43 iwe 350
lbf 46805-4

Ausgeschieden
im Jahr 2025

D1728751

Nanotechnology

Volume 4: Information Technology II

Edited by Rainer Waser

Related Titles

Nanotechnologies for the Life Sciences

Challa S. S. R. Kumar (ed.)

Volume 1: Biofunctionalization of Nanomaterials
2005
978-3-527-31381-5

Volume 2: Biological and Pharmaceutical Nanomaterials
2005
978-3-427-31382

Volume 3: Nanosystem Characterization Tools in the Life Sciences
2005
978-3-527-31383-9

Volume 4: Nanodevices for the Life Sciences
2006
978-3-527-31384-6

Volume 5: Nanomaterials - Toxicity, Health and Environmental Issues
2006
978-3-527-31385-3

Volume 6: Nanomaterials for Cancer Therapy
2006
978-3-527-31386-0

Volume 7: Nanomaterials for Cancer Diagnosis
2006
978-3-527-31387-7

Volume 8: Nanomaterials for Biosensors
2006
978-3-527-31388-4

Volume 9: Tissue, Cell and Organ Engineering
2006
978-3-527-31389-1

Volume 10: Nanomaterials for Medical Diagnosis and Therapy
2007
978-3-527-31390-7

Nanotechnology

Günter Schmid (ed.)

Volume 1: Principles and Fundamentals
2008
978-3-527-31732-5

Harald Krug (ed.)

Volume 2: Environmental Aspects
2008
978-3-527-31735-6

Rainer Waser (ed.)

Volume 3: Information Technology I
2008
978-3-527-31738-7

Rainer Waser (ed.)

Volume 3: Information Technology II
2008
978-3-527-31737-0

Viola Vogel (ed.)

Volume 5: Nanomedicine and Nanobiotechnology
2009
978-3-527-31736-3

Harald Fuchs (ed.)

Volume 6: Nanoprobes
2009
978-3-527-31733-2

Michael Grätzel, Kuppuswamy Kalyanasundaram (eds.)

Volume 7: Light and Energy
2009
978-3-527-31734-9

Lifeng Chi (ed.)

Volume 8: Nanostructured Surfaces
2009
978-3-527-31739-4

www.wiley.com/go/nanotechnology

G. Schmid, H. Krug, R. Waser, V. Vogel, H. Fuchs,
M. Grätzel, K. Kalyanasundaram, L. Chi (Eds.)

Nanotechnology

Volume 4: Information Technology II

Edited by Rainer Waser

WILEY-VCH Verlag GmbH & Co. KGaA

10 lbf 46905

The Editor

Prof. Dr.-Ing. Rainer Waser
RWTH Aachen
Institut für Elektrotechnik II
Sommerfeldstr. 24
52074 Aachen
Germany

Cover: Nanocar reproduced with kind permission of Y. Shirai/Rice University

UNIVERSITÄTSBIBLIOTHEK KIEL
FACHBIBLIOTHEK
INGENIEURWISSENSCHAFTEN

All books published by Wiley-VCH are carefully produced. Nevertheless, authors, editors, and publisher do not warrant the information contained in these books, including this book, to be free of errors. Readers are advised to keep in mind that statements, data, illustrations, procedural details or other items may inadvertently be inaccurate.

Library of Congress Card No.: applied for

British Library Cataloguing-in-Publication Data
A catalogue record for this book is available from the British Library.

Bibliographic information published by the Deutsche Nationalbibliothek
Die Deutsche Nationalbibliothek lists this publication in the Deutsche Nationalbibliografie; detailed bibliographic data are available in the Internet at http://dnb.d-nb.de.

© 2008 WILEY-VCH Verlag GmbH & Co. KGaA, Weinheim

All rights reserved (including those of translation into other languages). No part of this book may be reproduced in any form – by photoprinting, microfilm, or any other means – nor transmitted or translated into a machine language without written permission from the publishers. Registered names, trademarks, etc. used in this book, even when not specifically marked as such, are not to be considered unprotected by law.

Typesetting Thomson Digital, Noida, India
Printing betz-druck GmbH, Darmstadt
Binding Litges & Dopf Buchbinderei GmbH, Heppenheim

Printed in the Federal Republic of Germany
Printed on acid-free paper

ISBN: 978-3-527-31737-0

Contents

Nanotechnology. Volume 4: Information Technology II. Edited by Rainer Waser
Copyright © 2008 WILEY-VCH Verlag GmbH & Co. KGaA, Weinheim
ISBN: 978-3-527-31737-0

Preface

Beyond any doubt, Information Technology constitutes the area in which nanotechnology is most advanced. Since its origination during the 1960s, semiconductor technology as the driving force of information technology has advanced and continues to advance at an exponential pace. Today, semiconductor-based information technology penetrates almost all areas of contemporary society – and we are still only at the beginning of a new era. Within the coming decades completely new applications may emerge such as personal real-time translation systems, fully automatic navigation systems for cars, intelligent software agents for the internet, and autonomous robots to assist in our daily lives.

The main ingredient of the tremendous evolution of semiconductor circuitry has been the technological opportunity of ever-shrinking the minimum feature size in the fabrication of semiconductor chips. This led to a corresponding increase in the component density on the chips, decreasing energy consumption of the individual logic and memory cells, as well as higher clock frequencies and the development of multi-core architectures. All this added up to a doubling of the computer performance of chips approximately every 18 months, known as Moore's law. During the first decades of development, semiconductor technology was referred to as *microelectronics,* but this was changed to *nanoelectronics* a few years ago when the minimum feature size was reduced to below 100 nm. At about the same time, the component density has surpassed the one billion per chip mark and continues to progress at an unrestrained pace. Research areas related to nanoelectronics, however, comprise much more than simply the extension of current semiconductor technology to still smaller structures. More importantly, they cover the entire physics of nanosized objects with manifold properties that are unmatched in the macroscopic world and which might one day be exploited to store, to transmit, and to process information. In addition, they deal with technological approaches which are completely different to the *top-down* concept based on lithographical methods. The alternative *bottom-up* concept starts with the chemistry of, for example, organic molecules, nanocrystals, nanotubes, or nanowires, and strives for the self-organization of structures which can themselves act as assemblies of functional

Nanotechnology. Volume 4: Information Technology II. Edited by Rainer Waser
Copyright © 2008 WILEY-VCH Verlag GmbH & Co. KGaA, Weinheim
ISBN: 978-3-527-31737-0

devices. Furthermore, nanoelectronics research investigates completely new computational concepts and architectures.

This text on *Information Technology* within the series *Nanotechnology*, is divided into two volumes and covers the concepts of potential future advances of the semiconductor technology right up to their physical limits, as well as alternative concepts which might one day augment the semiconductor technology, or even replace it in designated areas. Some readers may be familiar with the book *Nanoelectronics and Information Technology* (Wiley-VCH, 2nd edition, 2005) which I have edited. Although the topic of the present book is quite similar, the target is somewhat different. While the first volume represents an advanced text book, the present two volumes emphasize encyclopedic reviews in-line with the concept of the series. Yet, wherever possible, I have strived for a complementarity of the topics covered in the two texts.

This volume covers three parts:

Part One – *Basic Principles and Theory* – includes chapters on the mesoscopic transport of electrons, single electron effects and processes dominated by the electron spin. Furthermore, the fundamental physics of computational elements and its limits are covered.

Part Two – *Nanofabrication Methods* – starts with the prospects of various optical and non-optical lithography techniques, describes the manipulation of nanosized objects by probe methods, and closes with chemistry- and biology-based bottom-up concepts.

Part Three – *High-Density Memories* – begins with an outlook at the future potential of current memories such as Flash and DRAM, attributes magnetoresistive and ferroelectric RAM, and reports about the perspectives of resistive RAM such as phase-change RAM and electrochemical metallization RAM.

Nanotechnology Volume 4 will cover the following topics:

- *Logic Devices and Concepts* – ranges from advanced and non-conventional CMOS devices and semiconductor nanowire device, via various spin-controlled logic devices, and concepts involving carbon nanotubes, organic thin films, as well as single organic molecules, to the visionary idea of intramolecular computation.
- *Architectures and Computational Concepts* – covers biologically inspired structures, and quantum cellular automata, and finalizes by summarizing the main principles and current approaches to coherent solid-state-based quantum computation.

There are many people to whom I owe acknowledgments. First of all, I would like to express my sincere thanks to the authors of the chapters, for their dedication, their patience, and their willingness whenever I requested modifications.

Next, I must pay tribute to the following colleagues (in alphabetical order) for critically reviewing the concept of the text and for their advice on topic and author selection: George Bourianoff (Intel Corp.), Ralph Cavin (Semiconductor Research Corp.), U-In Chung (Samsung Electronics), James Hutchby (Semiconductor Research Corp.), Christoph Koch (Caltech), Phil Kuekes (Hewlett Packard Research Laboratories), Heinrich Kurz (RWTH Aachen University), Rich Liu (Macronix Intl.

Ltd.), Hans Lüth (FZ Jülich), Siegfried Mantl (FZ Jülich), Tobias Noll (RWTH Aachen University), Stanley Williams (Hewlett Packard Research Laboratories), and Victor Zhirnov (Semiconductor Research Corp.).

Heartfelt thanks are due to Günther Schmid, editor of the series *Nanotechnology*, who invited and motivated me, and the staff of Wiley-VCH, in particular Gudrun Walter and Steffen Pauly, who supported me in every possible way.

Last – but certainly not least – I was greatly assisted by Dagmar Leisten, who redrew most of the original figures in order to improve their graphical quality, by Thomas Pössinger for his layout work and design ideas aiming at a more consistent appearance of the book, and by Maria Garcia for all the organizational work around such a project and for her sustained support.

Aachen, January 2008 *Rainer Waser*

List of Contributors

Dan A. Allwood
Department of Engineering Materials
University of Sheffield
Sheffield S10 5NA
United Kingdom
d.allwood@sheffield.ac.uk

Supriyo Bandyopadhyay
Department of Electrical and Computer Engineering
Virginia Commonwealth University
Richmond, VA 23284
USA
sbandy@vcu.edu

Thomas Bjørnholm
Nano-Science Center
University of Copenhagen
The H. C. Ørsted Institute
Universitetsparken 5
DK-2100 Copenhagen Ø
Denmark
tb@nano.ku.dk

Tomas Bryllert
Lund University
Solid State Physics
P.O. Box 118
S-221 00 Lund
Sweden
Present address:
Jet Propulsion Laboratories
California Institute of Technology
4800 Oak Grove Drive
Pasadena, CA 91109
California
USA

Russell P. Cowburn
Blackett Physics Laboratory
Imperial College London
Prince Consort Road
London SW7 2BW
United Kingdom
r.cowburn@imperial.ac.uk

André DeHon
Department of Electrical and Systems Engineering
200 South 33rd Street
Philadelphia, PA 19104
USA
andre@cs.caltech.edu; andre@acm.org

Nanotechnology. Volume 4: Information Technology II. Edited by Rainer Waser
Copyright © 2008 WILEY-VCH Verlag GmbH & Co. KGaA, Weinheim
ISBN: 978-3-527-31737-0

Linus Fröberg
Lund University
Solid State Physics
P.O. Box 118
S-221 00 Lund
Sweden
and
Qumat Technologies AB
St. Fiskaregatan 13E
S-222 24 Lund
Sweden

Akira Fujiwara
NTT Basic Research Laboratories
NTT Corporation
3-1 Morinosato- Wakamiya
Atsugi
Kanagawa 243-0198
Japan

Edward Goldobin
Physikalisches Institut-
Experimentalphysik II
Universität Tübingen
Auf der Morgenstelle 14
76076 Tübingen
Germany
gold@uni-tuebingen.de

Dan Hammerstrom
Electrical and Computer Engineering
Department
Maseeh College of Engineering and
Computer Science
Portland State University
1930 SW 4th Street
Portland, OR 97207
USA
strom@cecs.pdx.edu

Hiroshi Inokawa
Research Institute of Electronics
Shizuoka University
3-5-1 Johoku
Hamamatsu 432-8011
Japan

Hagen Klauk
Max Planck Institute for Solid-State
Research
Heisenbergstr. 1
70569 Stuttgart
Germany
H.Klauk@fkf.mpg.de

Raphael D. Levine
The Fritz Haber Research Center for
Molecular Dynamics
The Hebrew University of Jerusalem
Jerusalem 91904
Israel
and
Department of Chemistry and
Biochemistry
The University of California Los Angeles
Los Angeles, CA 90095-1569
USA
rafi@fh.huji.ac.il

Erik Lind
Lund University
Solid State Physics
P.O. Box 118
S-221 00 Lund
Sweden
and
Qumat Technologies AB
St. Fiskaregatan 13E
S-222 24 Lund
Sweden

Björn Lüssem
Sony Deutschland GmbH
Hedelfinger Str. 61
70327 Stuttgart
Germany
luessem@sony.de

Massimo Macucci
Dipartimento di Ingegneria
dell'Informazione
Università di Pisa
Via Caruso 16
I-56122 Pisa
Italy
massimo.macucci@iet.unipi.it

M. Meyyappan
Center for Nanotechnology
NASA Ames Research Center
Moffett Field, CA 94035
USA
meyya@orbit.arc.nasa.gov or
mmeyyappan@mail.arc.nasa.gov

Katsuhiko Nishiguchi
NTT Basic Research Laboratories
NTT Corporation
3-1 Morinosato- Wakamiya
Atsugi
Kanagawa 243-0198
Japan

Yukinori Ono
NTT Basic Research Laboratories
NTT Corporation
3-1 Morinosato-Wakamiya
Atsugi
Kanagawa 243-0198
Japan
ono@aecl.ntt.co.jp

Françoise Remacle
Department of Chemistry, B6c
University of Liège
B-4000 Liège 1
Belgium
fremacle@ulg.ac.be

Lothar Risch
Tizianstrasse 27
85579 Neubiberg
Germany
lothar-risch@gmx.de

Lars Samuelson
Lund University
Solid State Physics
P.O. Box 118
S-221 00 Lund
Sweden
lars.samuelson@ftf.lth.se

Yasuo Takahashi
Graduate School of Information Science
and Technology
Hokkaido University
Sapporo
Hokkaido 060-0814
Japan

Claes Thelander
Lund University
Solid State Physics
P.O. Box 118
S-221 00 Lund
Sweden
and
Qumat Technologies AB
St. Fiskaregatan 13E
S-222 24 Lund
Sweden

Martin Weides
Institute for Solid State Research (IFF)
and
CNI - Center of Nanoelectronic Systems for Information Technology
Forschungszentrum Jülich
52425 Jülich
Germany
m.weides@fz-juelich.de

Lars-Erik Wernersson
Lund University
Solid State Physics
P.O. Box 118
S-221 00 Lund
Sweden

I
Logic Devices and Concepts

Nanotechnology. Volume 4: Information Technology II. Edited by Rainer Waser
Copyright © 2008 WILEY-VCH Verlag GmbH & Co. KGaA, Weinheim
ISBN: 978-3-527-31737-0

1
Non-Conventional Complementary Metal-Oxide-Semiconductor (CMOS) Devices

Lothar Risch

1.1
Nano-Size CMOS and Challenges

The scaling of complementary metal-oxide-semiconductor (CMOS) is key to following Moore's law for higher integration densities, faster switching times, and reduced power consumption at reduced costs. In today's research laboratories MOSFETs with minimum gate lengths below 15 nm have already been demonstrated. An example of such a small transistor is shown in Figure 1.1a, where the transmission electron microscopy (TEM) cross-section shows a functional, fully depleted silicon-on-insulator (SOI) transistor with 14 nm gate length, 20 nm spacers using a 17 nm thin silicon layer and a 1.5-nm gate dielectric. The gate has been defined with electron-beam (e-beam) lithography. For the contacts, elevated source drain regions were grown with selective Si epitaxy to lower the parasitic resistance, and a high dose of dopants was implanted into the epi layer for source and drain. In Figure 1.1b, a TEM cross-section through the fin of a SONOS memory FinFET is shown with a diameter of 8 nm, surrounded by the ONO charge-trapping dielectric. As can be seen, many critical features in Si-MOSFETs are already in the range in the range of 1 to 20 nm.

However, achieving the desired performance gain in electrical parameters from scaling will in time become very challenging, as indicated in the International Technology Roadmap for Semiconductors (ITRS) by many red brick walls [1] (see Figure 1.2).

The three main limiting factors for a performance increase are related to physical laws. Gate leakage stops SiO_2 scaling (see Figure 1.3), while source drain leakage reduction needs higher channel doping and shallower junctions. However, this increases junction capacitance, junction leakage, gate-induced drain currents, reduces carrier mobility and increases parasitic resistance. Because of this, transistors with astoundingly small gate lengths down to 5 nm have been realized [2]; although these are the smallest MOSFETs produced until now, their performance is worse than that of a 20-nm device.

When considering memories, the situation is not much different, and for mainstream DRAM and Floating Gate Flash several constraints can be foreseen. For DRAM, the storage capacitance at small cell size and a low leakage cell transistor

Nanotechnology. Volume 4: Information Technology II. Edited by Rainer Waser
Copyright © 2008 WILEY-VCH Verlag GmbH & Co. KGaA, Weinheim
ISBN: 978-3-527-31737-0

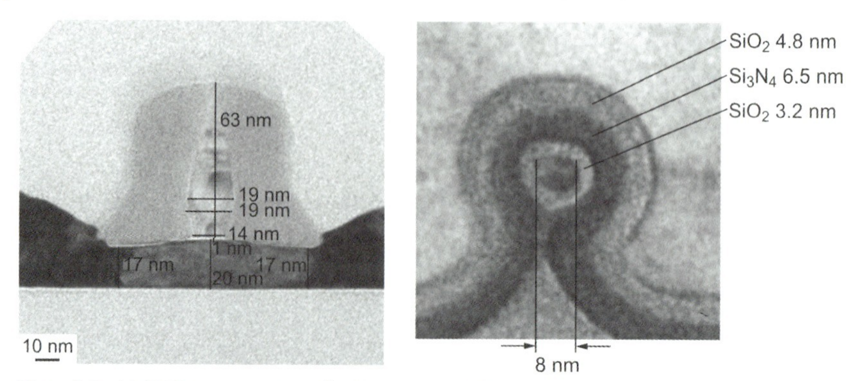

Figure 1.1 (a) A TEM cross-section of a 14-nm gate SOI transistor with raised source/drain (S/D) on 17 nm Si, $t_{ox} = 1.5$ nm. (b) TEM cross-section of a SONOS SOI FinFET across a 8-nm wire-type fin.

become a critical issue. For Floating Gate, the high drain voltages and scaling of the gate dielectric, as well as coupling to neighboring cells, are critical.

Therefore, on the way to better devices, two strategies are proposed by ITRS. The first strategy is to implement new materials as performance boosters. Among these are high-k dielectrics and metal gates, high-mobility channels and low-resistivity or

Year		04	07	10	13	16
Node [nm]		90	65	45	32	22
L_G [nm]	hp	37	25	18	13	9
	lop	53	32	22	16	11
	lstp	65	37	25	18	13
V_{dd} [V]	hp,lstp	1.2	1.1	1.0	0.9	0.8
	lop	0.9	0.8	0.7	0.6	0.5
I_{on} [mA/μm]	hp	1.1	1.5	1.9	2.05	2.4
	lop	0.53	0.57	0.77	0.78	0.92
	lstp	0.44	0.51	0.76	0.88	0.86
I_{off} [nA/μm]	hp	50	70	100	300	500
	lop	3	5	7	10	30
	lstp	0.01	0.025	0.06	0.08	0.1

Figure 1.2 ITRS 04 roadmap: gate lengths and currents for high performance, low operation power, low standby power.

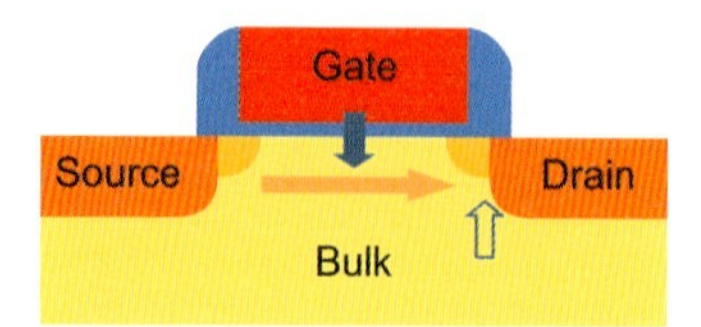

Figure 1.3 Scaling limits of scaled MOSFETs: source to drain, gate dielectric tunneling and junction leakage.

metal source drain junctions. This will lead to a remarkable improvement in the performance of transistors. The second strategy is to develop new device structures with better electrostatic control, such as fully depleted SOI and multi-gate devices. These can also be utilized in DRAMs as low leakage cell transistors, as well as in nanoscale non-volatile Flash memories.

1.2 Mobility Enhancement: SiGe, Strained Layers, Crystal Orientation

Carrier mobility enhancement of electron and holes provide the key to increase the on-currents without higher gate capacitance and without degrading the off-currents. Several methods have been developed, including SiGe heterostructures [3] with a higher hole mobility for the p-channel transistor. This is achieved by growing a thin epitaxial $Si_{1-x}Ge_x$ layer, where x is the Ge concentration, with a thickness of 5–10 nm for the channel region directly on Si (see Figure 1.4). On top of the SiGe layer a thin Si cap layer is deposited with a thickness of 3–5 nm, which is also used for the growth of the gate oxide. This forms a quantum well for the holes due to a step in the valence band of the Si/SiGe/Si heterostructure, with a depth of about 150 mV for a Ge content of 20%. The SiGe layer is under bi-axial compressive strain due to the smaller lattice constant of Si compared to SiGe (see Figure 1.4a). The mobility is enhanced because of the lower effective mass of the holes in SiGe and a split of the degenerated three-valence bands, thus reducing intervalley scattering. Compared to pure Si with a peak hole mobility of about 110 $cm^2\,Vs^{-1}$, with 0.25 Ge 210 $m^2\,Vs^{-1}$ have been achieved [4],

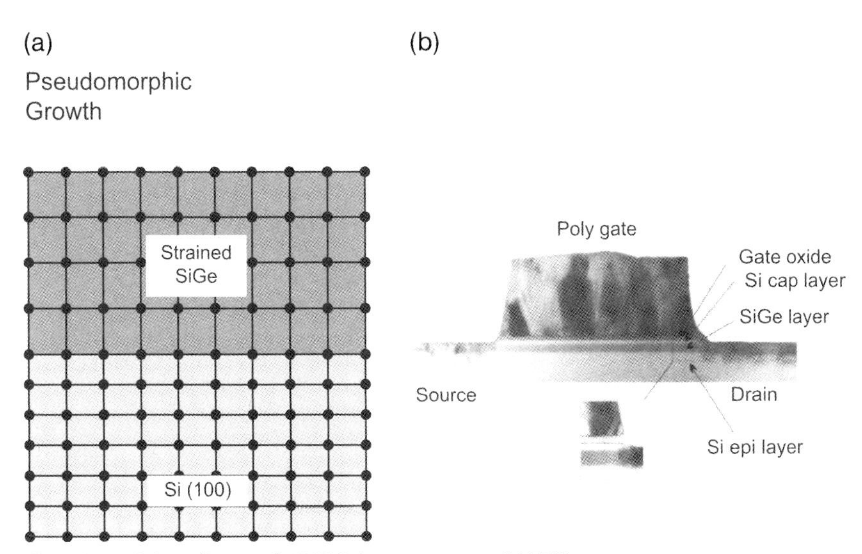

Figure 1.4 (a) Crystal lattice of a Si/SiGe heterostructure. (b) TEM cross-section of a p-channel MOSFET with a Si/SiGe/S quantum well.

extracted from MOSFET measurements. Whereas, the SiGe channel on Si is beneficial for the hole mobility, strained silicon offers both an improved electron and hole mobility, together with a surface channel [5]. The strain is created by a relaxed graded SiGe buffer layer, typically with a thickness of about 3 μm and a Ge concentration of 20–30%. A thin Si layer is grown on top of the relaxed SiGe layer in the range of 10 to 20 nm, which is now under biaxial tensile strain due to the larger lattice constant of the SiGe buffer layer.

Both techniques provide global bi-axial strain on the wafer and are based on Si/SiGe epitaxy. A critical issue here is the increased process complexity, the density of defects and wafer cost. Moreover, the implementation of tensile strain for the n-channel and compressive strain for the p-channel would be desirable, and is difficult to achieve with global strain. Therefore, local uni-axial strain techniques have now become mainstream for mobility enhancement, and these can provide tensile and compressive strain by depositing dedicated layers around the transistor. This method was introduced [6] for the 90-nm CMOS generation. In the n-channel transistor a nitride capping layer with tensile strain is used to improve the drive current by 10–15%. For the p-channel transistor, an embedded SiGe source drain region provides compressive strain and increases the drive current by 25%. TEM cross-sections of the n- and p-channel devices are shown in Figure 1.5 [6].

Another mobility-enhancement technique is based on the crystal orientation dependence of the mobility. Until now, the (1 0 0) surface of silicon wafers has been used with a channel orientation of the transistors in the <0 1 1> direction (see Figure 1.6). This is optimal for the electron mobility but will decrease the hole mobility, which is twice that at the (1 1 0) plane in the <1 0 0> direction. If (1 1 0) wafers are used or rotated (1 0 0) wafers by 45° with the channel in the <1 0 0> direction, the hole mobility is improved remarkably while electron mobility is reduced only moderately (see Figure 1.7) [7].

Therefore, another channel orientation is an effective means to increase p-channel performance, and an improvement of up to 15% has been reported [8]. Unfortunately, the embedded SiGe source drain regions with compressive strain have no remarkable influence in this crystal direction.

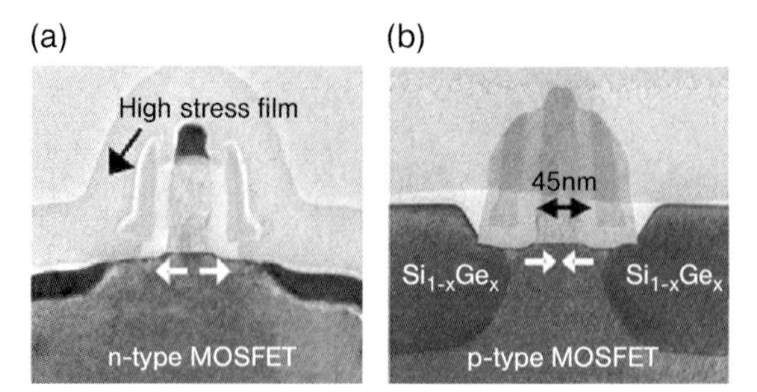

Figure 1.5 (a) 90-nm technology NMOS transistor with tensile stress nitride layer; (b) PMOS, showing heteroepitaxial SiGe source/drain inducing compressive strain [6] .

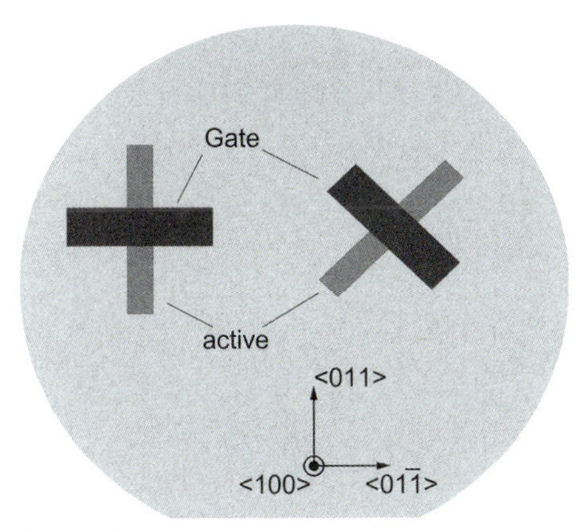

Figure 1.6 Crystal orientation and channel direction on (1 0 0) Si wafers.

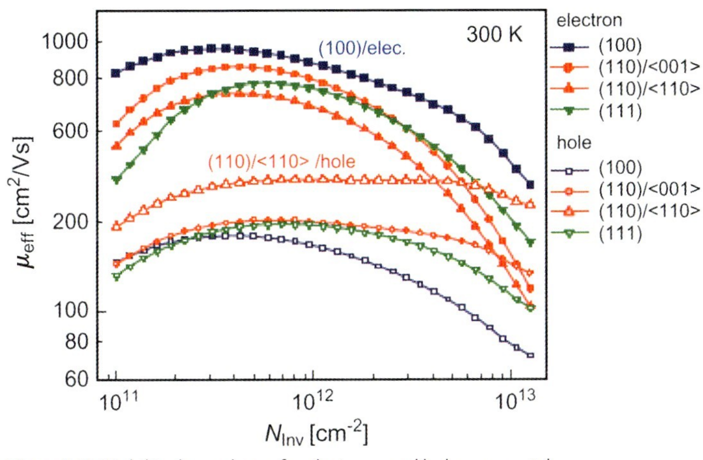

Figure 1.7 Mobility dependence for electrons and holes on crystal orientation and channel direction [7].

1.3
High-*k* Gate Dielectrics and Metal Gate

As indicated in the ITRS roadmap, scaling of the classical SiO_2 gate dielectric and increasing the gate capacitance in order to achieve higher drive currents reached completion at about 2 nm for low standby power, due to Fowler–Nordheim tunneling currents through the gate dielectric. By using nitrided oxides, the minimum thickness could be extended to about 1 nm for high-performance applications with a gate leakage current of about 10^3 A cm^{-2} [9]. The introduction of high-*k* dielectrics allows the use of thicker dielectric layers in order to reduce the tunneling currents at the same equivalent oxide thickness, or to provide thinner dielectrics for continuous scaling. Unfortunately,

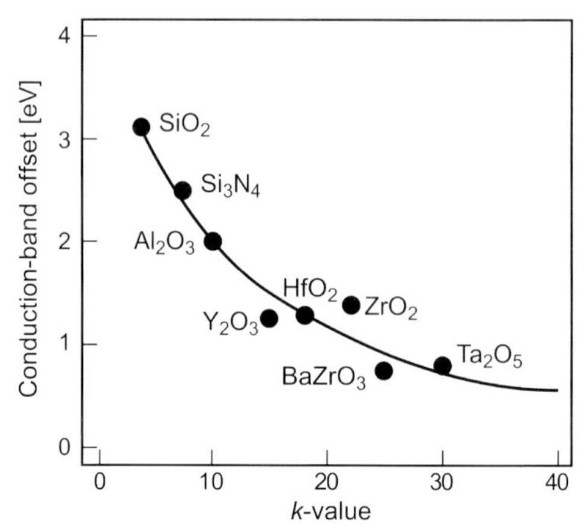

Figure 1.8 Conduction band offset versus k-value for different high-*k* materials [10].

all known high-*k* materials have a smaller bandgap than SiO_2. In Figure 1.8 the conduction band offset as a function of the dielectric constant is shown for different materials [10]. For the highest *k* materials such as Ta_2O_5 ($k = 30$) or TiO_2 ($k = 90$), the bandgap becomes too small and leads to increased gate leakage. Other critical issues are the growth of an interfacial layer during processing. Today, the most mature high-*k* dielectrics are based on Hf. Among these, HfO_2 ($k = 17$–25), HfSiO ($k = 11$) and HfSiON ($k = 9$–11), the latter are the more temperature-stable. An equivalent oxide thickness of below 1 nm has been demonstrated for these high-*k* materials [10]. Other candidates are ZrO_2 and La_2O_3 with dielectric constants between 20 and 30; however, the former is incompatible with a poly silicon gate and requires a metal gate.

For most high-*k* dielectrics a degradation of mobility is observed due to an increased scattering by phonons or a high fixed charge density at the interface. Especially for Al_2O_3, the hole mobility reduction is not acceptable. For the best Hf-based high-*k* dielectrics a 20% lower mobility has been achieved until now, compared to SiO_2.

Closely related to the high-*k* dielectric is a new gate material which avoids the depletion layer of poly silicon gates and the reaction of the high-*k* material with silicon at higher process temperatures. Moreover, metal gates offer the possibility of adjusting the threshold voltage with the workfunction of the gate material instead of doping in the channel, and this decreases the mobility at higher doping concentrations. The desired workfunctions for bulk with n+ poly and p+ poly silicon gates for low-power/high-performance applications with low doped transistor channels are shown in Figure 1.9.

Midgap-like materials such as TiN, TiSiN and W are suitable for n- and p-channel transistors with threshold voltages in the range of 300 to 400 mV, especially for fully depleted SOI or multi-gate transistors with lower channel doping concentrations. For optimized logic processes with low V_t transistors for high performance, in the range of 100 to 200 mV, dual metals with n+ and p+ poly-silicon-like workfunctions must be integrated. For n-channel transistors Ru is a candidate, and for p-channel Ta or RuTa alloys.

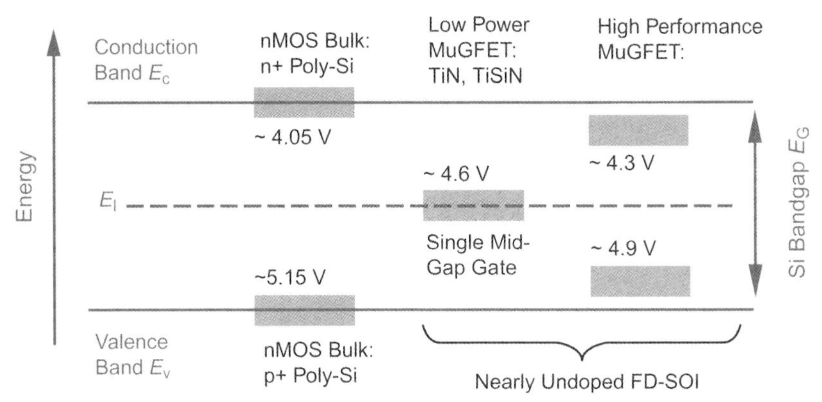

Figure 1.9 Desired workfunction for bulk and FD MOSFETS [24], Pacha ISSCC 2006.

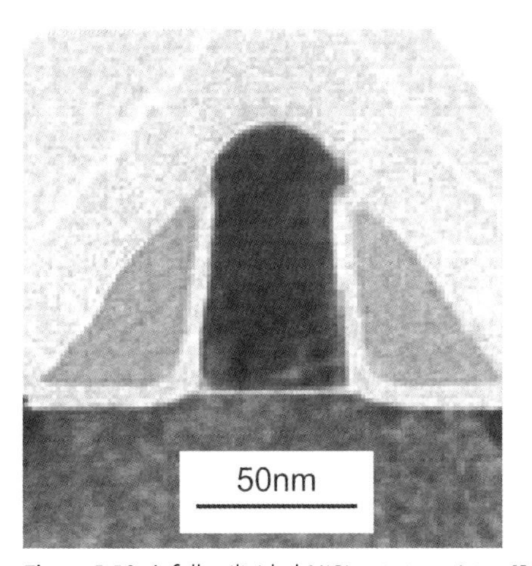

Figure 1.10 A fully silicided NiSi gate transistor [10].

Another gate matcrial option is a tunable workfunction, such as fully silicided NiSi implanted with As and B, or Mo implanted with N. Until now, a shift of the workfunction in a conduction band direction of 200 to 300 mV has been reported [11]. A cross-section of a 50-nm transistor with a fully silicided NiSi gate is shown in Figure 1.10. Here, two approaches have been pursued: the first approach, with Thin Poly, allows the simultaneous silicidation of the source/drain (S/D) and gate, while the second approach uses CMP, offers the independent silicidation of the S/D and gate, and also avoids the formation of thick silicides in the S/D [10].

1.4 Ultra-Thin SOI

Many of the device problems due to short channel effects are related to the silicon bulk. The SOI [12] uses only a thin silicon layer for the channel, which is isolated from

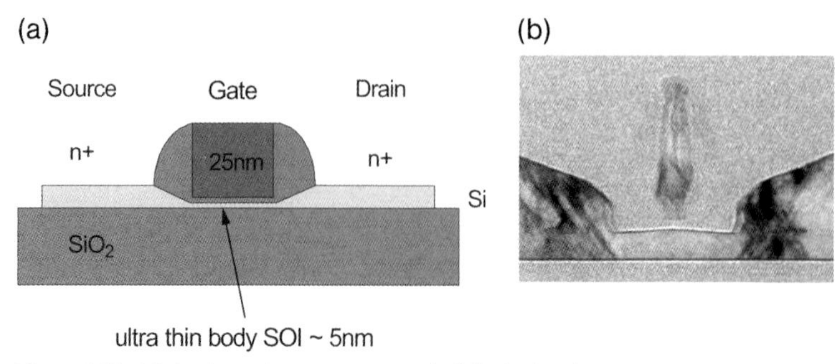

Figure 1.11 (a) A schematic cross-section of a fully depleted SOI transistor with a raised source drain. (b) TEM cross-section of a 12-nm gate fully depleted SOI transistor on 16-nm silicon.

the bulk by a buried oxide. Several companies producing semiconductors have already switched to SOI for high-performance microprocessors or low-power applications. Typically, the thickness of the Si layer is in the range of 50 to 100 nm, and the doping concentrations are comparable to those of bulk devices. This situation, which is referred to as *partially depleted SOI*, has several advantages, most notably a 10–20% higher switching speed. However, further down-scaling faces similar issues as the bulk, and here thinner Si layers [13], which lead to fully depleted channels, are of interest.

A schematic representation and a TEM cross-section of a thin-body SOI transistor with 12-nm gate length and 16-nm Si thickness on 100 nm buried oxide is shown in Figure 1.11. The gate has been defined with e-beam lithography while, for the contacts, raised source drain regions were grown with selective Si epitaxy and a high dose of dopands was implanted into the epi layer.

The experimental current–voltage (I–V) characteristics of n-channel SOI transistors with gate lengths down to 12 nm are shown in Figure 1.12. For gate lengths

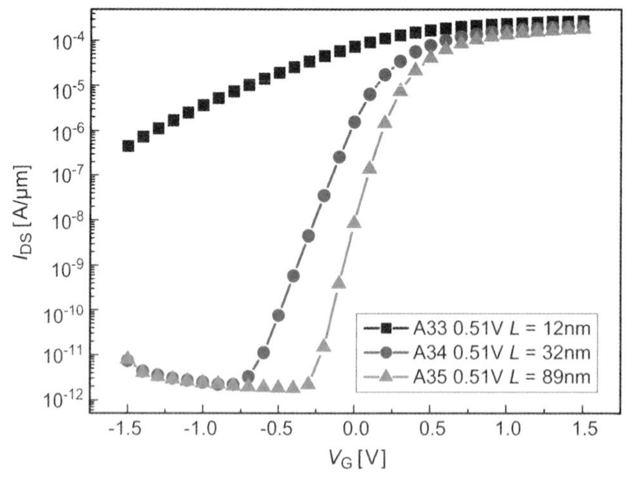

Figure 1.12 Experimental I–V characteristics of 89 to 12-nm SOI transistors on 16-nm silicon with undoped channel, n+ poly gate, $t_{ox} = 1.5$ nm.

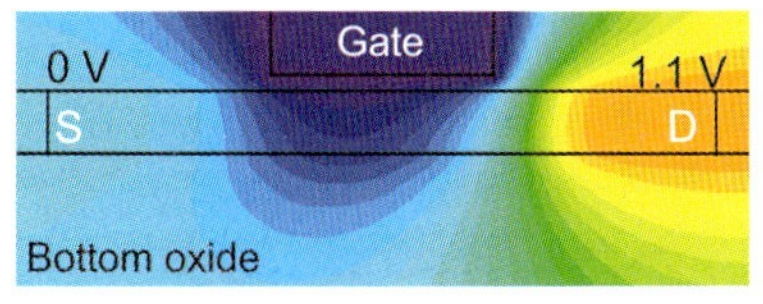

Figure 1.13 Potential distribution in a 30-nm single gate SOI transistor ($t_{Si} = 10$ nm, $t_{ox} = 2$ nm, $V_g = 0$ V, $V_d = 1.1$ V, midgap gate material).

>32 nm, subthreshold slopes of 65 mV dec^{-1} have been reached but, due to the still relatively thick Si body of 16 nm, short channel effects begin to increase below 30 nm gate length, and the transistors with 12 nm gate length cannot easily be turned off.

A two-dimensional (2-D) device simulation of the electrostatic potential of an SOI transistor with undoped channel and a thinner silicon body of 10 nm is shown in Figure 1.13 at a drain voltage of 1.1 V and a gate voltage of 0 V. For a gate length of 30 nm the gate potential controls the channel quite well. However, even with 10 nm Si thickness the potential barrier is slightly lowered at the bottom of the channel.

>This gives rise to an increase in the subthreshold slope as function of gate length, even for 5 nm Si thickness and 1 nm gate oxide (see the device simulation in Figure 1.16). A single-gate SOI exhibits the ideal subthreshold slope of 60 mV dec^{-1} down to about 50-nm gate lengths. In the gate length range of 50 to 20 nm, the turn-off characteristics are still good, and therefore ultra-thin SOI can provide a device architecture which is superior to that of bulk and suitable for the 32-nm node. A simple scaling rule for fully depleted SOI devices proposes a Si thickness of about one-fourth of the gate length in order to achieve good turn-off characteristics.

Whilst in these devices the channel was either low or undoped, this is not feasible in bulk devices because of the punch through from source to drain. The mobility of the charge carriers and the on-current is higher due to lower electric fields; this is shown graphically in Figure 1.14 for different channel doping concentrations. At a

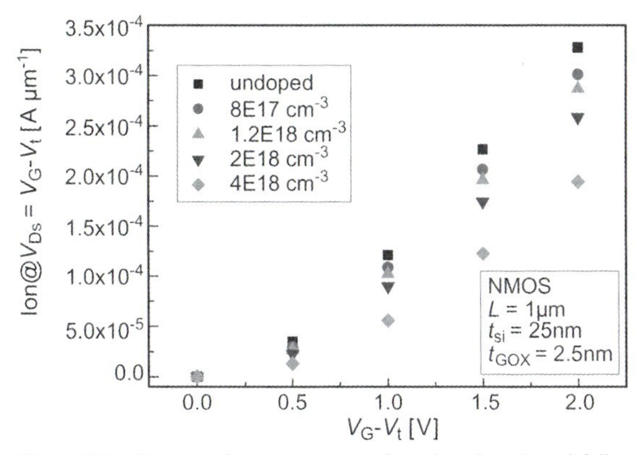

Figure 1.14 Measured on-currents at doped and undoped fully depleted SOI transistors at $V_g - V_t = 1$V.

gate voltage overdrive of 1 V the saturation current of the undoped transistor is twice that of the doped channel, at 4E18 cm^{-3} [14].

Moreover, without channel doping the Zener tunneling currents are reduced as well as electrical parameter variations, due to statistical fluctuations of the doping atoms.

1.5 Multi-Gate Devices

Further reduction of the gate length will require two or more gates for control of the channel, together with thin Si layers. The advantage of a multi-gate is to suppress the drain field much more effectively.

This is illustrated in Figure 1.15, by using the same simulation conditions as in Figure 1.13 and adding a bottom gate to the 30-nm SOI transistor. As shown in Figure 1.15, the electrostatic potential barrier is much higher than in the single-gate device. The better electrostatic control results in a steeper subthreshold slope; this can be seen in Figure 1.16, with a drift diffusion simulation of single- and double-gate transistors. A very thin Si thickness of 5 nm and a equivalent gate oxide thickness of 1 nm has been assumed, with a drain voltage of 1 V. Compared to the single gate, a

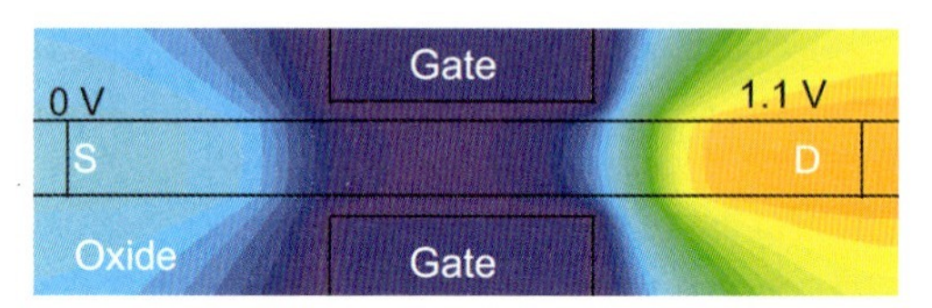

Figure 1.15 Electrostatic potential in a double-gate transistor with 30-nm gate length and 10-nm Si thickness; $V_g = 0$ V; $V_d = 1.1$ V; midgap gate material.

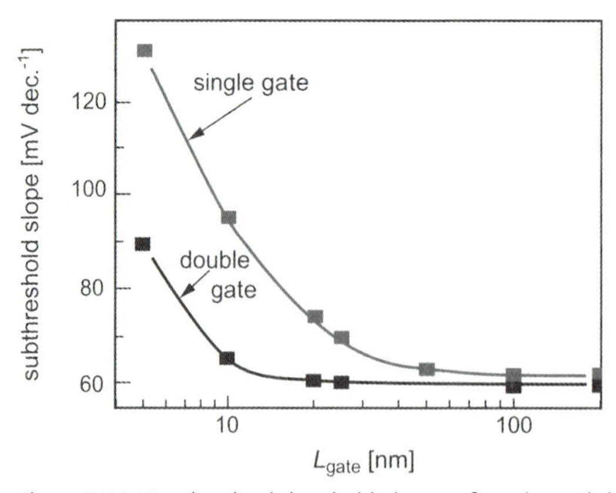

Figure 1.16 Simulated subthreshold slopes of single- and double-gate SOI transistors.

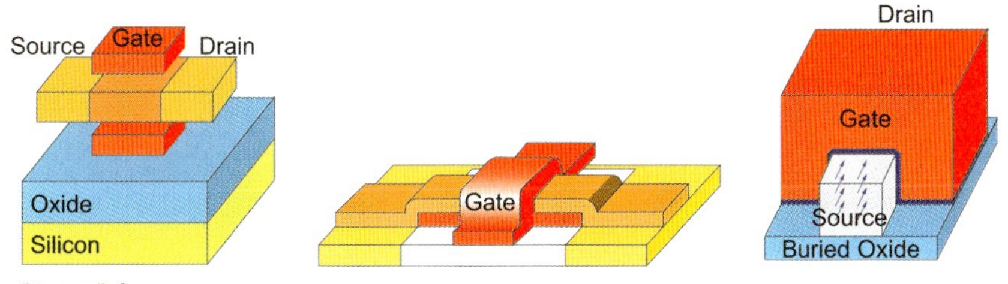

Figure 1.17 Three architectures for multi-gate devices. Left: Planar double-gate wafer-bonded [16]; Center: Gate all-around device [17]; Right: FinFET [18].

10-nm gate length and a subthreshold slopc of 65 mV dec^{-1} are predicted for a double gate, and even 5 nm seems feasible with a reasonable subthreshold slope.

The challenge for multi-gate transistors will be to develop a manufacturable process with self-aligned gates to S/D regions. Three promising concepts have been investigated within the EC project NESTOR [15]: the first was a planar double-gate SOI transistor, which uses wafer bonding [16]; the second was a gate all-around device, based on silicon-on-nothing (SON) [17]; and the third was a FinFET type [18] (see Figure 1.17).

1.5.1 Wafer-Bonded Planar Double Gate

The planar double-gate transistor is an evolution of the ultra-thin SOI transistor, with a top and a bottom gate being used for better control of the channel. Processing starts with the bottom gate, spacers and elevated S/D regions using a SOI wafer with a thin silicon layer (see Figure 1.18). The gate is then encapsulated with dielectric layers and planarized with chemical mechanical polishing (CMP). Next, a handle wafer with an oxide layer is bonded onto the wafer with the bottom gate. The bulk Si of the top wafer is then completely removed down to the buried oxide, which acts as an etch stop. After removal of the buried oxide, a gate dielectric is deposited on the thin Si layer. Finally, the top gate and metallization are processed as in a conventional transistor.

An atomic force microscopy (AFM) image of a double-gate transistor test-structure with two separated contacts for the bottom and top gate is depicted in Figure 1.19a, using e-beam litho and an alignment mark for the top gate. A TEM cross-section of the first devices with a p+ poly-Si top and a n+ poly-Si bottom gate for V_t adjustment is shown in Figure 1.19b [19].

Recently, functional double-gate transistors with a TiN metal gate and lengths down to 12 nm and 8 nm for the top and bottom gates have been demonstrated [16] (see Figure 1.20). The 20-nm devices show good short-channel characteristics with $S = 102\ \mathrm{mV\,dec^{-1}}$, an off-current in the range of $1\ \mu\mathrm{A}\,\mu\mathrm{m}^{-1}$, and an on-current of $1250\ \mu\mathrm{A}\,\mu\mathrm{m}^{-1}$.

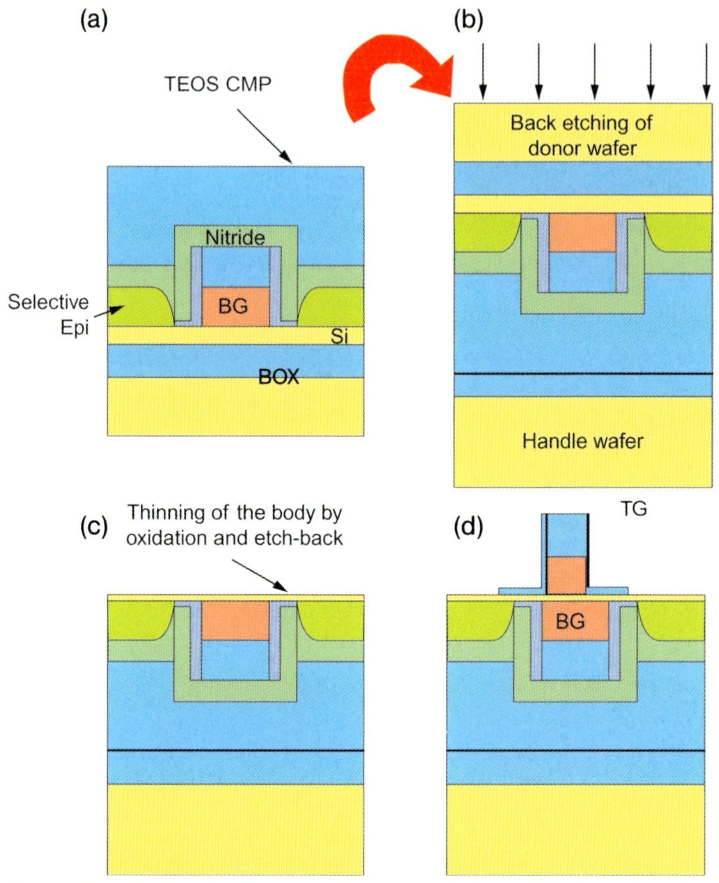

Figure 1.18 Process flow for a wafer-bonded double-gate transistor: Bottom gate, raised source drain and planarization, wafer bonding and back etch of Si bulk wafer, back etch Si channel, gate dielectric and top gate. BOX = Buried Oxide; BG = Buried Gate; TEOS = tetraethyl orthosilicate; CMP = chemical mechanical polishing.

1.5.2 Silicon-On-Nothing Gate All Around

The second approach for multi-gate architectures is based on silicon-on-nothing, as proposed by [20], which uses bulk Si wafers instead of SOI. A SiGe layer is grown with selective chemical vapor deposition (CVD) epitaxy and on top, non-selectively, a thin Si layer for the channel (see Figure 1.21). Next, the SiGe layer is removed by an isotropic etching process. The gate dielectric is then deposited around the silicon bridge, followed by the gate material, which is either poly-Si or a TiN metal gate. A 40-nm gate length and very thin Si channels down to 15 nm have been successfully

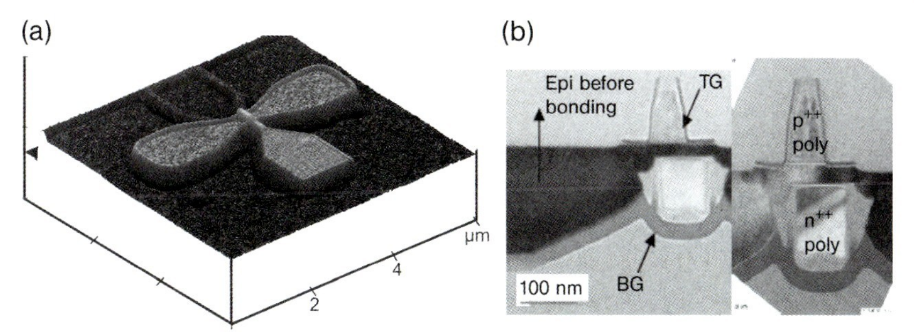

Figure 1.19 (a) AFM image of planar double-gate transistor with top and bottom gate. (b) TEM cross-section of planar double-gate transistor with n+/p+ poly gates.

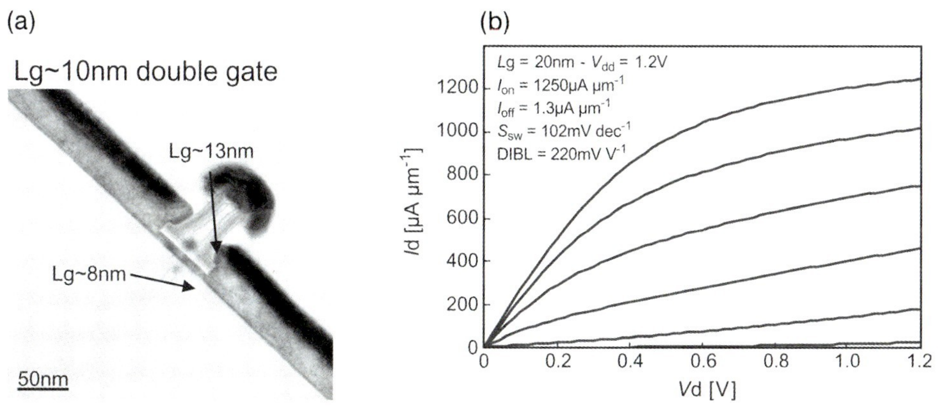

Figure 1.20 (a) TEM cross-section of a 10-nm double-gate transistor realized with wafer bonding [16]. (b) *I–V* characteristics of a 29-nm wafer-bonded double-gate device with a TiN metal gate [16].

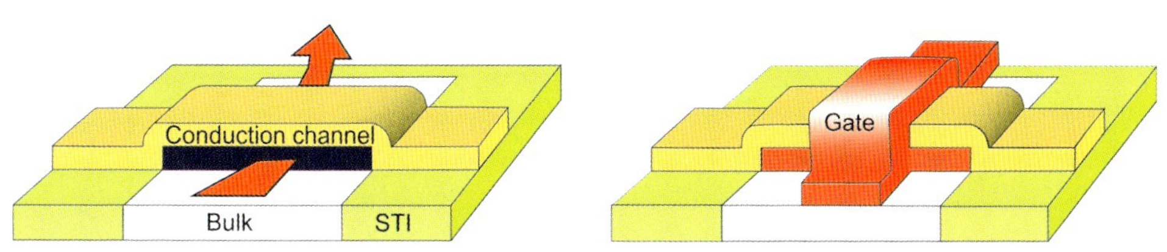

Figure 1.21 Gate all-around transistor processing based on silicon-on-nothing (SON) with a SiGe layer, which is removed for the gate [17].

fabricated [17]. Within the EC project NESTOR, devices with gate lengths of 25 nm have been achieved (see Figure 1.22a). These exhibit excellent short-channel characteristics, with $S = 70\,\text{mV}\,\text{dec}^{-1}$, DIBL = 11.8 mV, and high on currents of $1540\,\mu\text{A}\,\mu\text{m}^{-1}$ ($I_{\text{off}} = 2\,\mu\text{A}\,\mu\text{m}^{-1}$, $t_{\text{ox}} = 2\,\text{nm}$) at 1.2 V (see Figure 1.22b). As shown in

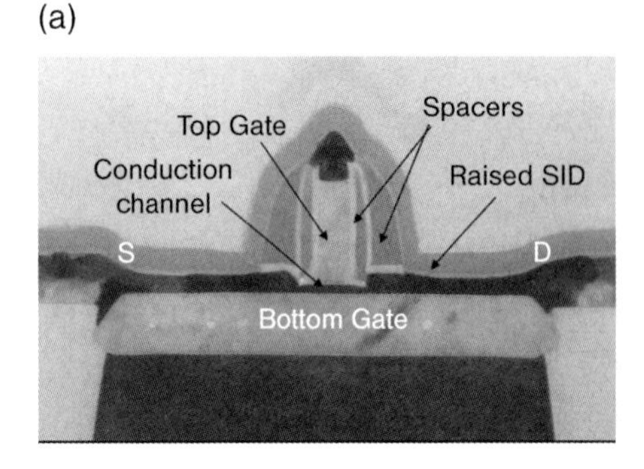

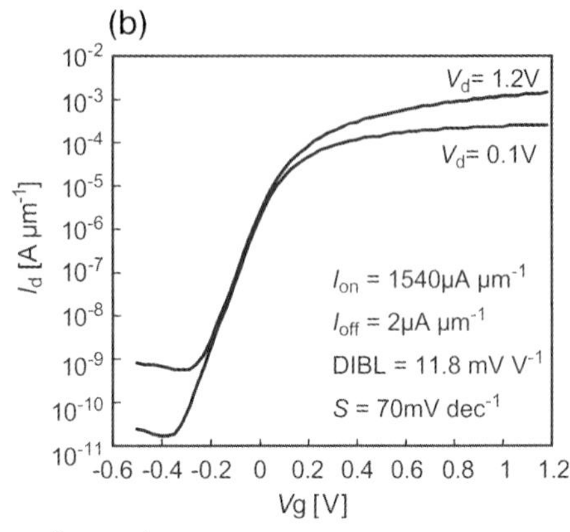

Figure 1.22 (a) TEM cross-section of 25-nm gate all-around SON transistor ($t_{ox} = 2$ nm; $t_{Si} = 10$ nm [15]. (b) Electrical characteristics of 25-nm gate all-around transistor [15].

Figure 1.22a, the bottom gate is still larger than the top gate. Ongoing studies have focused on a reduced bottom gate capacitance and a self-aligned approach.

Recently, multi-bridge transistors [21] have been reported using a similar type of SiGe layer etch technique for the fabrication of two or more channels stacked above each other and with an on-current of up to 4.2 mA μm^{-1} at 1.2 V.

1.5.3
FinFET

The FinFET [18, 22] can provide a double- or triple-gate structure with relatively simple processing (see Figure 1.23). First, the fin on SOI is structured with a tetraethylorthosilicate (TEOS) hardmask (Figure 1.23, left). A Si_3N_4 capping layer shields the top of the fin for a double-gate FinFET, and the same process flow can be used for triple-gate devices, without the capping layer. Next, a gate dielectric and the poly-Si gate are deposited and structured with litho and etching (Figure 1.23, center). The buried oxide provides an etch stop for the definition of the fin height. After this, a gate spacer is formed, raised source/drain regions are grown with epitaxy, and highly doped n+ or p+ regions implanted (Figure 1.23, right). The source/drain regions are enhanced using selective Si epitaxy to lower the sheet resistance. The facet of the Silicon epitaxy has been optimized to reduce the capacitance of drain to gate.

A TEM cross-section of a 20-nm tri-gate FinFET [23] is shown in Figure 1.24. Here, the top of the Si fin is also used for the channel, and no corner effects are observed at low fin doping concentrations. The fin and the gate layer have been processed with e-beam lithography. The smallest fin widths are in the range of 10 nm (see also Figure 1.30).

TEM cross-sections of a tri-gate device with larger fins of about 36 nm are also shown in Figure 1.24. The fin height is in the range of 35 nm, the corners are rounded by sacrificial oxidation, the gate dielectric is 2–3 nm SiO_2, and the poly gate surrounds the fin with a slight under-etch of the buried oxide.

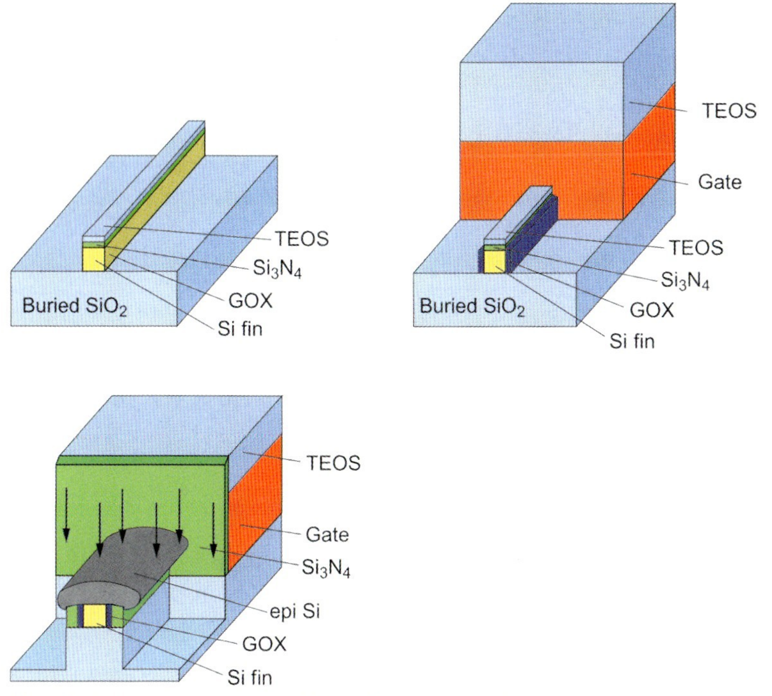

Figure 1.23 Process flow for a FinFET on buried oxide with a capping layer on top of the fin, a poly-Si gate, and raised source/drain regions with implantation. For details, see the text.

The measured I–V_g characteristics of n- and p-channel FinFETs with 20-nm and 30-nm gate length, respectively, are depicted in Figure 1.25. For the n-channel transistor a saturation current of 1.3 mA μm^{-1} (normalized by fin height) at an off-current of 100 nA μm^{-1} has been achieved at a gate voltage of 1.2 V, despite a relaxed

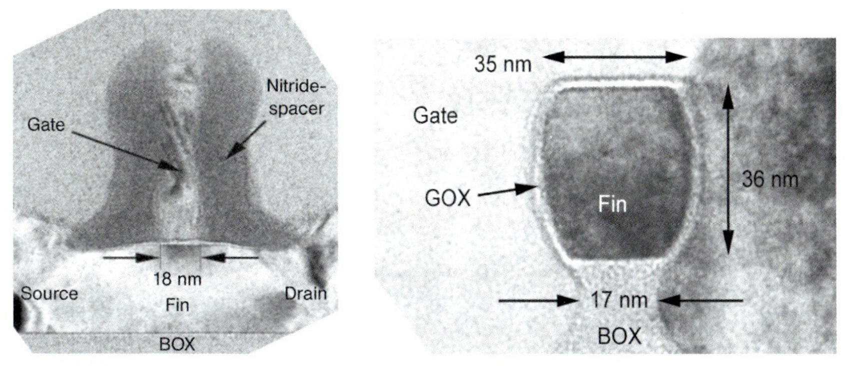

Figure 1.24 TEM cross-sections of a tri-gate FinFET on 100-nm buried oxide along and across the fin. Left: cross section along the fin; only the gate length is visible (18 nm). Right: cross section of the fin; on top it is 35 nm, height 36 nm, bottom ~17 nm.

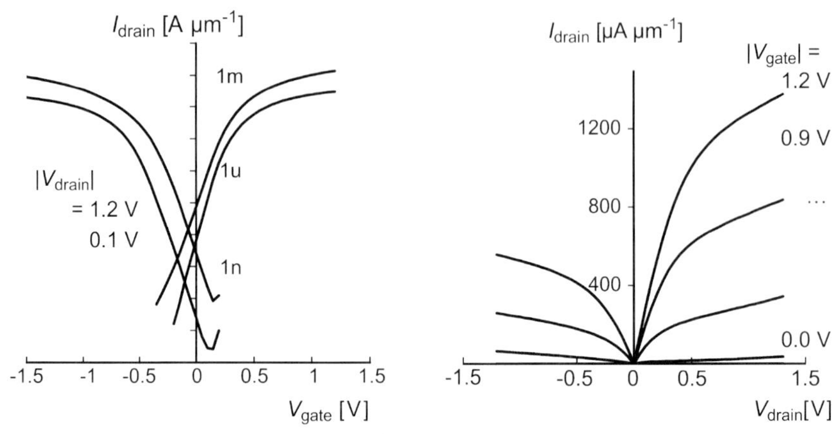

Figure 1.25 Measured I–V characteristics of 20-nm n-FinFETs (left) and 30-nm p-FinFETs (right) with $t_{ox} = 3$ nm (n) and 2 nm (p).

gate oxide thickness of 3 nm. For the p-channel, a high on current of 500 μA/μm and an off current in the range of 5 nA μm^{-1} is measured at 30-nm gate length. The FinFET has the advantage of self-aligned source and drain regions.

In Figure 1.25 the current was normalized on the height of a single fin. The electrical width of the device would be 2.2 times larger. For circuit applications, multi-fins are often needed in order to achieve higher drive currents (in Figure 1.26 the device has four fins) [24]. For a comparison with planar transistors it is important how many fins with height, width and pitch can be integrated on the same area as for the conventional device.

With respect to the switching time of multi-gate devices, the drive current together with the gate capacitance must be considered. Here, it was shown by simulation, that multi-gate devices can achieve 10–20% faster delay times compared to single-gate devices, mainly due to the better I_{off}/I_{on} ratio [25]. This was confirmed experimentally in Ref. [24] for inverter FO2 ring oscillators, where tri-gate FinFETs with TiSiN gate,

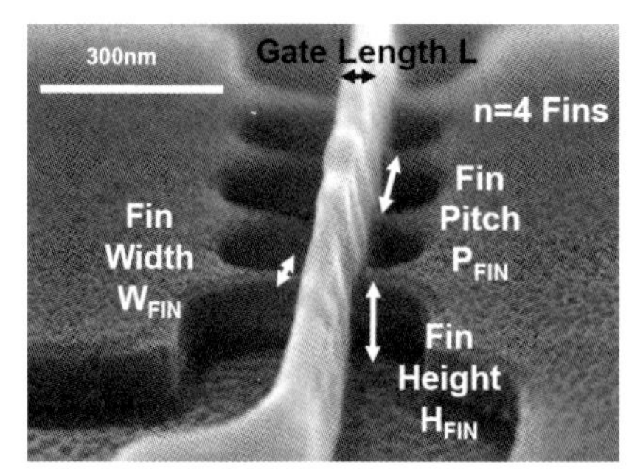

Figure 1.26 Scanning electron microscopy image of a multi-channel FinFET [24] with four fins on SOI. The gate length is 60 nm, fin width 30 nm, and pitch 200 nm.

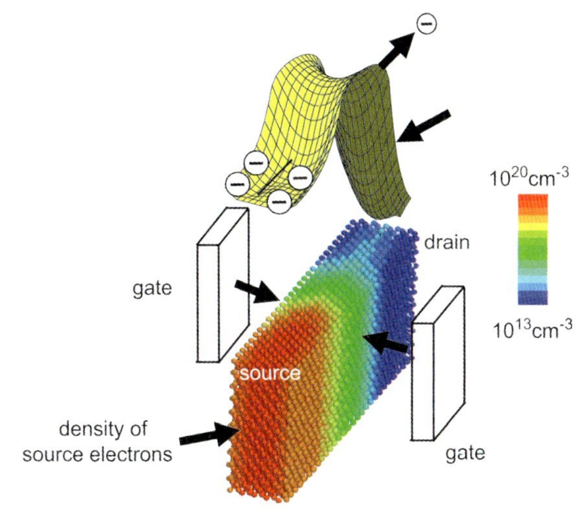

Figure 1.27 Atomistic simulation of a double-gate FinFET using the tight binding method.

55 nm length, and a low-doped channel achieved, with 21 ps, a much better speed performance than comparable planar MOSFETs in a 65 nm low-power CMOS technology, especially for sub-1 V power supply voltages.

1.5.4 Limits of Multi-Gate MOSFETs

The physical limit for the minimum channel length of multi-gate transistors has been investigated with 3-D quantum mechanical simulations using the tight binding method [26]. The device is composed of atoms in the silicon crystal lattice; the current can flow either by thermionic emission across the potential barrier of the channel, or directly via tunneling through the barrier from source to drain (see Figure 1.27).

In Figure 1.28, the simulated source drain current as a function of gate voltage is given with and without band to band tunneling for different gate lengths. An aggressive Si thickness of 2 nm and equivalent oxide thickness of 1 nm has been assumed. For gate lengths of 8 nm the tunneling contribution is on the order of the current over the potential barrier. At 4 nm the current is increased by two orders of magnitude by tunneling, but even 2-nm gates seem possible with off currents in the range of $\mu A\,\mu m^{-1}$, corresponding to ITRS hp specifications. Gate control is still effective and would achieve a subthreshold slope of about 140 mV dec^{-1}.

1.6 Multi-Gate Flash Cell

Multi-gate transistors are also very suitable for highly integrated memories with small gate lengths. Flash memory cells require thicker gate dielectrics than in logic

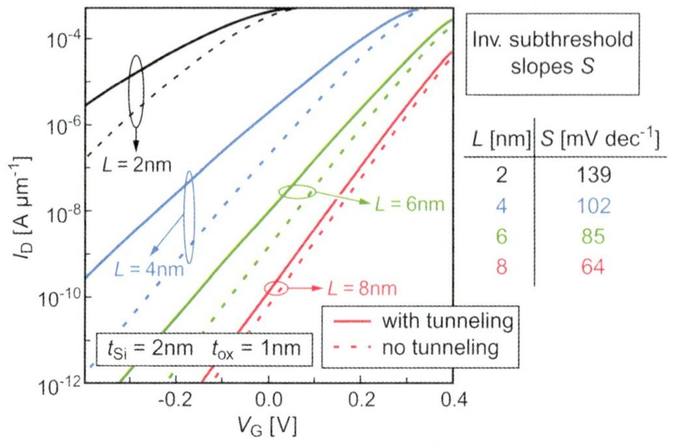

Figure 1.28 Thermionic and total current (+tunneling) of double-gate FinFETs simulated with the tight binding method [26].

applications, and therefore exhibit enhanced short channel effects. Currently, the most widely used Flash cell consists of a transistor with a floating gate [27] or a charge-trapping dielectric [28] sandwiched between the gate electrode and the channel region. A small amount of charge is transferred into the storage region either by tunneling or hot electron injection. This can be stored persistently and read out by a shift in the I–V_g characteristics. A schematic cross-section of a tri-gate FinFET memory transistor with improved electrostatic channel control compared to a planar device is shown in Figure 1.29, where a multilayer ONO gate dielectric around the fin serves as the storage element.

An experimentally realized memory structure [29] with a very small Si fin of 12 nm width and height of 38 nm is shown in Figure 1.30. The multilayer dielectric consisted of 3 nm SiO_2, 4 nm Si_3N_4 and 6.5 nm SiO_2.

The charge is uniformly injected into the nitride trapping layer by Fowler–Nordheim tunneling. The electrical function has been verified experimentally down to

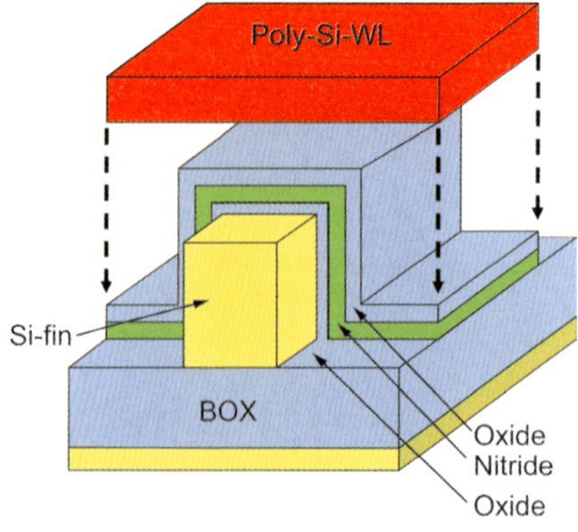

Figure 1.29 Schematic cross-section of a tri-gate charge-trapping memory field-effect transistor (FET).

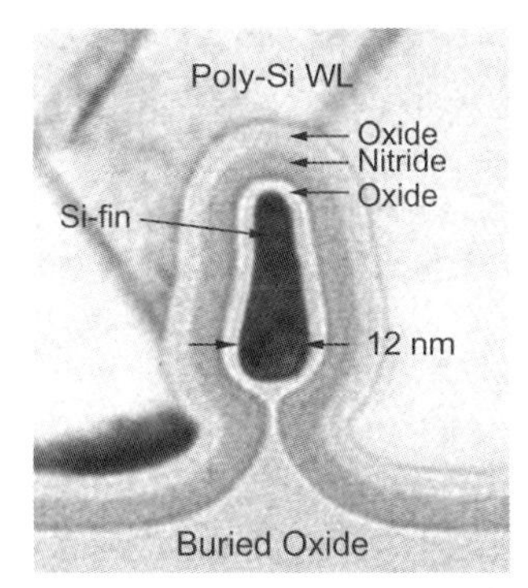

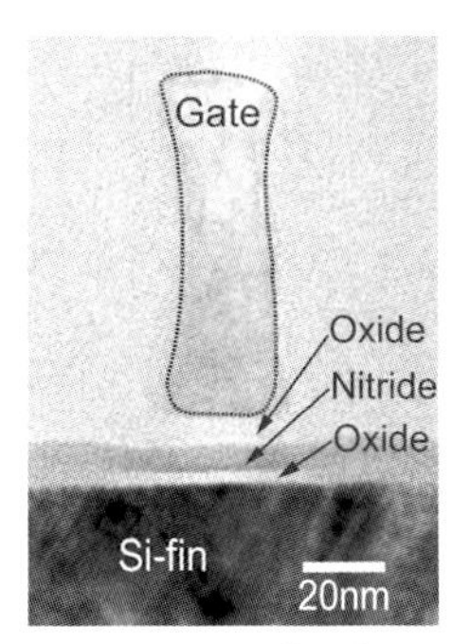

Figure 1.30 TEM cross-section of a tri-gate memory cell with 12 nm fin width and ONO dielectric.

20 nm gate length [29]. During an applied gate voltage of +12.5 V, 2 ms, electrons are injected and shift the I–V_g curves to positive voltages (write) (see Figure 1.31).

A V_t shift of about 4 V (write) was obtained using a fin width of 12 nm at gate lengths of 80 to 20 nm. The application of a negative gate voltage (erase) of 11 V, 2 ms, injects holes into the nitride layer or detraps electrons and shifts the I-Vg curves back to low V_t.

Due to the large V_t shift, multi-level storage becomes also feasible. Four levels with about 1 V separation have been programmed in the 40-nm memory transistor. The charge of one level corresponds to about 100 electrons.

The retention time for the charge-trapping dielectric has been tested, and a programming window of 3.6 V for single level was extrapolated after 10 years. Excellent retention properties between all levels are observed (see Figure 1.32).

If operated in a 4–5 F^2 high-density array such as NAND, the tri-gate cell would enable memory densities up to 32 Gbit at a die size of 130 mm^2 for the 25-nm node. A schematic NAND layout is shown in Figure 1.33. Finally, scaling is limited by the thickness of the two oxide–nitride–oxide layers, plus the minimal gate electrode thickness between the fins.

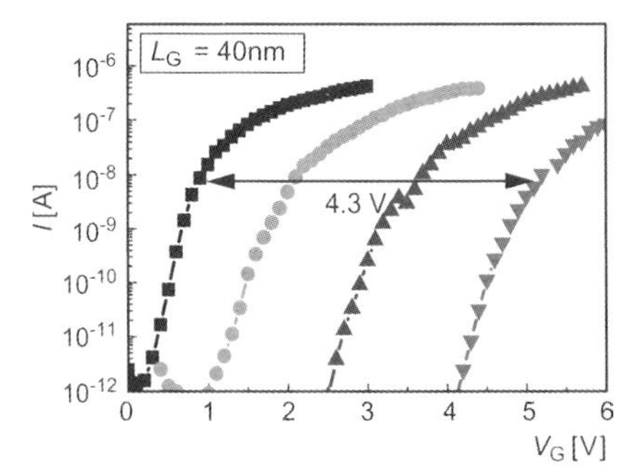

Figure 1.31 Write characteristics of a tri-gate memory cell with 40 nm gate length and multi-level operation. The different symbols represent write voltages between 0 V and 4.3 V.

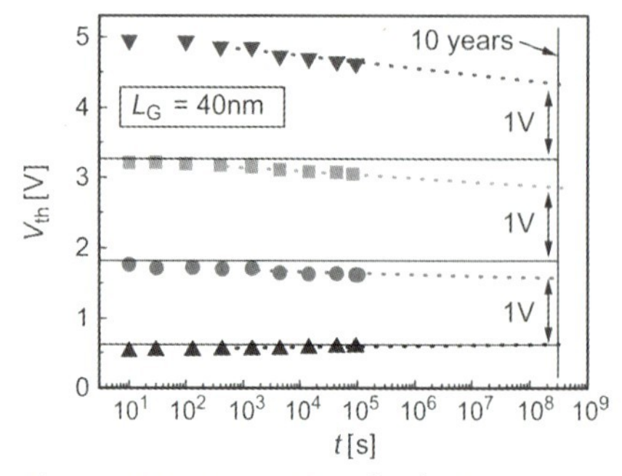

Figure 1.32 Retention time for the 40-nm tri-gate memory cell with oxide–nitride–oxide (ONO) dielectric at room temperature. The different symbols indicate the different write voltages of Figure 1.31.

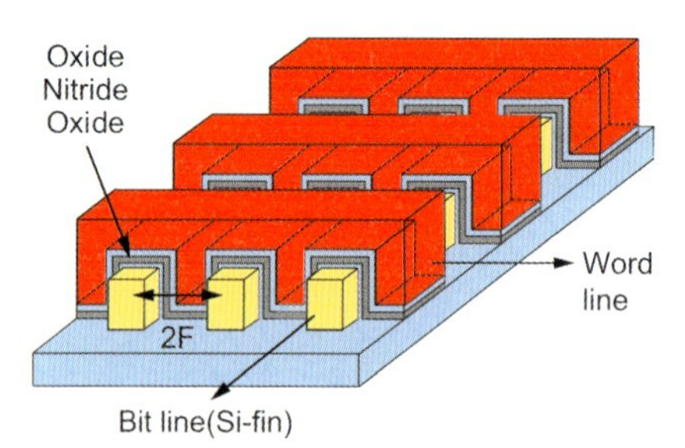

Figure 1.33 4 F^2 NAND array with tri-gate memory cells.

1.7
3d-DRAM Array Devices: RCAT, FinFET

For DRAMs, extremely low leakage current array devices below 1 fA per cell are required in order to avoid too-high charge losses during the refresh time interval. A contribution to the leakage current originates from the sub-V_t current of the cell transistor, while others are junction leakage and tunneling currents through the dielectric of the storage capacitor. With respect to the cell transistor, the channel doping cannot be increased in order to improve the turn-off characteristics, because of the electric field, which will initiate trap-assisted tunneling leakage currents [30] at $E > 0.5\,\text{MV}\,\text{cm}^{-1}$. Therefore, the planar DRAM cell transistors can be scaled down only to about 70 nm [30]. A schematic and a SEM cross-section of the 70-nm trench DRAM cell are shown in Figure 1.34.

For future applications, new cell transistor structures must be implemented. For stack DRAM cells, the transition to a recessed channel array transistor (RCAT) has already been reported for the 90- to 80-nm generation [31]. Such a device, with a U-shaped groove etched into silicon with a depth of about 200 nm, is shown in

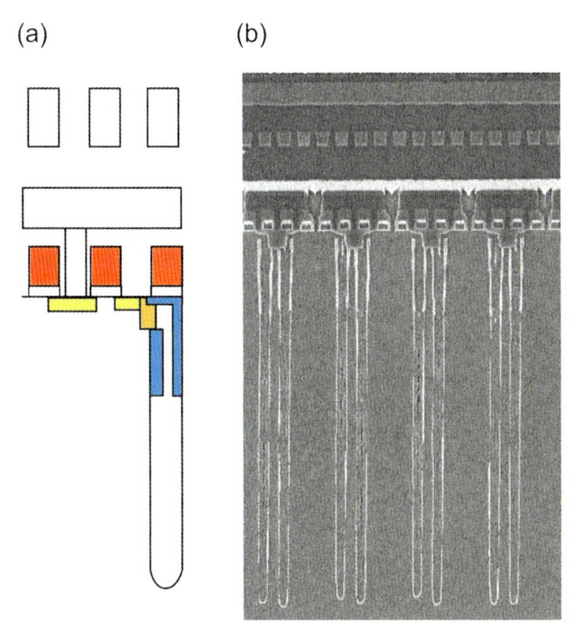

Figure 1.34 (a) Schematic cross-section of a trench DRAM cell with planar cell transistor and buried strap capacitor node contact [30]. The yellow rectangle indicates a n+ doped region in p-well; the red area is the gate (wordline); the blue is an isolation oxide. The other line is the bitline and the second wordline on top. (b) SEM cross-section of the 70-nm trench DRAM cell [30].

Figure 1.35 [32]. After gate dielectric growth, the groove is filled with the poly-silicon gate material. Bitline and storage node contacts are on the planar silicon. Such a structure is suitable for sub-70-nm generations because it provides a longer channel for lower I_{off} currents. In this Extended U-shape Device, a gate wrap-around the Si sidewalls with a depth of 6–10 nm increases the on-current and improves the subthreshold slope. The 3d device has been integrated into a 90-nm DRAM test array [32]. Simulation and measurement are shown in Figure 1.35b, with and without a corner device of about 6 nm. The subthreshold slope is in the range of 95 to 130 mV dec^{-1} at 85 °C, and the side gates enhance the on-current by 30%.

Reducing the width of the cell transistor to sublithographic dimensions and utilizing deeper vertical sidewalls leads to a fully depleted FinFET device with improved electrostatic control and increased on-currents [32]. A schematic cross-section in bitline direction of a trench cell with a FinFET array transistor, together with a SEM cross-section of a realized structure in 90 nm technology, are shown in Figure 1.36. The fin has a width of about 20 nm and a height of 50 nm. The transistor has been implemented using a local Damascene technique for fin and gate. The local gates are connected with a WSi wordline, which is also used for the gate layer of the planar transistors in the periphery circuits. The body is connected to the substrate and isolated to the neighboring fin with Sallow Trench Isolation.

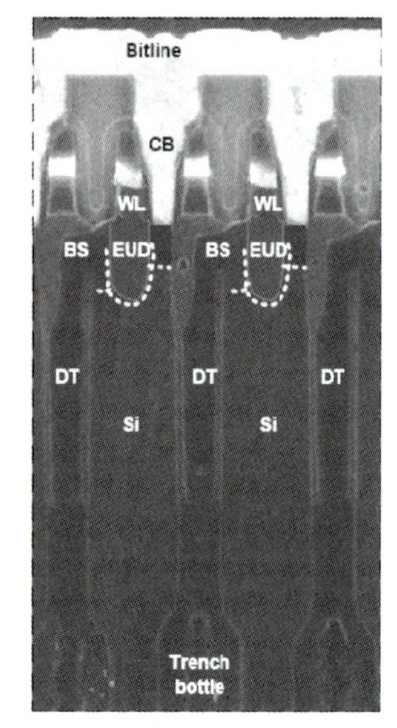

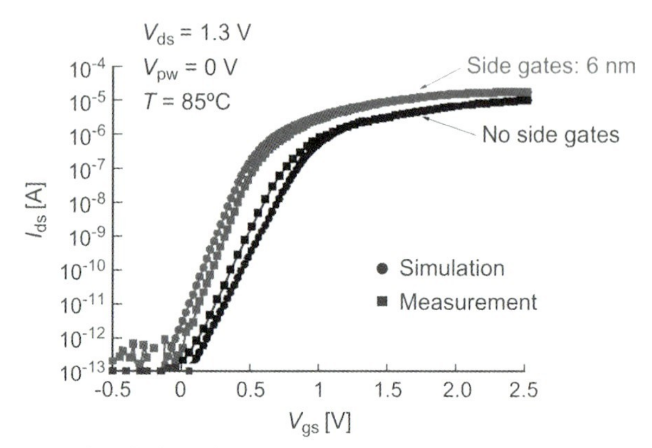

Figure 1.35 (a) SEM cross-section along bitline of a 90-nm trench DRAM cell with Extended U-Shape Device [32]. (b) Measurement and simulation of the I–V characteristics of the Extended U-Shape Device with and without 6-nm side gates at the corner [32].

A steep sub-threshold slope of 77 mV dec^{-1} without any drain-induced barrier lowering and back bias effect has been measured [32] in a 90-nm demonstrator (see Figure 1.37). According to simulations, the high I_{off}/I_{on} ratio will be maintained at least down to 40 nm, with a subthreshold slope of 89 mV dec^{-1} and without any remarkable influence of the adjacent trench cells, which can disturb the potential in the array device.

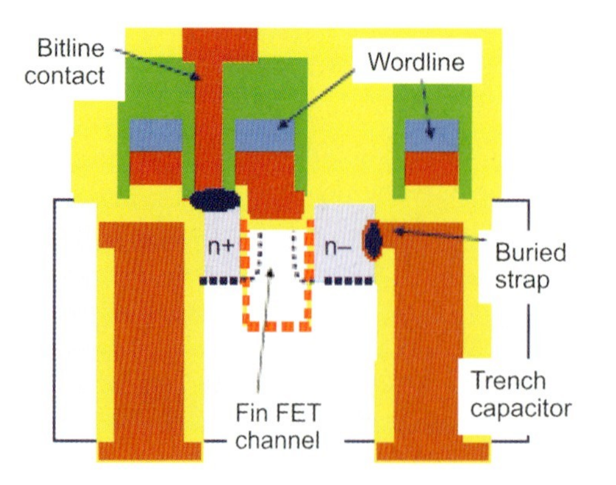

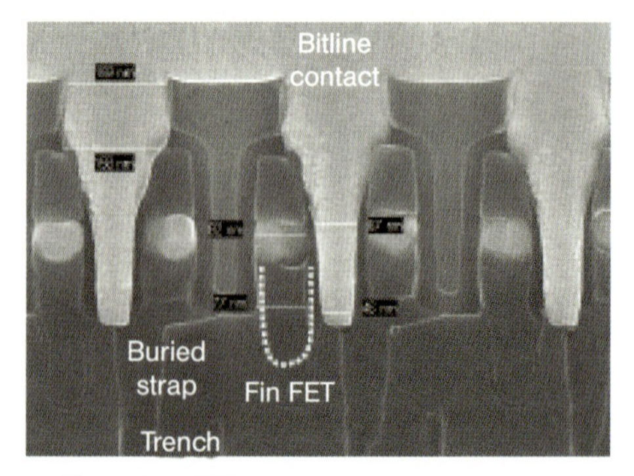

Figure 1.36 Trench DRAM with a FinFET-type cell transistor [32].

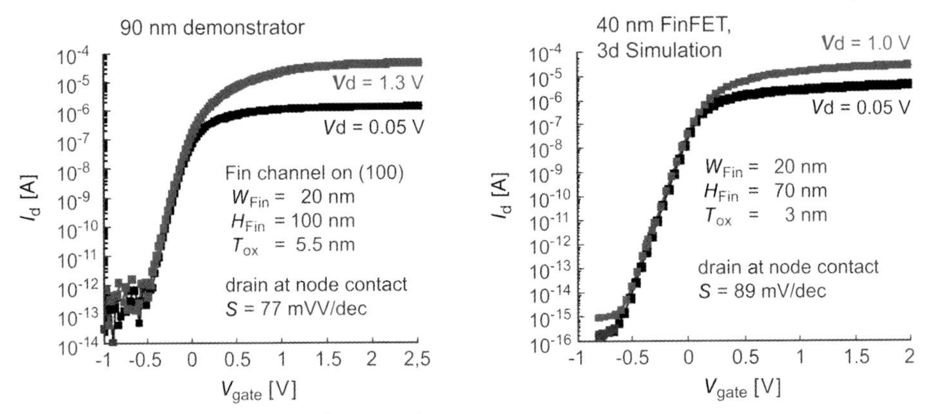

Figure 1.37 Measured FinFET array device *I–V* characteristics in a 90-nm trench cell demonstrator and simulation for 40 nm.

1.8 Prospects

Assuming that lithography tools such as Extreme Ultra-Violet will be available for the sub-45-nm technology nodes, it seems very likely that the scaling of Si CMOS will continue down to the 22-nm node, with the start of production in the year 2016, according to the ITRS roadmap. In this scenario – which is known as *More Moore* – technology costs must be reduced per chip from generation to generation, and performance must be increased. This will be expected especially for memories and microprocessors, and in order to fulfill these requirements more challenging new process modules, such as metal gate, high-*k* dielectrics, and strain will need to be integrated with high yield and in good time. On the other hand, conventional bulk CMOS may run into performance constraints below the 45-nm generation. Multi-gate devices with thin silicon channels and better electrostatic control may take over and will allow further downscaling, but with more complex processing. For DRAMs and Flash, the integration of such 3d transistors with very low leakage currents has already been started. Ultimately, beyond 10 nm the process tolerances and variability of the electrical parameters will become the most limiting factors. In addition, with the consistently good scaling potential of Si MOSFETs, many applications such as low-frequency RF, analogue, and powerFETs, displays and sensors do not require extremely small feature sizes. Therefore, additional functionality on the chip – referred to as *More than Moore* – will be another key trend.

Another important issue is the increasing research into new logic and memory devices. Among these are the 1d wire structures of Si, Ge or carbon with source, drain and gate, such as Si MOSFETs. These devices show similar *I–V* characteristics to Si (or even better), depending on the normalization of the current on the small width of the devices. However, the manufacturability and integration on a large scale has still to be proven, and the key for success would be the integration capability with Si CMOS.

With regards to memories, many promising new concepts have appeared, based on new materials such as the storage element. Among these are included non-volatile memories, with a large change in resistance, such as Phase-Change or Conductive Bridging. These memories can be combined very well with a Si access transistor and CMOS circuitry. With these evolutionary elements, non-conventional CMOS represents the most realistic approach for high-density logic and memories, and will undoubtedly represent the dominant technology of the nanoelectronics era.

Acknowledgments

The studies on SOI MOSFETs have been partly supported within the BMBF project HSOI, and Multi-Gate Devices within Extended CMOS and the EC Project NESTOR, IST-2001-37114. The author thanks the NESTOR partners for their courtesy, especially S. Deleonibus, T. Poiroux, P. Coronel, S. Harrison, N. Collaert, and Y. Ponomarev. Thanks are also expressed to the author's colleagues at Infineon/ Qimonda for their contributions, notably M. Alba, L. Dreeskornfeld, J. Hartwich, F. Hofmann, G. Ilicali, J. Kretz, E. Landgraf, T. Lutz, H. Luyken, W. Rösner, M. Specht, M. Staedele, C. Pacha, and W. Mueller.

References

1 ITRS Roadmap 2004 edition, http://public.itrs.net.

2 H. Wakabayashi, S. Yamagami, N. Ikezawa, A. Ogura, M. Narihiro, K.-I. Arai, Y. Ochiai, K. Takeuchi, T. Yamamoto, T. Mogami, Sub-10 nm planar-bulk-CMOS devices using lateral junction control (5 nm CMOS), *IEDM Technical Digest* 2003, 989.

3 D. K. Nayak, J. C. S. Park, K. Wang, K. P. MacWilliams, Enhancement-Mode Quantum-Well GexSi1-x PMOS, *IEEE-EDL* 1991, **12**, 154.

4 L. Risch, *et al.*, Fabrication and electrical characterization of Si/SiGe p-channel MOSFETs with a delta doped boron layer, Proceedings of ESSDERC, p. 465, 1996.

5 K. Rim, S. Koester, M. Hargrove, J. Chu, P. M. Mooney, J. Ott, T. Kanarsky, P. Ronsheim, M. Ieong, A. Grill, H.-S. P. Wong, Strained Si CMOS (SS CMOS) Technology, Proceedings VLSI Symposium, p. 59, 2001.

6 T. Ghani, M. Armstrong, C. Auth, M. Bost, P. Charvat, G. Glass, T. Hoffmann, K. Johnson, C. Kenyon, J. Klaus, B. McIntyre, K. Mistry, A. Murthy, J. Sandford, M. Silberstein, S. Sivakumar, P. Smith, K. Zawadzki, S. Thompson, M. Bohr, A 90 nm high volume manufacturing logic technology featuring novel 45 nm gate length strained silicon CMOS transistors, *IEDM Technical Digest* 2003, 978.

7 H. Irie, K. Kita, K. Kyuno, A. Toriumi, In-plane mobility anisotropy and universality under uni-axial strains in n- and p-MOS inversion layers on (1 0 0), (1 1 0), and (1 1 1) Si, *IEDM Technical Digest* 2004, 225.

8 H. Sayama, Y. Nishida, H. Oda, T. Oishi, S. Shimizu, T. Kunikiyo, K. Sonoda, Y. Inoue, M. Inishi, Effect of <1 0 0> channel direction for high performance SCE immune p-MOSFET with less than 0.15 μm gate length, *IEDM Technical Digest* 1999, 657.

9 B. Tavel, M. Bidaud, N. Emonet, D. Barge, N. Planes, H. Brut, D. Roy, J. C. Vildeuil, R. Difrenza, K. Rochereau, M. Denais, V. Huard, P. Llinares, S. Bruyère, C. Parthasarthy, N. Revil, R. Pantel, F. Guyader, L. Vishnubhotla, K. Barla, F. Arnaud, P. Stolk, M. Woo, Thin oxynitride solution for digital and mixed-signal 65 nm CMOS platform, *IEDM Technical Digest* 2003, **27.6**, 643.

10 S. De Gendt, Advanced Gate Stacks: high k and metal gates, 2004 IEDM Short Course 45 nm CMOS Technology.

11 J. Kedzierski, D. Boyd, P. Ronsheim, S. Zafar, J. Newbury, J. Ott, C., Jr. Cabral, M. Ieong, W. Haensch, Threshold voltage control in NiSi-gated MOSFETs through silicidation induced impurity segregation, *IEDM Technical Digest* 2003, **13.3**, 315.

12 J. P. Colinge, *SOI Technology: Materials to VLSI*, 2nd edition, Boston, MA, Kluwer, 1997.

13 B. Doris, M. Ieong, H. Zhu, Y. Zhang, M. Steen, W. Natzle, S. Callegari, V. Narayanan, J. Cai, S. H. Ku, P. Jamison, Y. Li, Z. Ren, V. Ku, D. Boyd, T. Kanarsky, C. D'Emic, M. Newport, D. Dobuzinsky, S. Deshpande, J. Petrus, R. Jammy, W. Haensch, Device design considerations for ultra-thin SOI MOSFETs, *IEDM Technical Digest* 2003, 631.

14 J. Hartwich, L. Dreeskornfeld, F. Hofmann, J. Kretz, E. Landgraf, R. J. Luyken, M. Specht, M. Staedele, T. Schulz, W. Rösner, L. Risch, Off-current adjustments in ultra-thin SOI MOSFETs, *Proceedings of ESSDERC*, p. 305, 2004.

15 EC Project NESTOR. IST-2001-37114.

16 M. Vinet, T. Poiroux, J. Widiez, J. Lolivier, B. Previtali, C. Vizioz, B. Guillaumot, Y. Letiec, P. Besson, B. Biasse, F. Allain, M. Casse, D. Lafond, J.-M. Hartmann, Y. Morand, J. Chiaroni, S. Deleonibus, High performance 10 nm bonded planar double metal gate CMOS transistors, *IEEE-EDL* May 2005, 317.

17 S. Harrison, P. Coronel, F. Leverd, R. Cerutti, R. Palla, D. Delille, S. Borel, S. Jullian, R. Pantel, S. Descombes, D. Dutartre, Y. Morand, M. P. Samson, D. Lenoble, A. Talbot, A. Villaret, S. Monfray, P. Mazoyer, J. Bustos, H. Brut, A. Cros, D. Munteanu, J.-L. Autran, T. Skotnicki, Highly performant double gate MOS-FET realized with SON process, *IEDM Technical Digest* 2003, 449.

18 X. Huang, W.-C. Lee, C. Kuo, D. Hisamoto, L. Chang, J. Kedzierski, E. Anderson, H. Takeuchi, Y.-K. Choi, K. Asano, V. Subramanian, T.-J. King, J. Bokor, C. Hu, Sub 50 nm FinFET, *IEDM Technical Digest* 1999, 67.

19 G. Ilicali, W. Weber, W. Rösner, L. Dreeskornfeld, J. Hartwich, J. Kretz, T. Lutz, J. P. Mazellier, M. Städele, M. Specht, J. R. Luyken, E. Landgraf, F. Hofmann, L. Risch, R. Käsmaier, W. Hansch, Planar double gate transistors with asymmetric independent gates, Proceedings International SOI Conference, 2005.

20 S. Monfray, D. Chanemougame, S. Borel, A. Talbot, F. Leverd, N. Planes, D. Delille, D. Dutartr, R. Palla, Y. Morand, S. Decombes, M.-P. Samsan, N. Vulliet, T. Sparks, A. Vandooren, T. Skotnicki, SON technological CMOS platform: Highly performant devices and SRAM cells, *IEDM Technical Digest* 2004, 635.

21 S.-Y. Lee, E.-J. Yoon, S.-M. Kim, C. W. Oh, M. Li, J.-D. Choi, K.-H. Yeo, M.-S. Kim, H.-J. Cho, S.-H. Kim, D.-W. Kim, D. Park, K. Kim, A novel sub 50 nm multi-bridge-channel MOSFET (MBCFET) with extremely high performance, 2004 Symposium on VLSI Technology, p. 200.

22 B. Yu, L. Chang, S. Ahmed, H. Wang, S. Bell, C.-Y. Yang, C. Tahery, C. Ho, Q. Xiang, T.-J. King, J. Bokor, C. Hu, M.-R. Lin, D. Kyser, FinFET scaling to 10 nm gate length, *IEDM Technical Digest* 2002, 251.

23 W. Roesner, E. Landgraf, J. Kretz, L. Dreeskornfeld, H. Schäfer, M. Städele, L. Risch, Nanoscale FinFETs for low-power applications, *Solid-State Electronics* 2004, **48**, 1819.

24 C. Pacha, K.v. Arnim, T. Schulz, W. Xiong, M. Gostkowski, G. Knoblinger, A. Marshall, T. Nirschl, J. Bertold, C. Russ, H. Gossner, C. Duvvury, P. Patruno, R. Cleavelin, K. Schruefer, Circuit design issues in multi-gate FET CMOS technologies, Proceedings ISSCC, 2006.

25 M. Städele, R. J. Luyken, M. Specht, G. Ilicali, W. Rösner, L. Risch, Speed considerations of fully depleted single and double gate SOI transistors, *Proceedings ULIS*, p. 87, 2005,

26 M. Städele, A. Di Carlo, P. Lugli, F. Sacconi, B. Tuttle, Atomistic tight-binding calculations for the transport in extremely scaled SOI devices, *IEDM Technical Digest* 2003, 229.

27 J.-H. Park, S.-H. Hur, J.-H. Lee, J.-T. Park, J.-S. Sel, J.-W. Kim, S.-B. Song, J.-Y. Lee, S.-J. Son, Y.-S. Kim, M.-C. Park, S.-J. Choi, U.-I. Chung, J.-T. Moon, K.-T. Kim, K. Kim, B.-L. Ryu, 8 Gb MLC NAND flash memory using 63 nm process technology, *IEDM Technical Digest* 2004, 873.

28 J. Willer, C. Ludwig, J. Deppe, C. Kleint, S. Riedel, J.-U. Sachse, M. Krause, R. Mikalo, E. Stein, V. Kamienski, S. Parascondola, T. Mikolajick, J.-M. Fischer, M. Isler, K.-H. Kuesters, I. Bloom, A. Shapir, E. Lusky, B. Eitan, 110 nm NROM technology for code and data flash products, 2004 Symposium on VLSI Technology, p. 76.

29 M. Specht, U. Dorda, L. Dreeskornfeld, J. Kretz, F. Hofmann, M. Staedele, R. J. Luyken, W. Rösner, H. Reisinger, E. Landgraf, T. Schulz, J. Hartwich, R. Kömmling, L. Risch, 20 nm tri-gate SONOS memory cells with multi-level operation, *IEDM Technical Digest* 2004, 1083.

30 J. Amon, A. Kieslich, L. Heineck, T. Schuster, J. Faul, J. Luetzen, C. Fan, C.-C. Huang, B. Fischer, G. Enders, S. Kudelka, U. Schroeder, K.-H. Kuesters, G. Lange, J. Alsmeier, A highly manufacturable deep trench based DRAM cell layout with a planer array device in a 70 nm technology, *IEDM Technical Digest* 2004, 73.

31 H. S. Kim, D. H. Kim, J. M. Park, Y. S. Hwang, M. Huh, H. K. Hwang, N. J. Kang, B. H. Lee, M. H. Cho, S. E. Kim, J. Y. Kim, B. J. Park, J. W. Lee, D. I. Kim, M. Y. Jeong, H. J. Kim, Y. J. Park, Kinam. Kim, An outstanding and highly manufacturable 80 nm DRAM technology, *IEDM Technical Digest* 2003, **17.2**, 411.

32 W. Mueller, G. Aichmayr, W. Bergner, E. Erben, T. Hecht, C. Kapteyn, A. Kersch, S. Kudelka, F. Lau, J. Luetzen, A. Orth, J. Nuetzel, T. Schloesser, A. Scholz, U. Schroeder, A. Sieck, A. Spitzer, M. Strasser, P.-F. Wang, S. Wege, R. Weiset, Challenges for the DRAM cell scaling to 40 nm, *IEDM Technical Digest* 2005, **14.1**.

Further Reading

Short Course on Silicon+: Augmented Silicon Technology, Organizer T.-J. King, IEDM, December 7, 2003.

Emerging Nano-Electronics: Scaling MOSFETs to the Ultimate Limits and Beyond-MOSFET Approaches, Organizers P. Zeitzoff, T. Mogami, VLSI Technology Short Course, June 14, 2004.

Advanced CMOS Devices on bulk and SOI: Physics, modeling and characterization, T. Poiroux, G. Le Carval, Short Courses ESSDERC, September 12, 2005.

Non Classical CMOS: Novel Materials, Novel Device Structures, and Technology Roadmap, H.-S. Philip Wong, Short Courses ESSDERC, September 12, 2005.

2
Indium Arsenide (InAs) Nanowire Wrapped-Insulator-Gate Field-Effect Transistor

Lars-Erik Wernersson, Tomas Bryllert, Linus Fröberg, Erik Lind, Claes Thelander, and Lars Samuelson

2.1
Introduction

Semiconductor nanowires [1–9] offer the possibility to form a new class of semiconductor device. Nanowire technology enables new material combinations and also the possibility to enhance the potential control in down-scaled channels using wrap-around gates. As the lateral dimensions of semiconductor materials are scaled down towards 100 nm and below (which can be easily achieved with the nanowire technology), fewer constraints become apparent in terms of lattice matching between materials. This opens the path to a heterogeneous materials integration that cannot be accomplished with conventional bulk semiconductor technology. For example, it has been shown that segments of InP can be incorporated in indium arsenide (InAs) nanowires [10], and that InAs nanowires can be grown on Si substrates [11], in spite of about 3.5% and 7% lattice mismatch, respectively. These material combinations cannot be synthesized in the bulk, nor with planar epitaxial techniques. The second advantage is related to the challenges that the technology is facing as the planar transistors are scaled down towards the 22 nm node and beyond. At this length scale, the transistors are more sensitive to short-channel effects related to the reduced potential control in the channel and the body of the devices. This is reflected in an increased output conductance and sub-threshold swing of the transistors that degrade the transistor performance. Dual gates, trigates and FinFETs have been demonstrated to reduce these issues. Taking the technology one step further is to completely surround the channel with a wrapped gate, and for this technology vertical nanowires are ideal.

Several groups have reported on the successful fabrication of vertical nanowire transistors [12–22]. Various implementations of Si transistors have been reported and, in particular, it has been shown that the wire geometry may be used to fabricate different advanced transistors with benefits in sub-threshold characteristics and switching behavior. The present authors' effort has been focused on the vertical

Nanotechnology. Volume 4: Information Technology II. Edited by Rainer Waser
Copyright © 2008 WILEY-VCH Verlag GmbH & Co. KGaA, Weinheim
ISBN: 978-3-527-31737-0

implementation of III–V transistors, and in the following sections are described the processing and characteristics of both long- and short-channel transistors. The importance is also demonstrated of introducing a high-k dielectric, its influence on the device characteristics, and the benefits of heterostructure design.

2.2 Nanowire Materials

In these studies, InAs has been the primary choice of material in the transistors. For various reasons, wrap-gate transistors based on silicon will naturally have a very strong standing, due primarily to the compatibility with silicon technology in general and also to the fact that Si nanowires can be made with diameters even <5 nm and yet still be conducting. InAs, on the other hand, shows very strong lateral confinement effects already for diameters around 30 nm, making very narrow uncapped InAs transistors depleted of charge-carriers and, in that sense, less promising. In contrast, n-type InAs has highly attractive material properties, with a reproducible Fermi-level pinning in the conduction-band, with a very high room-temperature mobility and ideal ohmic-contact properties. The remainder of this chapter focuses on the use of InAs as the active transistor channel material and the use of P-containing $InAs_{1-x}P_x$-alloys for enhancement of the transistor functionality and performance.

2.3 Processing

The nanowires used here are grown with chemical beam epitaxy (CBE), using patterned Au discs to locate the nanowire growth and to set the diameter of the nanowires (Figure 2.1). The ability to form well-defined matrices of nanowires is a key feature both for the post-growth device processing and for the transistor design, in that the number of wires determines the drive current and the transconductance of the nanowire transistor. The uniformity in length provides good starting conditions for uniform top contact formation. Typically, nanowire matrices ranging from 1×1 to 10×10 are used to form the vertical transistors in order to reach drive currents approaching 10 mA, but a nanowire transistor may be defined by anything from a single wire to, say, 10^4 wires. Outside the active transistor region, smaller arrays of nanowires are formed to create alignment markers for optical lithography in the post-growth processing described below. The seed for these wires are formed in the same seed and growth steps as the actual transistor wires.

After the growth, either long-channel or short-channel transistors may be formed by processing the transistors in two different ways, as shown in Figures 2.2 and 2.4, respectively [15–22]. For the long-channel transistors, the vertical nanowire matrix is first covered by a SiN_x gate-dielectric layer, followed by a sputtered Ti/Au gate metal that is covering the nanowires uniformly. The sample is spin-coated by a photoresist, which is back-etched to the desired gate length, after which the gate metal is

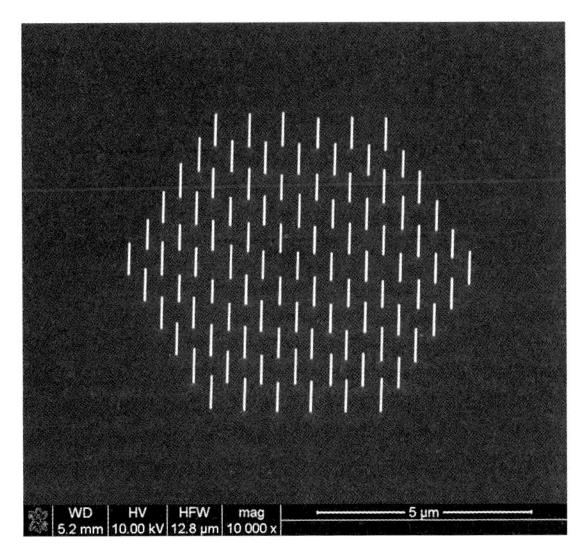

Figure 2.1 Scanning electron microscopy image of a nanowire matrix grown by chemical beam epitaxy.

selectively removed by wet-etching. After removal of the resist, the sample is covered by a second resist layer and the SiN_x is etched to open for formation of the drain Ti/Au top contact by evaporation. Finally, an airbridge is created by electroplating from the drain contact, and the resist is dissolved. With this technology, the transistor structure shown in Figure 2.3 is formed. While this technology provides good long-channel devices, in which fluctuations in the gate-length are less important due to the smaller relative change, is seems difficult to reproducibly scale the definition of the

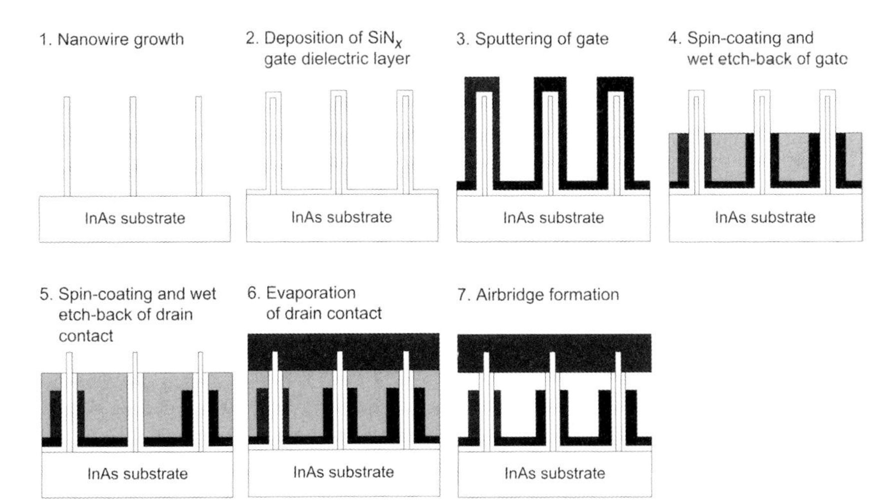

Figure 2.2 Processing scheme for long-channel transistors with sputtering and back-etching.

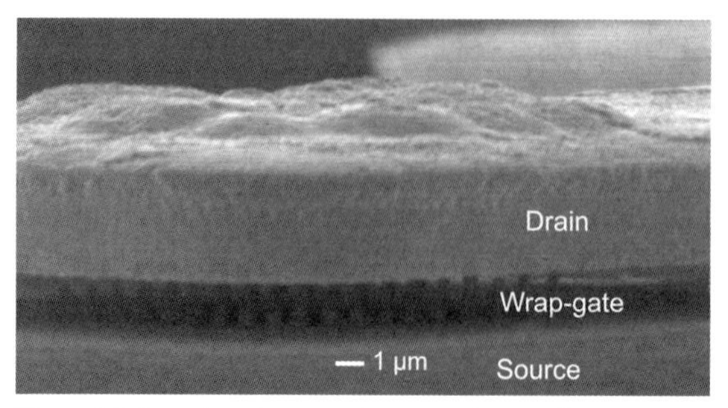

Figure 2.3 Scanning electron microscopy image of wrap-gate transistor with air-bridge drain contact [15].

gate-length below 100 nm when using this back-etch process. Instead, a direct evaporation method is used to form gates with a length below 100 nm, as described next.

For the processing of short-channel transistors, a direct evaporation of the gate metal has been developed. In this process, the metal gate is evaporated onto the SiN_x-covered nanowires, the main benefit of this approach being that the gate-length is determined by the thickness of the evaporated layer. This is in contrast to the previously described long-channel process, where it is set by the thickness of the back-etched polymer film. A scanning electron microscopy (SEM) image of a formed 80 nm-thick gate is shown in Figure 2.5. As can be seen in the image, an intimate contact is formed between the gate layer and the nanowire. Following gate formation, the drain contact is formed by spin-coating the sample with a resist and wet etching of

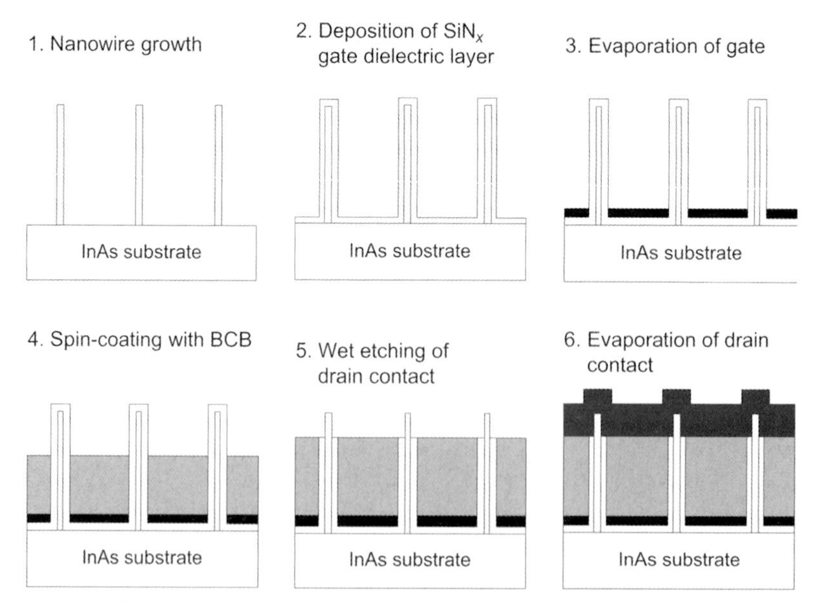

Figure 2.4 Processing scheme for short-channel transistors with evaporation of the gate [21]. BCB = benzocyclobutane.

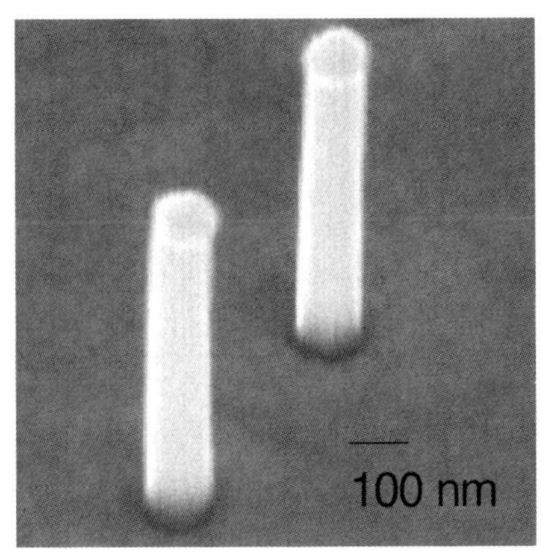

Figure 2.5 InAs nanowires coated with SiNx penetrating an evaporated Ti/Au gate electrode [15].

the tips of the nanowires that penetrate the organic film. Finally, an evaporated drain top contact is formed over the wires.

2.4 Long-Channel Transistors

The long-channel transistors have been characterized using the substrate as a grounded source contact. The data for a 10×10 nanowire matrix with a gate length of 800 nm, a wire diameter of 80 nm, and a thickness of 40 nm in the SiN_x-layer is shown in Figure 2.6. The transistor shows a good current saturation already at low drain biases $V_{sd} = 0.2$ V and a transconductance of about 1 mS. At larger drain biases ($V_{sd} > 1$ V) the transistor shows an increase in the drain current, most likely related to impact ionization processes in the InAs channel. In order to explain the transistor operation, the transistor structure is modeled as a cylindrical version of the planar metal-insulator-semiconductor field-effect transistor (MISFET), as shown in Figure 2.7. In the MISFET, the gate potential is used to deplete carriers in the channel and thus to modulate the conducting area in the cross-sectional core of the nanowire. As the gate length is a factor ten-fold larger than the diameter of the wire and the thickness of the dielectric, a good potential control is obtained in the channel, and this is reflected in the measured low-output conductance of the transistor.

The long-channel transistors show the expected $I_{sd} \sim V_g^2$ dependence, as demonstrated for a 40-nanowire transistor in Figure 2.8. From these data the threshold voltage is deduced to be $V_t = -0.16$ V. From the analytic fitting in Figure 2.8, values are deduced for the carrier concentration and the mobility ($N_d = 3 \times 10^{17}\,\mathrm{cm}^{-3}$, $\mu = 9600\,\mathrm{cm}^2\,\mathrm{V}^{-1}\,\mathrm{s}^{-1}$). The sub-threshold characteristics of the transistor are shown in Figure 2.9. At $V_{sd} = 0.2$ V an inverse sub-threshold slope of 100 mV decade^{-1} was measured, and a maximum current on-off ratio above 1000. To further verify the mode of operation in the transistor, the transistor characteristics were also simulated

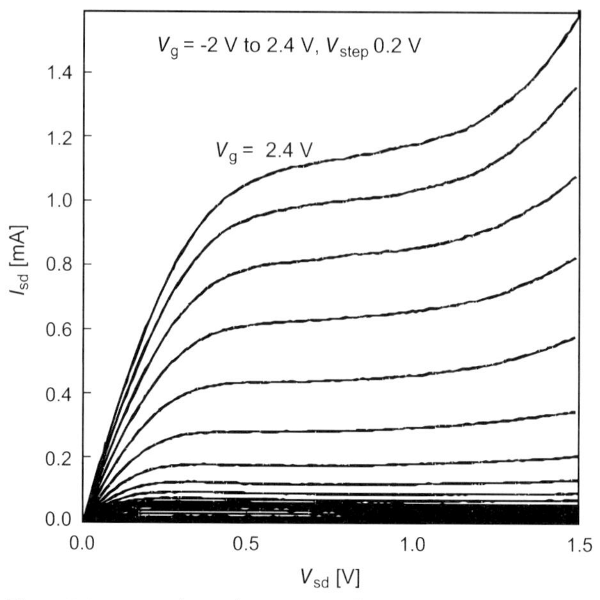

Figure 2.6 Measured *I–V* characteristics for an InAs long-channel NW-transistor [16].

using the Atlas device simulator created by Silvaco [16]. As these devices have a long channel length and a wide diameter, effects related to lateral quantization, doping fluctuations and ballistic transport may be omitted, and the transistors may be modeled within the drift-and-diffusion formalism. The simulated data in Figure 2.9 are obtained for $N_d = 2 \times 10^{17}\,\mathrm{cm}^{-3}$ and $\mu_e = 10\,000\,\mathrm{cm}^2\,\mathrm{V}^{1-1}\,\mathrm{s}^{-1}$, and reproduce the measured data both in the on-state and in the off-state. Thus, the measured data may be reproduced both by analytical modeling and by simulation in a MISFET model.

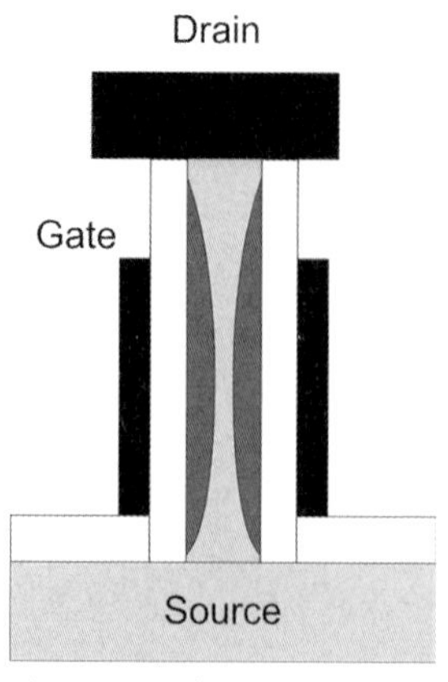

Figure 2.7 Schematic illustration of nanowire MISFET operation.

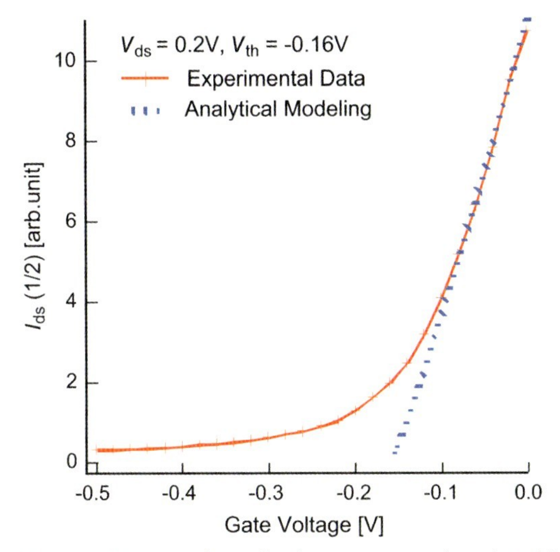

Figure 2.8 Measured transfer characteristics with analytical fitting to deduce the threshold voltage [16].

2.5 Short-Channel Transistors

Scaling is of importance to any FET-technology, and for the nanowire FET the scaling of both the gate length and nanowire diameter must be considered. The processing outlined above has been used the to fabricate transistors with 80 nm nominal gate length [19, 20]. During the growth of these nanowires, matrices with different nanowire diameters (70 and 55 nm) have been included on the same sample. Both types of nanowire transistor were processed in the same batch, and the transistor characteristics compared (see Figure 2.10). In both cases, good transistor characteristics were observed, with both transistors showing a limited current saturation, even

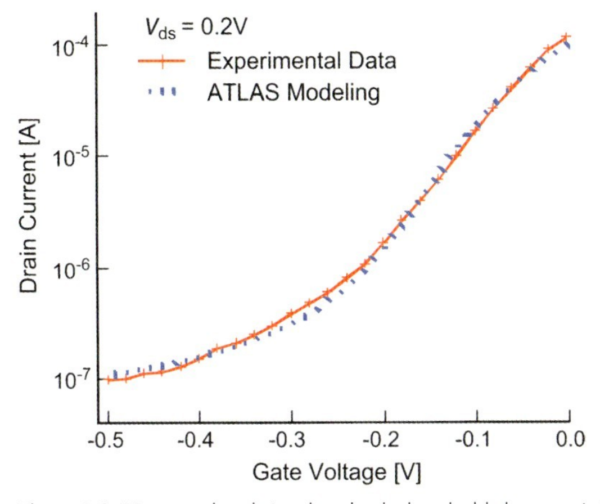

Figure 2.9 Measured and simulated sub-threshold characteristics for a long-channel transistor [15].

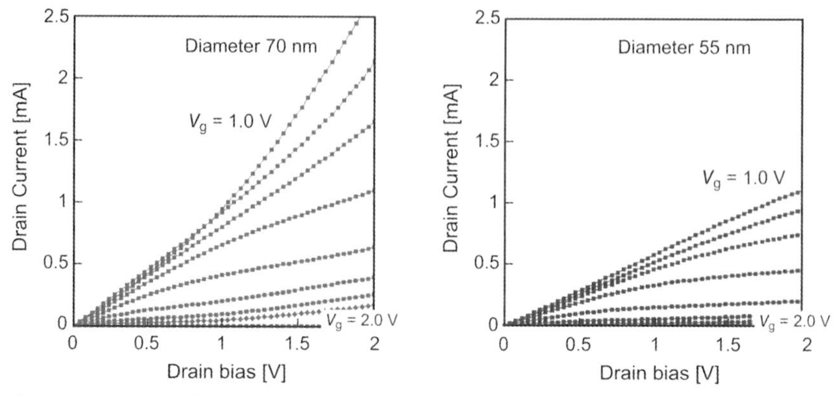

Figure 2.10 Measured transistor characteristics for 80 nm gate-length NW-transistor with 70 nm (left) and 55 nm (right) diameters [20].

at comparably large drain voltages. The increased saturation voltage arises from a series resistance in the 1 μm separation between the gate and the drain. For biases above 1 V, the larger-diameter transistors show a punch-through in the characteristics, a feature that was not observed in the narrower transistors that have a better potential control. When the drain current was normalized with the circumference of the nanowire, only a minor drive current reduction per gate width was observed as the diameter was reduced; this demonstrates good scalability in the technology.

In order to scale the gate length further, the relatively thick SiN_x dielectric was replaced with 10 nm HfO_2, a material with a higher dielectric constant (15) (Figure 2.11) [21, 22]. The HfO_2 was deposited using atomic layer deposition (ALD), which gives a uniform dielectric coverage and a very accurate thickness. A 100-nm layer of silicon oxide was also deposited, to act as a lower-*k* spacer layer between the InAs substrate and the wrap-gate. Next, a 50-nm Cr gate layer was formed by metal evaporation. Finally, a 100- to 200-nm-thick polymer layer was deposited on top of the gate to provide insulation between the gate and the drain contact. Despite a very short gate length (50 nm), considerably improved dc characteristics were observed compared to previous device designs. Transconductance values up to 0.8 S mm^{-1} were obtained ($V_{sd} = 1$ V), with an inversed sub-threshold slope around 100 mV dec^{-1}. The transconductance values were in this case normalized to the total nanowire circumference for the array. In addition to a gate swing of 0.5 V, an I_{on}/I_{off} ratio >1000 at $V_{sd} = 0.5$ V following the conventional definition [23] was observed, whereas a maximum I_{on}/I_{off} ratio above 10^4 was measured.

2.6
Heterostructure WIGFETs

The wrap-gate transistors show a good on-state characteristics, but even the long gate transistor characteristics suffer from a comparably large inverse sub-threshold slope (100 mV dec^{-1}) and a non-negligible off-state current. This is worse for the devices

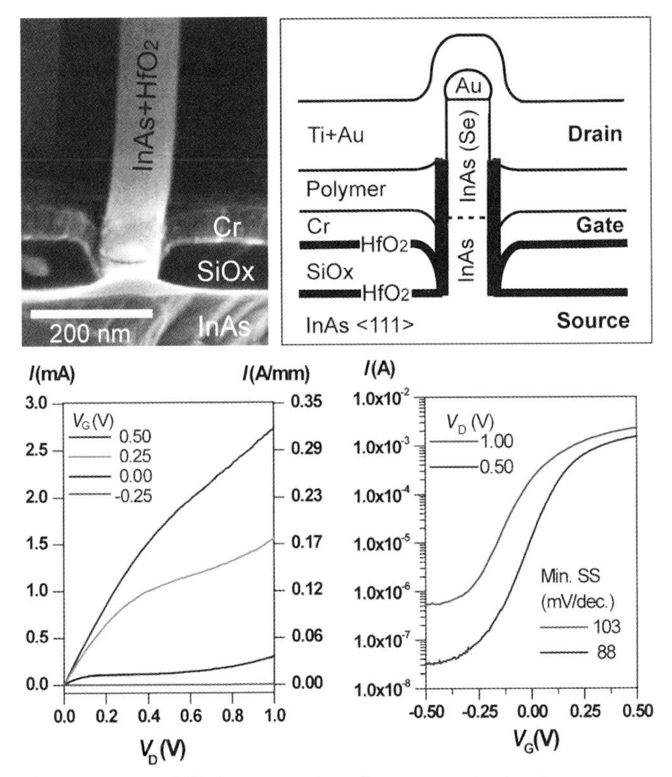

Figure 2.11 (Top left) A cross-section of a test sample showing the SiOx spacer layer, and the Cr wrap gate. (Top right) Illustration of the device design for an individual nanowire element in an array. (Bottom left) Output characteristics for a 61 nanowire wrap-gate array. (Bottom right): Sub-threshold characteristics for the same device for two different drive voltages [21].

with a short gate and a comparably thick (40 nm) gate-dielectrics. The transistors with high-*k* gate oxides also show effects related to the narrow InAs band gap that allows for impact ionization processes and thus creates a limited potential barrier in the off-state. The nanowire technology offers alternative transistor designs in that heterostructure segments may be incorporated into the transistor channel to alter the band gap in critical regions. A segment of InAsP was introduced into the InAs channel of a nanowire transistor and the role of the barrier in transistor performance subsequently investigated [24]. A 150 nm-long segment of InAsP was introduced into a 4 μm-long, 50 nm-diameter InAs nanowire grown by CBE. The nominal P content in the InAsP segment was 30%, and the conduction band barrier 180 meV. The nanowire was placed in a lateral geometry with a Si/SiO_2 back gate, where two drain contacts and one source contact was used in order to fabricate and evaluate transistors with the same geometry differing in only the InAsP barrier (Figure 2.12). Room-temperature data for the two types of transistor are shown graphically in Figure 2.13.

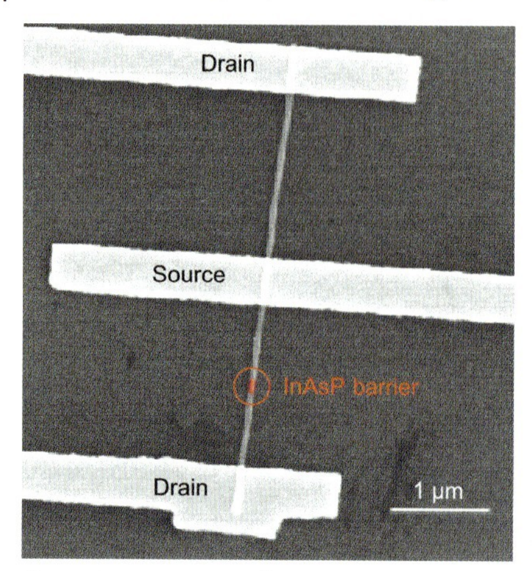

Figure 2.12 Scanning electron microscopy top image of lateral heterostructure nanowire transistor [23].

Both transistors showed good characteristics with current saturation and a decent drive current level. For a given bias condition ($V_{sd} = 0.3$ V and $V_g = 2.0$ V) the InAs transistor had a factor 2:1 higher drive current than the InAsP transistor. This was expected due to the introduction of the barrier. From the transfer characteristics, however, it should be noted that the current reduction was not related to a degradation in the transconductance, but rather to a shift in the threshold voltage. In fact, the measured transconductance remained constant, and for a fixed gate-overdrive the drive current was the same. When turning to the sub-threshold characteristics, major improvements were noted in both the inverse sub-threshold slope and the maximum I_{on}/I_{off} ratio as the barrier was introduced. Finally, temperature-dependent measurement of the current level was used to verify the presence of the barrier and to evaluate its height (Figure 2.14). The role of the barrier in this geometry is not only to block the off-current in the body of the wire, but also (and even more in this lateral geometry) to block the leakage current along the edges of the wires.

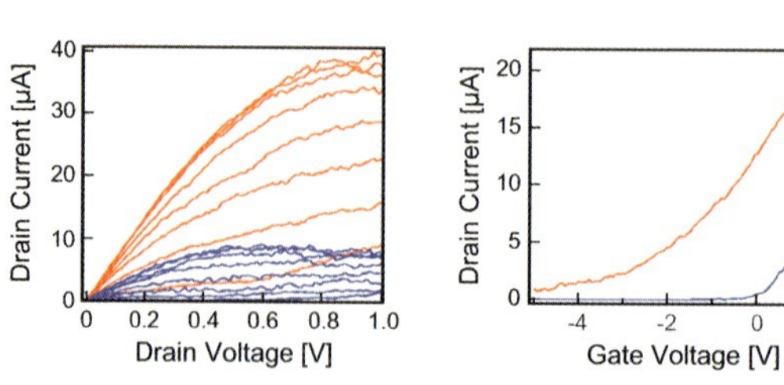

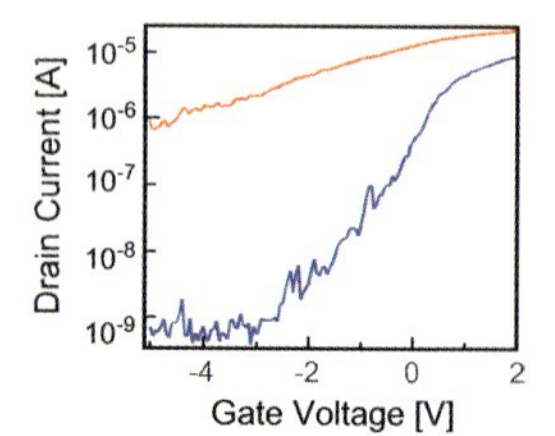

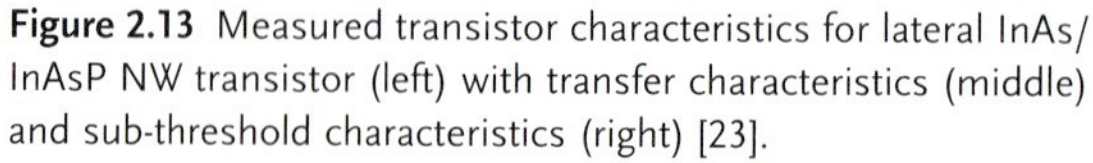

Figure 2.13 Measured transistor characteristics for lateral InAs/InAsP NW transistor (left) with transfer characteristics (middle) and sub-threshold characteristics (right) [23].

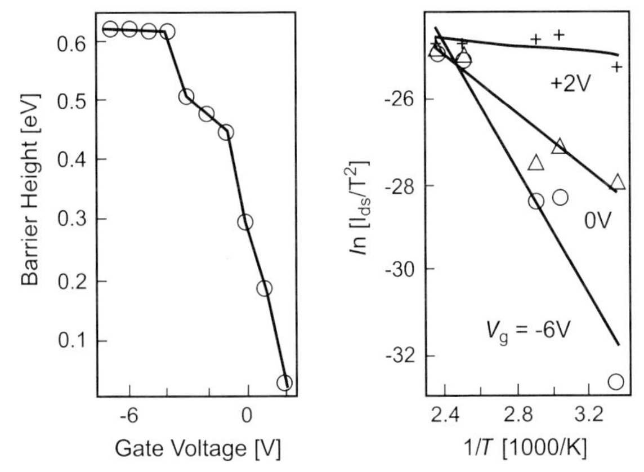

Figure 2.14 Deduced activation energies for varying gate bias (left) and the corresponding Arrhenius plot (right).

2.7 Benchmarking

It is of great value to perform an early evaluation of the potential in this new wrap-gate transistor technology. Hence, the performance of 100 nm gate-length transistors (structure shown in Figure 2.15) has been simulated and the characteristics evaluated according to the metrics of high-performance logic devices, including the gate delay

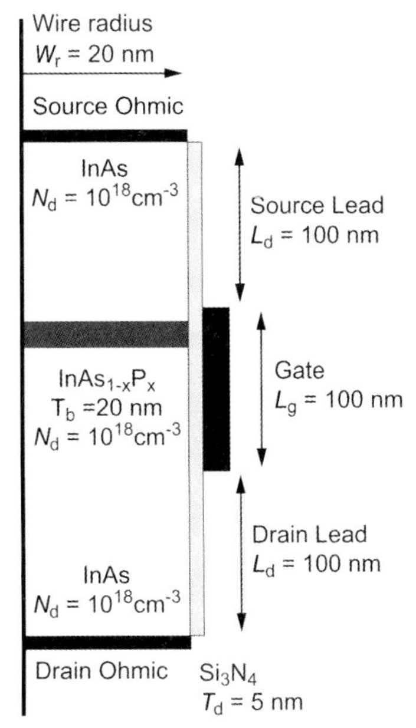

Figure 2.15 Schematic of nanowire geometry used for the bench-marking [25].

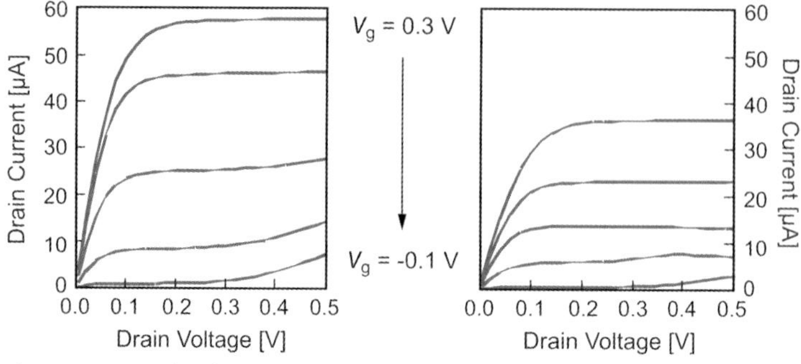

Figure 2.16 Simulated *I–V* characteristics for InAs nanowire wrap-gate transistor (left) and InAs/InAsP nanowire wrap-gate transistor (right) [25].

($\tau = C_{gg}V_{ds}/I_{on}$), the energy-delay-product ($\text{EDP} = \tau C_{gg}{V_{ds}}^2$), the current I_{on}/I_{off} ratio, and the inverse sub-threshold slope [25]. InAsP barriers with different P contents were further introduced into the InAs channel as a way of reducing the off-current, in analogy with the lateral devices previously described. In these simulations, the wire diameter was set to 40 nm and the doping level fixed to $1.0 \times 10^{18}\,\text{cm}^{-3}$ in order to obtain usable threshold voltages around $V_t = 0\,\text{V}$ and to avoid parasitic access resistance in the source and drain leads. The mobility was set to $10\,000\,\text{cm}^2\,\text{V}^{-1}\,\text{s}^{-1}$ and v_{sat} to $3 \times 10^5\,\text{m}\,\text{s}^{-1}$, in accordance with the fitting in Figure 2.9. The simulated output characteristics are shown in Figure 2.16 for a pure InAs channel and for a channel with a barrier of 20% P, respectively. Both types of transistor showed a good saturation already at about $V_{sd} = 0.1\,\text{V}$, which was a reflection of the low access resistance and the possibility of achieving excellent ohmic contacts to InAs. Obviously, the drive current for fixed gate bias was reduced by a factor 1.5 as the 20% barrier was introduced.

By evaluation of the transfer characteristics (Figure 2.17), this reduction in the drive current was found to be mainly related to a positive shift of the threshold

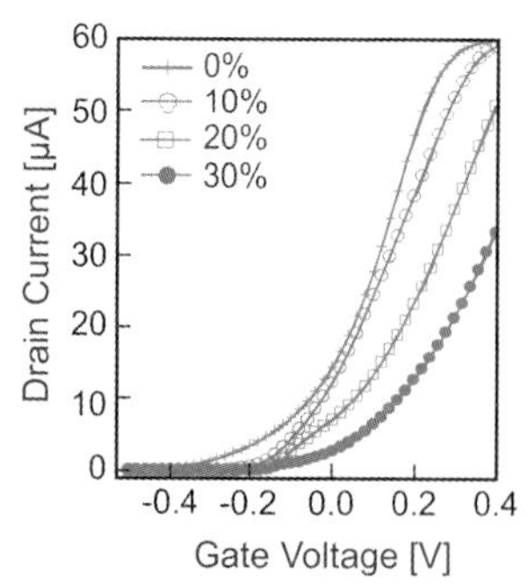

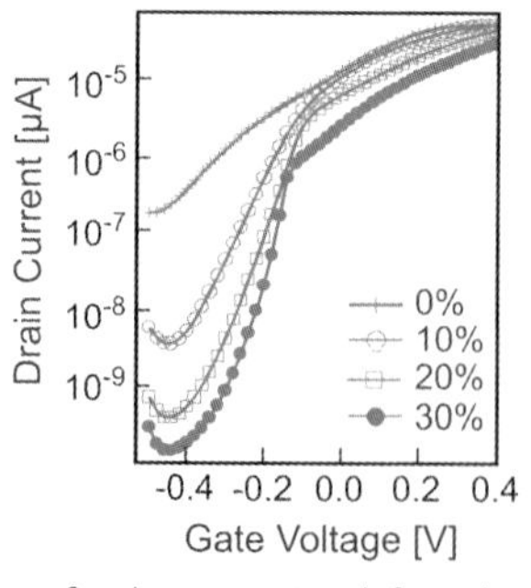

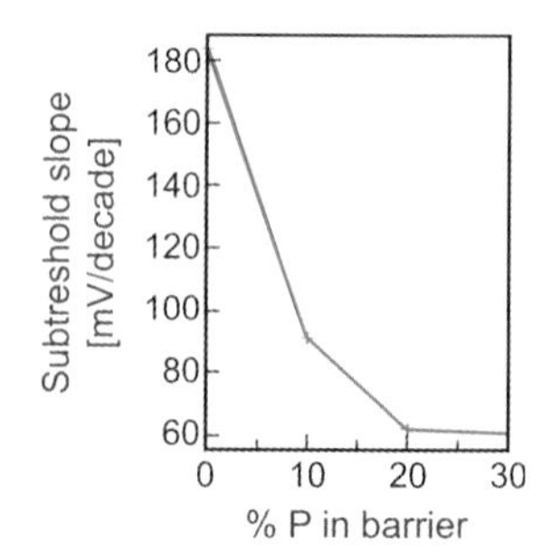

Figure 2.17 Simulated transfer characteristics (left), sub-threshold characteristics, and deduced inverse sub-threshold slope (right) for InAsP nanowire transistors with varying P content [24].

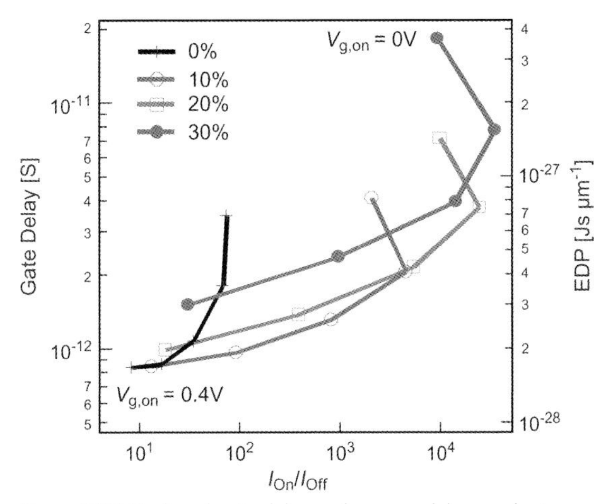

Figure 2.18 Deduced gate delay and energy–delay product as function of current on/off ratio for InAsP nanowire transistors with different P content and at varying bias conditions [25].

voltage, whilst there was only a minor degrading in the transconductance. On a logarithmic scale, it should be noted not only that the barrier provides a way to shift the threshold voltage, but also that the transition from the exponential to the linear characteristics becomes sharper, the current on/off ratio increases, and the inverse sub-threshold slope is reduced. All of these factors reduce the power consumption in circuits. The deduced critical metrics – that is, the gate delay, the energy delay product, the current on/off ratio, and the inverse sub-threshold swing – are shown in Figure 2.17 for various P contents in the barrier. As the threshold voltage shift with the barrier height, the evaluation is performed for various $V_{g,on}$, while the gate swing is kept constant at 0.5 V for all cases. The main point in Figure 2.18 is that the pure InAs channel provides the shortest gate delay due to the largest drive current, but also the lowest current I_{on}/I_{off} ratio related to the narrow band gap. While the introduction of the P barrier into the channel increases the minimum gate delay, it provides a way of increasing the I_{on}/I_{off} ratio to above 10^3 even for a gate swing of 0.5 V, a value required for many circuit applications. As an added benefit, the sharper transition between the on- and off-states further reduces the sensitivity to the choice of gate bias. These simulations show the potential in the technology and also demonstrate the benefit of introducing heterostructures into the nanowires.

2.8 Outlook

Based on the experimental results obtained to date, the question might be asked as to how far the nanowire FET technology may be developed? Critical issues for scaled devices are related to the growth of narrow nanowires with diameters of 10 to 30 nm and the processing of vertical gates on the 20 nm length scale. Based on experimental results, devices processed on these dimensions seem feasible in the near future. However, in order for these devices to be competitive it will be necessary for the drive current to be increased and the parasitics reduced. Likewise, good control of the carrier concentration in the channel and in the source and drain regions will be needed, as will an understanding and control of the interface properties in capped wires. The main benefit of the wire geometry – the possibility for heterostructure design in the axial and radial directions – may well prove to be the key when addressing these issues.

Acknowledgments

These studies were conducted within the Nanometer Structure Consortium at Lund University, with financial support from the Swedish Research Council (V.R.), the Swedish Foundation for Strategic Research (S.S.F.), the Knut and Alice Wallenberg Foundation (K.A.W.), and from the European Union via the project NODE 015783.

References

1 C. P. Auth, J. D. Plummer, Scaling theory for cylindrical, fully-depleted, surrounding-gate MOSFETs, *IEEE Electron Device Lett.* 1997,**18** (2), 74–76.

2 H. Takato, K. Sunouchi, N. Okabe, A. Nitayama, K. Hieda, F. Horiguchi, F. Masuoka, Impact of Surrounding Gate Transistor (SGT) for ultra-high-density LSIs, *IEEE Trans. Electron. Dev.* 1991, **38** (3), 573.

3 S. D. Suk, S.-Y. Lee, S.-M. Kim, E.-J. Yoon, M.-S. Kim, M. Li, C. W. Oh, K. H. Yeo, S. H. Kim, D.-S. Shin, K.-H. Lee, H. S. Park, J. N. Han, C. J. Park, J.-B. Park, D.-W. Kim, D. Park, B.-I. Ryu, High performance 5 nm radius twin silicon nanowire MOSFET (TSNWFET): fabrication on bulk Si wafer, characteristics, and reliability, *Int. Electron Devices Meeting Tech. Dig.* 2005, 735–738.

4 S. C. Rustagi, N. Singh, W. W. Fang, K. D. Buddharaju, S. R. Omampuliyur, S. H. G. Teo, C. H. Tung, G. Q. Lo, N. Balasubramanian, D. L. Kwong, CMOS inverter based on gate-all-around silicon-nanowire MOSFETs fabricated using top-down approach, *IEEE Electron Device Lett.* 2007, **28** (11), 1021–1024.

5 X. C. Jiang, Q. H. Xiong, S. Nam, *et al.*, InAs/InP radial nanowire heterostructures as high electron mobility devices, *Nano Lett.* 2007, **7** (10), 3214–3218.

6 J. Xiang, W. Lu, Y. J. Hu, Y. Wu, H. Yan, C. M. Lieber, Ge/Si nanowire heterostructures as high-performance field-effect transistors, *Nature* 2006, **441**, 489–493.

7 Y. Li, J. Xiang, F. Qian, *et al.*, Dopant-free GaN/AlN/AlGaN radial nanowire heterostructures as high electron mobility transistors, *Nano Lett.* 2006, **6** (7), 1468–1473.

8 P. Mohan, J. Motohisa, T. Fukui, Fabrication of InP/InAs/InP core-multishell heterostructure nanowires by selective area metal organic vapor phase epitaxy, *Appl. Phys. Lett.* 2006, **88**, 133105.

9 C. Thelander, P. Agarwal, S. Brongersma, *et al.*, Nanowire-based one-dimensional electronics, *Mater. Today* 2006, **9** (10), 28–35.

10 A. I. Persson, M. T. Björk, S. Jeppesen, L. Samuelson, J. B. Wagner, L. R. Wallenberg,

$InAs_{1.5-.5x}P_x$ nanowires for device engineering, *Nano Lett.* 2006, **6**, 403.
11 T. Mårtensson, J. B. Wagner, E. Hilner, A. Mikkelsen, C. Thelander, J. Stangl, B. J. Ohlsson, A. Gustafsson, E. Lundgren, L. Samuelson, W. Seifert, Epitaxial growth of indium arsenide nanowires on silicon using nucleation templates formed by self-assembled organic coatings, *Adv. Mater.* 2007, **19** (14), 1801–1806.
12 M. T. Björk, O. Hayden, H. Schmid, *et al.*, Vertical surround-gated silicon nanowire impact ionization field-effect transistors, *Appl. Physics Lett.* 2007, **90** (14), 142110.
13 O. Hayden, M. T. Björk, H. Schmid, *et al.*, Fully depleted nanowire field-effect transistor in inversion mode, *Small* 2007, **3** (2), 230–234.
14 H. T. Ng, J. Han, T. Yamada, P. Nguyen, Y. P. Chen, M. Meyyappan, Single crystal nanowire vertical surround-gate field-effect transistor, *Nano Lett.* 2004, **4** (7), 1247–1252.
15 T. Bryllert, L. Samuelson, L. E. Jensen, L.-E. Wernersson, Vertical high mobility wrap-gated InAs nanowire transistors, in: Proceedings, 63rd Device Research Conference, Santa Barbara, CA, USA, 2005.
16 L.-E. Wernersson, T. Bryllert, E. Lind, L. Samuelson, Wrap-gated InAs nanowire, in: Proceedings, Field Effect Transistor 2005 International Electron Device Meeting, December 5–7, IEDM Technical Digest, Washington DC, USA, pp. 265–268, 2005.
17 T. Bryllert, L.-E. Wernersson, L. E. Froberg, L. Samuelson, Vertical high-mobility wrap-gated InAs nanowire transistor, *IEEE Electron Device Lett.* 2006, **27** (5), 323–325.
18 T. Bryllert, L.-E. Wernersson, T. Löwgren, L. Samuelson, Vertical wrap-gated nanowire transistors, *Nanotechnology* 2006, **17** (11), 227–230.
19 L.-E. Wernersson, E. Lind, L. Samuelson, T. Löwgren, J. Ohlsson, Nanowire field-effect transistor, *Jap. J. Appl. Phys.* 2007, **46** (4B), 2629–2631.
20 T. Löwgren, J. Ohlsson, L. Samuelson, L.-E. Wernersson, Control of threshold voltage in 80 nm gate length InAs vertical nanowire WIGFETs, *Device Research Conference Tech. Digest* 2007, 165–166.
21 C. Thelander, L. E. Fröberg, C. Rehnstedt, L. Samuelson, L.-E. Wernersson, Vertical enhancement-mode InAs nanowire field-effect transistor with 50 nm wrap-gate, *IEEE Electron Device Lett.* 2008, **29**, 206–208.
22 C. Rehnstedt, C. Thelander, L. E. Fröberg, B. J. Ohlsson, L. Samuelson, L.-E. Wernersson, Drive current and threshold voltage control in vertical InAs wrap-gate transistors, *Electron Lett.* (accepted for publication) 2008, **44**, 236–237.
23 R. Chau, S. Datta, M. Doczy, B. Doyle, B. Jin, J. Kavalieros, A. Majumdar, M. Metz, M. Radosavljevic, Benchmarking nanotechnology for high-performance and low-power logic transistor applications, *IEEE Trans. Nanotechnol.* 2005, **4** (2), 153–158.
24 E. Lind, A. I. Persson, L. Samuelson, L.-E. Wernersson, Improved subthreshold slope in an InAs nanowire heterostructure field-effect transistor, *Nano Lett.* 2006, **6** (9), 1842–1846.
25 E. Lind, L.-E. Wernersson, InAsP/InAs nanowire heterostructure field effect transistors, *Device Res. Conf. Tech. Digest* 2006, 173–174.

3
Single-Electron Transistor and its Logic Application

Yukinori Ono, Hiroshi Inokawa, Yasuo Takahashi, Katsuhiko Nishiguchi, and Akira Fujiwara

3.1
Introduction

Complementary metal-oxide-semiconductor (CMOS) technology will face significant technological limitations shortly after 2010 [1], and intensive studies are currently being conducted in computational architecture, circuit design, and device fabrication to find ways to overcome this impending crisis. The major problem, especially for logic large-scale integrated circuits (LSIs), is that rapidly increasing power dissipation due to ever larger numbers of transistors and higher levels of interconnections is pushing CMOS circuits beyond their cooling limit. This points to the need for some drastic change in how LSIs are built, either at the system architecture or base device level, or both. Roughly speaking, achieving low-power operation of LSIs requires that both the total capacitance of circuits and the operation voltage are reduced, which means in turn that the number of electrons participating in the operation of some unit instruction must also be reduced. Single-electron transistors (SETs) [2–5], the characteristics of which are literally governed by the movement of single electrons, are considered to be the devices that will allow such a change. Their operation is basically guaranteed even when device size is reduced to the molecular level. Their performance, such as the peak-to-valley current ratio, improves as they become smaller. These properties are quite beneficial for large-scale integration. In addition, SETs are able not only to operate as simple switches but also to have high functionality. Many theoretical studies have been conducted to evaluate the possibility of building SET-based LSIs, and fundamental computational capability has already been proved.

In this chapter, after a brief explanation of SET operation principles and fabrication processes, some experimentally tested SET logic circuits will be introduced. In addition, the merits and demerits of the SET as a logic device will also be discussed, and some brief ideas proposed concerning SET logic circuits.

Nanotechnology. Volume 4: Information Technology II. Edited by Rainer Waser
Copyright © 2008 WILEY-VCH Verlag GmbH & Co. KGaA, Weinheim
ISBN: 978-3-527-31737-0

3.2
SET Operation Principle

Before considering the SEToperation principle, imagine a small conductive sphere or "island" floating over the ground. If one electron is taken from the ground and placed in the sphere, then there will be an increase in the electrostatic potential of the sphere. This is given by e/C, where C is the capacitance of the sphere to the ground and e is the elementary charge, 1.6×10^{-19} C. When the sphere – and hence C – is extremely small, the potential increase becomes significant. For example, for a nanometer-scale sphere having a capacitance C of say 1 aF (1×10^{-18} F), the increase in the potential e/C reaches 160 mV. This is much larger than the thermal noise voltage at room temperature, 25.9 mV. The potential increase prevents another electron from entering the sphere unless that electron has an energy larger than e^2/C. This phenomenon is called the *Coulomb blockade*. If the potential can be decreased by e/C, by applying an external bias, then a second electron can (but the third one cannot) enter the sphere. If this occurs by quantum-mechanical tunneling, then it is called *single-electron tunneling*. Any single-electron device, including the SET, has at least one small conductive island and its operation relies on the Coulomb blockade and single-electron tunneling.

The cross-sectional view and equivalent circuit of the SET is shown in Figure 3.1. The SET is, like a conventional transistor, a three-terminal device consisting of a source, drain, and gate. There is however an additional component, called the "Coulomb island", between the source and drain. The Coulomb island is also called

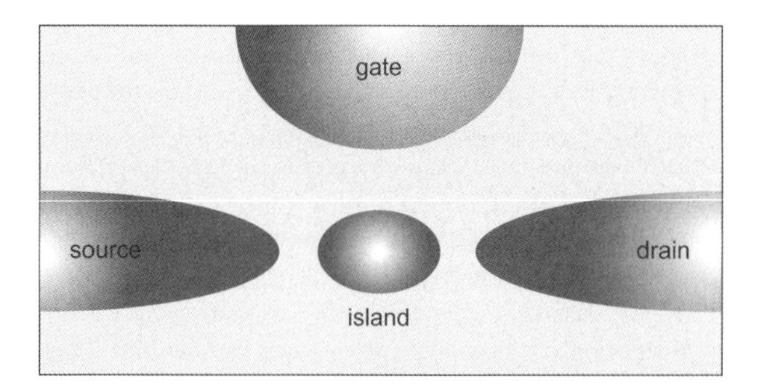

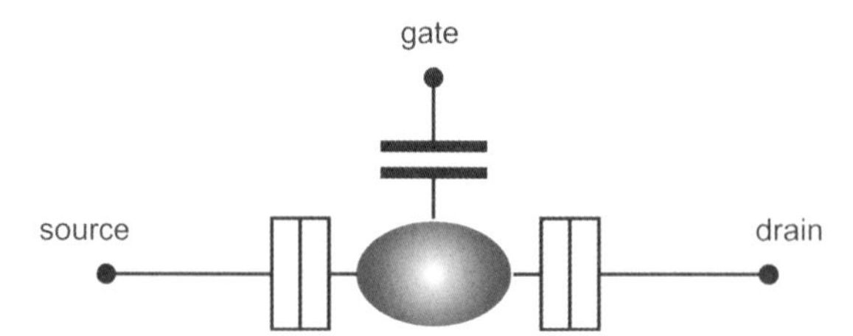

Figure 3.1 Cross-sectional view and equivalent circuit of the SET. In the equivalent circuit, double boxes indicate the tunnel junctions.

simply the island or a quantum dot. The Coulomb island must be conductive so that electrons can travel from the source to the drain via it. The role of this island is to capture/donate one electron from/to the source/drain, and otherwise to hold captured electrons. The region between the island and source (and also drain) must not be a good conductor; it must basically be insulating, so that electrons move to/from the island only by the tunneling. This region is called the *tunnel capacitor* or the *tunnel junction*. One the other hand, the region between the island and the gate should be insulating so as not to allow electrons to flow between them, as in a conventional transistor. In the equivalent circuit, the double box symbolizes a tunnel capacitor, which is a special capacitor that allows quantum mechanical tunneling of electrons, as mentioned above. The region sandwiched by the tunnel capacitors corresponds to the island, which is designated by an oval for visualization. The region between the island and the gate can be expressed as a normal capacitor.

Figure 3.2 explains what happens when the gate voltage is varied with a fixed small source/drain voltage. When a positive voltage V_g is applied to the gate, positive charges are induced there, whose number is given by $C_g V_g/e$, where C_g is the gate capacitance. Then, in order to minimize the free energy of the system, the SET tries to induce the same number of negative charges (i.e. electrons) in the island, and these electrons are conveyed from the source or drain through the tunnel junctions. If $C_g V_g/e$ is some integer N, the island obtains N electrons. After reaching this number, no more electron movement occurs. This is the *Coulomb blockade state* (the left equivalent circuit in Figure 3.2). When $C_g V_g/e$ is not an integer, for example, a half integer $N + 1/2$, the number of electrons in the island changes with time so that it becomes $N + 1/2$ *on average*. What actually occurs in the SET is as follows. First, one electron enters the island from the source and the number of electrons in the island

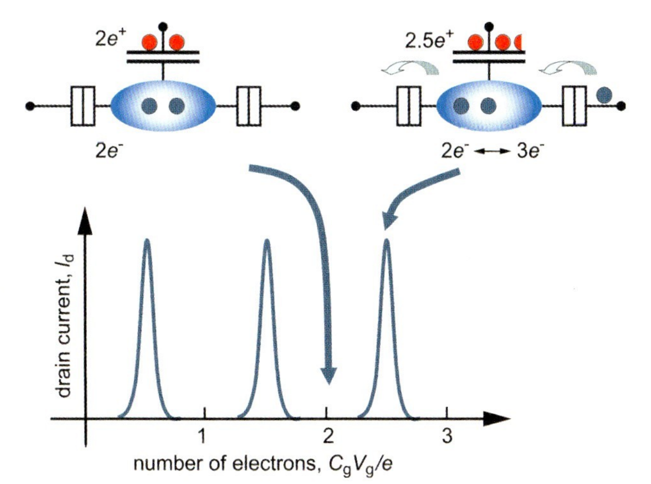

Figure 3.2 Drain current I_d as a function of number of charges $C_g V_g/e$ induced in the gate. C_g and V_g are the capacitance and gate voltage, respectively. The equivalent circuits explain the Coulomb-blockade state (left) and single-electron-tunneling state (right).

becomes $N+1$. Next, one electron emits to the drain from the island, resulting in N. This one-by-one electron transfer is repeated so that there is a net current between the source and drain. This is the *single-electron-tunneling state* (the right equivalent circuit in Figure 3.2). As a result, when the gate voltage is swept, the Coulomb-blockade state and the single-electron-tunneling state appears by turn, and the drain current versus gate voltage characteristics exhibit a repetition of sharp peaks, as shown in Figure 3.2. This is known as *Coulomb-blockade oscillation*. This ON-OFF characteristic indicates that the SET can function as a switching device.

For the complete description of the SET operation, consider the stability chart in the gate-voltage/drain-voltage plane in Figure 3.3. The rhombic-shaped regions colored in red are the region for the Coulomb-blockade state, and are known as *Coulomb diamonds*. Outside the Coulomb diamonds, the number of electrons in the island fluctuates between certain numbers. The degree of the fluctuation is determined by how far the voltage conditions are from the Coulomb diamonds. In the blue regions, the fluctuation is minimum – that is, the electron number changes only between two adjacent integers. These regions are for the single-electron-tunneling states. The shape and size of the Coulomb diamonds are determined only by the gate and junction capacitances. For example, the maximum drain voltage for the Coulomb blockade is given by e/C_Σ, where $C_\Sigma = C_g + C_d + C_s$ is the total capacitance of the island and C_d and C_s are junction capacitances at the drain and source. Each slope of the diamond is given by $-C_g/C_d$ and $C_g/(C_g + C_s)$.

A more detailed explanation of the SET operation can be found in textbooks [6, 7] and review articles [8, 9]. At this point, mention should be made of only one more item – which is what the SET *cannot* do. As explained above, the SET can convey electrons one by one, but the time interval of each transfer event is uncontrollable.

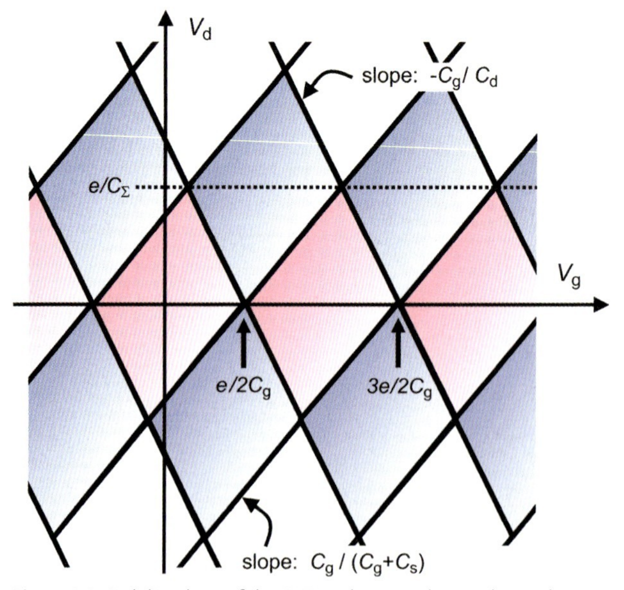

Figure 3.3 Stability chart of the SET in the gate/drain voltage plane.

This is because the transfer relies on tunneling, which is inherently stochastic. Therefore, it is difficult for the SET to transfer just one electron while preventing a second electron from being transferred. In other words, the transfer accuracy is quite low in the SET. In order to overcome this drawback, new single-electron devices have been invented and experimentally demonstrated. Sometimes called *single-charge-transfer devices*, these include the single-electron turnstile [10] and single-electron pump [11]. Although their structure is somewhat complicated (some of them possess two or more islands), they can transfer just one electron in one cycle of the gate clock, thereby providing high transfer accuracy. This function – the clocked single-electron transfer – is quite beneficial for implementing a certain level of logic architecture where one electron represents one bit. In this chapter, however, attention will be focused on the SET, and the single-charge-transfer devices and related logic styles will not be described in any detail. Very few experimental studies have been conducted on the single-charge-transfer logic circuits because of the difficulty of their fabrication. Hence, for single-charge-transfer logic, the reader is referred to review articles [8, 12].

3.3 SET Fabrication

When SETs are fabricated, two criteria should be borne in mind. First, the resistance of the tunnel junction must be sufficiently larger than the quantum resistance $R_q = h/e^2$ (~25.8 kΩ), where h is the Planck constant. Otherwise, the number of electrons in the island fluctuates because of the Heisenberg uncertainty principle. Because of this requirement, the current drivability is low, which is one major demerit of the SET as a logic element. Second, the energy for adding one electron to the island must be larger than the thermal energy. Otherwise, heated electrons tunnel through the barriers and the Coulomb blockade does not function. For example, as the temperature rises, each peak in Figure 3.2 broadens, and finally smears out. This relationship is expressed as $E \gg kT$, where E is the addition energy. The addition energy can be expressed in the form e^2/C_Σ, and thus a SET with a smaller island can operate at a higher temperature.

When the de Broglie wavelength of electrons is much smaller than the island size – which is the case of metal islands that are not too small ($\gg$1 nm) – charges are induced right at the island surface and E is determined only by the island size and the spatial configuration of the electrodes. However, when the de Broglie wavelength is comparable to the island size – typically as in the case of semiconductor islands with a nanometer size – the quantum size effect causes the kinetic energy of an electron to increase and hence E to increase. Thus, semiconductor islands can have larger addition energy than a metal island of the same size.

In order to explain how small the island must be made, Figure 3.4 shows the relationship between the island size (dot size) and the addition energy for a Si island embedded in SiO_2 dielectrics [13]. Both, the quantized level spacing and the charging energy (which can be defined as the addition energy for ideal metals) increase as the

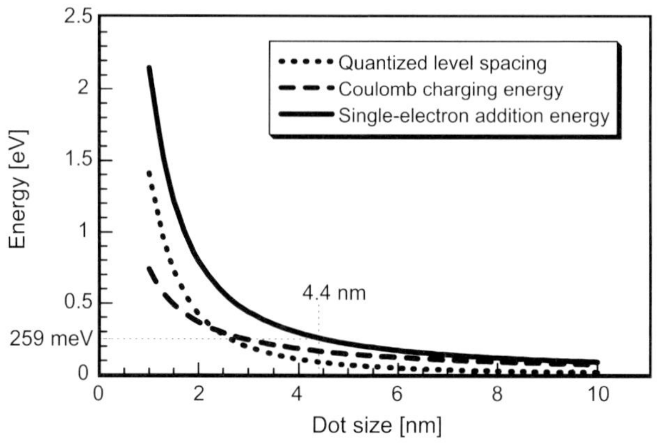

Figure 3.4 Relationship between dot size and addition energy [13].

island size decreases, and this leads to an increase in the resultant addition energy. If an addition energy 10 times larger than the room-temperature energy (25.9 meV) is required for the proper operation of a SET circuit, then an island as small as 4 nm is needed. The creation of such a small island and attaching tunnel junctions to it represents a technological challenge in SET fabrication. However, any conducting material can be used as long as the above criteria are satisfied, and an addition energy much larger than the room-temperature energy has already been demonstrated.

Historically, research into single-electron devices began with metals [4] and then expanded to semiconductors [14–18] and other materials, such as carbon nanotubes [19–22] and some molecules [23–28].

Among metals, a major material is aluminum, because its oxide functions as a good dielectric for tunnel junctions. Tunnel junctions are commonly made using Dolan's shadow evaporation technique [29]. The junction capacitance can be controlled in such structures to within 10% if the Al–AlO_x junctions have a relatively large capacitance of several hundred attofarads. Making smaller junctions is less easy, and thus electrical measurements with this material are commonly carried out below 1 K. However, it has been shown that making an extremely small SET, the island of which has an addition energy much larger than kT of room temperature, is possible [30].

Among semiconductors, Si is the most widely used material in research aimed at practical applications. Si SETs are commonly made on a certain type of Si substrate called silicon-on-insulator (SOI) [31]. In SOI substrate, a thin Si layer (typically 100–400 nm) is formed on a buried SiO_2 layer. Thinning the Si layer and reducing its size in the lateral direction by lithography enables small Si structures to be produced. It is possible to further miniaturize these structures by using thermal oxidation: Si is consumed during the oxidation, and thus the volume of the Si structures is reduced. With Si, it is relatively easy to make smaller islands compared with Al. Common measurement temperatures in Si SET research are 1 K to 100 K, and several groups have observed the Coulomb-blockade oscillation at higher temperatures (100–300 K) [13, 32–48].

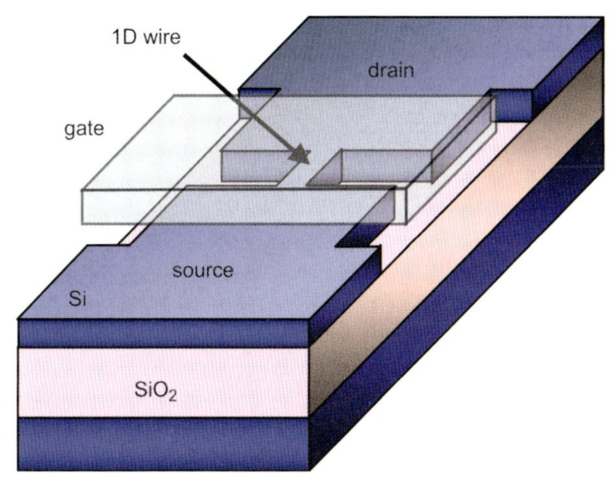

Figure 3.5 Basic device structure of PADOX SETs. The 1-D wire is converted to a Coulomb island and attaching tunnel junctions after thermal oxidation of the Si.

One reliable way of fabricating Si SETs is pattern-dependent oxidation, or PADOX [34, 49], and this has enabled the fabrication of a room-temperature-operating SET for the first time. It has been shown that the gate and junction capacitances are controllable even when their values are very small (a few attofarads) [34, 49, 50]. The PADOX method requires no special material for tunnel junctions as they are made of Si itself. PADOX exploits an oxidation-induced band modification [51], which makes it possible to produce a Coulomb island and tunnel junctions simultaneously during the gate oxidation step [52]. Figure 3.5 shows the basic structure for the PADOX SET. A one-dimensional (1-D) Si wire is converted to an island and tunnel junctions after thermal oxidation. As the name indicates, the final Si structure is dependent on the initial structures before oxidation. By elaborately designing the initial structures, a variety of SET configurations become possible [49, 53]. Figure 3.6 shows the drain current versus gate voltage characteristics of a PADOX SET measured at 27 K. Clear oscillation is observed. The PADOX method has contributed to the fabrication of many experimental SET circuits owing to its high controllability and the stability of the current characteristics.

Recent progress in the SET fabrication process has resulted in a very clear Coulomb blockade oscillation at room temperature; an example of this is shown in Figure 3.7, where a peak-to-valley ratio as large as 400 is achieved [48]. Although the working mechanisms underlining this excellent performance at room temperature are not satisfactorily understood at present, it is evident that room-temperature-operating SETs can be made. The focus of Si SET research is therefore moving from how to make room-temperature-operating SETs to how to control their size, which is still very difficult.

Carbon nanotubes are attractive for attaining small capacitance and thus high operation temperature because they have extremely small diameters of the order of a

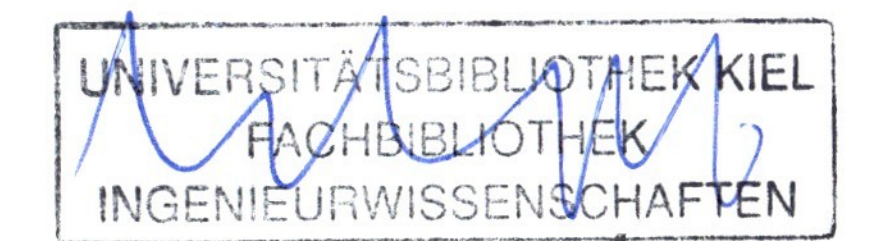
UNIVERSITÄTSBIBLIOTHEK KIEL
FACHBIBLIOTHEK
INGENIEURWISSENSCHAFTEN

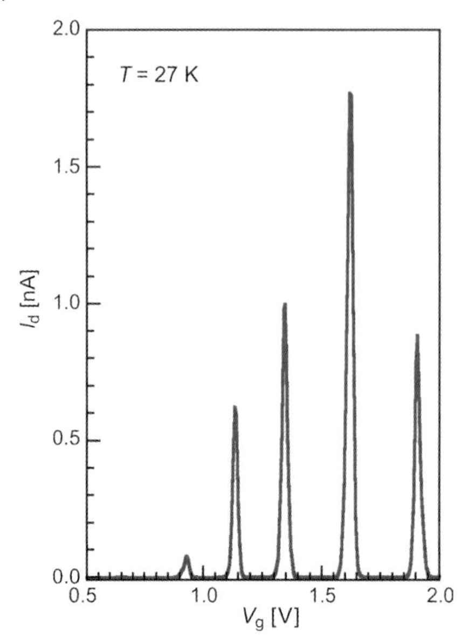

Figure 3.6 Current characteristics of a PADOX SET, measured at 27 K.

few nanometers. In the case of a 1-μm-long single-wall nanotube with a diameter of 1.4 nm suspended 100 nm above a ground plane, the addition energy E would be 8 meV, and this could be further increased by reducing the length. In fact, carbon nanotube SETs with E corresponding to this estimate have been reported [19, 20]. For practical applications, the nanotube diameter, chirality (i.e. electronic structure) and the locations of the tunnel junctions and nanotube itself must be controlled more precisely. Although these issues have already been partly addressed [21, 22], much further improvement is needed.

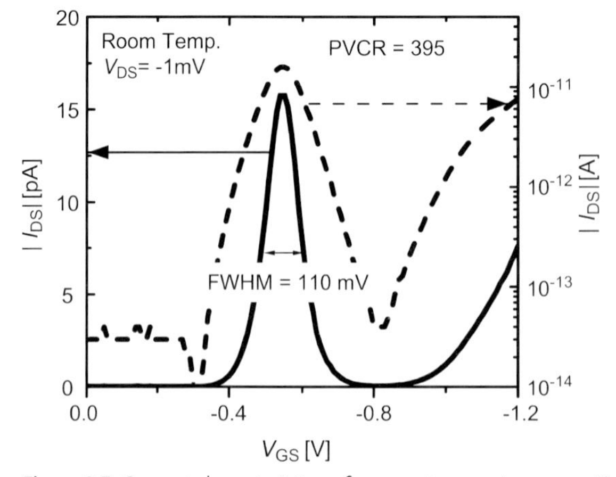

Figure 3.7 Current characteristics of a room-temperature-operating Si SET [48].

Fabrication methods that use molecules as building blocks are anticipated. In such methods, functions and characteristics are determined by chemical synthesis, without relying on lithographic techniques. Research into the charging effect or Coulomb blockade in a molecule began during the mid-1990s [23, 24], and more recently persuasive data showing conductance modulation by gate potential have been obtained [25–28]. Although the present understanding of transport in a molecule is improving, many issues of circuit integration, including architectural design, synthesis, and interfacing with external circuits, remain.

At this point, mention should be made of an infamous problem in SETs, known as the *background charge problem* [54]. Due to randomly distributed mobile and immobile charges in the dielectrics, the device characteristics may change over time and differ from one device to another. This is because SETs have a high sensitivity to charges due to their small size, and it makes the integration of SETs difficult. A typical case is seen in SETs made from metals and GaAs/AlGaAs heterostructures. For example, the characteristics of SEDs with Al–AlOx junctions change at least once a day. In order to stabilize such behavior, it may be necessary to wait for a long time after cooling down before measurements can be made [55]. The situation is similar for carbon nanotubes and some molecules, as these suffer from a large noise superimposed on the current characteristics, the origin of which is unknown.

The background charge problem is not specific to SETs, however, and may occur in any nanoscale *field-effect* device due to their high sensitivity to charges. In addition, the amount, location, and stability of the background charges are highly material- and process-dependent. In fact, it has already been shown that PADOX SETs have excellent long-term stability. The drift of the characteristics is less than $0.01e$ over a period of a week at cryogenic temperatures [56]. More practically, no noticeable change in the characteristics have been observed for more than eight years, during which time thermal cycling between room temperature and $\sim$20 K has occurred several times [57]. It has also been shown that the voltage at which the first Coulomb-blockade-oscillation peak appears is controllable [58]. These results demonstrate that PADOX SETs are not significantly influenced by slowly moving or immovable background charges, which indicates that the background charges problem is not intrinsic but rather can be solved. At present, no clear answers have been identified as to how seriously fewer fixed charges and/or faster motion of charges, which causes 1/f noise, will obstruct integration. However, it is believed that a circuit design with some degree of defect tolerance would relax the effects.

In summary, for room temperature operation, an island smaller than 10 nm is necessary. At present, Si is preferable for the SET fabrication from the viewpoints of operation stability and temperature. Some experimental data are available showing the control of the peak positions and the peak intervals in current characteristics. However, these parameters are still difficult to control in room-temperature-operating SETs. Also, there are no data showing the control of the resistance of room-temperature-operating SETs, and these points remain the subjects of future studies. A more complete description of the fabrication process for Si SETs can be found in Ref. [59].

3.4 Single-Electron Logic

Many logic styles have been proposed and analyzed for single-electron devices. Most of them can be categorized into two groups: charge-state logic; and voltage-state logic.

Charge-state logic [8, 12], which uses one electron to represent one bit, is highly specific to single-electron devices and might be in some sense the ultimate logic. Devices other than SETs, such as single-electron pumps, are building blocks. However, very few experimental studies have been reported regarding circuit operation based on this scheme because of the difficulty of the fabrication.

Voltage-state logic [60–88], which will be described here in detail, uses the SET as a substitute for the conventional MOS field-effect transistor (FET); hence it is referred to as SET logic. Although the circuit characteristics are predominated by the Coulomb blockade and single-electron tunneling, these phenomena are not directly employed for computation. Instead, the current produced by the sequence of single-electron tunnelings is used, and the bit is represented by the voltage generated by the accumulation of plural electrons. Actually, this is not a genuine single-electron logic because 10^1 to 10^3 electrons will be used in the operation. In many aspects, this logic is analogous to CMOS logic. The major advantage is that the accumulated technologies can be employed for CMOS circuit designs. However, the logic is not merely a copy of CMOS logic because the SET has completely different current characteristics from the MOSFET; that is, the current oscillates as a function of gate voltage. Some important elemental circuits such as an inverter, an exclusive-or (XOR) gate, a partial-sum/carry-out circuit, and an analog-to-digital converter, have been experimentally demonstrated [43, 46, 89–102]. Some of these will be introduced in the following three subsections.

3.4.1 Basic SET Logic

The voltage gain of the SET can be defined as for conventional transistors [5]. It is known from Figure 3.3 that when the drain voltage increases with a fixed gate voltage, a current begins to flow at the edge of the Coulomb diamond. As a result, the output drain voltage V_d for a fixed input drain current I_d exhibits a Coulomb diamond as a function of the gate voltage V_g. The measured characteristics for a Si SET are shown in Figure 3.8 (upper panel). The two slopes in the figure correspond to the inverting and non-inverting voltage gains G_I and G_{NI}. As shown in Figure 3.3, their values are determined by the capacitances as

$$G_I = C_g/C_d \tag{3.1}$$

$$G_{NI} = C_g/(C_g + C_s) \tag{3.2}$$

Although G_{NI} is always smaller than unity, G_I exceeds unity if $C_g > C_d$. Therefore, CMOS-like logic circuits can be prepared using SETs as substitutes for MOSFETs. From Eq. (3.1) it is clear that the SET must have a large C_g in order to obtain a high

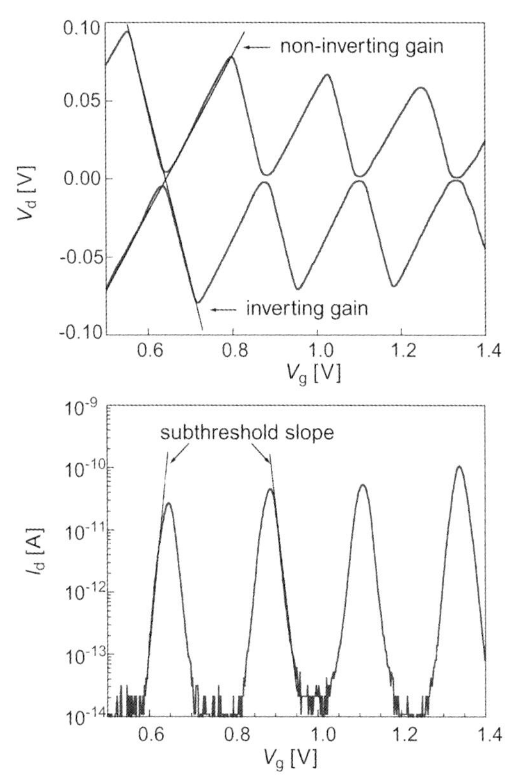

Figure 3.8 Electrical characteristics of a PADOX SET. Upper panel: Drain voltage V_d as a function of gate voltage V_g for a fixed drain current of ± 10 pA. Lower panel: Drain current I_d as a function of V_g for a fixed drain voltage of 10 mV. The measurement temperature was 27 K [50].

inverting gain, which in turn means that the total capacitance of the SET island will tend to increase. Therefore, the voltage gain and operating temperature are in a trade-off relationship and it is not easy to produce SETs with a larger-than-unity gain that operate at high temperatures. A G_I value larger than unity has been achieved in metal-based [103, 104], GaAs-based [105], and Si-based [48, 50, 106] SETs.

The current cut-off characteristics are determined by the subthreshold slope S in the I_d–V_g characteristics that rise and fall almost exponentially at the tails of the peaks. Figure 3.8 (lower panel) shows the output drain current I_d for a fixed input drain voltage V_d plotted as a function of V_g on a logarithmic scale. At a sufficiently low temperature and high tunnel resistance, S is given by

$$S = [d(\log_{10} I_d)/dV_g]^{-1} = \ln 10(C_\Sigma/C_g)kT/e \tag{3.3}$$

This equation is similar to that for a MOSFET. It also indicates that a high inverting voltage gain G_I is needed to obtain a steep subthreshold slope. Upon $C_s = C_d$, a G_I of 4 corresponds to a C_Σ/C_g of 1.5, or $S = 90\,\mathrm{mV\,dec^{-1}}$ at room temperature.

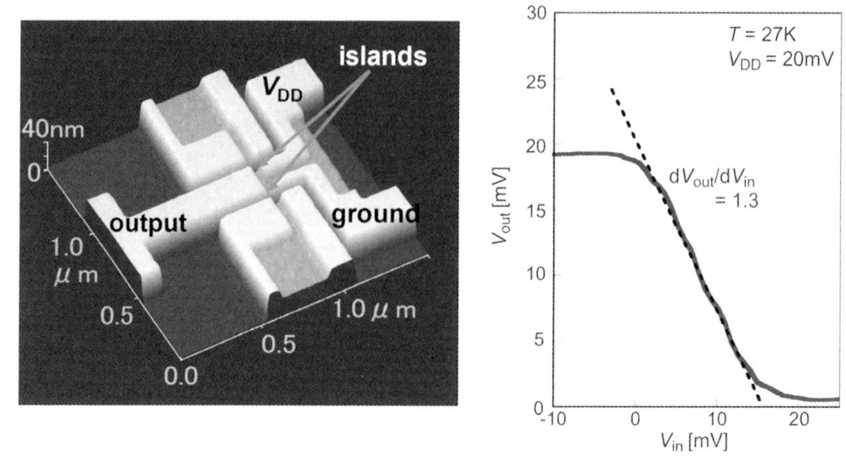

Figure 3.9 Si complementary single-electron inverter. Left: AFM image. Right: Input–output transfer curve measured with a power supply voltage V_{DD} of 20 mV. The measurement temperature was 27 K [89].

A logic circuit can be made by employing the above-mentioned voltage gain. The complementary inverter, which is the most fundamental logic element, was fabricated using Si SETs [89]. Figure 3.9 (left) shows an atomic force microscopy (AFM) image of an SET inverter made by Si. Two SETs with a voltage gain of about 2 were packed in a small (100×200 nm) area. As shown in Figure 3.9 (right), the input and output transfer curve attains both a larger than unity gain and a full logic swing at 27 K. Other complementary SET inverters, made from Al [90] and carbon-nanotube [91] have been reported, and resistive-loaded inverters have also been fabricated [92, 93].

3.4.2 Multiple-Gate SET and Pass-Transistor Logic

An important feature of SETs is that they can have *plural gates*. Such a multi-gate configuration enables the sum-of-products function to be implemented at the gate input level. That is, the total charges induced in the gates are expressed as $\Sigma C_{gi} V_{gi}$, where C_{gi} and V_{gi} are the gate capacitance and input voltage of the *i*-th gate. Provided that the gate input voltage V_{gi} is set to $e/2C_{gi}$, the SET is ON when the number of the ON gates is odd, and OFF when the number of the ON gates is even (Figure 3.10). This function is XOR. The SET XOR gate [65, 68] is a powerful tool for constructing arithmetic units such as adders and multipliers because XOR is nothing other than what is termed "half-sum", which is the lower order bit calculated by adding two one-bit binary numbers. The XOR gate has also been demonstrated experimentally using a Si dual-gate SET [94]. A scanning electron microscopy (SEM) image of the dual-gate SET is shown in Figure 3.11. The XOR function was confirmed in output drain current at 40 K, as shown in the figure.

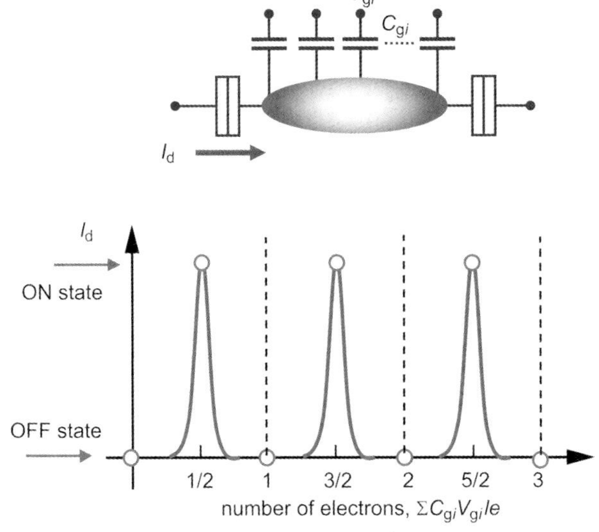

Figure 3.10 A multigate SET. Top: An equivalent circuit. Bottom: Current characteristics as a function of number of charges accumulated at the gates.

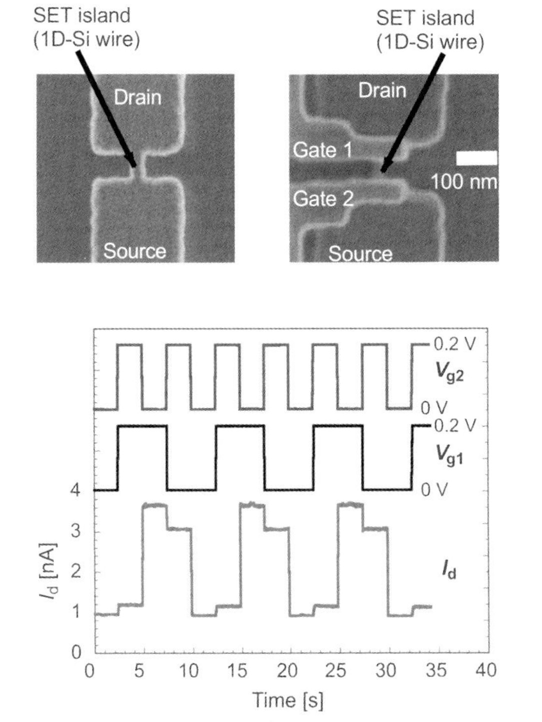

Figure 3.11 An experimental SET X-OR gate. Top: SEM images of the device before (left) and after (right) gate formation. Bottom: Output drain current for square-wave gate inputs [94].

In the multi-gate configuration, the gate capacitance for each gate is inherently smaller than that of the single-gate version. Therefore, it is more difficult to attain the larger-than-unity gain as the number of the gate increases. The CMOS-domino-type logic was proposed as a way of using SETs without a voltage gain [73]. A combinational logic circuit is built in a SET logic tree, where SETs are used as pull-down transistors. The point is that the tree is operated with a sufficiently small drain voltage in order to make the Coulomb blockade effective. The output signal is then amplified by using MOSFETs before being transferred to the next logic segment.

A single-electron pass-transistor logic, where SETs are used both as pull-up and pull-down transistors, has also been studied. The fundamental circuit of the single-electron pass-transistor logic was fabricated using PADOX SETs, and half-sum and carry-out for the half adder has been experimentally demonstrated [95, 96]. Figure 3.12 shows the equivalent circuits and the measurement data. Both half-sum and carry-out are correctly output at 25 K. What is significant here is that the gate and total capacitances, and even the peak positions of the used SETs, were well controlled for these operations. This is the first arithmetic operation ever performed by SET-based circuits. There have been attempts to construct logic elements [97–102] that operate based on the above-mentioned domino-type logic, pass-transistor logic, or the so-called binary-decision-diagram logic.

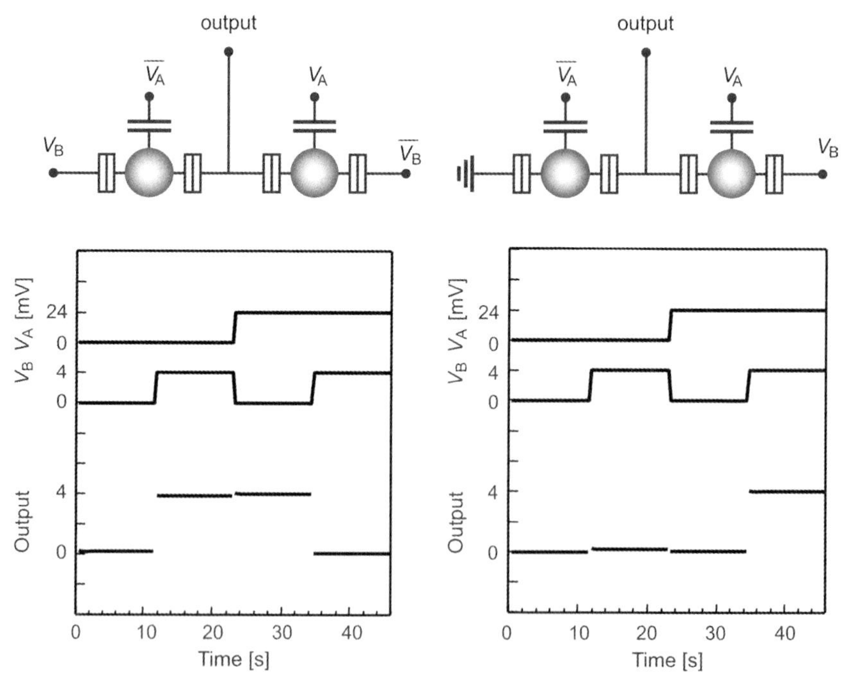

Figure 3.12 "Half-sum" and "carry-out" operations using SETs. In the equivalent circuits, V_A, V_B are inputs for addends and $\bar{V}_A$, $\bar{V}_B$ are their inverses [96].

3.4.3
Combined SET-MOSFET Configuration and Multiple-Valued Logic

In SETs, the applicable drain voltage is limited to a value smaller than e/C_Σ in order to maintain the Coulomb blockade. This may be an obstacle to driving a series of SETs and external circuits that require a high input voltage. A combined SET-MOSFET configuration has been proposed as a way to overcome this drawback [83, 85]. Figure 3.13(left) shows the equivalent circuit of the inverter based on this configuration. A MOSFET with a fixed gate bias V_{gg} is connected to the drain of a SET, and the inverter is driven by a constant current load, I_0. The MOSFET keeps the SET drain voltage sufficiently low, which helps to maintain the Coulomb blockade. As the drain voltage is almost independent of the output voltage, V_{out}, a large output voltage and voltage gain can be obtained.

The output voltage V_{out} and output resistance of the combined SET-MOSFET inverter are given by [86]:

$$V_{out} = -G_{m(SET)} R_{d(SET)} (1 + G_{m(MOS)} R_{d(MOS)}) V_{in}, \tag{3.4}$$

$$R_{out} = R_{d(MOS)} + (1 + G_{m(MOS)} R_{d(MOS)}) R_{d(SET)} \tag{3.5}$$

where $G_{m(SET)}$ is the transconductance of the SET, and $R_{d(SET)}$ and $R_{d(MOS)}$ are the drain resistances of the SET and MOSFET, respectively. The voltage gain of the SET is multiplied by that of the MOSFET, which means that the voltage gain of the SET-MOSFET inverter becomes very large because of the large voltage gain of the

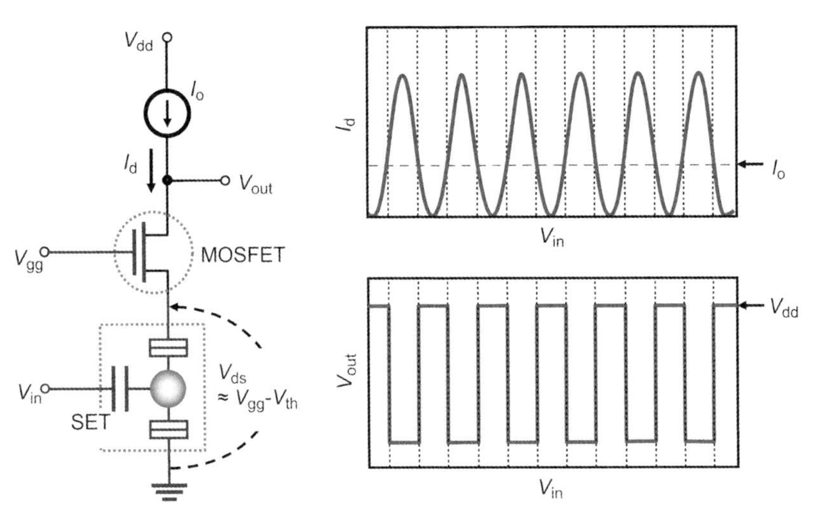

Figure 3.13 Left: Schematic of the universal literal gate comprising a SET, a MOSFET, and a constant-current load I_0. Right: $I_d - V_{in}$, and expected transfer ($V_{in} - V_{out}$) characteristics. $I_d - V_{in}$ characteristics are almost completely independent of V_{out} as the V_{ds} of the SET is kept nearly constant at ($V_{gg} - V_{th}$, the threshold voltage of the MOSFET [86].

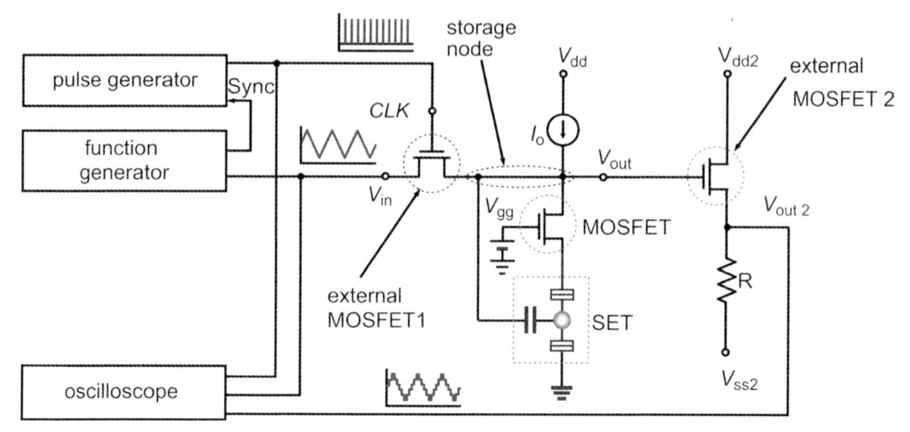

Figure 3.14 Measurement set-up for the single-electron quantizer [85]. CLK = clock.

MOSFET (see Figure 3.13, right). In fact, the measured voltage gain of the SET-MOSFET inverter was about 40 [86].

In this configuration, another important point is that the $I_d - V_g$ characteristics reflect the oscillatory I_d–V_g characteristics (Figure 3.13, right). This characteristic is referred to as the "universal literal", which is a basic unit for multiple-valued logic. Multiple-valued logics have potential advantages over binary logics with respect to the number of elements per function and operating speed. They are also expected to relax the interconnection complexity inside and outside of LSIs. These are advantageous, as they allow a further reduction in the power dissipation in LSIs and the chip sizes. However, success has been limited, partially because the devices that have been used (MOSFETs and negative-differential-resistance devices, such as resonant tunneling diodes) are inherently single-threshold or single-peak, and are not fully suited for multiple-valued logic. The oscillatory behavior seen in Figure 3.13 shows that the SET is suitable for implementing multiple-valued logic. By exploiting this behavior, a quantizer was fabricated. Figures 3.14 and 3.15 show the measurement set-up for the quantizer and the measured data, respectively. The triangular input V_{in} was successfully quantized into six levels.

3.4.4 Considerations on SET Logic

Many research groups have claimed that SETs could provide low-power circuits. In order to make clear the meaning of this claim, two parameters must first be discussed, namely information throughput I and the power density P. These parameters can be written in the following forms:

$$I = \alpha n f \tag{3.6}$$

$$P = E_{bit} n f \tag{3.7}$$

where n is the density of the binary switches, f the operating frequency, and E_{bit} the bit energy. A dimensionless parameter, α, was introduced which was referred to

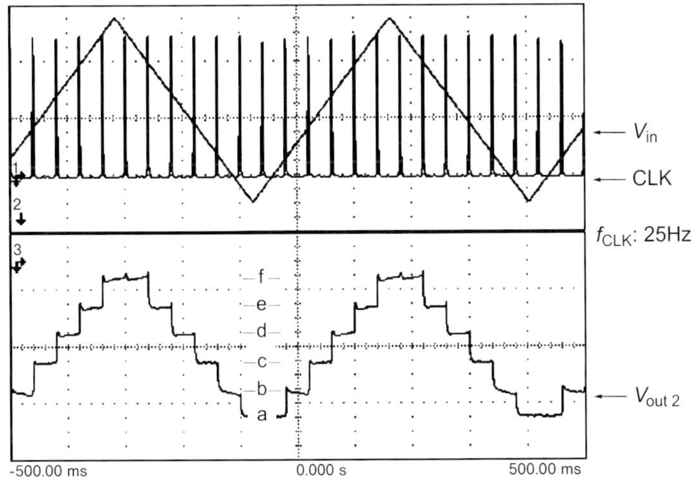

Figure 3.15 Experimental data for the single-electron quantizer [85]. CLK = clock.

as "functionality". The meaning of this parameter is simple; for example, if a transistor is available that can work more efficiently than a simple binary switch, then the information throughput can be increased. In such a case, the transistor has α larger than unity. Then, for low-power operations the aim would be to increase I while keeping P small. Therefore, an important parameter is the information throughput per power density, I/P, which is given by α/E_{bit}. A lower E_{bit} is better for larger I/P, but E_{bit} has a lower bound in order to avoid noise-induced bit errors. This minimum value is dependent on how many errors the system allows, and thus on the system architecture. However, the minimum E_{bit} will not change significantly unless the architecture is changed to an exotic version, like a neural network or fuzzy logic. Thus, the only option is to increase α. If the suggestion was to increase α on the device level, then there would be a need to depart from logic based on simple binary switches. This leads to a very important conclusion – that changing materials for the transistor channel, say, to carbon nanotubes or other molecules, is not the way to reduce dynamic power loss as long when transistors are used as binary switches. Hence, it is expected that the SET be an alternative device and would be highly functional, as shown previously in the chapter.

It should be mentioned, however, that the above discussion is rather too crude to draw any decisive conclusions. Actually, I/P does not include n and f, but indeed a large-scale integration and a fast device is needed in order to accomplish computation within acceptably short periods of time. Therefore, a more reasonable parameter may be $I^2/P\,(=\alpha^2 nf/E_{bit})$, and the size and the speed of the device need still to be discussed. At this time the static power loss should also be considered – that is, the loss due to leakage currents, which is independent of f. These points are discussed in the following sections.

It is important now to highlight once more the difference between voltage-state logic (or SET logic) and charge-state logic. The requirement for the addition energy – and hence for the island size – is different between the two. For charge-state logic, the addition energy should be sufficiently large so as to avoid bit errors caused by thermal noise. As charge-state logic uses single electrons to represent bit information, the bit energy is given by $(1/2)C_\Sigma V^2$, where $V = e/C_\Sigma$. This is in effect the addition energy. If the bit error requirements for CMOS LSIs are assumed, then the bit energy will have to be 10^2 larger than the thermal noise energy. This means that an additional energy 10^2 larger than the room-temperature energy is needed. From Figure 3.4 it is clear that an island as small as 1 nm is needed. SET logic, on the other hand, uses the voltage generated at output terminals to represent bits, as do CMOS circuits. The bit energy is therefore given by $(1/2)C_L V^2$, where C_L is the load capacitance. If the term $V = e/C_\Sigma$ is adopted for the power supply voltage of SET logic circuits, then the bit energy is C_L/C_Σ larger than the case for charge-state logic. Therefore, for SET logic, the addition energy requirement does not come from the bit error requirement but rather from the static power loss because a small addition energy causes the valley current (i.e. the OFF-current) to increase. There is no clear guideline as to how small the OFF-current should be, because the acceptable static power loss depends on the degree of the power-saving ability of the system. If the requirement for low-operating-power (LOP) applications are adopted – as stated in the International Technology Roadmap for Semiconductors (ITRS) – the source/drain OFF-state leakage current should be on the order of 10^{-9} A μm^{-1}, which corresponds to 10 pA for 10-nm SETs. This will be achievable. It is also necessary to have a large ON/OFF current ratio and, again, considering the requirement for LOP applications, a ratio of 10^5 is needed. From this ratio, the addition energy should be about 16 kT or larger, which was derived based on the standard theory of single-electron tunneling. This estimation does not consider the quantum leakage current, which becomes significant as the junction resistance approaches the quantum resistance. However, as long as the junction resistance is not very close to the quantum resistance, the above criteria for the addition energy will be a reasonable basis for later discussions. With this requirement, an addition energy E as large as 0.4 eV is required for room-temperature operation. This addition energy corresponds to the total island capacitance ($C_\Sigma = e^2/E$) of 0.4 aF, an excitation voltage (e/C_Σ) of 0.4 V and, from Figure 3.4, an island size of about 3 nm.

There are two time scales for evaluating SET speed: one is the intrinsic switching time, and the other for the circuit speed. The intrinsic switching time defines how fast the SET changes its states, and is determined by the RC time constant of the tunneling, $C_\Sigma R_j$, where R_j is the junction resistance. If it is assumed that $C_\Sigma = 0.4$ aF and $R_j = 1$ MΩ ($\sim 4R_q$), the switching speed will be 0.4 ps and thus the SET is a fairly fast switching device. The problem is that only one electron is moved by the switching event, and it thus takes a much longer time to change the state of the output terminal with a larger capacitance. The time for changing the state of the output terminal is determined by $C_L R_{SET}$, where R_{SET} is the resistance of the SET ($\sim 4R_j$). If it is assumed that $C_L = 100$ aF and $R_{SET} = 4$ MΩ, the time is 0.4 ns or 2.5 GHz. It will also be helpful to compare the SET current density with that of the present nMOS transistor (for LOP applications), which is about 600 μA μm^{-1}. Assuming that the size, resistance, and drain voltage of the SET are 3 nm, 4 MΩ, and 0.4 V, respectively, the SET current density will be 33 μA μm^{-1}. Although this is not fatally bad, it implies that the use of SETs is restricted to a local communication with relatively small load capacitances. Crudely speaking, the SET is inherently slower by the factor of at least 10^{-1} than FETs because the SET cannot operate with the resistance smaller than R_q, whereas FETs can.

The SET can be made very small, but this does not necessarily mean it will be the smallest. Ideally, a molecular-sized FET can be imagined, and could be made as small as the SET. Therefore, small size is not a major merit of the SET; rather, the main merit is that its operation is guaranteed even at the molecular level, and some parameters – such as the switching speed and current peak-to-valley ratio – can be improved owing to the reduced capacitance. At this point, it might be safe to say that SETs have no definite advantage over ultimately scaled-down MOSFETs from the viewpoint of the physical size itself.

Now, a return should be made to Eqs. (3.6) and (3.7). Considering the above arguments that SET size is comparable to that of ultimate future FETs, and that the circuit speed is 10^{-1} to 10^{-2} worse than its CMOS counterparts, the functionality α will need to be increased by 10^1 or more in order to make a drastic improvement in I/P while keeping I^2/P comparable to that for CMOS circuits.

Several ideas for improving the functionality have been reported. One is to use the SET XOR gates introduced in Section 3.4.2. Figure 3.16 shows the equivalent circuit of a full adder based on the SET XOR gate. A full adder can be constructed using six SETs, whereas this requires 28 MOSFETs in CMOS logic [106]. This can be interpreted as $\alpha = 4.7$. It has also been reported that, by integrating SET full adders, multi-bit adders can be constructed in a very area-efficient manner: there are no long wires in the carry-propagation path, which leads to fast operation in spite of a low drivability of the SET [80]. Another idea is based on the SET-MOSFET configuration introduced in Section 3.4.3. Based on this configuration, a SET logic gate family has been proposed [87, 88].

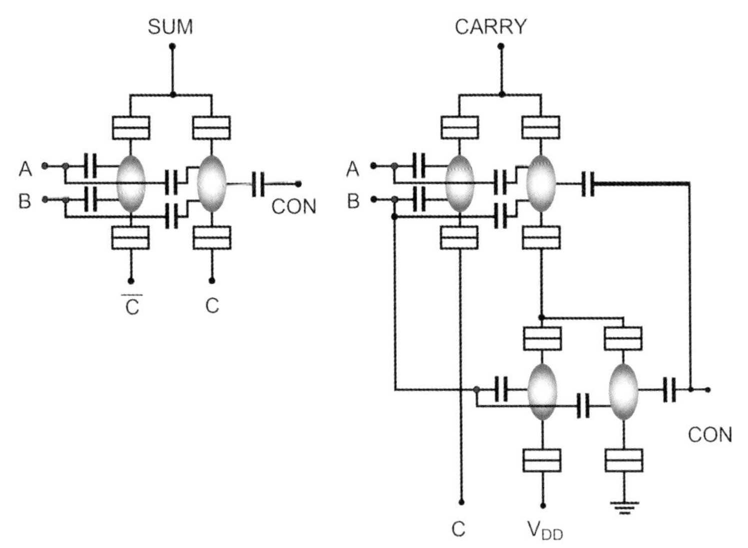

Figure 3.16 The SET full adder. A and B are addends and C is carry. CON is the control signal, which controls the phase of the Coulomb blockade oscillation [80].

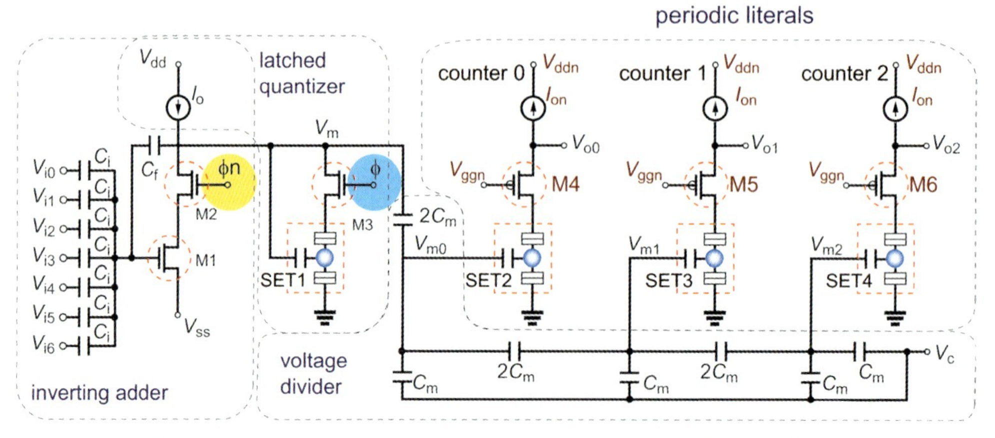

Figure 3.17 Circuit diagram of SET-based 7-3 counter. The circuit consists of five types of device: SETs (SET1–SET4), n-channel MOSFETs (M1–M3), p-channel MOSFETs (M4–M6), and constant current loads for the first stage and literals. No adjustment is required in the device parameters for devices of the same type. Clock ϕ and ϕn are complementary, and the multiple-valued data are latched when ϕ is high. V_{ddn} is set at a negative value to provide consistent voltage levels among the circuit blocks in and out of the counter [87].

These SET logic gates are useful for implementing binary logic circuits, multiple-valued logic circuits, and binary/multiple-valued mixed logic circuits in a highly flexible manner. As an example, a 7-3 counter is shown in Figure 3.17. This is a member of the M-N counters, which are generalized counters defined in the framework of Counter Tree Diagrams. Most adders, including those for redundant number systems, could be represented in this framework. The 7-3 counter can be constructed using four SETs and 10 FETs with some passive components, whereas 198 FETs are required in CMOS logic. The functionality α is of the order of 10^1 in this case.

The increase in α in the two examples is due to the application of the SET periodic function for implementing the operation "add". This is because the parity and the periodic function are the fundamentals of the arithmetic. These examples strongly suggest that the best use of the SETs will be in arithmetic units such as adders, while other arithmetic units such as multipliers are made from adders. Adders can be built by a repetition of relatively simple layouts, and require lesser amounts of long wiring, which will compensate for the low drivability of SETs. It is believed that, by pursuing this direction, a more efficient way to increase the functionality will be found. For this purpose, much larger-scale circuits should be investigated than have been studied to data.

In summary, the SET can function as a fairly fast switching device and, although the SET has low drivability, this will not prove fatal. The periodic function of the SET current characteristics suggests that it should be applied to arithmetic units such as adders and multipliers, which might reduce their dynamic power consumption. The most suitable applications for SET-based voltage-state circuits will be for LOP arithmetic units. However, there may be no merit in using SETs in terms of static power consumption, and some elaborate system architectures would be required to reduce this. For low standby power applications, charge-state logic circuits should pursued where, in principle, there is no leakage current.

3.5 Conclusions

Despite concern persisting with regards to reducing dynamic power consumption, this problem will not be solved simply by changing the raw materials of transistor production. One way of reducing power requirements is to use functional devices rather than simple switches, and among the large numbers of emerging devices the SET is one of the best functional units, on the basis of its unique current characteristics.

However, two critical problems remain when applying SETs to logic circuits. The first – basically technological – problem is to control the size of the nanometer-scale islands and attached tunnel junctions. During the early 1990s, very few investigators considered that SETs operating at room temperature could be fabricated, yet today they can be prepared with good ON-OFF current ratios. Moreover, their performance continues to improve. With this in mind it is likely that, in the future, a new technology will emerge for integrating millions of room-temperature-operating SETs. The second problem is to identify the "killer" applications for SETs, and this is a more fundamental and critical question. During recent years, much effort has been expended in designing SET circuits, and those which utilize the periodic nature of the SET current characteristics appear to show the greatest promise for the construction of LOP circuits. Nonetheless, further studies will be necessary to develop SET circuits that are sufficiently powerful to surpass CMOS circuits, or at least to replace a proportion of them. For this purpose, a collaboration among system architects, circuit designers, and process engineers is clearly called for.

References

1 J. D. Meindl (Ed.), *Proc. IEEE* 2001, 89.

2 D. V. Averin, K. K. Likharev, *J. Low Temp. Phys.* 1986, **62**, 345.

3 L. S. Kuzmin, K. K. Likharev, *J. Exp. Theoret. Physics Lett. Lett.* 1987, **45**, 495.

4 T. A. Fulton, G. J. Doran, *Phys. Rev. Lett.* 1987, **59**, 109.

5 K. K. Likharev, *IEEE Trans. Magn.* 1987, **23**, 1142.

6 D. V. Averin, K. K. Likharev, *Mesoscopic Phenomena in Solids*, Chapter 6, B. L. Altshuler, P. A. Lee, R. A. Webb (Eds.), Elsevier, Amsterdam, 1991.

7 H. Grabert, M. H. Devoret (Eds.), *Single Charge Tunneling*, Plenum, New York, 1992.

8 K. K. Likharev, *Proc. IEEE* 1999, **87**, 606.

9 M. A. Kastner, *Rev. Mod. Phys.* 1992, **64**, 849.

10 L. J. Geerligs, V. F. Anderegg, P. A. M. Holweg, J. E. Mooij, *Phys. Rev. Lett.* 1990, **64**, 2691.

11 H. Pothier, P. Lafarge, P. F. Orfila, C. Urbina, D. Esteve, M. H. Devoret, *Physica B* 1991, **169**, 573.

12 Y. Ono, A. Fujiwara, K. Nishiguchi, H. Inokawa, Y. Takahashi, *J. Appl. Phys.* 2005, **97**, 031101.

13 M. Saitoh, N. Takahashi, H. Ishikuro, T. Hiramoto, *Jpn. J. Appl. Phys.* 2001, **40**, 2010.

14 J. H. F. Scott-Thomas, S. B. Field, M. A. Kastner, H. I. Smith, D. A. Antoniadis, *Phys. Rev. Lett.* 1989, **62**, 583.

15 U. Meirav, M. A. Kastner, S. J. Wind, *Phys. Rev. Lett.* 1990, **65**, 771.

16 C. de Graaf, J. Caro, S. Radelaar, V. Lauer, K. Heyers, *Phys. Rev. B* 1991, **44**, 9072.

17 H. Matsuoka, T. Ichiguchi, T. Yoshimura, E. Takeda, *Appl. Phys. Lett.* 1994, **64**, 586.

18 D. Ali, H. Ahmed, *Appl. Phys. Lett.* 1994, **64**, 2119.

19 M. Bockrath, D. H. Cobden, P. L. McEuen, N. G. Chopra, A. Zettl, A. Thess, R. E. Smalley, *Science* 1997, **275**, 1922.

20 S. J. Tans, M. H. Devoret, H. Dai, A. Thess, R. E. Smalley, L. J. Geerlings, C. Dekker, *Nature* 1997, **386**, 474.

21 K. Ishibashi, D. Tsuya, M. Suzuki, Y. Aoyagi, *Appl. Phys. Lett.* 2003, **82**, 3307.

22 K. Matsumoto, S. Kinoshita, Y. Gotoh, K. Kurachi, T. Kamimura, M. Maeda, K. Sakamoto, M. Kuwahara, N. Atoda, Y. Awano, *Jpn. J. Appl. Phys.* 2003, **42**, 2415.

23 H. Nejoh, *Nature* 1991, **353**, 640.

24 V. Mujica, M. Kemp, A. Roitberg, M. Ratner, *J. Chem. Phys.* 1996, **104**, 7296.

25 N. B. Zhitenev, H. Meng, Z. Bao, *Phys. Rev. Lett.* 2002, **88**, 226801.

26 J. Park, A. N. Pasupathy, J. I. Goldsmith, C. Chang, Y. Yaish, J. R. Petta, M. Rinkoski, J. P. Sethna, H. D. Abruna, P. L. McEuen, D. C. Ralph, *Nature* 2002, **417**, 722.

27 W. Liang, M. P. Shores, M. Bockrath, J. R. Long, H. Park, *Nature* 2002, **417**, 725.

28 S. Kubatkin, A. Danilov, M. Hjort, J. Cornil, J.-L. Bredas, N. Stuhr-Hansen, P. Hedegard, T. Bjørnholm, *Nature* 2003, **425**, 698.

29 G. J. Doran, *Appl. Phys. Lett.* 1977, **31**, 337.

30 Y. A. Pashkin, Y. Nakamura, J. S. Tsai, *Appl. Phys. Lett.* 2000, **76**, 2256.

31 J.-P. Colinge (Ed.), *Silicon-on-insulator technology: Materials to VLSI*, 2nd edn., Kluwer Academic Publishers, Boston, 1997.

32 Y. Takahashi, M. Nagase, H. Namatsu, K. Kurihara, K. Iwadate, Y. Nakajima, S. Horiguchi, K. Murase, M. Tabe, *International Electron Devices Meeting, Technical Digest, p.* 938, IEEE, Piscataway, NJ, 1994.

33 Y. Takahashi, M. Nagase, H. Namatsu, K. Kurihara, K. Iwadate, Y. Nakajima, S. Horiguchi, K. Murase, M. Tabe, *Electronics Lett.* 1995, **31**, 136.

34 Y. Takahashi, H. Namatsu, K. Kurihara, K. Iwadate, M. Nagase, K. Murase, *IEEE Trans. Electron Devices* 1996, **43**, 1213.

35 E. Leobandung, L. Guo, Y. Wang, S. Y. Chou, *Appl. Phys. Lett.* 1995, **67**, 938.

36 E. Leobandung, L. Guo, S. Y. Chou, *Appl. Phys. Lett.* 1995, **67**, 2338.

37 H. Ishikuro, T. Fujii, T. Saraya, G. Hashiguchi, T. Hiramoto, T. Ikoma, *Appl. Phys. Lett.* 1996, **68**, 3585.

38 H. Ishikuro, T. Hiramoto, *Appl. Phys. Lett.* 1997, **71**, 3691.

39 H. Ishikuro, T. Hiramoto, *Appl. Phys. Lett.* 1999, **74**, 1126.

40 D. H. Kim, J. D. Lee, B.-G. Park, *Jpn. J. Appl. Phys.* 2000, **39**, 2329.

41 D. H. Kim, S.-K. Sung, K. R. Kim, J. D. Lee, B.-G. Park, B. H. Choi, S. W. Hwang, D. A. Park, *IEEE Trans. Electron Devices* 2002, **49**, 627.

42 D. H. Kim, S.-K. Sung, K. R. Kim, J. D. Lee, B.-G. Park, *J. Vac. Sci. Technol.* 2002, **B20**, 1410.

43 K. Uchida, J. Koga, R. Ohba, A. Toriumi, *IEEE Trans. Electron Devices* 2003, **50**, 1623.

44 K. R. Kim, D. H. Kim, J. D. Lee, B.-G. Park, *Appl. Phys. Lett.* 2004, **84**, 3178.

45 M. Saitoh, T. Hiramoto, *Appl. Phys. Lett.* 2004, **84**, 3172.

46 M. Saitoh, H. Harata, T. Hiramoto, *Jpn. J. Appl. Phys.* 2005, **44**, L338.

47 H. Harata, M. Saitoh, T. Hiramoto, *Jpn. J. Appl. Phys.* 2005, **44**, L640.

48 K. Miyaji, M. Saitoh, T. Hiramoto, *Appl. Phys. Lett.* 2006, **88**, 143505.

49 Y. Ono, Y. Takahashi, K. Yamazaki, M. Nagase, H. Namatsu, K. Kurihara, K. Murase, *IEEE Trans. Electron Devices* 2000, **47**, 147.

50 Y. Ono, K. Yamazaki, Y. Takahashi, *IEICE Trans. Electron.* 2001, **E-84C**, 1061.

51 K. Shiraishi, M. Nagase, S. Horiguchi, H. Kageshima, M. Uematsu, Y. Takahashi, K. Murase, *Physica* 2000, **E 7**, 337.

52 S. Horiguchi, M. Nagase, K. Shiraishi, H. Kageshima, Y. Takahashi, K. Murase, *Jpn. J. Appl. Phys.* 2001, **40**, L29.

53 A. Fujiwara, Y. Takahashi, K. Murase, M. Tabe, *Appl. Phys. Lett.* 1995, **67**, 2957.

54 A. B. Zorin, F.-J. Ahlers, J. Niemeyer, T. Weimann, H. Wolf, V. A. Krupenin, S. V. Lotkhov, *Phys. Rev. B* 1996, **53**, 13682.

55 W. H. Huber, S. B. Martin, N. M. Zimmerman, *Proceedings of Experimental Implementation of Quantum Computation*, p. 176, Rinton Press, Princeton, 2001.

56 N. M. Zimmerman, W. H. Huber, A. Fujiwara, Y. Takahashi, *Appl. Phys. Lett.* 2001, **79**, 3188.

57 Y. Takahashi, Y. Ono, A. Fujiwara, K. Shiraishi, M. Nagase, S. Horiguchi, K. Murase, *Proceedings of Experimental Implementation of Quantum Computation*, p. 183, Rinton Press, Princeton, 2001.

58 A. Fujiwara, M. Nagase, S. Horiguchi, Y. Takahashi, *Jpn. J. Appl. Phys.* 2003, **42**, 2429.

59 Y. Takahashi, Y. Ono, A. Fujiwara, H. Inokawa, *J. Phys.: Condens. Matter* 2002, **14**, R995.

60 J. R. Tucker, *J. Appl. Phys.* 1992, **72**, 4399.

61 M. I. Lutwyche, Y. Wada, *J. Appl. Phys.* 1994, **75**, 3654.

62 M. Kirihara, N. Kuwamura, K. Taniguchi, C. Hamaguchi, *Ext. Abstracts 1994 International Conference on Solid State Devices and Materials*, p. 328, Business Center for Academic Societies Japan, Tokyo, 1994.

63 H. Fukui, M. Fujishima, K. Hoh, *Jpn. J. Appl. Phys.* 1995, **34**, 1345.

64 A. N. Korotkov, R. H. Chen, K. K. Likharev, *J. Appl. Phys.* 1995, **78**, 2520.

65 R. H. Chen, A. N. Korotkov, K. K. Likharev, *Appl. Phys. Lett.* 1996, **68**, 1954.

66 S. Amakawa, H. Fukui, M. Fujishima, K. Hoh, *Jpn. J. Appl. Phys.* 1996, **35**, 1146.

67 M. Fujishima, H. Fukui, S. Amakawa, K. Hoh, *IEICE Trans. Electron.* 1997, **E80-C**, 881.

68 M.-Y. Jeong, Y.-H. Jeong, S.-W. Hwang, D.-M. Kim, *Jpn. J. Appl. Phys.* 1997, **36**, 6706.

69 H. Iwamura, M. Akazawa, Y. Amemiya, *IEICE Trans. Electron.* 1998, **E81-C**, 42.

70 S. Amakawa, H. Majima, H. Fukui, M. Fujishima, K. Hoh, *IEICE Trans. Electron.* 1998, **E81-C**, 21.

71 M. Kirihara, K. Nakazato, M. Wagner, *Jpn. J. Appl. Phys.* 1999, **38**, 2028.

72 M. Akazawa, K. Kanaami, T. Yamada, Y. Amemiya, *IEICE Trans. Electron.* 1999, **E82-C**, 1607.

73 K. Uchida, K. Matsuzawa, A. Toriumi, *Jpn. J. Appl. Phys.* 1999, **38**, 4027.

74 S. Shimano, K. Masu, K. Tsubouchi, *Jpn. J. Appl. Phys.* 1999, **38**, 403.

75 Y. Takahashi, A. Fujiwara, Y. Ono, K. Murase, *Proceedings 30th IEEE International Symposium on Multi-Valued Logic*, p. 411, IEEE, Los Alamitos, CA, 2000.

76 K. Uchida, K. Matsuzawa, J. Koga, R. Ohba, S. Takagi, A. Toriumi, *Jpn. J. Appl. Phys.* 2000, **39**, 2321.

77 K. Uchida, J. Koga, R. Ohba, A. Toriumi, *IEICE Trans. Electron.* 2001, **E84-C**, 1066.

78 M.-Y. Jeong, B.-H. Lee, Y.-H. Jeong, *Jpn. J. Appl. Phys.* 2001, **40**, 2054.

79 K.-T. Liu, A. Fujiwara, Y. Takahashi, K. Murase, Y. Horikoshi, *Jpn. J. Appl. Phys.* 2002, **41**, 458.

80 Y. Ono, H. Inokawa, Y. Takahashi, *IEEE Trans. Nanotechnol.* 2002, **1**, 93.

81 Y. S. Yu, J. H. Oh, S. W. Hwang, D. Ahn, *Electronics Lett.* 2002, **38**, 850.

82 Y. Mizugaki, P. Delsing, *Jpn. J. Appl. Phys.* 2001, **40**, 6157.

83 H. Inokawa, A. Fujiwara, Y. Takahashi, *Appl. Phys. Lett.* 2001, **79**, 3618.

84 H. Inokawa, Y. Takahashi, *IEEE Trans. Electron Devices* 2003, **50**, 455.

85 H. Inokawa, A. Fujiwara, Y. Takahashi, *IEEE Trans. Electron Devices* 2003, **50**, 462.

86 H. Inokawa, A. Fujiwara, Y. Takahashi, *Jpn. J. Appl. Phys.* 2002, **41**, 2566.

87 H. Inokawa, Y. Takahashi, K. Degawa, T. Aoki, T. Higuchi, *IEICE Trans. Electron.* 2004, **E87-C**, 1818.

88 K. Degawa, T. Aoki, T. Higuchi, H. Inokawa, Y. Takahashi, *IEICE Trans. Electron.* 2004, **E87-C**, 1827.

89 Y. Ono, Y. Takahashi, K. Yamazaki, M. Nagase, H. Namatsu, K. Kurihara, K. Murase, *Appl. Phys. Lett.* 2000, **76**, 3121.

90 C. P. Heij, P. Hadley, J. E. Mooij, *Appl. Phys. Lett.* 2001, **78**, 1140.

91 D. Tsuya, M. Suzuki, Y. Aoyagi, K. Ishibashi, *Jpn. J. Appl. Phys.* 2005, **44**, 1588.

92 F. Nakajima, K. Kumakura, J. Motohisa, T. Fukui, *Jpn. J. Appl. Phys.* 1999, **38**, 415.

93 K. Nishiguchi, S. Oda, *Appl. Phys. Lett.* 2001, **78**, 2070.

94 Y. Takahashi, A. Fujiwara, K. Yamazaki, H. Namatsu, K. Kurihara, K. Murase, *Appl. Phys. Lett.* 2000, **76**, 637.

95 Y. Ono, Y. Takahashi, *International Electron Devices Meeting, Technical Digest*, p. 297, IEEE, Piscataway, NJ, 2000.

96 Y. Ono, K. Yamazaki, M. Nagase, S. Horiguchi, K. Shiraishi, Y. Takahashi, *Microelectron. Eng.* 2001, **59**, 435.

97 N. J. Stone, H. Ahmed, *Electronics Lett.* 1999, **35**, 1883.

98 A. Fujiwara, Y. Takahashi, K. Yamazaki, H. Namatsu, M. Nagase, K. Kurihara, K. Murase, *IEEE Trans. Electron Devices* 1999, **46**, 954.

99 N. Takahashi, H. Ishikuro, T. Hiramoto, International Electron Devices Meeting, Technical Digest, p. 371, IEEE, Piscataway, NJ, 1999.

100 S. Kasai, H. Hasegawa, *IEEE Electron Device Lett.* 2002, **23**, 446.

101 F. Nakajima, Y. Miyoshi, J. Motohisa, T. Fukui, *Appl. Phys. Lett.* 2003, **83**, 2680.

102 Y. Miyoshi, F. Nakajima, J. Motohisa, T. Fukui, *Appl. Phys. Lett.* 2005, **87**, 033501.

103 G. Zimmerli, R. L. Kautz, J. M. Martinis, *Appl. Phys. Lett.* 1992, **61**, 2616.

104 E. H. Visscher, S. M. Verbrugh, J. Lindeman, P. Hadley, J. E. Mooij, *Appl. Phys. Lett.* 1995, **66**, 305.

105 Y. Satoh, H. Okada, K. Jinushi, H. Fujikura, H. Hasegawa, *Jpn. J. Appl. Phys.* 1995, **38**, 410.

106 R. A. Smith, H. Ahmed, *Appl. Phys. Lett.* 1997, **71**, 3838.

107 R. Zimmermann, W. Fichtner, *IEEE J. Solid-State Circuits* 1997, **32**, 1079.

4
Magnetic Domain Wall Logic

Dan A. Allwood and Russell P. Cowburn

4.1
Introduction

The integrated circuit which, during recent years, has become the basis of modern digital electronics, functions by making use of electron charge. However, electrons also possess the quantum mechanical property of spin, which is responsible for magnetism. New "spintronic" technologies seek to make use of this electron spin, sometimes in conjunction with electron charge, in order to achieve new types of device. Several spintronic devices are currently being developed that outperform traditional electronics. Often, this results from an increased functionality, which means that a single spintronic element performs an operation that requires several electronic elements.

Different approaches to spintronics have been developed by the semiconductor and magnetism communities. Although there have been some very impressive demonstrations of spin-polarized charge transport and ferromagnetism in cooled semiconductors [1–3], the lack of a reliable room-temperature semiconductor ferromagnet has hampered their application. Within the magnetism community, however, considerable success has been achieved at room temperature by using common ferromagnetic materials such as $Ni_{81}Fe_{19}$ (Permalloy). This approach offers the benefits of low power operation, non-volatile data storage (no power required), and a high tolerance of both impurities and radiation.

A great success of electronics has been the ability to use groups of transistors for performing Boolean logic operations. Each type of operation has a particular relationship between its input and output states, each of which can take the value "1" or "0". These relationships are shown in the "truth tables" in Table 4.1 for the Boolean NOT, AND, and OR logic operations. Logical NOT has a single input and a single output, with the output having the opposite state of the input. Logical AND has two independent inputs and has a single output that is "1" for an input combination of "11" and "0" otherwise. Conversely, a logical OR output is "0" for an input combination of "00" and "1" otherwise. Importantly, a suitable combination of NOT

Nanotechnology. Volume 4: Information Technology II. Edited by Rainer Waser
Copyright © 2008 WILEY-VCH Verlag GmbH & Co. KGaA, Weinheim
ISBN: 978-3-527-31737-0

Table 4.1 Truth tables of some common Boolean logic functions.

NOT		AND			OR		
Input A	Output B	Input A	Input B	Output C	Input A	Input B	Output C
0	1	0	0	0	0	0	0
		0	1	0	0	1	1
1	0	1	0	0	1	0	1
		1	1	1	1	1	1

and AND operations, or NOT and OR operations, allows any computation to be performed. The CMOS architecture of NOT and AND logic gates and other circuit elements are listed in Table 4.2. Logical "1" and "0" are represented by the presence or absence of electrical charge, usually measured as either a high and low (zero) voltage. Signal splitting (fan-out) and signal cross-over is achieved by appropriate routing of wire tracks, although for signal cross-over this requires complex fabrication in three dimensions.

Magnetic logic seeks to perform the functions necessary for a logic system with ferromagnetic metals to make use of the advantages that these materials offer. One approach has been to use magnetic/non-magnetic/magnetic tri-layer structures known as *magnetic tunnel junctions* (MTJs) [4, 5]. These have an electrical resistance that depends on the relative orientation of magnetization of the two magnetic layers, and are commonly used in magnetic random access memory (MRAM) [6, 7]. Logic gates made from MTJs [8–12] can perform the logic operations described in Table 4.1, as well as others such as NAND, NOR and XOR. Alternatively, MTJs may be used to provide a switchable bias to CMOS transistors so that a single logic gate may be capable of performing one of two logic operations, say logical AND and OR, as desired [13]. Rather than the MTJ switching on every logic operation, it would simply toggle when the CMOS logic gate operation needs to be switched. This could dramatically increase the logic density of *field-programmable gate arrays* (FPGAs), which are used as flexible alternatives to printed circuit boards.

The other major approach to performing logic operations with magnetic materials has been to propagate magnetic solitons. This can be achieved using chains of isolated nanoscale dots, each element separated by a few tens of nanometers from its neighbors [14–17]. Each dot has uniform magnetization, but the proximity of adjacent dots means that they undergo magnetostatic interactions so that an ordered magnetization configuration is achieved. The same effect is achieved with a row of freely-rotating macroscopic bar magnets. However, defects in the magnetic ordering can be created in the chain of dots. These defects are magnetic solitons, and can be propagated through dot chains by the application of a suitable magnetic field. Furthermore, junctions of dots can be used to perform various Boolean logic functions [17]. The major challenge with this technology is to control the magnetostatic interactions between different pairs of dots in order to improve reliability and device yield.

Table 4.2 Common electronic circuit symbols and the equivalent CMOS and domain wall logic devices.

Symbol	CMOS Circuit	Domain Wall Logic Circuit
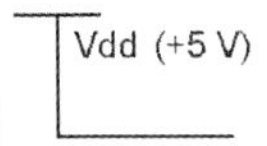 Vdd (+5 V)	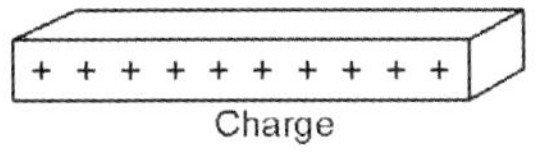 + + + + + + + + + + Charge	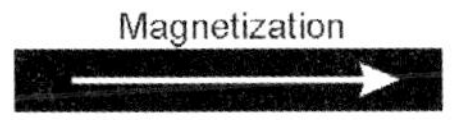 Magnetization
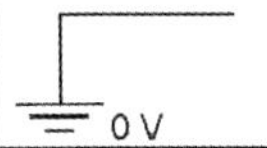 0 V	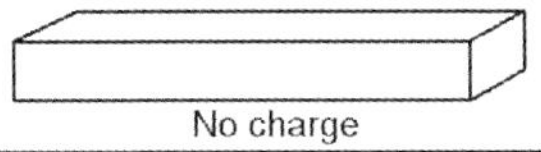 No charge	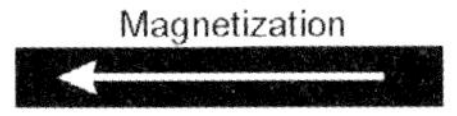 Magnetization
Fan-out 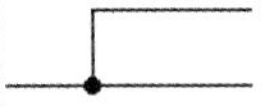	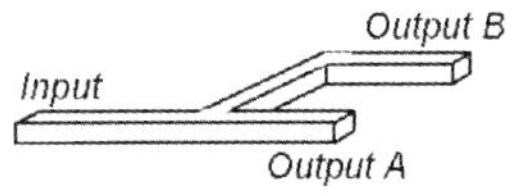 Input, Output B, Output A	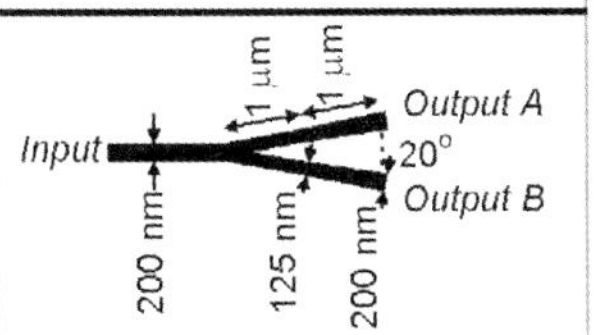 Input, 1 µm, 1 µm, Output A, 20°, Output B, 200 nm, 125 nm, 200 nm
Cross-over 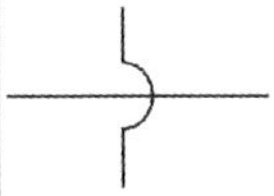	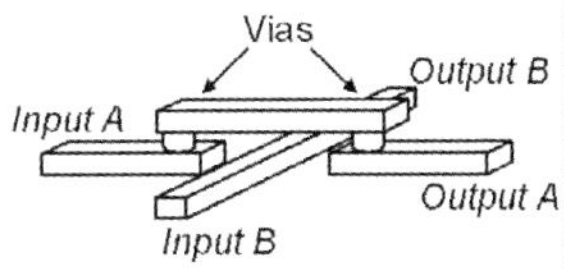 Vias, Input A, Output B, Output A, Input B	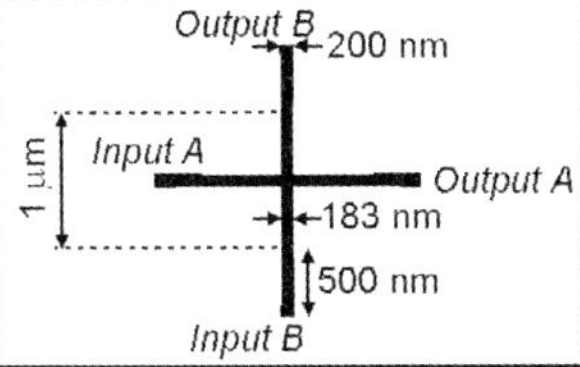 Output B, 200 nm, 1 µm, Input A, Output A, 183 nm, 500 nm, Input B
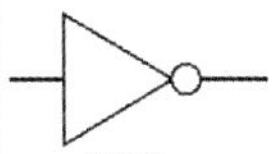 NOT	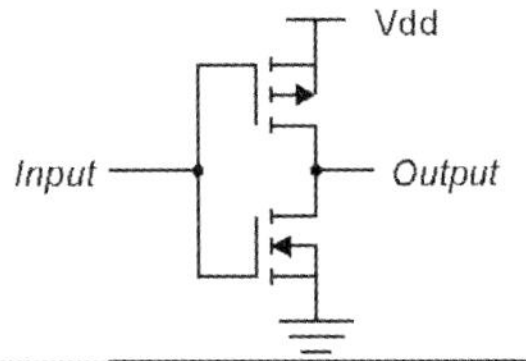 Vdd, Input, Output	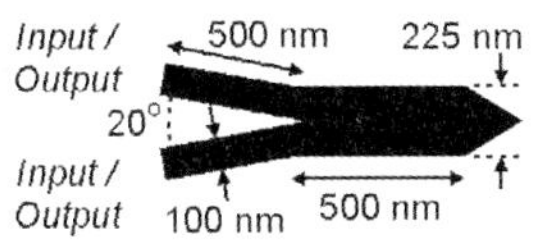 Input / Output, 500 nm, 225 nm, 20°, Input / Output, 100 nm, 500 nm
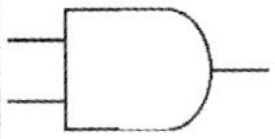 AND	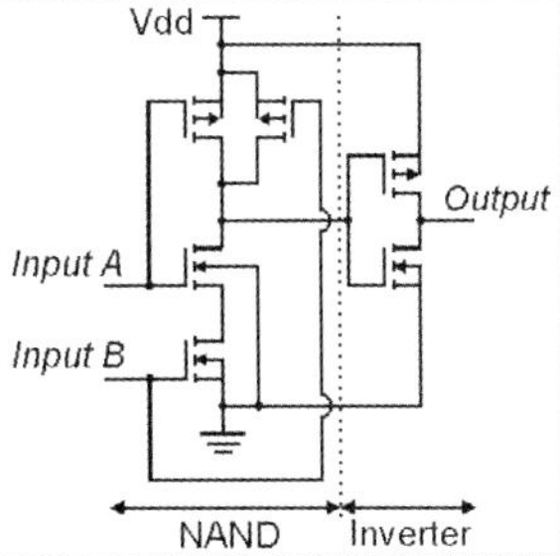 Vdd, Output, Input A, Input B, NAND, Inverter	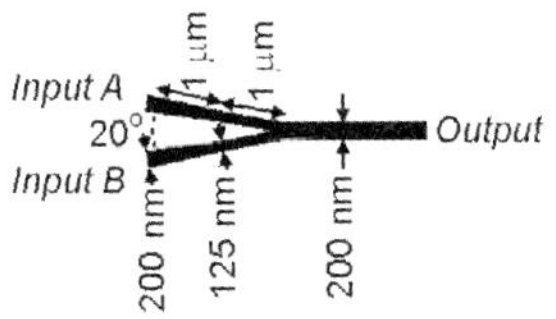 Input A, 1 µm, 1 µm, 20°, Output, Input B, 200 nm, 125 nm, 200 nm
NOT-fan hybrid 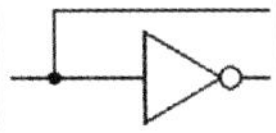	Sequential placement of fan and NOT elements above	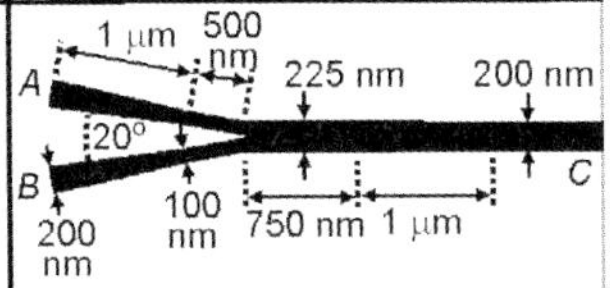 1 µm, 500 nm, 225 nm, 200 nm, A, 20°, B, C, 200 nm, 100 nm, 750 nm, 1 µm
Data input	Various	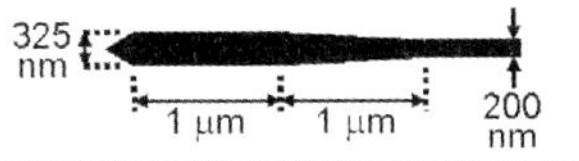 325 nm, 1 µm, 1 µm, 200 nm

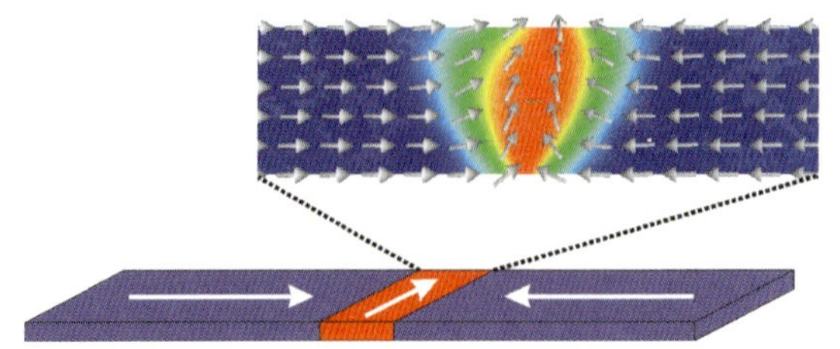

Figure 4.1 Schematic diagram of a transverse 'head-to-head' domain wall in a planar Permalloy nanowire 200 nm wide and 5 nm thick. The expanded region shows a numerically calculated and more detailed view of a domain wall's magnetic structure.

Alternatively, circuits made from planar magnetic nanowires can be used, with wires typically 100 to 250 nm wide and 5 to 10 nm thick. The *shape anisotropy* (geometry) of these wires creates a magnetic easy axis in the wire long axis direction that defines the stable orientations of magnetization (Figure 4.1). This system with two opposite stable magnetization orientations is ideal for representing logical "1" and "0" (see Table 4.2). Where opposite magnetizations meet they are separated by a transition region through which magnetization rotates by 180° (Figure 4.1). This is another form of a magnetic soliton, and is called a *domain wall*. For the wire dimensions relevant here, domain walls are typically approximately 100 nm wide. Domain walls can be moved by applying magnetic fields, and it is this ability which is exploited in magnetic domain wall logic. Domain walls travel down sections of nanowire between nanowire junctions where logic operations are performed. Crucially, the influence of nanowire imperfections on domain wall propagation is very significantly reduced compared with the propagation of solitons in interacting dots. Furthermore, very little power is required either to propagate a domain wall or to perform a logic operation, compared to the lowest power CMOS equivalents or magnetic alternatives. This combination makes magnetic domain wall logic a robust, low-power logic technology. The remainder of this chapter is devoted to explaining how magnetic domain wall logic functions, and what the future prospects of the technology might be.

4.2 Experimental

All of the magnetic structures shown here are fabricated by focused ion beam (FIB) milling [18] of 5 nm-thick Permalloy films. The films were thermally evaporated onto Si(0 0 1) substrates with a native oxide present in a chamber with a base pressure $<10^{-7}$ torr. FIB milling used 30 keV Ga^+ ions which were focused to a diameter of ~7 nm at the substrate. The Ga^+ ions sputter the magnetic material and scatter within the film and substrate to implant at lateral distances up to 40 nm from the spot center [19]. Both of these processes lead to a loss of ferromagnetic order in the Permalloy film and allow nanostructures to be defined. A $150 \times 150\,\mu m$ square of magnetic material is cleared around each nanowire circuit to allow optical analysis of the nanostructure without interference from the surrounding film. Sample images are obtained from secondary electron emission during a single FIB scan.

Magnetization measurements are performed using a magneto-optical Kerr effect (MOKE) magnetometer in the longitudinal MOKE configuration [20]. The instrument has a ~5 μm spatial resolution, given by the focused laser spot diameter, and is sensitive to single magnetization reversal events in individual nanowires [21]. The magnetometer also includes a facility for mapping the sample susceptibility using MOKE signals in order to select particular regions of a structure for measurement. In-plane magnetic fields are applied to samples at a frequency of 27 Hz using a quadrupolar electromagnet. This two-dimensional (2-D) field is characterized by orthogonal fields H_x and H_y, with amplitudes H_x^0 and H_y^0, respectively. The directions of H_x and H_y are defined in the images of each structure.

4.3 Propagating Data

Probably the most common measurement in magnetism is to determine the major hysteresis loop of a magnetic material. Figure 4.2a shows one such measurement from a 100 nm-wide Permalloy wire with field H_x applied along the wire length (along the magnetic easy axis) [22]. The sharpness of the transitions indicates that magnetization reversal most likely occurs by domain wall nucleation from one wire end, rapid propagation through the wire, and annihilation at the other wire end.

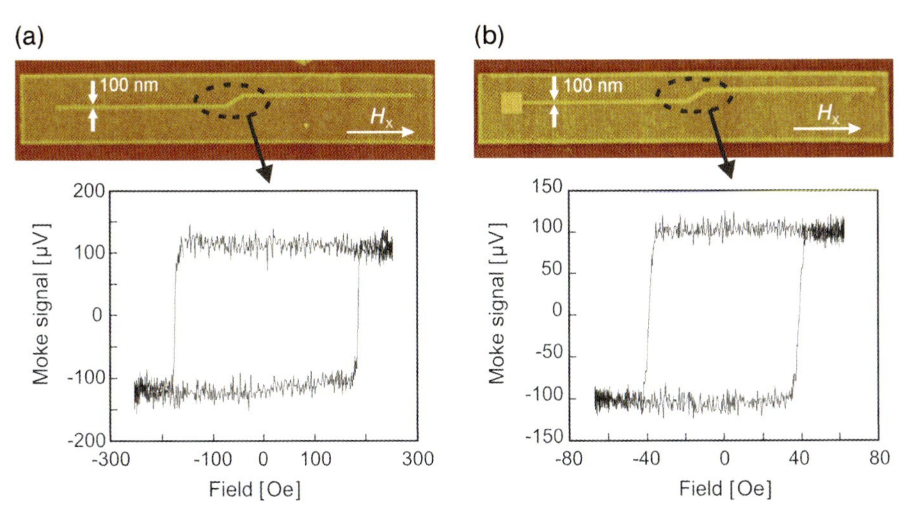

Figure 4.2 Atomic force microscopy images and magneto-optically measured magnetization hysteresis curves from: (a) 100 nm-wide wire and (b) 100 nm-wide wire with a 1 μm × 1 μm square domain wall "injection pad" [22]. Measuring either side of the kink does not change the observed switching field. The direction of the magnetic field H_x is indicated in both images.

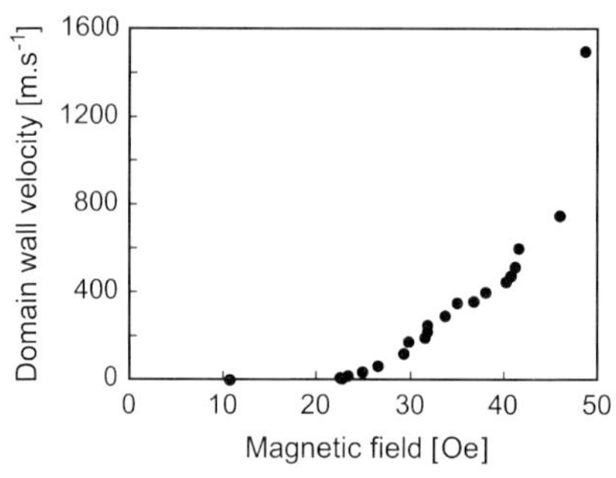

Figure 4.3 Measured domain wall velocity in a 200 nm-wide, 5 nm-thick $Ni_{81}Fe_{19}$ wire as a function of magnetic field along the wire long axis [23].

For this wire, magnetization reversal occurs at a coercive field $H_c = 180$ Oe. Unwanted domain wall nucleation from wire ends, junctions and corners must generally be avoided in logic circuits, as this will corrupt any existing data. It is imperative, therefore, domain walls can be introduced and propagated at magnetic fields lower than, in this case, 180 Oe. Figure 4.2b shows a structure with a similar wire to that in Figure 4.2a, but now with a 1 μm × 1 μm square pad attached to one end. MOKE measurement of the wire shows that $H_c = 39$ Oe. This reduction in H_c is a result of the square pad undergoing magnetization reversal at $H_c = 26$ Oe (not shown) before a domain wall is injected into the wire at $H_x = 39$ Oe. Different regions of the wire all have the same coercive field, even beyond the 30° kink, showing that domain walls can propagate in wires and through changes of wire direction at fields significantly lower than nucleation fields. To quantify this low-field propagation more precisely, the domain wall velocity was measured in a 200 nm-wide Permalloy wire (Figure 4.3), in an experiment described elsewhere [23]. Measured domain wall velocities exceeded 1500 m s^{-1} for certain fields applied along the wire long axis. Other studies [24–27] have shown that domain wall velocity does not increase continually with field, but rather reaches a maximum value before reducing at higher fields. Interestingly, domain wall propagation is still observed at fields as low as 11 Oe [23], albeit with very low velocities of 0.01 m s^{-1}. The data in Figures 4.2 and 4.3 indicate that nanowire devices operating by domain wall propagation will require 11 Oe $< H_x <$ 180 Oe, although these field values will change once wire junctions are introduced.

Simply using a unidirectional field will not allow domain walls to be separated reliably and, hence, normal data streams containing both "1"s and "0"s cannot be propagated. Instead, use is made of the orthogonal fields H_x and H_y to create a magnetic field vector that rotates in the plane of the sample to control domain wall propagation around smooth 90° wire corners. An important rule for understanding the nanowire circuits is that domain walls will propagate around corners of the same sense of rotation as the applied field – that is, a clockwise rotating field will lead to domain walls traveling around corners clockwise. In a correctly designed nanowire circuit, the sense of field rotation will define a unique direction of domain wall propagation and, hence, data flow. This is an essential feature of a Boolean logic

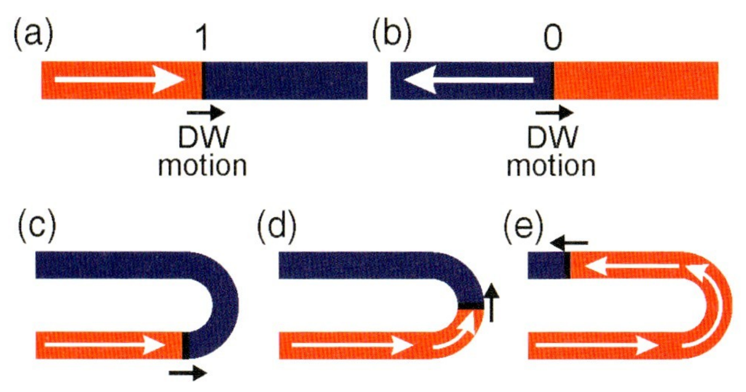

Figure 4.4 Schematic diagrams showing the definition of: (a) logical "1" and (b) logical "0" in magnetic domain wall (DW) logic. Panels (c–e) show sequentially how the wire magnetization changes when a domain wall propagates around a 180° wire corner.

system. Interestingly, the direction of data flow in magnetic domain wall logic can, in principle, be reversed simply by reversing the sense of field rotation.

At this point it must be considered how binary data are represented in magnetic domain wall logic. It was mentioned in Section 4.1 how the two opposite magnetization directions supported in magnetic nanowires can be used to represent the logic states "1" and "0". However, care must be taken here with the definition chosen to use with magnetic nanowire circuits, as the wires can change directions. Figure 4.4a and b show, schematically, two similar magnetic wires with opposite magnetizations containing a single domain wall. A simple approach would be to say that magnetization pointing to the right represents logical "1", and that pointing to the left represents logical "0". However, Figure 4.4c–e shows the magnetization following a domain wall that propagates around a 180° wire corner. In the final situation (Figure 4.4e), the magnetization is continuous up to the domain wall, meaning that there are no changes in logic state up to this point. However, the absolute directions of magnetization are opposite on either side of the turn, and so the simple definition cannot then be valid. Instead, the choice is made to define data representation in terms of the direction of magnetization relative to the direction of domain wall motion. In Figure 4.4c–e, the magnetization following the domain wall is always oriented in the direction of domain wall motion, so the logic state represented remains unchanged. This robust definition allows measurements from different parts of logic circuits to be interpreted correctly.

4.4 Data Processing

The NOT-gate was the first domain wall logic device to be introduced [28–30], and is foundational to the development of all other logic elements. Figure 4.5 shows the geometry of a NOT-gate, and illustrates its principle of operation. The NOT-gate is a junction formed by two wires. For a given field rotation, one wire will act as the input

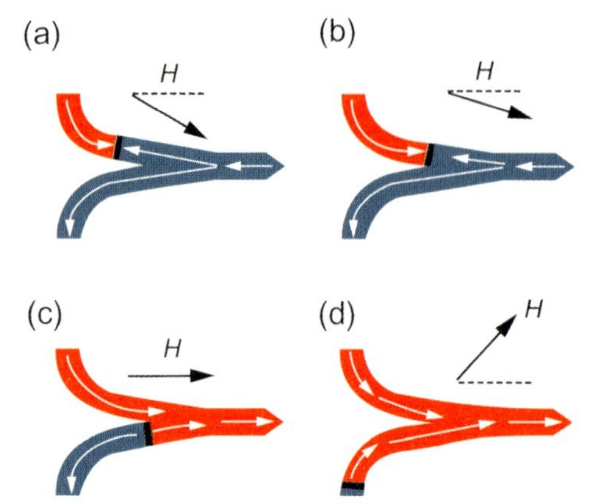

Figure 4.5 Schematic diagrams showing the operating principle of a magnetic nanowire NOT-gate [28]. The black arrows represent the instantaneous magnetic field vector, while the white arrows show the wire's internal magnetization configuration. A domain wall is shown as the black line that separates the oppositely magnetized (red and blue) magnetic domains.

and the other wire as the output. A small central "stub" which emerges from the wire junction is an important part of the device as it ensures there is sufficient shape anisotropy to maintain a magnetization component in the direction in which the stub points. The dimensions for an optimized NOT-gate design are given in Table 4.2. Under a rotating magnetic field, H, a domain wall enters the NOT-gate input wire (Figure 4.5a) before reaching the wire junctions (Figure 4.5b). The magnetization following the domain wall points in the direction of domain wall propagation. Provided that there is sufficient field, the domain wall expands over the junction and splits in two, with one domain wall traveling along the central stub, leaving the stub magnetization reversed, and the other free to propagate on the NOT-gate output wire (Figure 4.5c). As the field continues to rotate, the domain wall in the output wire leaves the NOT-gate. The magnetization following the domain wall is now pointing away from the direction of domain wall motion. The magnetization on either side of the wire junction is reversed, and the device has inverted the input logical state. The reversal in magnetization means that the inversion process would be expected to require a one-half cycle of field.

Figure 4.6a shows a structure containing a NOT-gate fabricated in a square loop of magnetic wire that has not been used to test the operation of the logic device [28–30]. The loop provides feedback to the NOT-gate by joining the output wire to the input wire. Having a single inverter in the loop guarantees that at least one domain wall will be present [28–30] and removes the need, at this stage, for explicit data input. If the principle of operation for a NOT-gate described above is correct, it should be possible to predict the switching period of the NOT-gate/loop structure in terms of field cycles. As mentioned above, a domain wall will take a one-half field cycle to propagate through the NOT-gate junction. Traveling around the 360° wire loop will then add another field cycle. So, magnetization reversal would be expected

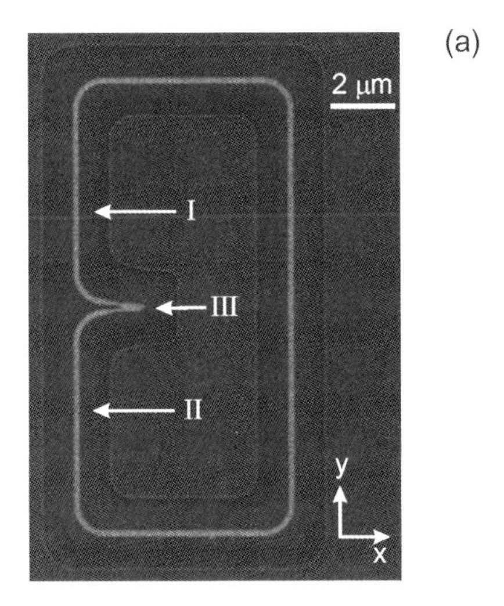

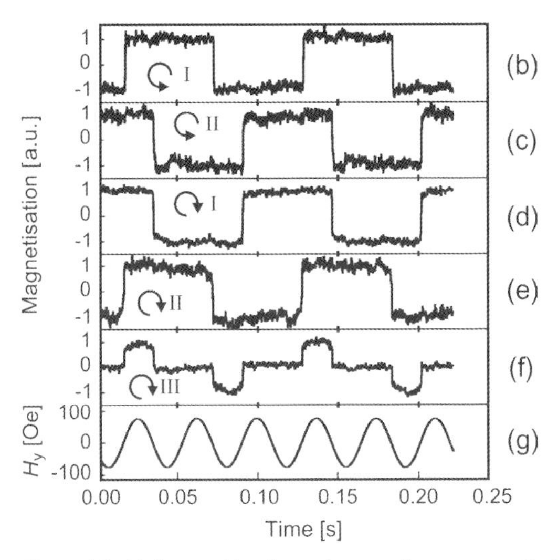

Figure 4.6 (a) Focused ion beam image of a ferromagnetic NOT-gate and feedback loop [29]. Also shown are the *x*- and *y*-directions, and the measurement regions I, II and III where the ~5 μm diameter magnetometer laser spot was positioned. (b–e) Measured MOKE signals during application of an in-plane rotating magnetic field ($H_x^0 = 77$ Oe, $H_y^0 = 74$ Oe) for anti-clockwise field rotation at measurement position (b) I and (c) II, and clockwise field rotation at measurement position (d) I and (e) II. (f) A measured trace from position III with a clockwise field rotation; (g) the *y*-component of the rotating field.

to be seen at any position in the loop every 3/2 field cycles, giving a switching period of three field cycles.

Figure 4.6b shows a time-averaged MOKE trace obtained from position I on the structure (Figure 4.6a) under anti-clockwise field conditions [29]. As expected, a

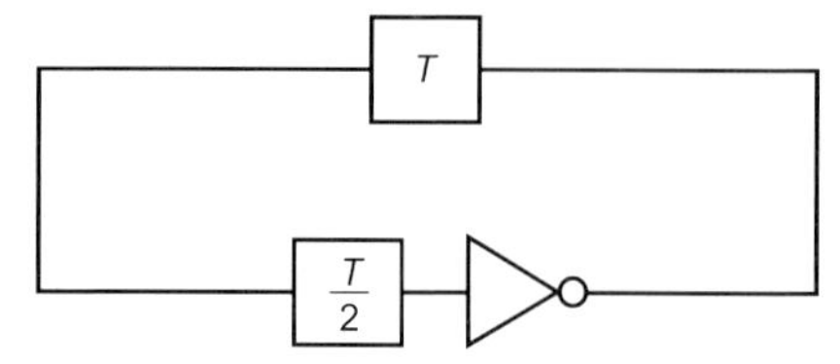

Figure 4.7 An equivalent electronic circuit used to model the ferromagnetic NOT-gate and feedback loop. The square boxes correspond to time delay elements, where *T* is the periodic time of the applied rotating magnetic field.

three-field cycle switching period was observed, indicating that the wire junction is performing as a NOT-gate. To validate this further, Figure 4.6c shows a MOKE trace obtained from position II, the other side of the NOT-gate (Figure 4.6a), with the same field conditions. This trace is inverted compared to that in Figure 4.6b, except for a one-half field cycle delay, consistent with the operation of a NOT-gate outlined above. As the NOT-gate has an equal number of input and output wires that have identical geometry, it may be operated reversibly. Figure 4.6d and e show measurements from positions I and II, respectively but now under a clockwise-rotating field. The observed phase relationship indicates that position II has now become the input and position I the output. Figure 4.6f shows a MOKE trace that is obtained with the laser spot positioned over the NOT-gate (position III, Figure 4.6a). A domain wall is observed to enter and leave the NOT-gate to correlate with the traces shown in Figure 4.6d and e. An important aspect of this initial demonstration is that the applied field acts as both power supply and clock to the magnetic circuit. The structure in Figure 4.6a can be thought of as having the equivalent electronic circuit shown in Figure 4.7. Electronic invertors do not have a delay in terms of a clock cycle, so a buffer must be introduced with a delay of $T/2$, where T represents a clock period. Another buffer with a delay of T is then introduced within a feedback loop to represent the time spent by a domain wall propagating around the wire loop. The signal from this circuit will replicate those observed in Figure 4.6.

One of the advantages of having input and output wires of identical forms is that logic gates can be directly connected together. Figure 4.8a shows a magnetic shift register circuit made of 11 NOT-gates within a wire loop [28]. The expected switching period can again be calculated by $11 \times 1/2$ field cycles for domain wall propagation through NOT-gates plus one field cycle for the loop to give a magnetization reversal every 6.5 field cycles, or a switching period of 13 field cycles. Figure 4.8b shows the MOKE trace obtained from the structure, in which the 13-field cycle period is clearly observed. The measurement was obtained over 30 min of averaging, indicating that almost 10^5 logical NOT operations were successfully performed. If there was a problem with a domain wall propagating through a NOT-gate on just one occasion, the resultant phase difference introduced would be clearly visible in the time-averaged trace (Figure 4.8b). Although in this case the topology of the shift register structure has been used to ensure the presence of a single domain wall, it will be seen later (in Section 4.5) how similar shift registers can support complex data sequences. The one-half cycle delay for domain wall propagation creates a natural buffer between

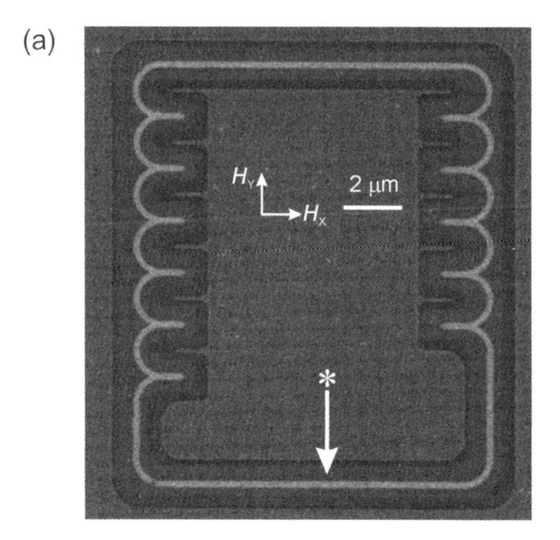

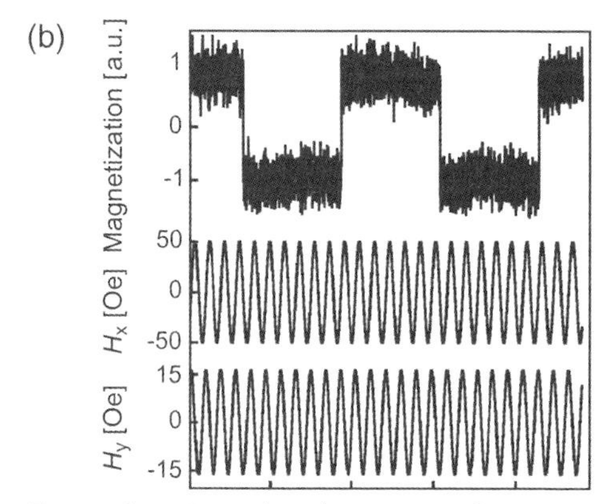

Figure 4.8 (a) Focused ion beam image of a magnetic ring including 11 NOT junctions, where the asterisk indicates the position of MOKE analysis [28]. The directions of the magnetic field components, H_x and H_y, are also indicated. (b) MOKE analysis of an identical structure within a clockwise rotating magnetic field ($H_x^0 = 50$ Oe and $H_y^0 = 15$ Oe).

data bits, removing the need for any complex circuitry, such as the flip-flop circuits that are commonly used in electronic memories.

Characterizing the performance of magnetic domain wall NOT-gates is essential for design optimization [29] and for integrating them with other types of nanowire junction. Here, structures similar to that in Figure 4.6a were used to assess a NOT-gate's operation as a function of in-plane rotating field amplitudes H_x^0 and H_y^0. Three types of operation are observed:

- When the field amplitudes are too low, domain walls experience pinning at the NOT-gate junction, and this leads either to no switching for very low fields or else de-phasing of the time-averaged MOKE trace when domain walls are pinned even once.

- At high fields, additional domain wall pairs are nucleated in the structure and the magnetization reversal has a single field cycle period.
- At intermediate fields the three-field cycle operation described above is observed.

It should be noted that, after domain wall nucleation is observed at high fields, it is necessary to reduce the number of domain walls back to one by applying field conditions for occasional de-pinning [29]. This allows domain wall pairs to meet and annihilate. Figure 4.9 shows the resulting phase diagram describing NOT-gate operation as a function of field. There are two phase boundaries present, one

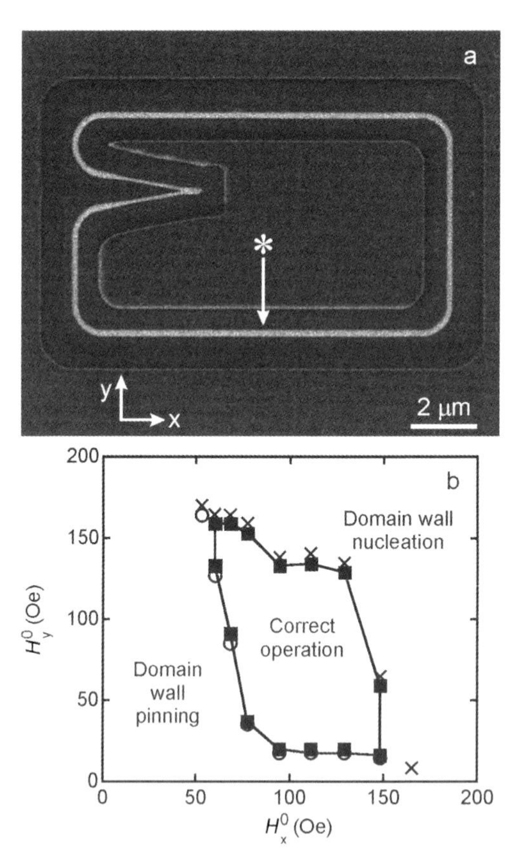

Figure 4.9 (a) Focused ion beam (FIB) image of NOT-gate and feedback loop [29]. The magnetic wire is the light-gray line, and all other features are a result of the FIB milling. Also shown are the measurement position (denoted by "*") and the *x*- and *y*-directions. (b) Experimentally determined phase diagram showing operation of the NOT-gate/feedback loop structure shown in (a) as a function of the rotating magnetic field component amplitudes H_x^0 and H_y^0. × = nucleating; ■ = correct operation; ○ = domain wall pinning. The region bounded by the solid line is the operating region for this device.

separating domain wall pinning from correct operation, and the other separating correct operation from domain wall nucleation. The two boundaries meet to define an area of field phase space in which the NOT-gate operates correctly. Figure 4.9 is taken for a NOT-gate with the optimized dimensions given in Table 4.2.

The other circuit elements that are required for a realistic logic system are a majority gate, signal fan-out, and signal cross-over. The NOT-gate operating phase diagram (Figure 4.9) provides a useful and necessary reference for comparing the performance of these additional elements to ensure compatibility. Figure 4.10a–c shows three structures used for testing the operating fields of majority gate junctions [31]. Each junction has two input wires and one output wire, with the structures having (a) no, (b) one, and (c) two input wires terminated by a domain wall "injection pad". The low field at which domain walls are introduced from an injection pad means that they provide a means of testing majority gate junction operation

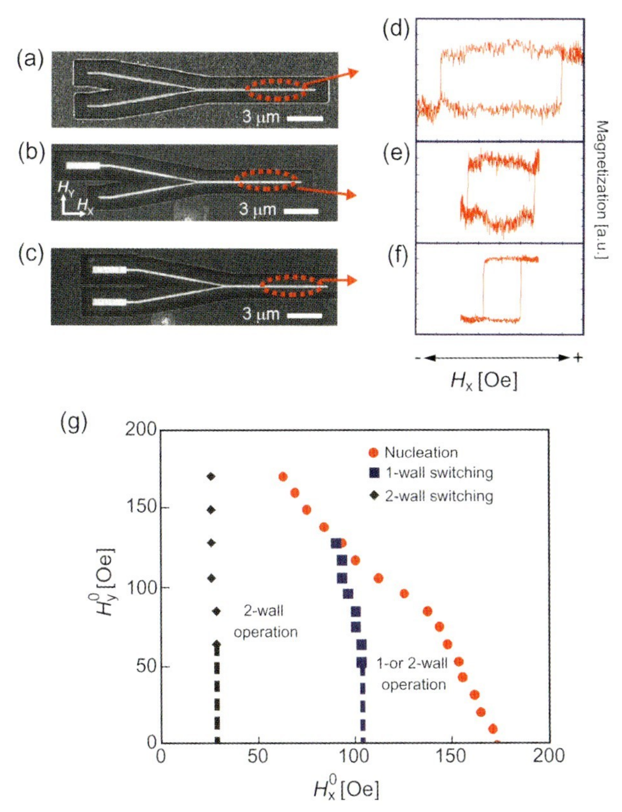

Figure 4.10 (a–c) Focused ion beam images of majority gate test structures with (a) zero, (b) one, and (c) two input wires connected to a 3 μm × 600 nm domain wall "injection pad" [31]. The directions of H_x and H_y are indicated in panel (b). (d–f) MOKE hysteresis loops from the output wires of panels (a)–(c), respectively. (g) Experimentally determined operating phase diagram of the majority gates in an in-plane rotating magnetic field as a function of the field amplitudes H_x^0 and H_y^0.

when 0, 1, or 2 domain walls are present in the input wires. Clearly, the output arm switching fields (Figure 4.10d–f) reduce as the number of domain walls present at the junction increases. In terms of magnetization dynamics, it is interesting that a single domain wall appears capable of expanding across a junction before propagating through the output wire, and that two domain walls are able to interact to enable very low output wire switching fields. Figure 4.10g shows a more detailed analysis of the operation of optimized majority gate junctions (dimensions given in Table 4.2) as a function of the in-plane rotating field amplitudes. The different input conditions now lead to two field-space regions of operation, depending on the number of domain walls present before switching. Crucially, comparison with Figure 4.9 shows that there is overlap between the NOT-gate operating region and both operating regions of the majority gate. The question remains, however, whether to use field amplitudes from the lower-field operating region of the majority gate, or the higher. The answer is to use both. For a majority gate aligned parallel to H_x, field conditions $H_x^0 = 120$ Oe and $H_y^0 = 50$ Oe will mean that the output wire will switch whenever there is just one domain wall present. This corresponds to an input condition of "01" or "10". Clearly, this should not happen for either an AND-gate or an OR-gate. However, by examining the truth tables for each (see Table 4.1), it becomes obvious that a "10" input should always lead to a "0" output for an AND-gate and a "1" output for an OR-gate. However, H_x need not be symmetric; instead, a dc field H_x^{DC} can additionally be applied to bias H_x so that for one sense of H_x the majority gate reverses with a domain wall in either input, while in the other sense of H_x the majority gate requires domain walls in both input wires. The AND/OR function of the gate is then selected by the polarity of H_{dc}.

Signal fan-out and signal cross-over junctions were developed in a similar manner [32], with optimized geometries shown in Table 4.2. Figure 4.11a shows a circuit that integrates all of the structures necessary for performing logic operations: a NOT-gate, a majority gate, two signal fan-out junctions and a signal cross-over element [33]. An anti-clockwise rotating field with amplitudes of $H_x^0 = 75$ Oe and $H_y^0 = 88$ Oe was used with $H_x^{DC} = -5$ Oe (Figure 4.11b) in order to circulate domain walls in an anti-clockwise direction and select logical AND operation for the majority gate. The NOT-gate/loop is similar to those discussed above, and will contain a single domain wall and a magnetization switching period of three field cycles, as before. The difference from Figure 4.11a, however, is that a fan-out structure is incorporated within the loop to split a domain wall each time it is incident on the junction. Part of the domain wall will continue propagating around the loop, while the other part exits the loop to the rest of the circuit. When used in this manner, the NOT-gate/loop acts as a three-field cycle period signal generator for testing the circuit. A domain wall that exits the NOT-gate/loop is then split again at a second fan-out junction. MOKE measurement at position I in Figure 4.11a shows that the three-field cycle period from the NOT-gate loop is preserved through both fan-out junctions (Figure 4.11b, trace I). The domain walls from the second fan-out junction now have separate paths before reaching the AND-gate inputs. The domain wall that passes position I simply has to propagate through two 90° corners and some straight wire sections. The resulting half-field cycle delay between domain walls passing position 1 and arriving at the AND-gate is indicated by trace II in Figure 4.11b. The other domain wall from

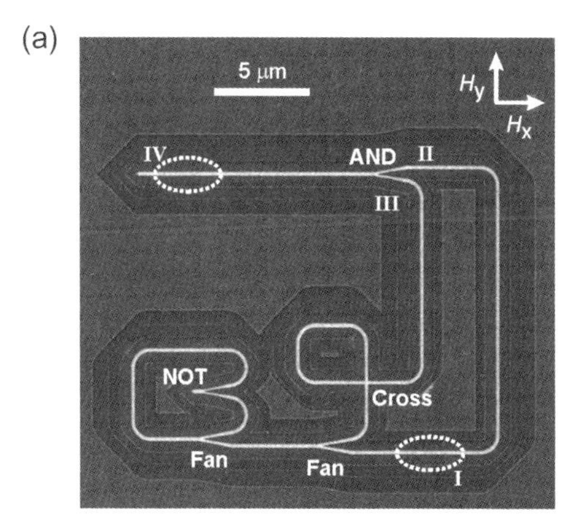

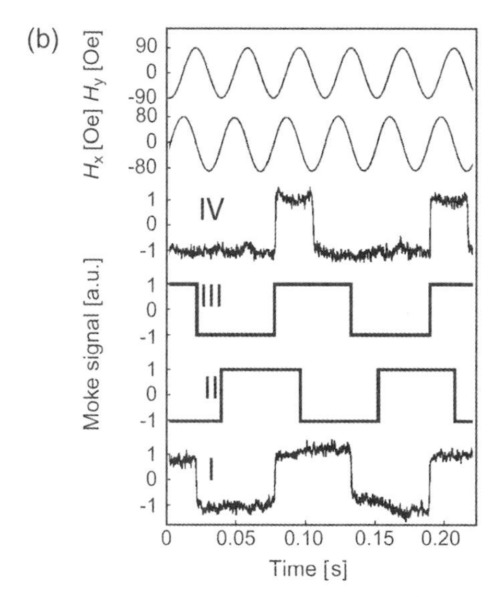

Figure 4.11 (a) Focused ion beam image of a magnetic nanowire circuit containing one NOT-gate, one AND-gate, two fan-out junctions and a cross-over junction [33]. MOKE measurements were made at positions I and IV, while positions II and III denote the inputs to the AND-gate. Also indicated are the directions of field components, H_x and H_y. (b) MOKE traces describing the operation of the magnetic circuit within an anti-clockwise rotating field with $H_x^0 = 75$ Oe, $H_y^0 = 88$ Oe and $H_x^{DC} = -5$ Oe. Experimental MOKE measurements from positions I and IV of the circuit are shown. Traces II and III are inferred from trace I, and show the magnetization state of the AND-gate's input wires.

the fan-out junction has to negotiate a cross-over junction and an additional 360° loop before arriving at an AND-gate input at position III (Figure 4.11a). The inclusion of the loop tests the operation of the cross-over element and will create a one-field cycle delay between domain walls arriving at positions II and III, as indicated in the inferred trace III in Figure 4.11b. Measurement at position IV in Figure 4.11a shows that the output is high only when both inputs are high, showing that the majority gate is operating correctly as an AND-gate. Furthermore, this demonstrates that all four of the element types can operate under identical field conditions in a single circuit.

4.5 Data Writing and Erasing

In the previous section, domain walls were introduced to nanowire junctions either by using topological constraints of a nanowire circuit or domain wall injection from a large area pad. These are both valid methods for developing logic elements, although a method of entering user-defined data is still required to create a viable logic system. Here, an element for data input is presented that is integrated with a domain wall shift register [33]. Furthermore, it is shown how data can be deleted from the shift register.

The design of the optimized data input element is shown in Table 4.2. Figure 4.12 shows the operating phase diagram of this element, obtained from simple test structures, overlaid with that of a NOT-gate. A single phase boundary for the data input element bisects the NOT-gate field operating area. Above the phase boundary, a domain wall is nucleated from the data input element, whereas below the phase boundary no magnetic reversal occurs. Two sets of field amplitudes can then be identified for operating both NOT-gates and the data input element. Below the data input element phase boundary are the *read* or *no-write* field conditions of

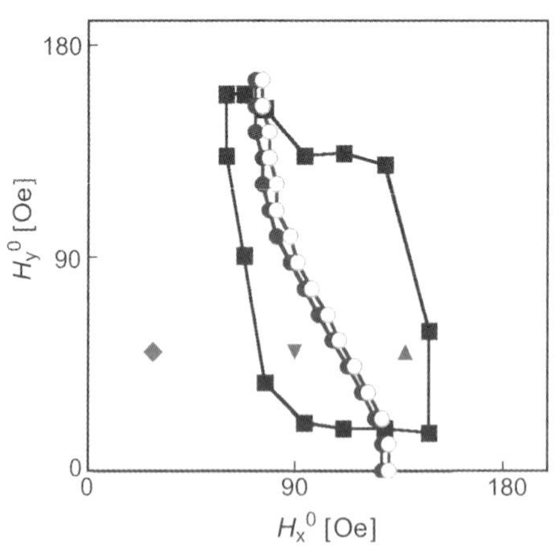

Figure 4.12 Operating field phase diagram of optimized NOT-gate and data input elements. Symbols represent the limits of the NOT-gate operating region (■), maximum field for no domain wall injection (●) and minimum field for reliable domain wall injection (○) from the data input element, and selected *write* field (▲), *read/no-write* field (▼) and *erase* field (◆) conditions. The lines are provided only as guides to the eye.

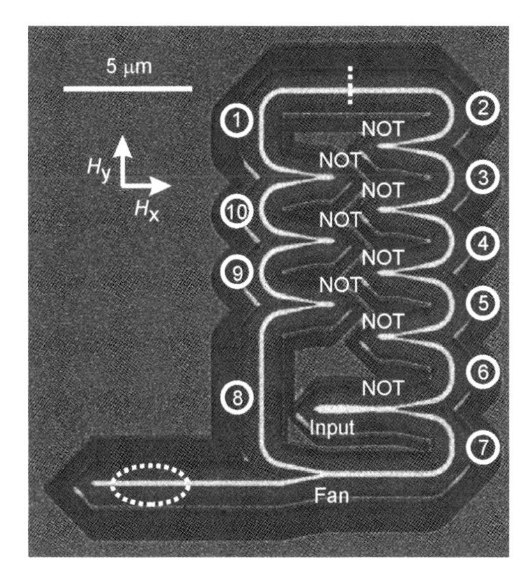

Figure 4.13 Focused ion beam image of a continuous shift register made of eight NOT-gates, a fan-out junction and a data input element connected to the central wire of one of the NOT-gates [33]. The shift register is divided into ten labeled cells, each separated from its neighbors either by a wire junction (NOT or fan-out) or a straight horizontal wire (indicated by dotted line). The field directions H_x and H_y are shown; the position of magneto-optical measurement is indicated by the dotted ellipse.

$H_x^{\text{no-write}} = 90$ Oe and $H_y^0 = 50$ Oe, and above the phase boundary are the *write* field conditions of $H_x^{\text{write}} = 138$ Oe and $H_y^0 = 50$ Oe (Figure 4.12).

Figure 4.13 shows an image of a shift register containing eight NOT-gates and one fan-out junction [33]. In addition, one NOT-gate has a data input element attached to its central stub. The fan-out element provides a monitor arm for MOKE measurement, as used in Section 4.4 above. The shift register can be divided into ten cells, each capable of holding a single domain wall and separated from its neighbors by a total of 180° of wire turn. Due to topological restrictions, domain walls can only be introduced or removed in pairs. Therefore, a data bit is represented by the presence or absence of a domain wall pair, so the shift register in Figure 4.13 contains five data bits.

Figure 4.14a–d shows, schematically, the operating principle of the data input element connected to the NOT-gate [33]. Initially, no domain walls are present and the two connecting wires to the NOT-gate have opposite magnetizations (Figure 4.14a). As the field rotates, the *write* field amplitude is used (Figure 4.14b) so that a domain wall is nucleated at the end of the data input element. This domain wall will propagate to the NOT-gate junction, where it will split into domain walls DW 1 and DW 2, one in each of the input/output wires (Figure 4.14b). As the field rotates further (Figure 4.14c), both domain walls follow the field rotation and propagate clockwise around corners. DW 1 propagates away from the NOT-gate, while DW 2 returns to the junction (Figure 4.14c). Finally, the field rotates to be oriented 180° from when nucleation occurred, but now with *no-write* conditions (Figure 4.14d). DW 1 has

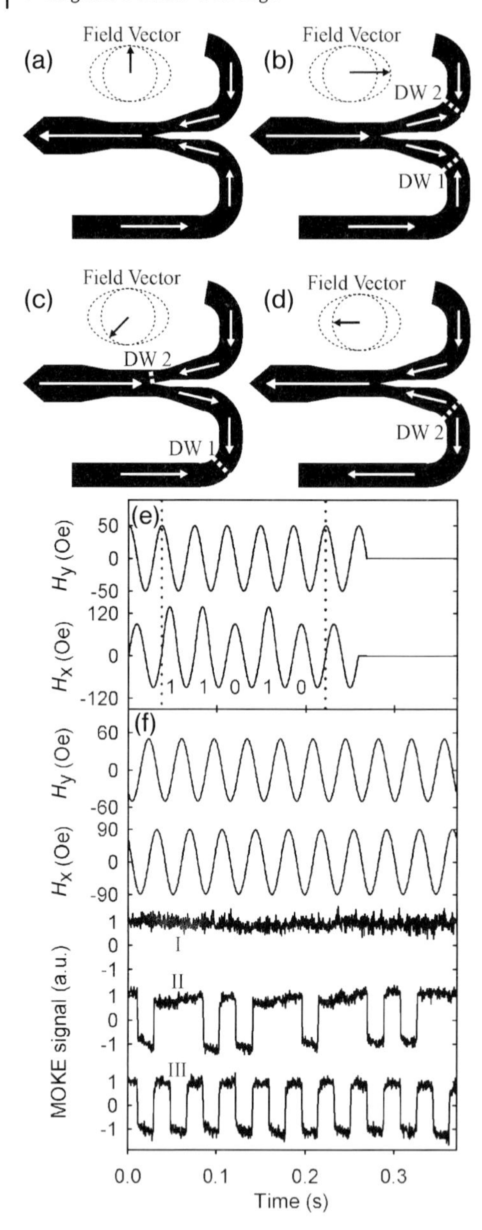

Figure 4.14 (a–d) Schematic diagrams describing the operation of a data input element [33], including instantaneous field vectors (black arrows), magnetization directions (white arrows) and position of domain walls (white dotted line). (e) *Write* field pattern with field amplitudes $H_x^{\text{no-write}} = 90$ Oe, $H_x^{\text{write}} = 138$ Oe and $H_y^0 = 50$ Oe. The five-bit sequence "11010" is generated during the interval between the dotted lines. (f) MOKE measurements ($H_x^0 = 90$ Oe and $H_y^0 = 50$ Oe) from the shift register in Figure 4.13 in reset configuration (trace I), after applying the *write* field pattern shown in (e) (trace II), and after a 1.85-ms duration half-sinusoid field pulse of amplitude $H_x^0 = 234$ Oe (trace III).

propagated out of the section shown in Figure 4.14, while DW 2 has propagated through the NOT-gate, to leave the NOT-gate and data input element's magnetization back in their initial configuration (Figure 4.14d). One single half-cycle of *write* field conditions has created a pair of domain walls – that is, a single data bit has been written.

Figure 4.14e shows a field sequence that is used to write data to the shift register in Figure 4.13. A combination of *write* and *no-write* field conditions is used to write the five-bit data sequence "11010". Time-averaged MOKE measurements were performed during continual application of the *read* field conditions. Trace I in Figure 4.14f shows the MOKE signal obtained prior to the application of the *write* field sequence in Figure 4.14e. No transitions are observed, meaning that no domain walls are present. After a single application of the *write* field sequence, the MOKE signal changes to show that pairs of domain walls are propagating continuously around the shift register (Figure 4.14f, trace II). Crucially, the pattern of domain wall pairs matches the original input data sequence of "11010", although the phase of the measurement is such that the MOKE trace starts part-way through this sequence. Note that in this case logical "1" is represented by a low MOKE signal, due to the 180° wire turn between the data input element and the measurement position. This observation confirms the principle of operation for a data input element outlined above. Delays of an hour between writing and successfully reading data have been seen, demonstrating the intrinsic non-volatility of the data storage. The whole shift register can be filled with domain walls, destroying any data present, by applying an *over-write* half-sinusoid field pulse of amplitude $H_x^0 = 243$ Oe and 1.85 ms pulse length (Figure 4.14f, trace III).

Individual domain wall pairs can also be removed from the shift register in Figure 4.13. This represents a selective bitwise delete operation. Almost all of the ten cells shown in Figure 4.13 are separated from their neighbors by a nanowire junction. The exceptions are cells 1 and 2, which are separated by a straight section of wire. Domain walls require *read* field conditions to propagate successfully through the shift register. However, when *erase* field conditions of $H_x^{\text{erase}} = 24$ Oe and $H_y^0 = 50$ Oe (see Figure 4.12) are used, domain walls cannot overcome the pinning potentials associated with the nanowire junctions. The only possible domain wall motion will be between cells 1 and 2, where there are no wire junctions. Figure 4.15a shows the field sequence for erasing a single pair of domain walls. The first half-cycle has *erase* field conditions, so the only domain wall propagation will be from cell 1 to cell 2. All other domain walls will remain pinned at the junctions between cells. The next half-cycle has *read* field amplitudes, so all domain walls will propagate forward by one cell, with the exception of the pair of domain walls in cell 2 which will meet and annihilate. The second full field cycle continues with *read* field conditions to move all domain walls on by two cells and allowing the field sequence to be repeated on the next domain wall pair. Figure 4.15b shows MOKE traces obtained from the shift register following an *over-write* half-sinusoid pulse and between 0 and 5 erase field sequences (Figure 4.15a). The MOKE traces have a five-cycle period and each erase sequence removes a pair of domain walls, validating the operating principle described above.

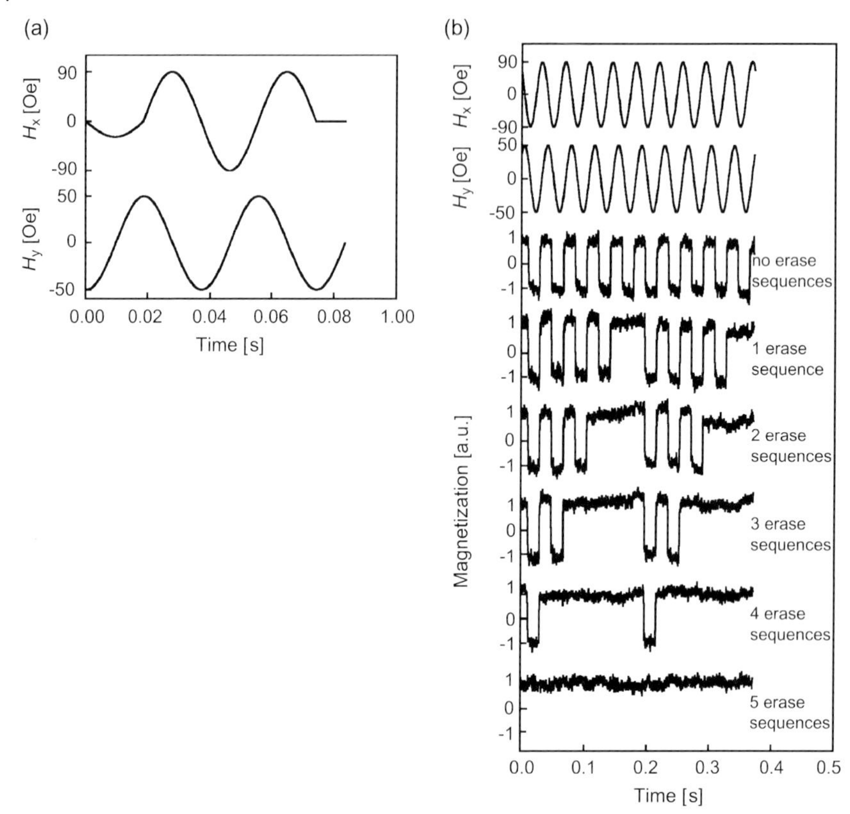

Figure 4.15 (a) *Erase* field sequence ($H_x^0 = 90$ Oe, $H_x^{erase} = 24$ Oe, $H_y^0 = 50$ Oe) applied to the structure in Figure 4.13. (b) *Read* field sequence $H_x^0 = 90$ Oe, $H_y^0 = 50$ Oe) and MOKE signals measured from the structure in Figure 4.13 following a saturating pulse and between zero to five 5 *erase* field sequences, as indicated.

4.6 Outlook and Conclusions

Domain wall logic is not a contender for a wholesale replacement of CMOS microelectronics. CMOS is a highly mature technology with many advantages, and still has many years of scaling available to it. The limited operational speed of domain wall logic does not render it suitable for many applications. However, a strong trend in microelectronics which is expected to apply to the relationship between CMOS and to many other areas of nanotechnology in the future, is to combine multiple technologies on a single platform: the System on Chip (SoC).

So – what does domain wall logic do well? First, it provides a high level of functionality to relatively simple structures. To implement an AND gate in CMOS would take six transistors, but domain wall logic achieves this simply by bringing two nanowires together. Similarly, the other high-level properties that have been highlighted in this chapter – such as input–output isolation and signal/power gain – are all intrinsic to the nanowire and do not have to be explicitly created.

The power dissipation per logic gate is extremely low. Microelectronic engineers usually measure dissipation from a gate by the power–delay product; that is to say, the product of how much power is dissipated multiplied by how long the gate takes to process a single function. The units of this quantity are energy, corresponding to the energy dissipated during the evaluation of the function performed by the gate. The power–delay product of CMOS depends on the size of the devices. Hence, in order to compare like with like, a 200 nm minimum feature size CMOS value of 10^{-2} pJ is considered [34]. On very general magnetic grounds, it can be said that an upper bound for the power–delay product for domain wall logic is $2M_S VH$, where M_S is the saturation magnetization of the magnetic material, V is the volume of magnetic material in a gate, and H is the amplitude of the applied field. Applying the parameters for a typical 200 nm domain wall logic gate gives 10^{-5} pJ – that is, 1000 times lower than the equivalent CMOS device. Because of the inefficiencies inherent in the generation of high-speed magnetic fields (see above), this does not necessarily mean that domain wall logic chips will not consume much power. What it does mean, however, is that the waste heat will be generated from the global field generator and not from the logic devices themselves. This is of particular relevance if the devices are to be stacked into three-dimensional (3-D), neural-like circuits. The two key technical difficulties in doing this in CMOS are: (i) distributing the power and clock to everywhere inside the volume of network; and (ii) extracting the waste heat from the center of the network so that the device does not melt. It is believed that domain wall logic is an excellent choice of primitive for 3-D architectures.

Non-volatility comes as standard. In a world of mobile computing and portable (or even wearable) devices, the concept of "instant-on" is becoming increasingly important. Users accept that devices cannot be expected to operate when there is no power, but as soon as the power becomes available they want the device to be ready, and not have to undergo a long boot process, or to have forgotten what it was doing when the power last failed. As there are currently very few non-volatile memory technologies available which can be embedded directly into CMOS, a data transfer process is usually required between a high-speed, volatile memory register in the heart of the CMOS logic and an off-chip, low-speed, non-volatile store where the state variables of the system are stored. With domain wall logic, all of this becomes redundant. Provided that the rotating field is properly controlled so that it stops gracefully as power fails, and does not apply intermediate levels of field leading to data corruption, the domain wall logic circuit should simply stop and retain all of its state variables. Then, as soon as the power returns, the logic continues from where it left off.

Domain wall logic can make use of redundant space on top of CMOS. Because no complex heterostructures are required, the logic elements can sit in a single layer fabricated as a Back End Of Line process after the CMOS has been laid down. This can improve the efficiency of the underlying CMOS by farming out some space-consuming task to the domain wall logic on top. As this space was never accessible to CMOS itself anyway, it all counts as a gain.

Being metals, the basic computational elements of domain wall logic are automatically radiation-hard, and so are suitable for use in either space or military applications.

Domain wall logic is very good at forming high-density shift registers that could be used as non-volatile serial memory for storing entire files, and so would not require high-speed random access. The hard disk drive and NAND Flash devices – for example, as used to store photographs in a digital camera – are examples of non-volatile serial memory. At present, both of these devices are 2-D in form, but registers made from domain wall logic elements have the potential to be stacked into three dimensions, without incurring extra wiring complexity, as the data and power can be transmitted remotely through magnetic fields (see above). In a hard disk drive the data are stored as rows of magnetic domains, and this would remain the same in a domain wall logic serial memory. What would differ is that, in a hard disk, the domains are mechanically rotated on their disk underneath a static sensor, whereas in domain wall logic the domains themselves would move under the action of an externally applied magnetic field along static domain wall conduits, potentially stacked into an ultrahigh-density, 3-D array.

Acknowledgments

The research studies described in this chapter were funded by the European Community under the Sixth Framework Programme Contract Number 510993: MAGLOG. The views expressed are solely those of the authors, and the other Contractors and/or the European Community cannot be held liable for any use that may be made of the information contained herein. D.A.A. acknowledges the support of an EPSRC Advanced Research Fellowship (GR/T02942/01).

References

1 T. Dietl, H. Ohno, F. Matsukura, J. Cibert, D. Ferrand, *Science* 2000, **287**, 1019.
2 H. Ohno, D. Chiba, F. Matsukura, T. Omiya, E. Abe, T. Dietl, Y. Ohno, K. Ohtani, *Nature* 2000, **408**, 944.
3 Y. Ohno, D. K. Young, B. Beschoten, F. Matsukura, H. Ohno, D. D. Awschalom, *Nature* 1999, **402**, 790.
4 G. A. Prinz, *Science* 1998, **282**, 1660.
5 S. A. Wolf, D. D. Awschalom, R. A. Buhrman, J. M. Daughton, S. von Molnar, M. L. Roukes, A. Y. Chtchelkanova, D. M. Treger, *Science* 2001, **294**, 1488.
6 S. A. Wolf, D. Treger, A. Chtchelkanova, *MRS Bulletin* 2006, **31**, 400.
7 R. W. Dave, G. Steiner, J. M. Slaughter, J. J. Sun, B. Craigo, S. Pietambaram, K. Smith, G. Grynkenich, M. DeHerrera, J. Åkerman, S. Tehrani, *IEEE Trans. Magn.* 2006, **42**, 1935.
8 R. Richter, L. Bar, J. Wecker, G. Reiss, *Appl. Phys. Lett.* 2002, **80**, 1291.
9 A. Ney, C. Pampuch, R. Koch, K. H. Ploog, *Nature* 2003, **425**, 485.
10 C. Pampuch, A. Ney, R. Koch, *Europhys. Lett.* 2004, **66**, 895.
11 G. Reiss, H. Brückl, A. Hütton, H. Koop, D. Meyners, A. Thomas, S. Kämmerer, J. Schmalhorst, M. Brzeska, *Phys. Stat. Sol. A* 2004, **201**, 1628.

12 D. Meyners, K. Rott, H. Brückl, G. Reiss, J. Wecker, *J. Appl. Phys.* 2006, **99**, 023907.

13 W. C. Black, B. Das, *J. Appl. Phys.* 2000, **87**, 6674.

14 R. P. Cowburn, M. E. Welland, *Science* 2000, **287**, 1466.

15 R. P. Cowburn, *Phys. Rev. B* 2002, **65**, 092409.

16 M. C. B. Parish, M. Forshaw, *Appl. Phys. Lett.* 2003, **83**, 2046.

17 A. Imre, G. Csaba, L. Ji, A. Orlov, G. H. Bernstein, W. Porod, *Science* 2006, **311**, 205.

18 G. Xiong, D. A. Allwood, M. D. Cooke, R. P. Cowburn, *Appl. Phys. Lett.* 2001, **79**, 3461.

19 D. Ozkaya, R. M. Langford, W. L. Chan, A. K. Petford-Long, *J. Appl. Phys.* 2002, **91**, 9937.

20 A. Hubert, R. Schäfer, *Magnetic Domains. The Analysis of Magnetic Microstructures*, Springer-Verlag, Berlin, 1998.

21 D. A. Allwood, G. Xiong, M. D. Cooke, R. P. Cowburn, *J. Phys. D Appl. Phys.* 2003, **36**, 2175.

22 R. P. Cowburn, D. A. Allwood, G. Xiong, M. D. Cooke, *J. Appl. Phys.* 2002, **91**, 6949.

23 D. Atkinson, D. A. Allwood, G. Xiong, M. D. Cooke, C. C. Faulkner, R. P. Cowburn, *Nature Mater.* 2003, **2**, 85.

24 Y. Nakatani, A. Thiaville, J. Miltat, *Nature Mater.* 2003, **2**, 521.

25 Y. Nakatani, A. Thiaville, J. Miltat, *J. Magn. Magn. Mater.* 2005, **290–291**, 750.

26 D. G. Porter, M. J. Donahue, *J. Appl. Phys.* 2004, **95**, 6729.

27 G. S. D. Beach, C. Nistor, C. Knutson, M. Tsoi, J. L. Erskine, *Nature Mater.* 2005, **4**, 741.

28 D. A. Allwood, G. Xiong, M. D. Cooke, C. C. Faulkner, D. Atkinson, N. Vernier, R. P. Cowburn, *Science* 2002, **296**, 2003.

29 D. A. Allwood, G. Xiong, M. D. Cooke, C. C. Faulkner, D. Atkinson, R. P. Cowburn, *J. Appl. Phys.* 2004, **95**, 8264.

30 X. Zhu, D. A. Allwood, G. Xiong, R. P. Cowburn, P. Grütter, *Appl. Phys. Lett.* 2005, **87**, 062503.

31 C. C. Faulkner, D. A. Allwood, M. D. Cooke, G. Xiong, D. Atkinson, R. P. Cowburn, *IEEE Trans. Magn.* 2003, **39**, 2860.

32 D. A. Allwood, G. Xiong, R. P. Cowburn, *J. Appl. Phys.* 2007, **101**, 024308.

33 D. A. Allwood, G. Xiong, C. C. Faulkner, D. Atkinson, D. Petit, R. P. Cowburn, *Science* 2005, **309**, 1688.

34 R. Waser, *Nanoelectronics and Information Technology*, Wiley VCH, Weinheim, 2003.

5
Monolithic and Hybrid Spintronics

Supriyo Bandyopadhyay

5.1
Introduction

An electron has three attributes: mass; charge; and spin. An electron's mass is too small to be useful for practical applications, but the charge is an enormously useful quantity that is utilized universally in every electronic device extant. The third attribute – spin – has played mostly a passive role in such gadgets as magnetic disks and magneto-electronic devices, where its role has been to affect the magnetic or the electrical properties in useful ways – for example, in the giant magnetoresistance devices used to read data stored in the magnetic disks of laptops and Apple iPods. Only recently has a conscious effort been made to utilize spin – either singly or in conjunction with the charge degree of freedom – to store, process, and transmit information. This field is referred to as modern "spintronics".

There are two distinct branches of spintronics:

- *Hybrid spintronics:* these devices are very much conventional electronic devices, as information is still encoded in the charge (ultimately detected as voltage or current), but spin augments the functionality of the device and *may* improve device performance. Examples of hybrid spintronic devices are *spin field effect transistors* (SPINFETs) [1] and *spin bipolar junction transistors* (SBJTs) [2], where information is still processed by modulating the charge current flowing between two terminals via the application of either a voltage or a current to the third terminal. The process by which the third terminal controls the voltage or current is spin-mediated – hence the term "spin transistors".
- *Monolithic spintronics:* here, charge has no direct role whatsoever. Rather, the information is encoded entirely in the spin polarization of an electron, which may be made to have only two *stable* values: "upspin" and "downspin", by placing the electron in a static magnetic field. "Upspin" will correspond to polarizations anti-parallel to the magnetic field, while "downspin" will be parallel. These two

Nanotechnology. Volume 4: Information Technology II. Edited by Rainer Waser
Copyright © 2008 WILEY-VCH Verlag GmbH & Co. KGaA, Weinheim
ISBN: 978-3-527-31737-0

polarizations can encode binary bits 0 and 1 for digital applications. Toggling a bit merely requires flipping the spin, *without any physical movement of charge*. It has recently been argued that as no charge motion (or current flow) is required, there can be tremendous energy savings during switching [3]. As a result, monolithic spintronic devices are far more likely to yield low-power signal processing units than are hybrid spintronic devices. An example of monolithic spintronic devices is the Single Spin Logic (SSL) paradigm that is described in Section 5.3.

In this chapter, the two most popular hybrid spintronic devices – the SPINFET and the SBJT – will be described, and evidence provided that neither device is likely to produce significant advantages in terms of speed or power dissipation over conventional charge-based transistors. The concept of single spin logic (SSL) will then be discussed, and its significant advantages in power dissipation over SPINFET or SBJT outlined. SSL also has significant advantages over any charge-based paradigm where charge, rather than spin, is used as the state variable to encode information. Finally, it will be shown that the *maximum* energy dissipation in switching a bit in SSL is the Landauer–Shannon limit of $kT\ln(p)$ per bit operation, where $1/p$ is the bit error probability. Some gate operations dissipate less energy than this because of *interactions* between spins, which may reduce dissipation [4], because many spins function collectively, as a single unit, to effect gate operation. This collective, cooperative dynamics is conducive to energy efficiency. Any discussion of adiabatic switching [5], which can reduce energy dissipation even further, is avoided as it is very slow, error-prone, and therefore impractical. The discussion of devices in non-equilibrium statistical distribution, where energy dissipation can be reduced below the Landauer–Shannon limit [6] is also avoided, simply because energy is required to maintain the non-equilibrium distributions over time, and that energy must be dissipated in the long term. The final section includes a very brief engineer's perspective on spin-based quantum computing (included at the request of the editor).

5.2
Hybrid Spintronics

Hybrid spintronic devices are those where spin is used to "enhance" the performance of charge but does not itself play a direct role in storing, processing or communicating information. The two most popular hybrid spintronic devices are the SPINFET and the SBJT.

5.2.1
The Spin Field Effect Transistor (SPINFET)

A schematic representation of the SPINFET, as proposed in the seminal studies of Ref. [1], is shown in Figure 5.1a. This device exactly resembles a conventional metal-

(a)

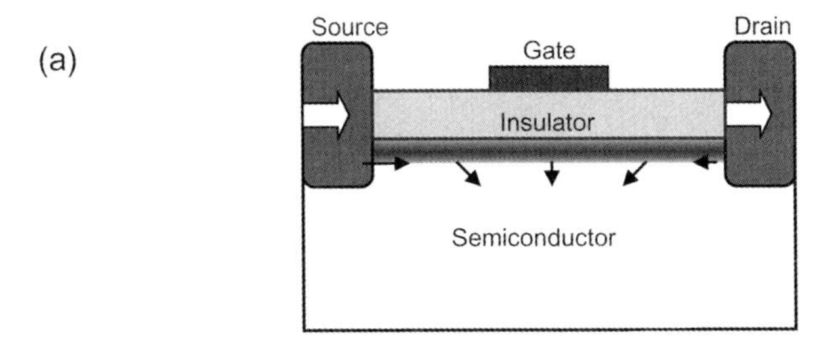

(b)

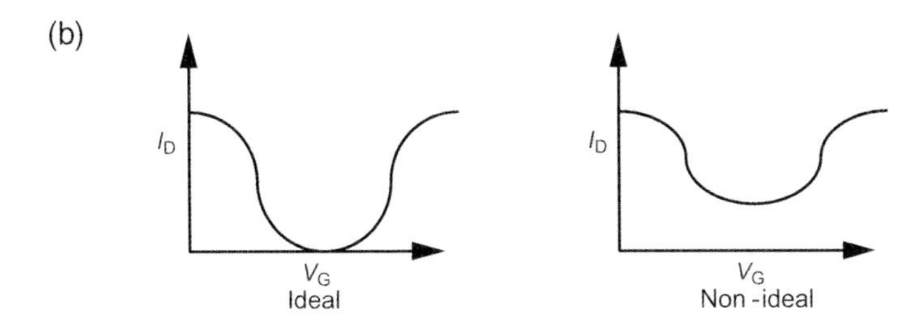

(c)

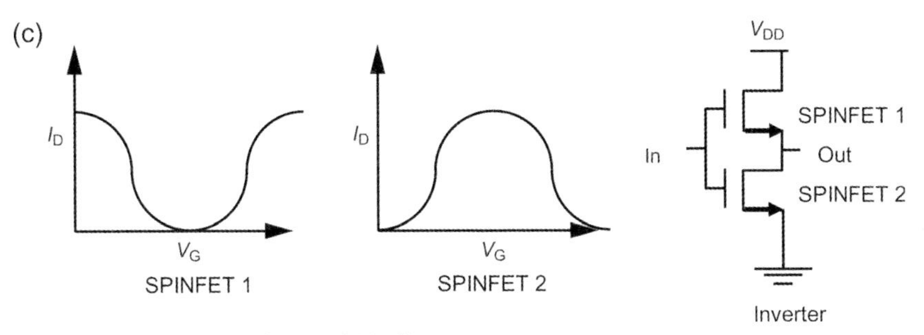

Figure 5.1 (a) Schematic of a spin field effect transistor (SPINFET). (b) Ideal and non-ideal transfer characteristic of a SPINFET. (c) Transfer characteristics of two SPINFETs with different threshold shifts and realization of a CMOS-analog inverter by connecting these two SPINFETs in series.

oxide-semiconductor-field-effect-transistor (MOSFET), except that the source and drain contacts are ferromagnetic. It will be assumed that the channel is strictly one-dimensional (1-D) (quantum wire), and only the lowest transverse subband is occupied by electrons. Both, source and drain contacts are magnetized so that their magnetic moments are parallel and point along the direction of current flow ($+x$-direction). As a result, when the source-to-drain voltage is turned on, the ferromagnetic source injects carriers into the channel with $+x$ polarized spins (the majority spins in the ferromagnet). It will also be assumed that the spin injection efficiency is 100% so that *only* majority spins ($+x$-polarized spins) are injected from the source contact, and absolutely no minority spin (i.e. $-x$-polarized spin) is injected. Immediately after injection into the channel, all spins are polarized along the $+x$ direction. When the gate voltage is switched on, it induces an electric field in the y-direction that causes a Rashba spin–orbit interaction [7] in the channel. This spin–orbit interaction acts like an effective magnetic field in the z-direction (which is the direction mutually perpendicular to the electron's velocity in the channel and the gate electric field). This pseudo-magnetic field B_{Rashba} causes the spins to precess in the x-y plane, as they travel towards the drain. The angular frequency of spin precession (which is essentially the Larmor frequency) is given by $\Omega = eB_{\text{Rashba}}/m^*$, where e is the electronic charge and m^* is the effective mass of the carrier. The pseudo-magnetic field B_{Rashba} depends on the magnitude of the gate voltage and the carrier velocity along the channel according to

$$B_{\text{Rashba}} = \frac{2(m^*)^2 a_{46}}{e\hbar^2} E_y v_x \tag{5.1}$$

where a_{46} is a material constant, E_y is the gate electric field, and v_x is the electron velocity.[1)]

The spatial rate of spin precession is

$$\frac{d\phi}{dx} = \frac{d\phi}{dt}\frac{1}{\frac{dx}{dt}} = \Omega/v_x = 2\frac{m^* a_{46}}{\hbar^2} E_y \tag{5.2}$$

which is *independent* of the carrier velocity and depends only on the gate voltage (or gate electric field). The total angle by which the spins precess in the x-y plane as they travel through the channel from source to drain is

$$\Phi = \frac{2m^* a_{46}}{\hbar^2} E_y L \tag{5.3}$$

where L is the channel length. This angle is independent of the carrier velocity and therefore is the *same* for every electron, regardless of its initial velocity or scattering history in the channel. If the gate voltage (and E_y) is of such magnitude that Φ is an odd multiple of π, then *every* electron has its spin polarization anti-parallel to the drain's magnetization when it arrives at the drain. These electrons are blocked by the drain, and therefore the source to drain current falls to zero. Here, it has been assumed that the drain is a perfect spin filter that allows only majority spins to transmit, while completely blocking the minority spins. Without a gate voltage, the

1) Some authors assume incorrectly that the Rashba field B_{Rashba} is proportional to wavevector k_x and not the velocity v_x. This makes a difference since, in the presence of the Rashba interaction, $v_x = \hbar k_x/m* \pm \eta/\hbar$, where η is the strength of the Rashba interaction. Following the derivation in this chapter, the reader can easily convince herself/himself that the SPINFET would not work as claimed if B_{Rashba} were proportional to wavevector k_x and not the velocity v_x. As the magnetic field associated with spin–orbit interaction is proportional to $\frac{\vec{v} \times \vec{E}}{2c^2}$ (where c is the speed of light in vacuum and $\vec{E}$ is the electric field seen by the electron), it is obvious that B_{Rashba} should be proportional to v_x and not to k_x.

spins do not precess[2] and the source to drain current is non-zero. Thus, the gate voltage causes current modulation via spin precession and this realizes transistor action.[3] This device is also briefly discussed in Chapter 3 in Volume III of this series.

It should be clear that (in this device) although "spin" plays the central role in current modulation, it plays no direct role in information handling. Information is still encoded in "charge" which carries the current from the source to the drain. The transistor is switched between the "on" and "off" states by changing the current with the gate potential, or by controlling the motion of charges in space. The role of spin is only to provide an alternate means of changing the current with the gate voltage. Thus, this device is a quintessential hybrid spintronic device.

5.2.1.1 The Effect of Non-Idealities

The operation of the SPINFET described above is an idealized description. In a real device, there will be many non-idealities. First, there will be a magnetic field along the channel because of the magnetized ferromagnetic contacts. This will cause problems, as it will add to B_{Rashba} and the total effective magnetic field will be $|\vec{B}_{\text{Rashba}} + \vec{B}_{\text{channel}}|$, which is no longer linearly proportional to carrier velocity v_x. As a result, the precession rate in space will no longer be given by Eq. (5.2)[4] and will *not* be independent of the carrier velocity (or energy). Therefore, at a finite temperature, different electrons having different velocities due to the thermal spread in carrier energy, or because of different scattering history, will suffer different amounts of precession Φ. As a result, when the current drops to a minimum, not all spins at the drain end will have their polarizations exactly anti-parallel to the drain's magnetization. Those that do not, will be transmitted by the drain and contribute to a *leakage current* in the off state [8]. This is extremely undesirable as it decreases the ratio of on- to off-current and will lead to standby power dissipation when the device is off.

A more serious problem is that the magnetic field changes the energy dispersion relations in the channel. In Figure 5.2, the energy dispersion relation (energy versus wavevector) is shown schematically with and without the channel magnetic field [9]. Without the magnetic field, the Rashba interaction lifts the spin degeneracy at any non-zero wavevector, but each spin-split band still has a fixed spin quantization axis (meaning that the spin polarization in each band is always the same and independent of wavevector) (Figure 5.2a). The spin polarizations in the two bands are anti-parallel and the eigenspinors in the two bands are orthogonal. Because of this orthogonality, there can be no scattering between the two bands. Electrons can scatter elastically

2) Even without the gate voltage, there is obviously some Rashba interaction in the channel due to the electric field associated with the hetero-interface. This field exists because the structure lacks inversion symmetry along the direction perpendicular to the hetero-interface. This vestigial interaction will cause some spin precession even at zero gate voltage, but this effect is simply equivalent to causing a fixed threshold shift.

3) The device described here is a "normally on" device. If the magnetizations of the source and drain are anti-parallel instead of parallel, or if the spin polarizations in the source and drain contacts have *opposite* signs (e.g. iron and cobalt), then the device will be a "normally off" device.

4) In Eq. (5.2), B_{Rashba} must be replaced by $|\vec{B}_{\text{Rashba}} + \vec{B}_{\text{channel}}|$ in Ω.

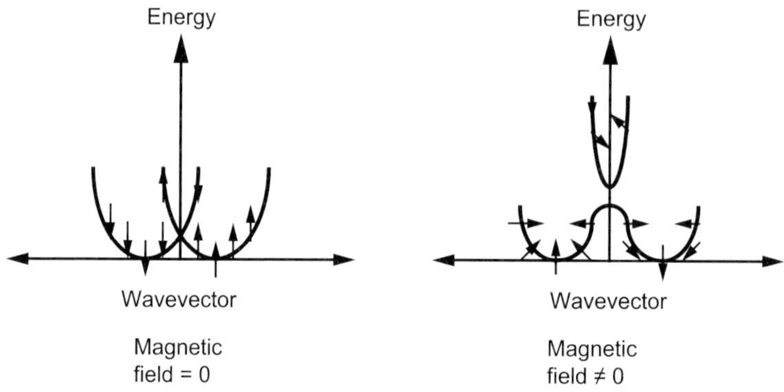

Figure 5.2 Schematic energy dispersion relationships for electrons in the channel of a one-dimensional SPINFET channel, with and without an axial magnetic field. The arrowheads indicate the spin polarization of a carrier in the corresponding wavevector state.

or inelastically only within the same band, but this does not alter the spin polarization since every state in the same band has exactly the same spin polarization. However, if a magnetic field is present in the channel, then the spin polarizations in both bands become *wavevector-dependent* and neither subband has a fixed spin polarization. Two states in two subbands with different wavevectors[5)] will have different spin polarizations that are not completely anti-parallel (orthogonal). Therefore, the matrix element for scattering between them is non-zero, which means that there is finite probability that an electron can scatter between them. Therefore, any momentum randomizing scattering event (due to interactions with impurities or phonons) will rotate the spin as the initial and final states have different spin polarizations. This rotation is random in time or space as the scattering event is random; therefore, it will cause spin relaxation. This is a new type of spin relaxation, and it is introduced solely by the channel magnetic field [10]. It is similar to the Elliott–Yafet spin relaxation mechanism [11] in the sense that it is associated with momentum relaxation. Any such spin relaxation in the channel will randomize the spin polarizations of electrons arriving at the drain and thus give rise to a significant leakage current. Therefore, the channel magnetic field causes leakage current in two different ways, both of which are harmful.

In Ref. [1], where the ideal SPINFET was analyzed, it was assumed that there is no spin relaxation in the channel. The transfer characteristic shown Figure 5.1b, which shows zero leakage drain current in the OFF-state, is predicated on this assumption. A question might arise as to whether the usual spin relaxation mechanisms are operative in the channel even without a channel magnetic field. For the ideal SPINFET, the answer is in the negative. The two spin relaxation mechanisms of concern are the Elliott–Yafet mode [11] and the D'yakonov–Perel' mode [12]. The

5) Eigenspinors in the two bands having the same wavevector are still orthogonal.

former is absent if the eigenspinors are wavevector-independent (as is the case with the ideal SPINFET), and the D'yakonov–Perel' mode is absent if carriers occupy only a single subband [13]. Thus, in the ideal 1-D SPINFET, there can be no significant spin relaxation (even if there is scattering due to interactions with non-magnetic impurities and phonons). If any spin relaxation occurs, it will be due to hyperfine interactions with nuclear spins. Since such interactions are very weak, they can be ignored for the most part. However, if there is an axial magnetic field in the channel, then all this changes and scattering with non-magnetic impurities or phonons will cause spin relaxation (and therefore a large leakage current). Consequently, it is extremely important to eliminate the channel magnetic field.

There are two ways to eliminate (or reduce) the channel magnetic field. One way is to magnetize the contacts in the y-direction instead of the x-direction. Since B_{Rashba} is in the z-direction, it makes no difference as to whether the spins are initially polarized in the x- or y-direction, as they precess in the x-y plane. The SPINFET works just as well if the source and drain contacts are magnetized in the $+y$ direction instead of the $+x$-direction. The advantage is that the magnetic field lines emanating from one contact, and sinking in the other, are no longer directed along the channel. Consequently, the channel magnetic field will be a fringing field, which is much weaker.

A more sophisticated approach is to play off the Dresselhaus spin–orbit interaction [14] against the channel magnetic field. This spin–orbit interaction is present in any zinc-blende semiconductor that lacks crystallographic inversion symmetry. In a 1-D channel oriented along the [100] crystallographic direction, the Dresselhaus spin–orbit interaction gives rise to a pseudo-magnetic field along the channel (x-axis), just as the Rashba spin–orbit interaction gives rise to a pseudo-magnetic field perpendicular to the channel (in the z-direction). The Dresselhaus field $B_{\text{Dresselhaus}}$ can be used to offset the channel magnetic field due to the contacts. Since $B_{\text{Dresselhaus}}$ depends on the carrier velocity v_x, the ensemble average velocity $\langle v_x \rangle$ (which is the Fermi velocity for a degenerate carrier concentration) can be tuned with a backgate to make $B_{\text{Dresselhaus}}$ equal and opposite to the channel magnetic field, thereby offsetting the effect of the channel field. This was the approach proposed in Ref. [8].

In Figure 5.1b, the transfer characteristic (drain current versus gate voltage) is shown for an ideal SPINFET and a non-ideal SPINFET (with a channel magnetic field), ignoring any fixed threshold shift caused by a non-zero Rashba interaction in the channel at zero gate voltage. It should be noted that the transfer characteristic is "oscillatory" and therefore non-monotonic. As a result, the transconductance ($\partial I_D/\partial V_G$), where I_D is the drain current and V_G is the gate voltage, can be either positive or negative depending on the value of V_G (gate bias). A fixed threshold shift can be caused in any SPINFET by implanting charges in the gate insulator. Imagine now that there are two ideal SPINFETs with transfer characteristics, as shown in Figure 5.1c. These two can be connected in series to behave as a complementary metal oxide semiconductor (CMOS) like inverter where an appreciable current flows only during switching. This would be a tremendous advantage as CMOS like operation can be achieved with just n-type SPINFETs where the majority carriers are electrons. In conventional CMOS technology, both an n-type and a p-type device would have normally have been needed. Here, we need only n-type devices. However, all this advantage is defeated if there is significant leakage current flowing through the SPINFET when it is "off". Thus, the leakage current is a rather serious issue and care must be taken to eliminate it as much as possible.

5.2.1.2 The SPINFET Based on the Dresselhaus Spin–Orbit Interaction

The Dresselhaus spin–orbit interaction can also be gainfully employed to realize a different kind of SPINFET [15]. In a 1-D channel, the strength of the Dresselhaus interaction depends on the physical width of the channel. If the 1-D channel is defined by a split-gate, then the voltages on the split gate can be varied to change the channel width and therefore the strength of the Dresselhaus interaction. As with the Rashba interaction, the Dresselhaus interaction also causes spins to precess in space at a rate independent of the carrier velocity because it gives rise to a pseudo-magnetic field $B_{\mathrm{Dresselhaus}}$ along the x-axis. The spins precess in the y-z plane. By varying the split gate voltage it is possible to change $B_{\mathrm{Dresselhaus}}$ and the precession rate, and therefore the angle Φ, by which the spins precess as they traverse the channel from source to drain. As the split gate voltage can be used to modulate the source to drain current, transistor action can be realized. A schematic representation of a Dresselhaus-type SPINFET is shown in Figure 5.3.

The advantage of the Dresselhaus-type SPINFET over the Rashba-type is that, in the former, there is never a strong magnetic field in the channel due to contacts [15], as $B_{\mathrm{Dresselhaus}}$ is in the x-direction and therefore the ferromagnetic contacts must be magnetized in the y-z plane (see Figure 5.3). By contrast, B_{Rashba} is in the z-direction, and therefore in the Rashba-type device the ferromagnetic contacts must be magnetized in the x-y plane. As mentioned above, the Rashba-type device could avoid a strong channel magnetic field if the contacts were to be magnetized in the y-direction, but this is difficult to do as the ferromagnetic layer thickness in the y-direction is much smaller than that in the x- or z-directions. Thus, the Rashba-type device will typically have some magnetic field in the channel, while the Dresselhaus-type device will not, except for the fringing fields. This feature eliminates many of the problems

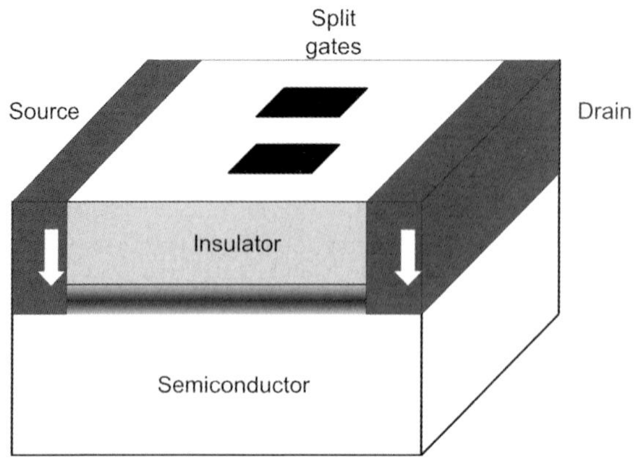

Figure 5.3 Schematic of a SPINFET based on the Dresselhaus spin–orbit interaction.

associated with the channel magnetic field in the Dresselhaus-type SPINFET, and could lead to a reduced leakage current. A comparison between the Rashba-type and Dresselhaus-type SPINFETs is provided in Ref. [15].

5.2.2 Device Performance of SPINFETs

In the world of electronics, the universally accepted benchmark for the transistor device is the celebrated metal-oxide-semiconductor-field-effect-transistor (MOSFET) which has been – and still is – the "workhorse" of all circuits. Therefore, the SPINFET must be compared with an equivalent MOSFET to determine if there are any advantages to utilizing spin. Surprisingly, in spite of many papers extolling the perceived merits of SPINFETs, this elementary exercise was not carried out until recently. When an ideal SPINFET was compared with an equivalent ideal MOSFET at low temperatures [16], the results were quite illuminating.

According to Ref. [1], the switching voltage necessary to turn a 1-D Rashba-type SPINFET from on to off, or vice versa, is given by

$$V_{switching}^{SPINFET} = \frac{\pi\hbar^2}{2m^* L\xi} \tag{5.4}$$

where m^* is the effective mass, L is the channel length (or source-to-drain separation), and ξ is the rate of change of the Rashba interaction strength in the channel per unit change of the gate voltage. This is given by [16]:

$$\xi = \frac{\hbar^2}{2m^*} \frac{\Delta(2E_g + \Delta)}{E_g(E_g + \Delta)(3E_g + 2\Delta)} \frac{2\pi e}{d} \tag{5.5}$$

where E_g is the bandgap of the channel semiconductor, Δ is the spin orbit splitting in the valence band of the semiconductor, e is the electronic charge, and d is the gate insulator thickness. If $d = 10$ nm,[6] then it can be calculated that in an InAs channel, $\xi = 10^{-28}$ C-m. This is the theoretical value, but an actual measured value is much less than this [17]. The compound InAs has strong spin–orbit interaction and therefore is an ideal material for SPINFETs.

Now, imagine that the same structure is used as a MOSFET. Then, the switching voltage that turns the MOSFET device off (depletes the channel of mobile carriers) is E_F/e, where E_F is the Fermi energy in the channel.[7] Thus, the ratio of the switching

6) The ideal semiconductor is a narrow gap semiconductor such as InAs, which has strong Rashba spin–orbit interaction. In that case, the gate insulator will probably be AlAs, which is reasonably lattice-matched to InAs. Because the conduction band offset between these two materials is not too large, a minimum of 10 nm gate insulator thickness may be necessary to prevent too much gate leakage.

7) An accumulation mode MOSFET has been assumed that is "normally-on". It has also been assumed that the normal channel carrier concentration is large enough that $E_F \gg kT$, where E_F is the Fermi energy and kT is the thermal energy. Therefore, the comparison is strictly valid at very low temperatures. It is believed that, at higher temperatures, the fundamental conclusions from this comparison will not be significantly altered.

voltages is

$$\frac{V_{switching}^{SPINFET}}{V_{switching}^{MOSFET}} = \frac{\pi\hbar^2 e}{2m^* L\xi E_F} \tag{5.6}$$

In order to maintain single subband occupancy in an InAs 1-D channel of reasonable width, E_F must be less than $\sim$3 meV. Therefore, from Eq. (3.6) it is found that the SPINFET will have a lower threshold voltage than a comparable MOSFET (at low temperature) only if its channel length exceeds 2.4 μm! Thus, no submicron SPINFET has any advantage over a comparable MOSFET in terms of switching voltage or dynamic power dissipation during switching. Currently, MOSFETs with 90 nm channel length are in production [18]. For a SPINFET with this channel length to have a lower threshold voltage than a comparable MOSFET, the channel must be made from a material in which the product $m^*\xi$ is 26-fold larger than it is in InAs, assuming that E_F is still 3 meV. Such materials are not currently available, but of course could become available in the future. The unfortunate spoiler is that materials which have large effective mass also tend to have weak spin–orbit interaction, which makes it difficult to increase the product $m^*\xi$.

One issue that requires some thought here is that the switching voltage of a MOSFET depends on E_F – and therefore the carrier concentration in the channel – while the switching voltage of a SPINFET depends on the channel length. It is not clear which quantity is easier to control in batch processing, but that would determine which device has an advantage in terms of threshold variability and, ultimately, of yield.

5.2.3 Other Types of SPINFET

Slightly different types of SPINFET ideas have also been reported in the literature, with names such as "Non-ballistic SPINFET" [19] or the "Spin Relaxation Transistor" [20–22].

5.2.3.1 The Non-Ballistic SPINFET

The channel of the so-called "non-ballistic SPINFET" has a two-dimensional (2-D) electron gas, like an ordinary MOSFET. Unlike in a 1-D structure (quantum wire), the spin split bands in a 2-D structure (quantum well or 2-D electron gas) do not have a fixed spin quantization axis (meaning that the spin eigenstates are wavevector-dependent; recall Figure 5.2b), even if there is no magnetic field. The only exception to this situation is when the Rashba and Dresselhaus interactions in the channel have exactly the same strength. In that case, each band has a fixed spin quantization axis, and the spin eigenstate in either band is wavevector-independent.

In the non-ballistic SPINFET, the Rashba interaction is first tuned with the gate voltage to make it exactly equal to the Dresselhaus interaction, which is gate voltage-independent in a 2-D electron gas. This makes the spin eigenstates wavevector-independent. Electrons are then injected into the channel of the transistor from a ferromagnetic source with a polarization that corresponds to the spin eigenstate in one of the bands. All carriers enter this band. As the spin eigenstate is wavevector-independent, any momentum relaxing scattering in the channel, which will change the electron's wavevector, will not alter the spin polarization (recall the discussion in Section 5.2.1.1). Scattering can couple two states within the same band, but not in two different subbands, as the eigenspinors in two different subbands are orthogonal. Therefore, regardless of how much momentum-relaxing scattering takes place in the channel, there will be no spin relaxation via the Elliott–Yafet mode. There will also be no D'yakonov–Perel' relaxation as it can be shown that the pseudo-magnetic field due to the Rashba and Dresselhaus interactions will be aligned exactly along the direction of the eigenspin. As all spins are initially injected in an eigenstate, they will always be aligned along the pseudo-magnetic field. Consequently, there will be no spin precession which would have caused D'yakonov–Perel' relaxation. As no major spin relaxation mechanism is operative, the carriers at the source will arrive at the drain with their spin polarization intact. If the drain ferromagnetic contact is magnetized parallel to the source magnetization, then all these carriers will exit the device and contribute to current. In order to change the current, the gate voltage is detuned; this makes the Rashba and Dresslhaus interaction strengths unequal, thereby making the spin eigenstates in the channel wavevector-dependent. In that case, if the electrons suffer momentum-relaxing collisions due to impurities, defects, phonons, and surface roughness, their spin polarizations will also rotate and this will result in spin relaxation. Thus, the carriers that arrive at the drain no longer have all their spins aligned along the drain's magnetization. Consequently, the overall transmission probability of the electrons decreases, and the current drops. This is how the gate voltage changes the source-to-drain current and produces transistor action.

One simple way of viewing the transistor action is that when the Rashba and Dresselhaus interactions are balanced, the channel current is 100% spin polarized (all carriers arriving at the drain have exactly the same spin polarization). However, when the two interactions are unbalanced, then the spin polarization of the current decreases owing to spin relaxation. The spin polarization can decrease to zero – *but no less than zero* – which means that, on average, 50% of the spins will be aligned and 50% anti-aligned with the drain's magnetization when the minimum spin polarization is reached. The "aligned" component in the current will transmit and the "anti-aligned" component will be blocked. Thus, the minimum current (off-current) of this transistor *is only one-half* of the maximum current (on-current). As the maximum ratio of the on-to-off conductance is only 2, this device is clearly unsuitable for most – if not all – mainstream applications. A recent simulation has shown that the on-to-off conductance ratio is only about 1.2 [23], which precludes use in any fault-tolerant circuit.

The situation can be improved dramatically if the source and drain contacts have *anti-parallel* magnetizations, instead of parallel. In that case, when the Rashba and Dresselhaus interactions are balanced, the transmitted current will be exactly zero, but when they are unbalanced, it is non-zero. Therefore, the on-to-off conductance becomes infinity. However, there is a caveat. This argument pre-supposes that the ferromagnetic contacts can inject and detect spins with 100% efficiency, meaning that only the majority spins are injected and transmitted by the ferromagnetic source and drain contacts, respectively. That has never been achieved, and even after more than a decade of research the maximum spin injection efficiency demonstrated to date at room temperature is only about 70% [24]. That means

$$\frac{I_{\mathrm{maj}} - I_{\mathrm{min}}}{I_{\mathrm{maj}} + I_{\mathrm{min}}} = 0.7 \tag{5.7}$$

where $I_{\mathrm{maj(min)}}$ is the majority (minority) spin component of the current. If the spin injection efficiency is only 70%, then 15% of the injected current is due to minority spins. These minority spins will transmit through the drain, even when the Rashba and Dresselhaus interactions are balanced. Thus, the off-current is 15% of the total injected current ($I_{\mathrm{maj}} + I_{\mathrm{min}}$), whereas the on-current is still at best 50% of the total injected current. Therefore, the on-to-off ratio of the conductance is $0.5/0.15 = 3.3$, which is not much better than 2.[8)] Consequently, achieving a large conductance ratio is very difficult, particularly at room temperature when the spin injection efficiency tends to be small. In order to make the conductance ratio 10^5 – which is what today's transistors have – the spin injection efficiency must be 99.999% at room temperature. This is indeed a tall order, and may not be possible in the near term. If the off (leakage) current is nearly one-third of the on-current (which is what it will be with present-day technology), then the standby power dissipation will be intolerable and the noise margin unacceptable.

The device described in Ref. [20] is identical to that in Ref. [19], and therefore the same considerations apply.

5.2.3.2 **The Spin Relaxation Transistor**

The proposed "spin relaxation transistor" [21, 22] is very similar to the non-ballistic SPINFET. With zero gate voltage, the spin–orbit interaction in the channel is either weak or non-existent, which makes the spin relaxation time very long. When the gate voltage is turned on, the spin–orbit interaction strength increases, which makes the spin relaxation time short. Thus, with zero gate voltage, the spin polarization in the drain current is large (maximum 100%), while with a non-zero gate voltage it is small (minimum 0%). This device cannot be any better than the non-ballistic SPINFET. If the drain and source magnetizations are parallel, then it is easy to see that in the "on" state, the transmitted current is at best 100% of the injected current, while in the "off" state, it is no less than 50% of the injected current (the same arguments as in Section 5.2.3.1 apply). Therefore, the on-to-off conductance ratio is less than 2. With anti-parallel magnetizations in the source and drain contacts, the off-current approaches 0, and the conductance ratio approaches infinity, but only if the contacts inject and detect spin with 100% efficiency. If the injection efficiency is only 70%

8) Here it has been assumed that the drain is a perfect spin filter, or equivalently, the spin detection efficiency at the drain is 100%.

then, following the previous argument, the conductance ratio is no more than 3.3 [25]. Therefore, this device, too, is unsuitable for mainstream applications.

5.2.4 The Importance of the Spin Injection Efficiency

In every proposal for SPINFETs discussed here [1, 15, 19–22] there has always been the tacit assumption that spin injection and detection efficiencies at the source/channel and drain/channel interfaces are 100%. This is of course unrealistic. It can be shown easily that the ratio of on- to off-conductance of SPINFETs of the type proposed in Refs. [1, 15] is

$$r = \frac{1+\xi_S\xi_D}{1-\xi_S\xi_D} \tag{5.8}$$

where ξ_S is the spin injection efficiency at the source/channel interface and ξ_D is the spin detection efficiency at the drain/channel interface. For SPINFETs of the types proposed in Refs. [19–22],

$$r = \frac{1}{1-\xi_S\xi_D} \tag{5.9}$$

Therefore, the spin injection/detection efficiency ξ is critical in order to obtain a large enough value of r. If the spin injection and detection efficiencies fall from 100% to 90%, the conductance on-off ratio drops from infinity to 5.2! Therefore, without a very high spin injection efficiency, approaching 100%, none of the SPINFETs will have a sufficiently high on-to-off conductance ratio to be of much use for anything. Thus, a closer look at spin injection efficiency is clearly required.

5.2.4.1 Spin Injection Efficiency

The science of "spin injection efficiency" (what controls it, what improves it, what does it depend on, etc.) is controversial (see Chapter 3 in Volume III), and there is scant agreement among research groups in this field. However, to the authors' knowledge nobody has claimed in the open literature that the spin injection efficiency can be 100% (even theoretically), particularly at elevated temperatures, except for Ref. [26]. These authors hold that, as there has been rapid experimental progress in improving spin injection efficiency over the past seven years (as of this writing), 100% efficient spin injection at room temperature should be around the corner. This optimism is not shared by the present author, since the two mechanisms suggested [26] as possible routes to achieving 100% spin injection efficiency are to use 100% spin-polarized half-metallic ferromagnets as spin injectors, and spin-selective barriers. Unfortunately, there are no 100% spin-polarized half-metals at any temperature above 0 K because of phonons and magnons [27], and even at 0 K the 100% spin polarization is destroyed by surfaces and inhomogeneities [27]. Thus, 100% spin-polarized half-metals simply do not exist. The best spin-selective barriers are resonant tunneling devices [28] that have spin-resolved energy levels. When the carrier energy is resonant with one of these levels, only the corresponding spin transits through the barrier. However, at any temperature exceeding 0 K, the thermal spread in the electron energy will cause injected electrons to tunnel through *both* levels if the level separation is less than kT (which it usually is). Therefore, both spins will be transmitted. This happens even if the spin levels themselves are not broadened by $\sim kT$ because of weak spin–phonon coupling. As long as both spins are transmitted, the spin "selection" suffers and that makes the spin injection efficiency considerably less than 100%.

To summarize, as yet there is no known method to suggest – even theoretically – the possibility of 100% spin injection efficiency. As a result, all SPINFETs discussed in this chapter suffer from the malady of a low on-to-off conductance ratio, and this alone may make them non-competitive with MOSFETs.

5.2.5 Spin Bipolar Junction Transistors (SBJTs)

The SBJT is identical with the normal bipolar junction transistor, except that the base is ferromagnetic and it has a non-zero spin polarization. The conduction energy band diagram of a heterojunction n-p-n SBJT is shown in Figure 5.4. Assuming that the carrier concentration in the base is non-degenerate, so that Boltzmann statistics apply, the spin polarization in the base is

$$\alpha_b = \tanh(\Xi/2kT) \tag{5.10}$$

where 2Ξ is the energy splitting between majority and minority spin bands in the base. Based on a small signal analysis, it was shown that the voltage and current gains afforded by the SBJT is about the same as a conventional BJT [29], but the short-circuit current gain β has a dependence on the degree of spin polarization in the base which can be altered with an external magnetic field using the Zeeman effect. Thus, the external magnetic field can act as a "fourth terminal", and this can lead to non-linear circuits such as mixers/modulators. For example, if the ac base current is a sinusoid with a frequency ω_1 and the magnetic field is an ac field which is another sinusoid with frequency ω_2, then the collector current will contain frequency components

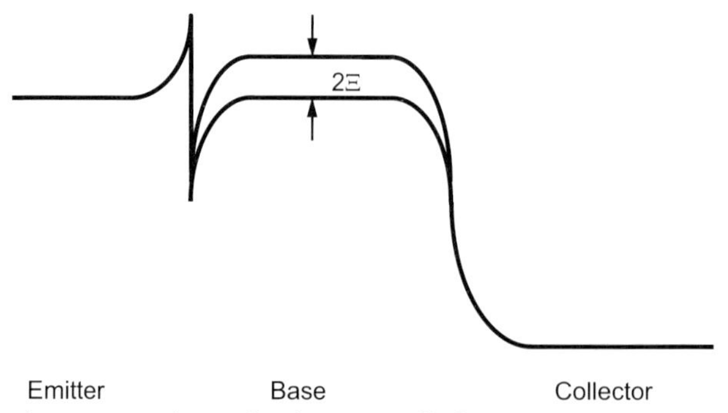

Figure 5.4 Conduction band energy profile for an n-p-n spin bipolar junction transistor (SBJT). The base is spin-polarized (ferromagnetic) and the spin splitting energy in the base is 2Ξ.

$\omega_1 \pm \omega_2$. This is one example where "spin" augments the role of "charge", making the SBJT another "hybrid spintronic" device.

5.2.6
The Switching Speed

The switching delay of any of the SPINFETs discussed above is limited by the transit time of carriers through the channel (or base). This is entirely due to the fact that information is encoded by charge (or current), and therefore the charge transit time is the bottleneck that ultimately limits the switching speed. Thus, hybrid spintronic devices do not promise any better speed than their charge-based counterparts.

5.3
Monolithic Spintronics: Single Spin Logic

At this point the discussion centers on "monolithic spintronics" where charge plays no role whatsoever and spin polarization is used to store, process and transmit information. In 1994, the idea was proposed of "single spin logic" (SSL) where a single electron acts as a binary switch and its two orthogonal (anti-parallel) spin polarizations encode binary bits 0 and 1 [30]. Switching between bits is accomplished by simply flipping the spin *without physically moving charges*. To the author's knowledge, this is the first known logic family (classical or quantum) based on single electron spins.

5.3.1
Spin Polarization as a Bistable Entity

The first step in SSL is to make the spin polarization of an electron a *bistable* quantity that has only two stable "values" that will encode the bits 0 and 1. In charge-based electronics, the state variables representing digital bits (voltage, current or charge), are not bistable but are continuous variables. So, why is the spin polarization required to be bistable? The reason is that in the world of electronics, there are analog-to-digital converters that can convert a continuous variable (analog signal) to a discrete variable (digital signal). More importantly, logic gates act as amplifiers with power gain and can automatically restore digital signal at logic nodes [31] if the signal is corrupted by noise. There are no equivalent analog-to-digital converters for spin polarization and no spin amplifiers, and therefore *Nature* must be relied upon to digitize spin polarization and make signal degeneration impossible. This can happen if *Nature* permits only two values of spin polarization – that is, it inherently makes it "bistable". That can be accomplished by placing an electron in a static magnetic field. The Hamiltonian describing a single electron in a magnetic field is

$$H = (\vec{p} - q\vec{A})^2/2m^* - (g/2)\mu_B \vec{B} \cdot \vec{\sigma} \tag{5.11}$$

where $\vec{A}$ is the vector potential due to the magnetic flux density $\vec{B}$, μ_B is the Bohr magneton, g is the Lande g-factor, and $\vec{\sigma}$ is the Pauli spin matrix. If the magnetic field is directed in the x-direction ($\vec{B} = B\hat{x}$), then diagonalization of the above Hamiltonian immediately produces two mutually orthogonal eigenspinors [1,1] and [1, − 1] which are the $+x$ and $-x$-polarized spins – that is, states whose spin quantization axes are parallel and anti-parallel to the x-directed magnetic field. Thus, the spin quantization axis (or spin polarization) has only two stable values and therefore becomes a *binary* variable. The "down" (parallel) or "up" (anti-parallel) states can encode logic bits 0 and 1, respectively.

5.3.2
Stability of Spin Polarization

Although the binary bits 0 and 1 can be encoded in the two anti-parallel spin polarizations, there remains a problem in that random bit flips caused by coupling of spins to the environment will cause bit errors and corrupt the data. The probability of a "bit flip" within a clock cycle is $1 - e^{-\frac{T}{<\tau>}}$, where T is the clock period and $<\tau>$ is the mean time between random spin flips. In order to make this probability small, it must be ensured that $<\tau> \gg T$.

If the host for the spin is a "quantum dot", then indeed $<\tau>$ can be quite long. In InP quantum dots, the single electron spin flip time has been reported to exceed 100 µs at 2 K [32]. More recently, several experiments have been reported claiming spin flip times (or so-called longitudinal relaxation time, T_1) of several milliseconds, culminating in a recent report of 170 ms in a GaAs quantum dot at low temperature (see the last reference in Ref. [33]). An extremely surprising result is that spin relaxation time $<\tau>$ in organic semiconductors can be incredibly long. It was found that the spin relaxation time in tris(8-hydroxyquinolinolate aluminum) – popularly known as Alq_3 – can be as long as 1 s at 100 K [34]. If the clock frequency is 5 GHz, then the clock period T is 200 ps, which is 5×10^9 times smaller than the spin flip time. Therefore, the probability that an unintentional spin flip will occur between two successive clock pulses is $1 - e^{-1/(5 \times 10E9)} = 2 \times 10^{-10}$, which can be handled by modern error correction algorithms [35].

Typical error probabilities encountered in today's integrated circuits range from 10^{-10} to 10^{-9}. If a 5-GHz clock is used and an error probability $1/p$ of 10^{-9} is required, the spin flip time needs to be only 200 ms, which is fairly easy to achieve today at low temperatures (77 K).

5.3.3
Reading and Writing Spin

"Spin" has one major disadvantage compared to "charge". Whereas, charge is extremely easy to read (or measure) with voltmeters, ammeters, electrometers, and so on, and extremely easy to write (or inject) with voltage and current sources, spin is much more difficult to read or write. Although reading is more difficult than writing, even single spins have been "read" recently. The following section discusses reading and writing.

5.3.3.1 Writing Spin

Spin bits in SSL are represented by the spin polarizations of single electrons, with each electron being hosted in a quantum dot. There are exotic methods of "writing" spin bits in quantum dots [3], but the easiest and conceptually most simple is to use local magnetic fields generated with inductive loops. These fields will orient the spins in the target dots along the direction of the field and write the chosen bit (see Chapter 4 in Volume III of this series). Local magnetic fields are generated on chip in magnetic random access memory (MRAM), but not with the spatial resolution required in the case of SSL. However, carbon nanotube-based inductive loops may be up to the task in the near future.

The next task is to estimate what amount of energy will be dissipated during the "writing" operation. Previously [3], the idea of writing spin using Rabi oscillation, as is done in electron spin resonance spectroscopy, was proposed. That mode of writing will not dissipate any energy whatsoever, but has no error tolerance as an ac magnetic pulse must be applied for precisely the correct duration. However, if overshoot or undershoot occurs, an error will be incurred that can build up and ultimately cause a bit-writing error. Error-free writing will require some dissipation, though this can be arbitrarily small.

5.3.3.2 Reading Spin

Reading spin bits is more difficult than writing spin bits. Here, the aim is to ascertain the spin polarization of a single electron in a solid (quantum dot), a feat which has been accomplished recently using three different methods [36–38]. The technique reported in Ref. [37] is eminently suitable for application in electronics.

5.3.4 The Universal Single Spin Logic Gate: The NAND Gate

The basic idea behind implementing logic gates in SSL is to engineer the interactions between input and output spin bits in such a way that the input–output relationship represents the "truth table" of the desired logic gate. This approach can be illustrated by showing how a NAND gate can be realized. The NAND gate is a universal gate with which any arbitrary combinational or sequential logic circuit may be implemented, and it is realized with a linear chain of three electrons in three quantum dots (see Figure 5.6). It will be assumed that only nearest-neighbor electrons interact via exchange as their wavefunctions overlap. Second nearest-neighbor interactions are negligible as exchange interaction decays exponentially with distance.

For NAND gate implementation, the leftmost and rightmost spins in Figure 5.5 must be regarded as the two "inputs bits", and the center spin as the corresponding "output bit". Assume that the downspin state ($\downarrow$) represents bit 1, and the upspin state ($\uparrow$) is bit 0. The global magnetic field, defining the spin quantization axes, is in the direction of "downspin". It has been shown recently, using a Heisenberg Hamiltonian to describe the 3-spin array, that as long as the Zeeman splitting energies caused by the local magnetic fields writing bits in the input dots is much larger than the

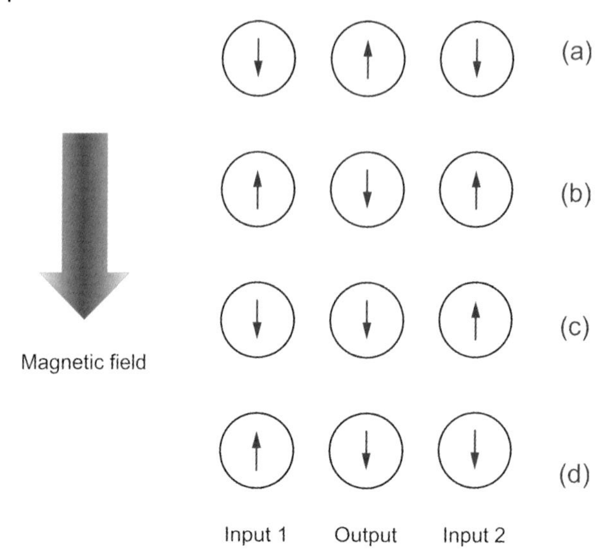

Figure 5.5 Single spin realization of the NAND gate. (a) When two inputs are [1 1]; (b) when two inputs are [0 0]; (c) when two inputs are [1 0]; and (d) when two inputs are [0 1]. Reproduced from Ref. 3 with permission from American Scientific Publishers: http://www.aspbs.com.

exchange coupling strength between neighboring dots, the ground-state spin configurations (determined by the directions of the local magnetic fields) are precisely those shown in Figure 5.5 [39]. In other words, if the input bits are written with local magnetic fields and the array is allowed to relax to the ground state in the presence of the local magnetic fields, then the output bit conforms to the diagrams in Figure 5.5a–d. It is evident that these configurations represent the truth table of the NAND gate:

Input 1	Input 2	Output
1	1	0
0	1	1
1	0	1
1	1	1

Therefore, if there is a 3-spin array, with nearest-neighbor exchange coupling, placed in a global magnetic field, and local magnetic fields align the spins in the peripheral dots to desired input bits, the output bit in the central dot will always be the NAND function of the input bits according to the truth table above, as long as two conditions are fulfilled:

- The array is in the thermodynamic ground state.

- The Zeeman splitting in the input dots caused by the local magnetic fields writing input data is much larger than the exchange coupling strength, which is roughly the energy difference between the triplet and singlet states in two neighboring dots.

Independent quantum mechanical calculations to confirm the working of the NAND gate were carried out by Molotkov and Nazin [40], while further investigations in this area have been conducted by Bychkov and coworkers [41].

Once the NAND gate is realized, only one other component is needed to implement any arbitrary combinational or sequential Boolean logic circuit. That element is a "spin wire" (with fan out) which will ferry spin logic signal from one stage to another *unidirectionally*. A spin wire (see Figure 5.6) consists of a linear array of quantum dots with clock pads between them. When the clock signal at a given pad is high, the potential barrier between the two surrounding dots is lowered and these dots are exchange-coupled because their wavefunctions overlap. This makes the spin states in these dots anti-parallel [42]. Therefore, by sequentially clocking the barriers, the spin bit can be replicated in every other dot, thus moving the spin signal along unidirectionally. Both, Sarkar et al. [43] and Bose et al. [44] have implemented a large array of combinational and sequential logic circuits using this approach. Two examples of their investigations are shown in Figure 5.7. The issue of unidirectionality and clocking will be revisited in Section 5.3.7.

5.3.5 Bit Error Probability

The NAND gate operates by relaxation to the thermodynamic ground state. It is the natural tendency of any physical system to gravitate towards the minimum energy

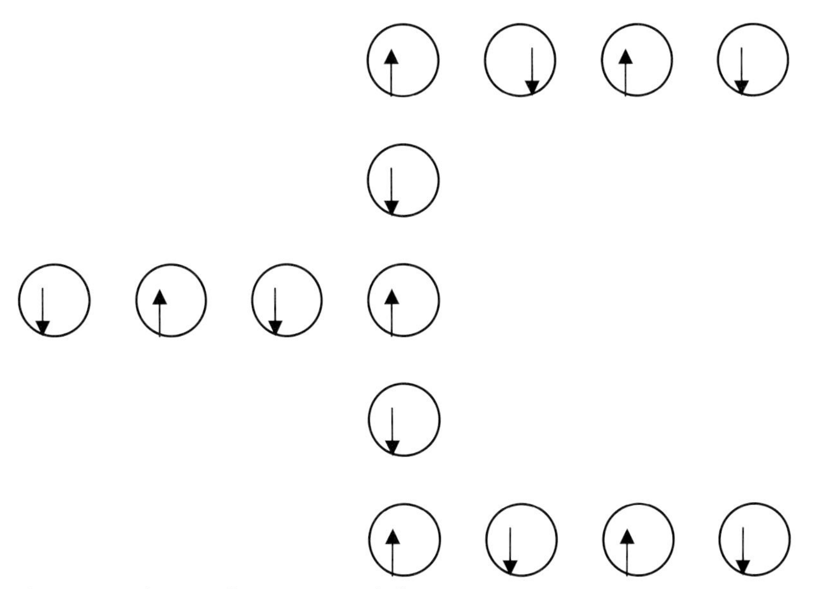

Figure 5.6 Realization of a "spin wire" with fan out.

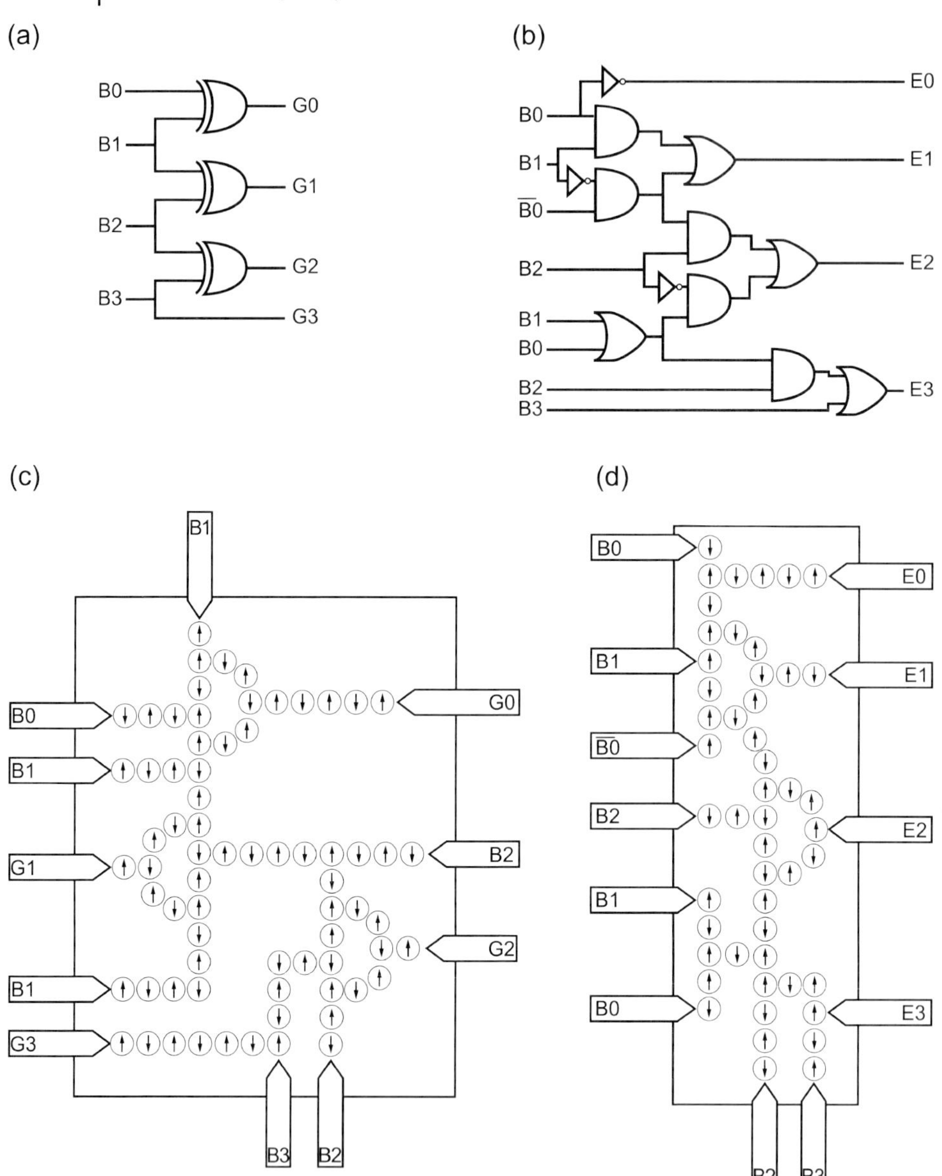

Figure 5.7 Single spin realizations of code converters. (a) Logic diagram for binary to Gray code converter; (b) logic diagram for binary to Excess-3 converter; (c) SSL realization of binary-to-Gray-code converter; and (d) SSL realization of binary-to-Excess-3 converter. The clock pads between successive cells are not shown for the sake of clarity. The input binary code is '1010'. Note that adjacent cells have anti-parallel polarizations indicating anti-ferromagnetic ordering. Reproduced from Ref. 43, with permission from Institute of Solid State Physics, Chernogolovka, Russia.

state (ground state), this being the law of thermodynamics. However, when a system achieves the ground state it need not stay there forever, as noise and fluctuations can take it to an excited state. If that happens and the NAND gate strays from the ground state, the results will be in error. This error probability is calculated next.

The NAND gate reaches ground state by exchanging phonons with the thermal environment (phonon bath). This brings it into thermodynamic equilibrium with the surrounding thermal bath. In that case, the occupation probability of any eigenstate of the gate will be given by Fermi–Dirac statistics. The ground-state occupation probability is $1/[\exp(E_{\text{ground}} - E_F)/kT + 1]$ (where E_F is the Fermi energy) and the excited state occupation probability is $1/[\exp(E_{\text{excited}} - E_F)/kT + 1]$. As the occupation probability of the ground state is *not* unity, the gate does not always work correctly with 100% certainty – in other words, the error probability $1/p$ is never zero. However, it does decrease with increasing energy separation between the excited and ground states.

The error probability $1/p$ is the sum of the ratios of the probabilities of being in the excited and ground state, summed over all excited states. This quantity is approximately $\sum_{\text{excited states}} e^{-\left(E_{\text{excited}} - E_{\text{ground}}\right)/kT}$, if the Fermi–Dirac statistics are approximated with Boltzmann statistics. It transpires that, in the case of the NAND gate, the second and higher excited states are far above in energy than the first excited state [39]. Therefore, only the first excited state E_1 can be retained in the sum above, and hence $E_1 - E_{\text{ground}} = kT\ln(p)$. It was also shown rigorously in Ref. [39] that $E_1 - E_{\text{ground}}$ is: (i) $4J - 2Z$ when the input bits are [1 1]; (ii) $4J + 2Z$ when the input bits are [0 0]; and (iii) $2Z$ when the input bits are [0 1] or [1 0]. Here, J is the exchange coupling strength between two neighboring dots, and $2Z$ is the Zeeman splitting energy in any dot due to the global magnetic field. Therefore, in order to attain an error probability of $1/p$ at a temperature T, it must be ensured that: (a) $2Z = kT\ln(p)$; and (b) $4J - 2Z = kT\ln(p)$, or $2J = kT\ln(p)$. The maximum values of J or Z are usually limited by technological constraints; for example, J is usually limited to 1 meV in gate-defined quantum dots [45]. These limits will then determine the maximum temperature of operation if a certain error probability $1/p$ is insisted on (this issue will be revisited in Section 5.3.10).

5.3.6
Related Charge-Based Paradigms

A similar idea for implementing logic gates using single electron charges confined in "quantum dashes" was proposed by Pradeep Bakshi and coworkers in 1991 [46]. There, logic bits were encoded in bistable charge polarizations of elongated quantum dots known as "quantum dashes". Coulomb interaction between nearest-neighbor quantum dashes pushes the electrons in neighboring dashes into antipodal positions, making the ground-state charge configuration "anti-ferroelectric", just as the exchange interaction in the present case tends to make the ground-state spin configuration almost "anti-ferromagnetic". Three Coulomb-coupled quantum dashes would realize a NAND gate in a way very similar to what was described here.

Bakshi's idea inspired a closely related idea known as "quantum cellular automata" [47], which uses a slightly different host, namely a four- or five-quantum dot "cell" instead of a quantum dash to store a bit. Here too the charge polarization of the cell is bistable and encodes the two logic bits. The only difference from Ref. [46] is that coulomb interaction makes the ground-state charge configuration ferroelectric, instead of anti-ferroelectric.

In the schemes of Refs. [46, 47] it is difficult to implement only nearest-neighbor interactions, at the exclusion of second-nearest-neighbor interactions, mainly because the Coulomb interaction is long range. The interaction in Refs. [46, 47] drops off as a polynomial of distance, but never exponentially with distance, unless strong screening can be implemented. If second-nearest-neighbor interactions are not much weaker than their nearest-neighbor counterparts, the ground-state charge configuration is weakly stable and not sufficiently robust against noise. In this respect, the spin-based approach has an advantage. As exchange interaction is short range (it always drops off exponentially with distance), it is much easier to make the second-nearest-neighbor interactions considerably weaker than the nearest-neighbor interactions.

A second issue is that in Refs. [46, 47], there is internal charge movement within each cell during switching, causing "eddy currents". This is a source of dissipation that is absent in the spin-based paradigm, as there is never any charge movement.

5.3.7
The Issue of Unidirectionality

The "unidirectional" propagation of logic signal was briefly mentioned in Section 5.3.5. This is a vital issue as the input signal should always influence (and determine) the output signal, but not the other way around. Unidirectionality is an important requirement for logic circuits [31]. A transistor is inherently unidirectional as there is *isolation* between its input and output terminals that guarantees unidirectionality. Therefore, it is easy to make logic circuits with transistors. In SSL, there is unfortunately *no isolation* between the input and output of the logic gate as exchange interaction is "bidirectional". Consider just two exchange coupled spins in two neighboring quantum dots; they will form a singlet state and therefore act as a natural NOT gate if one spin is the input and the other is the output (the output is always the logic complement of the input) [42]. However, exchange interaction, being bidirectional, cannot distinguish between which spin is the input bit and which is the output. The input will influence the output just as much as the output influences the input, and the master–slave relationship between input and output is lost. As the input and output are indistinguishable, it becomes ultimately impossible for logic signal to flow *unidirectionally* from an input stage to an output stage, and not the other way around. This issue has been discussed at length elsewhere [30, 48].

It is of course possible to forcibly impose unidirectionality in some (but not all) cases by holding the input cell in a fixed state until the desired output state is produced in the output cell. In that case, the input signal itself enforces unidirectionality because it is a symmetry-breaking influence. This approach was actually used to demonstrate a "magnetic cellular automaton", where the input enforced unidirectionality and produced the correct output [49]. However, there are problems with this approach. First, it can only work for a small number of cells before the influence of the input dies out. Second – and more important – the input cannot be changed until the final output has been produced, since otherwise the correct output *may not be produced at all.* That makes such architectures *non-pipelined* and therefore unacceptably slow. There may also be additional problems associated with random errors when this approach is employed; these are discussed in Ref. [50].

5.3.8 Unidirectionality in Time: Clocking

If unidirectionality cannot be imposed in *space*, then it must be imposed in *time*. This is accomplished by using clocking to activate successive stages sequentially in time [51, 52]. This strategy is well known and routinely adopted in bucket-brigade devices, such as charge-coupled-device (CCD) shift registers,[9)] where a push clock and a drop clock are used to lower and raise barriers between neighboring devices and thus steer a charge packet unidirectionally from one device to the next. The same can be done in single-spin circuits. A gate pad will be delineated between every two neighboring quantum dots, and a clock signal applied to this pad. During the positive clock cycle, a positive potential will appear over the potential barrier, thus isolating two neighboring quantum dots; this will lower the barrier temporarily to exchange-couple the two spins and result in the two spins assuming anti-parallel polarizations [42]. Then, during the negative clock cycle, the barrier is raised again to decouple the two spins. In this way, pairs of spin bits can be coupled sequentially in time, and the logic information transferred unidirectionally from one dot to the next in a bucket-brigade fashion. It has been shown previously [52] that a single-phase clock does not work, and that a three-phase clock is required to carry out this task.

The clocking circuit, however, introduces additional dissipation. The energy dissipated in the clock pad is $(1/2)CV^2$ if the clock pulse is applied non-adiabatically, and much less if applied adiabatically [6]. So, what should be the value of V? This is determined by noise considerations. The noise voltage on a capacitor is [53]:

$$U_n = \sqrt{\frac{kT}{C}} \tag{5.12}$$

and Ref. [53] prescribes that $V = 12\ U_n$ for reasonable error rates at clock frequencies over 10 GHz. In that case, the energy dissipated in the clock pads is $72\,kT$ if

9) CCDs also have no inherent unidirectionality. There, a push clock and a drop clock are used to steer charge packets from one device to the next. See, for example, D. K. Schroeder, in: G. W. Neudeck, R. F. Pierret (Eds.), *Advanced MOS Devices, Modular Series in Solid State Devices*, Chapter 3, Addison-Wesley, Reading, MA, **1987**.

applied non-adiabatically. If the clock signal is applied from a sinusoidal voltage source of amplitude V, then the energy dissipated per clock cycle is

$$E_{diss} = \frac{1}{2} C V^2 \frac{\omega}{\omega_{RC}} \tag{5.13}$$

where $\omega_{RC} = 1/RC$. The above formula holds if the clock circuit is modeled as a capacitor C in series with a resistor R representing the resistance in the charging path. Assuming that $R = 1\,\mathrm{k}\Omega$ and $C = 1\,\mathrm{aF}$, $\omega_{RC} = 10^{15}\,\mathrm{rad\,s^{-1}}$. If the clock frequency is 5 GHz, then $\omega = 3.45 \times 10^{10}\,\mathrm{rad\,s^{-1}}$. Therefore, $E_{diss} = 2.5 \times 10^{-3} kT$, which is negligible.

When clock pads are used, the most attractive feature of SSL is removed, namely the *absence* of interconnects (or "wires") between successive devices. "Wireless" exchange interaction plays the role of physical wires to transmit signals between neighboring devices, but in order to transmit "unidirectionally", each stage will need to be clocked and this requires a physical interconnect. A clock pad must be placed between pairs of quantum dots and wires must be attached to them to ferry the clock signal. Of course, a clock signal is also needed in traditional digital circuits involving CMOS, so that it is not an additional burden. Nevertheless, it still detracts from the appeal of a "wireless architecture", or the so-called "quantum-coupled architecture".

5.3.9
Energy and Power Dissipation

Most likely, by merely examining Figure 5.5, the reader can understand that the maximum energy dissipated when the gate switches between any two states is $2Z$, which is the Zeeman splitting in the output dot caused by the global magnetic field (this result was proved rigorously in Ref. [39]). Since it was shown in Section 5.3.5 that $2Z = kT\ln(p)$, the *maximum* energy dissipated during switching is $kT\ln(p)$, which was expected from the Landauer–Shannon result. The interesting point however is that the energy dissipated can be *less* than $kT\ln(p)$, depending on the initial and final states – that is, depending on the old and new input bit strings. If the gate switches from the state in Figure 5.5c to that in Figure 5.5a, the energy dissipated is actually $(2/3)\,kT\ln(p)$, while if it switches from the state in Figure 5.5b to that in Figure 5.5c, the energy dissipated is $(1/3)kT\ln(p)$ [39]. The reductions by factors of $1/3$ and $2/3$ are due to *interactions* between spins. Some implications of "interactions" were discussed in Ref. [4] in the context of reducing energy dissipation. What really happens here is that the three spins act collectively in unison as a single unit, because of the exchange-coupling between them, which reduces the total energy dissipation.

The maximum energy dissipation occurs when the gate switches from the state in Figure 5.5b to that in Figure 5.5a. That energy is $kT\ln(p)$. With $p = 10^{-9}$, this energy is $\sim 21\,kT$. By contrast, modern-day logic gates dissipate more than 50 000 kT when they switch [54].

5.3.10 Operating Temperature

In Section 5.3.5, the following result was established:

$$2J = kT\ln(p)$$
$$2Z = kT\ln(p)$$

The maximum value of J in semiconductor quantum dots is $\sim$1 meV [45], while in molecules it is about 6 meV [55]. Therefore, if an error probability p of 10^{-9} is required, $T = 1.1$ K if semiconductor quantum dots are employed. Room-temperature operation with this error probability will require J to be 270 meV, which is clearly unattainable at present.

Since $Z = J$ from the above relationships, the strength of the global magnetic field required can be estimated. That strength is found by setting $2Z = g\mu_B B_{\text{global}} = 2J = 2$ meV. If a material is used with a g-factor of 15 (e.g. InAs), then $B_{\text{global}} = 2.3$ Tesla, which is very reasonable.

As the energy separation between the spin levels ($g\mu_B B_{\text{global}}$) is 2 meV and kT at 1.1 K = 0.1 meV, it might be considered that a low temperature of 1.1 K was required in order to make the thermal broadening of the spin split levels much less (in this case, 20-fold less) than the level separation. However, this line of thinking would be entirely wrong, as the low temperature was needed for a small error probability ($1/p = 10^{-9}$), and *not* because there was a need to reduce the thermal broadening of levels. Spin, unlike charge, does not couple strongly to phonons, and therefore the thermal broadening of spin levels is much less than kT. If this were not the case, then electron spin resonance experiments at microwave frequencies could not be carried out at room temperature.

Unfortunately, this fact is not often understood or appreciated. For example, in Ref. [56] the authors state (contradicting Ref. [6]) that: "Energy barriers for a spin system play the same role as for a charge transfer system. The barrier must be large enough to make the different bit states distinguishable in a thermal environment". This argument is at best half-correct. For charge-based representation, the energy barrier separating logic states indeed need to be large enough that the different bit states are distinguishable in a thermal environment, since charge couples very strongly to the thermal environment, so that the thermal broadening of the energy states is $\sim kT$. This is *not* true for spin which, unlike charge, does *not* couple strongly with the thermal environment, so that the spin energy states are not broadened by $\sim kT$. As a result, the energy barrier separating logic states (i.e. $|g\mu_B B|$) can be much less than kT if bit error probability were not a concern. The energy barrier merely needs to be equal to $kT\ln(p)$, which would be *less than* kT if an error probability $1/p > 1/e = 0.367$ can be handled.

5.3.11 Energy Dissipation Estimates

It is now possible to estimate how much energy is dissipated in various actions, assuming $T = 1.1$ K:

Clocking	Bit flip
$2.48 \times 10^{-3} kT$ (sinusoidal) $= 4 \times 10^{-26}$ Joules ~ 0 (adiabatic)	$kT \ln(p)$ $[1/p = 10^{-9}]$ $= 3 \times 10^{-22}$ Joules

With a bit density of 10^{11} cm^{-2}, the dissipation per unit area is 3×10^{-22} Joules $\times$ 5 GHz $\times$ 10^{11} cm^{-2} = 0.15 W cm^{-2}. In comparison, the Pentium IV chip, with a bit density three orders of magnitude smaller, dissipates about 50 W cm^{-2} [57]. SSL dissipates 300 times less power with a bit density three orders of magnitude larger.

5.3.12
Other Issues

In charge-based devices such as MOSFETs, logic bits 0 and 1 are encoded by the *presence* and *absence* of charge in a given region of space. This region of space could be the channel of a MOSFET. When the channel is filled with electrons, the device is "on" and stores the binary bit 0. When the channel is depleted of electrons, the device is "off" and stores the binary bit 1. Switching between bits is accomplished by the physical motion of charges in and out of the channel, and this physical motion consumes energy.[10)] There is no easy way out of this problem since charge is a "scalar" quantity and therefore has only a *magnitude*. Bits can be demarcated only by a difference in the magnitude of charge or, in other words, by the relative presence and absence of charge. Therefore, to switch bits the magnitude of charge must be changed, and this invariably causes motion of the charges with an associated current flow. Spin, on the other hand, has a polarization which can be thought of as a "pseudo-vector" with two directions: "up" and "down". Switching between bits is accomplished by simply flipping the direction of the pseudo-vector with no change in magnitude of anything. As switching can be accomplished without physically moving charges (and causing a current to flow), spin-based devices could be inherently much more energy-efficient than charge-based devices, a point which was highlighted in Ref. [3].

It is because of this reason that SSL is much more energy efficient than the Pentium IV chip.

5.4
Spin-Based Quantum Computing: An Engineer's Perspective

Quantum computing is based on the idea of encoding information in a so-called "qubit", which is very different from a classical bit. It is a coherent superposition of

10) In fact, the physical motion of charges causes a current I to flow with an associated I^2R dissipation (where R is the resistance in the path of the current).

classical bits 0 and 1:

$$\begin{aligned} qubit &= a|0\rangle + b|1\rangle \\ |a|^2 + |b|^2 &= 1 \end{aligned} \tag{5.14}$$

The coefficients a and b are complex numbers, and the phase relationship between them is important. The "qubit" is the essential ingredient in the quantum Turing machine first postulated by Deutsch to elucidate the nuances of quantum computing [58].

The power of quantum computing accrues from two attributes: *quantum parallelism* and *entanglement*. Consider two electrons whose spin polarizations are made bistable by placing them in a magnetic field. If the system is classical, these two electrons can encode only two bits of information: the first one can encode either 0 or 1 (downspin or upspin), and the second one can do the same. By analogy, N number of electrons can encode N bits of information, as long as the system is classical.

But now consider the situation when the spin states of the two electrons are quantum mechanically entangled so that the two electrons can no longer be considered separately, but rather as one coherent unit. In that case, there are four possible states that this two-electron system can be in: both spins "up"; first spin "up" and the second spin "down"; the first spin "down" and the second spin "up"; and both spins "down". The corresponding "qubit" can be written as:

$$\begin{aligned} qubit &= a|\uparrow\uparrow\rangle + b|\uparrow\downarrow\rangle + c|\downarrow\uparrow\rangle + d|\downarrow\downarrow\rangle \\ |a|^2 + |b|^2 + |c|^2 + |d|^2 &= 1 \end{aligned} \tag{5.15}$$

Obviously, this system can encode four information states, as opposed to two. By analogy, if N qubits can be quantum mechanically entangled, then the system can encode 2^N bits of information, as opposed to simply N. This becomes a major advantage if N is large. Consider the situation when $N = 300$. There is no way in which a classical computer can be built that can handle 2^{300} bits of information as that number is larger than the number of electrons in the known universe. However, if just 300 electrons could be taken and their spins entangled (a very tall order), then 2^{300} bits of information could be encoded. Thus, entanglement bestows on a quantum computer tremendous information-handling capability.

The above must not be misconstrued to imply that a quantum computer is a "super-memory" that can store 2^N bits of information in N physical objects such as electrons. When a bit is "read", it always collapses to either 0 or 1. In Eq. (5.17), the probability that the qubit will be read as 0 is $|a|^2$, and the probability that it will be read as 1 is $|b|^2$. As either a 0 or a 1 is always read, a quantum computer allows *access* to no more than N bits of information. Thus, it is no better than a classical memory – in fact, it is worse! Because of the probabilistic nature of quantum mechanics, a stored bit will sometimes be read as 0 (with probability $|a|^2$) and sometimes as 1 (with probability $|b|^2 = 1 - |a|^2$). If repeated measurements are made of the stored bits, the exact same result will never be achieved every time, and thus the quantum system is not even a reliable memory.

The power of *entanglement* does not result in a super memory, but it is utilized in a different way, and is exploited in solving certain types of problem super efficiently.

Two well-known examples are Shor's quantum algorithm for factorization [59] and Grover's quantum algorithm for sorting [60]. Factorization is an *NP*-hard problem, but using Shor's algorithm, the complexity can be reduced to *P*. Grover's algorithm for sorting has a similar advantage. Suppose that there is a need to sort an *N*-body ensemble to find one odd object. By using the best classical algorithm, this will take $N/2$ tries, but using Grover's algorithm it will take only $\sqrt{N}$ tries. Thus, entanglement yields super-efficient algorithms that can be executed in a quantum processor where qubits are entangled. That is the advantage of quantum computing.

5.4.1 Quantum Parallelism

In classical computing, "parallelism" refers to the parallel (simultaneous) processing of different information in different processors. Quantum parallelism refers to the simultaneous processing of different information or inputs in the *same* processor. This idea, due to Deutsch, refers to the notion of evaluating a function once on a superposition of all possible inputs to the function to produce a superposition of outputs. Thus, all outputs are produced in the time taken to calculate one output classically. Of course, not all of these outputs are accessible since a measurement on the superposition state of the output will produce only one output. However, it is possible to obtain certain joint properties of the outputs [61], and that is a remarkable possibility.

Quantum parallelism may be illustrated with an example. Consider the situation when M inputs $x_1, x_2, x_3, \ldots x_M$ are provided to a computer, and their functions $f(x_1)$, $f(x_2), \ldots f(x_M)$ are to be computed. The results are then fed to another computer to calculate the functional $F(f(x_1), f(x_2), \ldots f(x_M))$.

With a classical computer, $f(x_1), f(x_2), \ldots f(x_M)$ will be calculated serially, one after the other. However, with a quantum computer, the initial state will be prepared as a superposition of the inputs:

$$|I\rangle = \frac{1}{\sqrt{M}}(|x_1\rangle + |x_2\rangle + \ldots |x_M\rangle) \tag{5.16}$$

This input will then be fed to a quantum computer and allowed to evolve in time to produce the output

$$|O\rangle = \frac{1}{\sqrt{M}}(|f(x_1)\rangle + |f(x_2)\rangle + \ldots |f(x_M)\rangle) \tag{5.17}$$

The output $|O>$ has been obtained in the time required to perform a single computation. Now, if the functional $F(f(x_1), f(x_2), \ldots f(x_M))$ can be computed from $|O>$, then a quantum computer will be extremely efficient. This is an example where "quantum parallelism" can speed up the computation tremendously.

There are two questions, however. First, can the functional be computed from a knowledge of the superposition of various $f(x_i)$ and not the individual $f(x_i)$s? The answer is "yes", but for a small class of problems – the so-called *Deutsch–Josza* class of problems which can benefit from quantum parallelism. Second, can the functional be computed correctly with unit probability? The answer is "no". However, if the answer obtained in the first trial is wrong (hopefully, the computing entity can distinguish right from wrong answers), then the experiment or computation is repeated until the right answer is obtained. The probability of getting the right answer within k iterations is $(1 - p^k)$ where p is the probability of getting the wrong answer in any iteration. The mean number of times the experiment should be repeated to get the correct answer is $M^2 - 2M - 2$.

The following is an example of the Deutsch–Josza class of problems. For integer $0 \leq x \leq 2L$, given that the function $f_{k(x)} \in [0,1]$ has one of two properties – (i) either $f_{k(x)}$ is independent of x; or (ii) one-half of the numbers $f_{k(0)}, f_{k(1)}, \ldots f_{k(2L-1)}$ are zero – determine which type the function belongs to, using the fewest computational steps.

The most efficient classical computer will require $L + 1$ evaluations whereas, according to Deutsch and Josza, a quantum computer can solve this problem with just two iterations.

5.4.2
Physical Realization of a Qubit: Spin of an Electron in a Quantum Dot

The all-important question that stirs physical scientists and engineers is: Which system is most appropriate to implement a qubit? It must be one where the phase relationship between the coefficients a and b in Eq. (5.17) are preserved for the longest time. Charge has a small phase-coherence time which saturates to about 1 ns as the temperature is lowered towards 0 K because of coupling to zero point motion of phonons [62]. Spin has a much longer phase-coherence time as it does not couple to phonons efficiently. As a result, the phase-coherence time may not rapidly decrease with increasing temperature. Measurements of the phase-coherence time (also called the transverse relaxation time, or T_2 time) in an ensemble of CdS quantum dots has recently been shown actually to increase with increasing temperature [63]. It is believed that this is because the primary phase relaxation mechanism for electron spins in these quantum dots is hyperfine interactions with nuclear spins. The nuclear spins are increasingly depolarized with increasing temperature, and that leads to an actual increase in the electron's spin coherence time with increasing temperature. Therefore, it is natural to encode a qubit by the coherent superposition of two spin polarizations of an electron.

In 1996, the idea was proposed of encoding a "qubit" by the spin of an electron in a quantum dot [42]. A simple spin-based "quantum inverter" was designed which utilized two exchange-coupled quantum dots. This was not a universal quantum gate, but relied on quantum mechanics to elicit the Boolean logic NOT function. The spin of the electron was used as a qubit. To the present author's knowledge, this was the first instance where the spin of an electron in a quantum dot was used to implement a qubit in a gate. This idea was inspired by single spin logic and so in many ways SSL could be regarded as the progeny of spin-based quantum computing.

Unlike SSL – which is purely classical and does not require "phase coherence" – quantum computing relies intrinsically on phase coherence and is therefore much more delicate. While the phase coherence of single spins can be quite long, it is doubtful that several entangled spins will have sufficiently long-lived phase coherence to allow a significant number of computational steps to be carried out, particularly when gate operations are performed on them. At the time of writing, the realization of practical scalable quantum processors based on electron spins appears to be decades away.

5.5 Conclusions

In this chapter, an attempt has been made to distinguish between the two approaches to spintronics, namely "hybrid spintronics" and "monolithic spintronics". It is unlikely that the hybrid approach will bring about significant advances in terms of energy dissipation, speed, or any other metric. The monolithic approach, on the other hand, is more difficult, but also more likely to produce major advances. The SSL idea revisited here is a paradigm that may begin to bear fruit with the most recent advances in manipulating the spins of single electrons in quantum dots [64–71]. This is a classical model and does not require the phase coherence that is difficult to maintain in solid-state circuits. There is also no requirement to "entangle" several bits, but rather a need to exchange-couple two bits pairwise. Thus, SSL is much easier to implement than quantum processors based on single electron spins.

Acknowledgments

The author acknowledges useful discussions with Profs. Marc Cahay and Supriyo Datta. These studies were supported by the National Science Foundation under grant ECCS-0608854, and by the Air Force Office of Scientific Research under grant FA9550-04-1-0261.

References

1 S. Datta, B. Das, *Appl. Phys. Lett.* 1990, **56**, 665.

2 (a) J. Fabian, I. Zutic, S. Das Sarma, *Appl. Phys. Lett.* 2004, **84**, 85; (b) M. E. Faltte, Z. G. Yu, E. Johnston-Halperin, D. D. Awschalom, *Appl. Phys. Lett.* 2003, **82**, 4740.

3 S. Bandyopadhyay, *J. Nanosci. Nanotechnol.* 2007, **7**, 168.

4 S. Salahuddin, S. Datta, *Appl. Phys. Lett.* 2007, **90**, 093503.

5 (a) V. V. Zhirnov, R. K. Cavin, J. A. Hutchby, G. I. Bourianoff, *Proc. IEEE* 2003, **91**, 1934; (b) R. K. Cavin, V. V. Zhirnov, J. A. Hutchby, G. I. Bourianoff, *Fluctuations Noise Lett.* 2005, **5**, C29. See also Chapter 4 in Volume III of this series.

6 D. Nikonov, G. I. Bourianoff, P. Gargini, www.arXiv.org/cond-mat/0605298.

7 E. I. Rashba, *Sov. Phys. Semicond.* 1960, **2**, 1109.

8 S. Bandyopadhyay, M. Cahay, *Physica E* 2005, **25**, 399.

9 M. Cahay, S. Bandyopadhyay, *Phys. Rev. B* 2003, **68**, 115316.

10 M. Cahay, S. Bandyopadhyay, *Phys. Rev. B* 2004, **69**, 045303.

11 R. J. Elliott, *Phys. Rev.* 1954, **96**, 266.

12 (a) M. I. D'yakonov, V. I. Perel', *Sov. Phys. JETP* 1971, **33**, 1053; (b) M. I. D'yakonov, V. I. Perel', *Sov. Phys. Solid State* 1972, **13**, 3023.

13 S. Pramanik, S. Bandyopadhyay, M. Cahay, *IEEE Trans. Nanotech.* 2005, **4**, 2.

14 G. Dresselhaus, *Phys. Rev.* 1955, **100**, 580.

15 S. Bandyopadhyay, M. Cahay, *Appl. Phys. Lett.* 2004, **85**, 1814.

16 S. Bandyopadhyay, M. Cahay, *Appl. Phys. Lett.* 2004, **85**, 1433.

17 J. Nitta, T. Takazaki, H. Takayanagi, T. Enoki, *Phys. Rev. Lett.* 1997, **78**, 1335.

18 Suman Datta, Intel Corporation, private communication.

19 J. Schliemann, J. C. Egues, D. Loss, *Phys. Rev. Lett.* 2003, **90**, 146801.

20 X. Cartoixa, D. Z.-Y. Ting, Y.-C. Chang, *Appl. Phys. Lett.* 2003, **83**, 1462.

21 (a) K. C. Hall, W. H. Lau, K. Gundogdu, M. E. Flatte, T. F. Boggess, *Appl. Phys. Lett.* 2003, **83**, 2937; (b) K. C. Hall, K. Gundogdu, J. L. Hicks, A. N. Kocbay, M. E. Flatte, T. F. Boggess, K. Holabird, A. Hunter, D. H. Chow, J. J. Zink, *Appl. Phys. Lett.* 2005, **86**, 202114.

22 K. C. Hall, M. E. Flatte, *Appl. Phys. Lett.* 2006, **88**, 162503.

23 E. Safir, M. Shen, S. Siakin, *Phys. Rev. B* 2004, **70**, 241302(R).

24 G. Salis, R. Wang, X. Jiang, R. M. Shelby, S. S. P. Parkin, S. R. Bank, J. S. Harris, *Appl. Phys. Lett.* 2005, **87**, 262503.

25 S. Bandyopadhyay, M. Cahay, www.arXiv.org/cond-mat/0604532.

26 M. E. Flatte, K. C. Hall, www.arXiv.org/cond-mat/0607432.

27 P. A. Dowben, R. Skomski, *J. Appl. Phys.* 2004, **95**, 7453.

28 T. Koga, J. Nitta, H. Takayanagi, S. Datta, *Phys. Rev. Lett.* 2002, **88**, 126601.

29 S. Bandyopadhyay, M. Cahay, *Appl. Phys. Lett.* 2005, **86**, 133502.

30 S. Bandyopadhyay, B. Das, A. E. Miller, *Nanotechnology* 1994, **5**, 113.

31 D. A. Hodges, H. G. Jackson, *Analysis and Design of Digital Integrated Circuits*, 2nd edition, McGraw-Hill, New York, 1988, p. 2.

32 M. Ikezawa, B. Pal, Y. Masumoto, I. V. Ignatiev, S. Yu. Verbin, I. Ya. Gerlovin, *Phys. Rev. B* 2005, **72**, 153302.

33 (a) R. Hanson, L. H. Willems van Beveren, J. M. Elzerman, W. J. M. Naber, G. H. L. Koppens, L. P. Kouwenhoven, L. M. K. Vandersypen, *Phys. Rev. Lett.* 2005, **94**, 196802; (b) M. Kroutvar, Y. Ducommun, D. Heiss, M. Bichler, D. Schuh, G. Abstreiter, J. J. Finley, *Nature* 2004, **432**, 81; (c) S. Amasha, K. MacLean, I. Radu, D. M. Zumbuhl, M. A. Kastner, M. P. Hanson, A. C. Gossard, www.arXiv.org/cond-mat/0607110.

34 S. Pramanik, C.-G. Stefanita, S. Patibandla, S. Bandyopadhyay, K. Garre, N. Harth, M. Cahay, *Nature Nanotech.* 2007, **2**, 216.

35 E. Knill, *Nature* 2005, **434**, 39.

36 D. Rugar, R. Budakian, H. J. Mamin, B. H. Chui, *Nature* 2004, **430**, 329.

37 J. M. Elzerman, R. Hanson, L. H. Willems van Beveren, B. Witkamp, L. M. K. Vandersypen, L. P. Kouwenhoven, *Nature* 2004, **430**, 431.

38 M. Xiao, I. Martin, E. Yablonovitch, H. W. Jiang, *Nature* 2004, **430**, 435.

39 (a) H. Agarwal, S. Pramanik, S. Bandyopadhyay, *New J. Phys.* 2008, **10**, 015001. (b) S. Bandyopadhyay, M. Cahay, *An Introduction to Spintronics*, CRC Press, Boca Raton, 2007, Chapter 13.

40 (a) S. N. Molotkov, S. S. Nazin, *JETP Lett.* 1995, **62**, 273; (b) S. N. Molotkov, S. S. Nazin, *Phys. Low Dim. Struct.* 1997, **10**, 85; (c) S. N. Molotkov, S. S. Nazin, *Zh. Eksp. Teor. Fiz.* 1996, **110**, 1439.

41 A. M. Bychkov, L. A. Openov, I. A. Semenihin, *JETP Lett.* 1997, **66**, 298.

42 (a) S. Bandyopadhyay, V. P. Roychowdhury, Proceedings International Conference on Superlattice Microstructures, Liege, Belgium 1996; (b) S. Bandyopadhyay, V. P. Roychowdhury, *Superlatt. Microstruct.* 1997, **22**, 411.

43 S. K. Sarkar, T. Bose, S. Bandyopadhyay, *Phys. Low Dim. Struct.* 2006, **2**, 69.

44 T. Bose, S. K. Sarkar, S. Bandyopadhyay, *IEE Proc. - Circuits, Dev. Syst.* 2007, **1**, 194.

45 (a) D. V. Melnikov, J.-P. Leburton, *Phys. Rev. B* 2006, **73**, 155301; (b) J.-P. Leburton, private communication.

46 (a) P. Bakshi, D. Broido, K. Kempa, *J. Appl. Phys.* 1991, **70**, 5150; (b) K. Kempa, D. A. Broido, P. Bakshi, *Phys. Rev. B* 1991, **43**, 9343; The logic implementations were made available to the author in early 1992 by P. Bakshi in a private communication.

47 C. S. Lent, P. D. Tougaw, W. Porod, G. H. Bernstein, *Nanotechnology* 1993, **4**, 49.

48 S. Bandyopadhyay, V. P. Roychowdhury, D. B. Janes, in: M. A. Stroscio, M. Dutta (Eds.), *Quantum-Based Electronic Devices and Systems*, World Scientific, Singapore, 1998, Chapter 1.

49 R. P. Cowburn, M. E. Welland, *Science* 2000, **287**, 1466.

50 M. Anantram, V. P. Roychowdhury, *J. Appl. Phys.* 1999, **85**, 1622.

51 S. Bandyopadhyay, V. P. Roychowdhury, *Jpn. J. Appl. Phys.* 1996, **35** (Part 1), 3350.

52 S. Bandyopadhyay, *Superlatt. Microstruct.* 2005, **37**, 77.

53 L. B. Kish, *Phys. Lett. A* 2002, **305**, 144.

54 The International Technology Roadmap for Semiconductors published by the Semiconductor Industry Association. http://public.itrs.net/.

55 C. F. Hirjibehedin, C. P. Lutz, A. J. Heinrich, *Science* 2006, **312**, 1021.

56 C. S. Lent, M. Liu, Y. Lu, *Nanotechnology* 2006, **17**, 4240.

57 P. P. Gelsinger, Proceedings, IEEE International Solid State Circuits Conference, IEEE Press, 2001.

58 D. Deutsch, *Proc. Royal Soc. London, Ser. A* 1985, **400**, 97.

59 P. W. Shor, in: Proceedings 37th Annual Symposium Foundations of Computer Science, IEEE Computer Society Press, p. 56, 1996.

60 (a) L. K. Grover, *Phys. Rev. Lett.* 1997, **79**, 325; (b) L. K. Grover, *Phys. Rev. Lett.* 1997, **79**, 4709.

61 R. Josza, *Proc. Royal Soc. London Ser. A* 1991, **435**, 563.

62 P. Mohanty, E. M. Q. Jariwalla, R. A. Webb, *Phys. Rev. Lett.* 1997, **78**, 3366.

63 S. Pramanik, B. Kanchibotla, S. Bandyopadhyay, Proceedings IEEE NANO 2006 Conference, Cincinnati, 2006.

64 M. Ciorga, A. S. Sachrajda, P. Hawrylak, C. Gould, P. Zawadzki, S. Jullian, Y. Feng, Z. Wasilewski, *Phys. Rev. B* 2000, **61**, R16315.

65 M. Piero-Ladriere, M. Ciorga, J. Lapointe, P. Zawadzki, M. Korukusinski, P. Hawrylak, A. S. Sachrajda, *Phys. Rev. Lett.* 2003, **91**, 026803.

66 C. Livermore, C. H. Crouch, R. M. Westervelt, K. L. Campman, A. C. Gossard, *Science* 1996, **274**, 1332.

67 T. H. Oosterkamp, T. Fujisawa, W. G. van der Wiel, K. Ishibashi, R. V. Hijman, S. Tarucha, L. P. Kouwenhoven, *Nature* 1998, **395**, 873.

68 A. W. Holleitner, R. H. Blick, A. K. Huttel, K. Eberl, J. P. Kotthaus, *Science* 2001, **297**, 70.

69 N. J. Craig, J. M. Taylor, E. A. Lester, C. M. Marcus, M. P. Hanson, A. C. Gossard, *Science* 2004, **304**, 565.

70 R. Hanson, B. Witkamp, L. M. K. Vandersypen, L. H. W. van Beveren, J. M. Elzerman, L. P. Kouwenhoven, *Phys. Rev. Lett.* 2003, **91**, 196802.

71 J. R. Petta, A. C. Johnson, J. M. Taylor, E. A. Laird, A. Yacoby, M. D. Lukin, C. M. Marcus, M. P. Hanson, A. C. Gossard, *Science* 2005, **309**, 2180.

6
Organic Transistors

Hagen Klauk

6.1
Introduction

Organic transistors are metal-insulator-semiconductor (MIS) field-effect transistors (FETs) in which the semiconductor is not an inorganic crystal, but a conjugated organic material. The fact that organic materials can be semiconductors may initially surprise, as most organic materials encountered today are excellent insulators. The fundamental property that leads to electrical conduction in carbon-based solids is conjugation – that is, the presence of alternating single and double bonds between neighboring carbon atoms (see Figure 6.1). Conjugation results in the delocalization of the π-electrons over the entire molecule – or at least over the conjugated portion of the molecule – and this allows electronic charge to be transported along the molecule.

Electrical conductivity in conjugated organic materials has been studied for a century, yet a complete picture of the charge transport physics in organics is still evolving. A central observation is that the intermolecular bonds in organic solids are not covalent bonds, but much weaker van der Waals interactions. As a consequence, electronic states are not delocalized over the entire solid, but are localized to a single molecule (or the conjugated portion of the molecule). Charge transport through organic solids is therefore limited by trapping in localized states, and likely involves some form of hopping between molecules. This means that the carrier mobilities in organic semiconductors are expected to be much smaller than the mobilities in inorganic semiconductor crystals.

In fact, carrier mobilities observed in organic solids vary greatly depending on the choice of material, its chemical purity, and the degree of molecular order in the solid (which determines the orbital overlap between neighboring molecules). Semiconducting polymers that arrange in amorphous films when prepared from solution usually have room-temperature mobilities in the range of 10^{-6} to $10^{-3}\,cm^2\,V^{-1}\,s^{-1}$. (For comparison, the carrier mobilities in single-crystalline silicon are near $10^3\,cm^2\,V^{-1}\,s^{-1}$.) Through molecular engineering, improved purification, and better control of the film deposition (so that charges can be transferred more easily

Nanotechnology. Volume 4: Information Technology II. Edited by Rainer Waser
Copyright © 2008 WILEY-VCH Verlag GmbH & Co. KGaA, Weinheim
ISBN: 978-3-527-31737-0

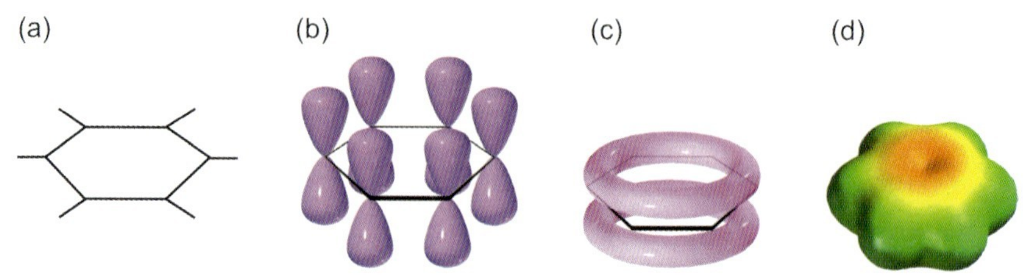

Figure 6.1 The concept of delocalized electrons in a conjugated molecule. (a) The carbon–carbon and carbon–hydrogen σ bonds in benzene. (b) The p orbital on each carbon can overlap with two adjacent p orbitals. (c) The clouds of electrons above and below the molecular plane. (d) The electrostatic potential map of benzene, showing that all the carbon–carbon bonds have the same electron density. (Figure adapted from: P. Y. Bruice, *Organic Chemistry*, Pearson Education, Upper Saddle River, NJ, USA.)

between molecules), the mobilities of conjugated polymers can be increased to about $0.5\,cm^2\,V^{-1}\,s^{-1}$. Small-molecule materials, on the other hand, often spontaneously arrange themselves into semicrystalline films when deposited by vacuum sublimation, and this results in room-temperature mobilities as large as $7\,cm^2\,V^{-1}s^{-1}$. Reports on carefully prepared single crystals of ultrapurified naphthalene suggest that mobilities at cryogenic temperatures can be as large as $400\,cm^2\,V^{-1}\,s^{-1}$ [1].

It is generally agreed that no single transport model can account for this wide a range of observed mobilities. Also, the temperature dependence of the mobility can be quite different for different organic materials, and in some cases different temperature-dependent mobility behavior has been observed even for the same organic material. Consequently, several different models for charge transport in organics have been proposed.

The model of *variable-range hopping* (VRH) assumes that carriers hop between localized electronic states by tunneling through energy barriers, and that the probability of a hopping event is determined by the hopping distance and by the energy distribution of the states. Specifically, carriers either hop over short distances with large activation energies, or over long distances with small activation energies. Since the tunneling is thermally activated, the mobility increases with increasing temperature. With increasing gate voltage, carriers accumulated in the channel fill the lower-energy states, thus reducing the activation energy and increasing the mobility. As Vissenberg and Matters have shown [2], the tunneling probability depends heavily on the overlap of the electronic wave functions of the hopping sites, which is consistent with the observation that the carrier mobility is significantly greater in materials with a larger degree of molecular ordering. Thus, the mobility is dependent on temperature, gate voltage, and molecular ordering (see Figure 6.2). The variable-range hopping model is usually discussed in the context of low-mobility amorphous semiconductor films (with room-temperature mobility less than about $10^{-2}\,cm^2\,V^{-1}\,s^{-1}$).

The *multiple trapping and release* (MTR) model adapted for organic transistors by Gilles Horowitz and coworkers [3] is based on the assumption that most of the charge carriers in the channel are trapped in localized states, and that carriers cannot move directly from one state to another. Instead, carriers are temporarily promoted to an

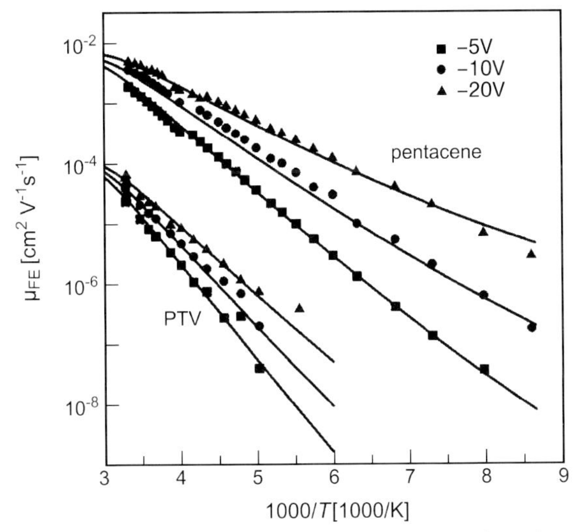

Figure 6.2 Temperature-dependent and gate voltage-dependent carrier mobility in a disordered polythienylenevinylene (PTV) film and in a solution-processed pentacene film. (Reproduced with permission from Ref. [2].)

extended-state band in which charge transport occurs. The number of carriers available for transport then depends on the difference in energy between the trap level and the extended-state band, as well as on the temperature and on the gate voltage (see Figure 6.3). The MTR model is generally considered to apply to materials with a significant degree of molecular ordering, such as polycrystalline films of small-molecule and certain polymeric semiconductors – that is, to materials which show room-temperature mobilities approaching or exceeding $0.1\,cm^2\,V^{-1}\,s^{-1}$. Carrier mobilities this large are not easily explained in the framework of the VRH model. On the other hand, the existence of an extended-state transport band in organic semiconductors as postulated by the MTR model is still the subject of debate.

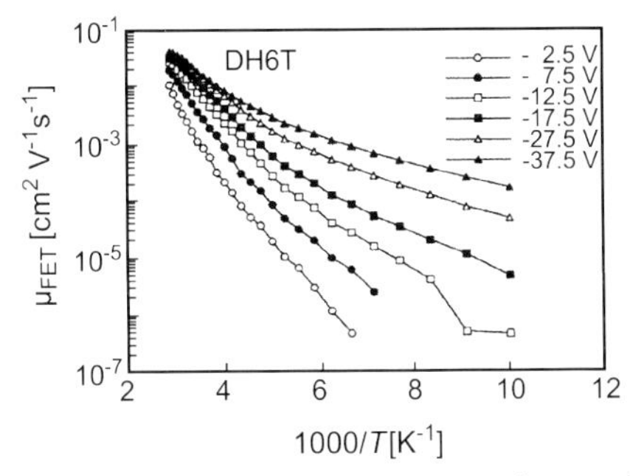

Figure 6.3 Temperature-dependent and gate voltage-dependent carrier mobility in a vacuum-deposited polycrystalline dihexylsexithiophene (DH6T) film. (Reproduced with permission from Ref. [3].)

6.2 Materials

Organic semiconductors essentially are available as two types: *conjugated polymers* and *conjugated small-molecule materials*. The prototypical semiconducting polymer is polythiophene (Figure 6.4a). While genuine polythiophene is insoluble and thus difficult to deposit in the form of thin films, alkyl-substituted polythiophenes, such as poly(3-hexylthiophene) (P3HT) (Figure 6.4b) have excellent solubility in a variety of organic solvents, and thin films are readily prepared by spin-coating, dip-coating, drop-coating, screen printing, or inkjet printing. Polythiophene was one of the first polymers to be used for organic transistors [4, 5], and remains one of the most popular semiconductors in organic electronics.

Generic polythiophenes usually form amorphous films with virtually no long-range structural order, very short π-conjugation length, and consequently poor mobilities, typically below $10^{-3}\,cm^2\,V^{-1}\,s^{-1}$. Obtaining usefully large mobilities in the polythiophene system requires highly purified derivatives which have been specifically synthesized to allow the molecules to self-organize into crystalline structures with a high degree of molecular order. An early example of such an engineered polythiophene is regioregular (RR) head-to-tail (HT) P3HT, initially synthesized by McCullough and coworkers in 1993 [6] and first employed for transistor fabrication by Bao and colleagues in 1996 [7]. In RR-HT-P3HT, the strong interactions between the regularly oriented alkyl side chains lead to a three-dimensional (3-D) lamellar structure in which the thienylene moieties along the polymer backbone are held in coplanarity (see Figure 6.5). The coplanarity of the thienylene moieties greatly increases the extent of π-conjugation, one consequence of which is a substantially increased carrier mobility (0.05 to $0.1\,cm^2\,V^{-1}\,s^{-1}$) compared with regiorandom P3HT (less than $10^{-3}\,cm^2\,V^{-1}\,s^{-1}$).

The microstructure of regioregular P3HT, its dependence on the degree of regioregularity, molecular weight, and deposition conditions, and the relationship between microstructure and carrier mobility, has been studied in detail by Sirringhaus and coworkers [8]. These authors found that the orientation of the microcrystalline lamellar domains with respect to the substrate surface is influenced by the molecular weight (i.e. the average polymer chain length), by the degree of

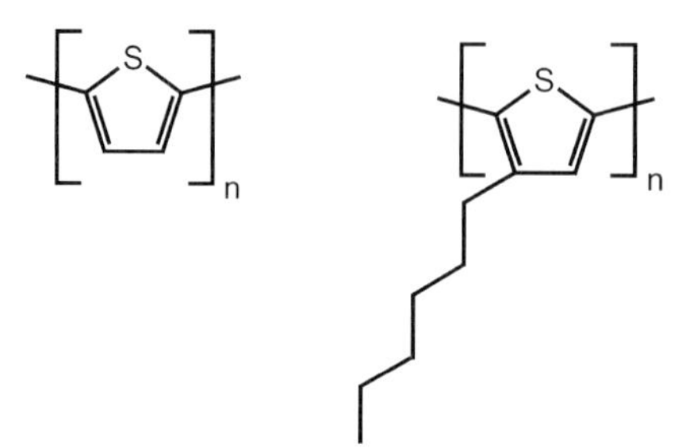

Figure 6.4 The chemical structures of polythiophene (left) and poly(3-hexylthiophene) (P3HT) (right).

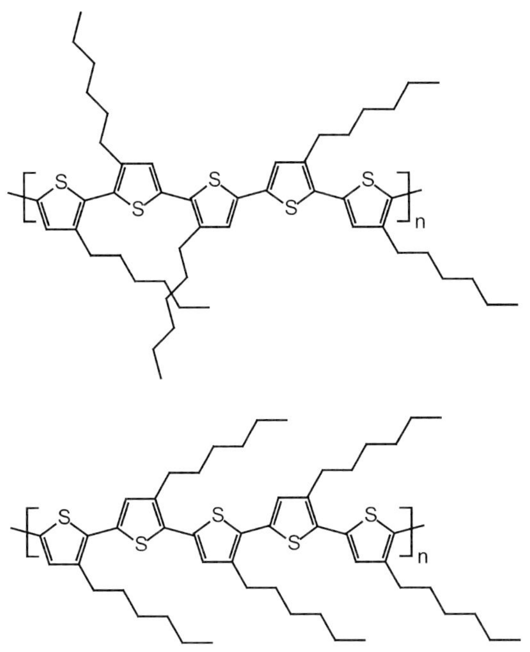

Figure 6.5 Schematic representation of regiorandom P3HT (left) and regioregular P3HT (right).

regioregularity, and by the deposition conditions (i.e. whether the film formation occurred quickly or slowly). The formation of ordered lamellae leads to a substantial overlap of the π-orbitals of neighboring molecules, but only in the direction perpendicular to the lamella plane. As a result, charge carrier transport and mobility in ordered P3HT films is highly anisotropic. In field-effect transistors (FETs), current flows parallel to the substrate, so the orientation of the lamellae with respect to the substrate surface is critical for the electrical performance of the transistors. Sirringhaus et al. were able to show that the transistor-friendly orientation of the lamellae (shown in Figure 6.6a) can be induced by selecting a polymer with a high degree of regioregularity (see Figure 6.6b) and, to a lesser extent, by choosing deposition conditions that favor a slow crystallization of the film.

Unfortunately, the large extent of π-conjugation in regioregular P3HT also leads to a significantly reduced ionization potential that makes the material very susceptible to photoinduced oxidation. This explains the commonly observed instability of P3HT transistors when operated in ambient air without encapsulation. A successful route to environmentally more stable self-organizing, high-mobility polythiophene derivatives was devised by Ong and coworkers [9]. This group recognized that the strategic placement of unsubstituted moieties along the polymer backbone and the resulting torsional deviations from coplanarity would reduce the effective π-conjugation length sufficiently to increase the ionization potential (and thus greatly improve oxidation resistance and environmental stability) while compromising the mobility only slightly, if at all. One particularly successful material which has emerged from this line of study is poly(3,3'''-didodecylquaterthiophene), better

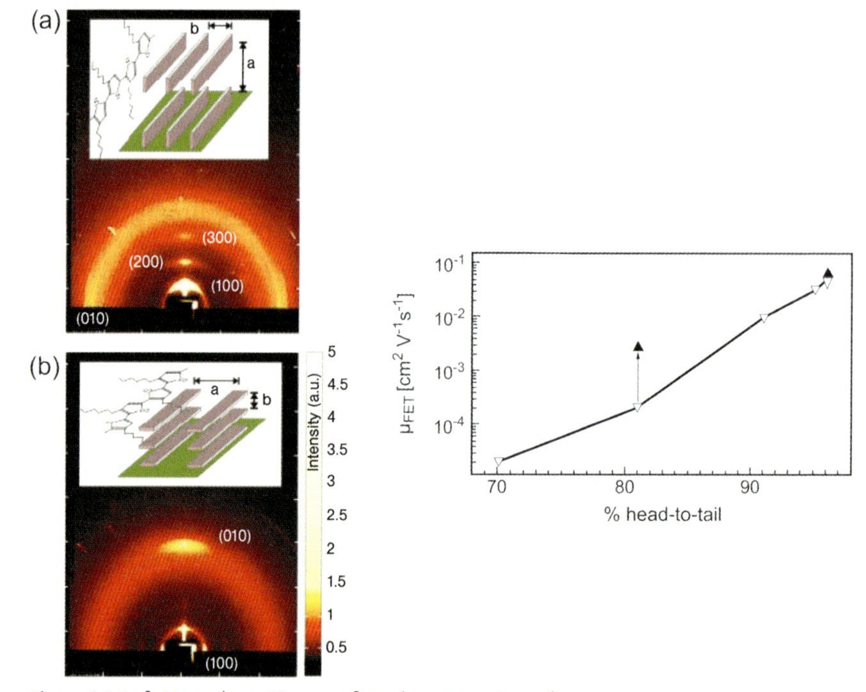

Figure 6.6 Left: Upon deposition on a flat substrate, regioregular poly(3-hexylthiophene) (P3HT) forms ordered lamellar domains, the orientation of which depends on the degree of regioregularity, molecular weight, and deposition conditions. Right: Relationship between the degree of regioregularity (quantified as the head-to-tail ratio) and the carrier mobility of P3HT transistors. (Reproduced with permission from Ref. [8].)

known as PQT-12 (see Figure 6.7). PQT-12 has shown air-stable carrier mobilities as large as $0.2\,cm^2\,V^{-1}\,s^{-1}$, and has been employed successfully in the fabrication of functional organic circuits and displays.

To further improve the performance and stability of alkyl-substituted polythiophenes, researchers at Merck Chemicals incorporated thieno[3,2-*b*]thiophene moieties into the polymer backbone [10] (see Figure 6.8). The effect of this is two-fold:

- The delocalization of carriers from the fused aromatic unit is less favorable than from a single thiophene unit, so the effective π-conjugation length is further reduced and the ionization potential becomes even larger than for polyquaterthiophene.
- The rotational invariance of the thieno[3,2-*b*]thiophene in the backbone promotes the formation of highly ordered crystalline domains with an extent not previously seen in semiconducting polymers. The molecular ordering is induced by annealing the spun-cast films in their liquid-crystalline phase and subsequent crystallization upon cooling. Carrier mobilities of $0.5\,cm^2\,V^{-1}\,s^{-1}$ have been reported.

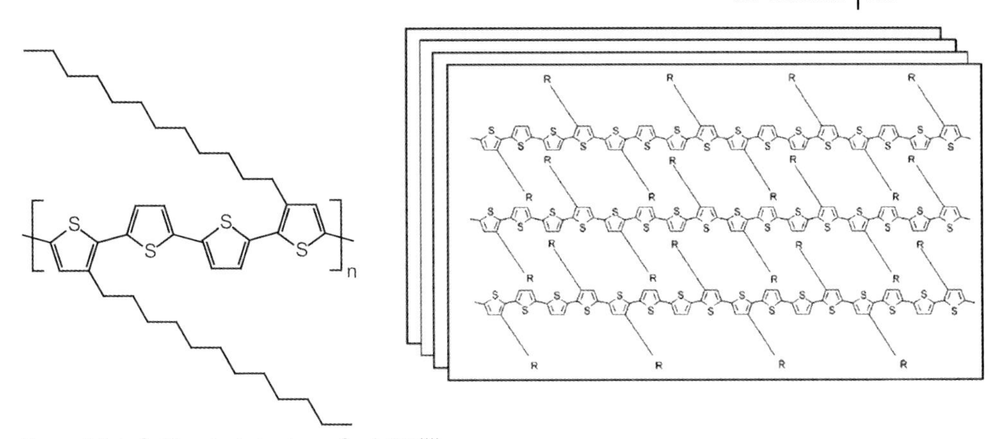

Figure 6.7 Left: Chemical structure of poly(3,3′′′-didodecylquaterthiophene) (PQT-12). Right: A schematic representation of the lamellar π-stacking arrangement. (Reproduced with permission from Ref. [9].)

Figure 6.8 The chemical structure of poly(2,5-bis(3-alkylthiophen-2-yl)thieno[3,2-*b*]thiophene (PBTTT).

Among the small-molecule organic semiconductors, the most widely studied systems include pentacene, sexithiophene, copper phthalocyanine, and the fullerene C_{60} (see Figure 6.9). Many small-molecule materials are insoluble in common organic solvents, but they often can be conveniently deposited using vacuum-deposition methods, such as thermal evaporation or organic vapor-phase deposition. In most cases, small-molecule organic semiconductors readily self-organize into semicrystalline films with a significant degree of molecular order (see Figure 6.10).

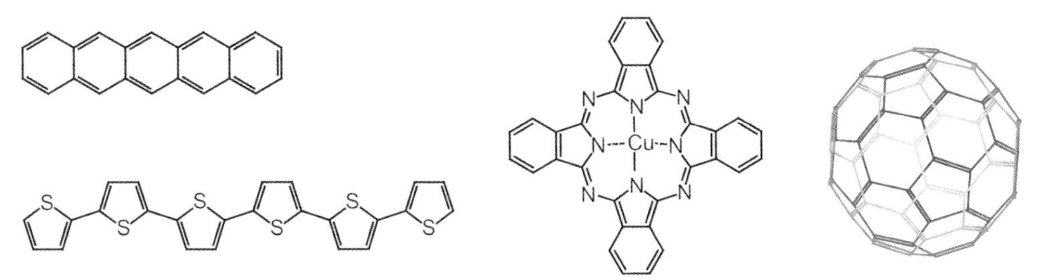

Figure 6.9 Chemical structures of pentacene (top left), sexithiophene (bottom left), copper phthalocyanine (center), and the fullerene C_{60} (right).

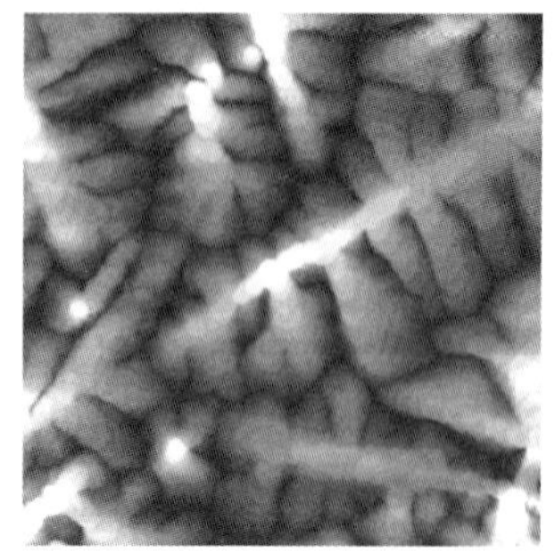

Figure 6.10 Pentacene self-organizes into an edge-to-face, or herringbone structure, forming semicrystalline films when deposited by evaporation onto amorphous substrates (left part of figure from: J. E. Anthony et al., Engineered pentacenes, in: *Organic Electronics*, Wiley-VCH, 2006.)

The use of vacuum-grown films of conjugated small-molecule materials for organic transistors was pioneered during the late 1980s by Madru and Clarisse (using metal phthalocyanines [11, 12]) and by Horowitz and Garnier (using oligothiophenes [13, 14]). Initial carrier mobilities were around $10^{-3}\,cm^2\,V^{-1}\,s^{-1}$, but these quickly improved to about $0.1\,cm^2\,V^{-1}\,s^{-1}$. In 1996, Jackson predicted and demonstrated that the carrier mobility of many organic semiconductors could be substantially improved by growing the films on low-energy surfaces [15, 16]. Inorganic dielectrics, such as silicon dioxide, are usually characterized by large surface energies favoring two-dimensional (2-D) growth of the first organic layer. Two-dimensional film growth typically results in large crystalline grains, and it was long believed that this was desirable to achieve good transistor performance. The surface energy of inorganic dielectrics is readily reduced by covering the surface with a self-assembled monolayer (SAM) of a methyl-terminated alkylsilane, such as octadecyltrichlorosilane (OTS). Organic film growth on low-energy SAM surfaces is distinctly three-dimensional, with much smaller grains and significantly more grain boundaries, yet the transistor mobilities were found to be significantly larger (by as much as an order of magnitude) compared with the large-grain films on high-energy surfaces. One explanation for this apparent discrepancy is that 2-D growth results in voids between disconnected grains, reducing the effective channel width of the transistor, and that such voids are efficiently filled when 3-D growth is favored [17, 18]. Carrier mobilities on high-energy surfaces (such as bare oxides) peak around $0.5\,cm^2\,V^{-1}\,s^{-1}$, while mobilities on low-energy surfaces (SAM-treated oxides or polymer dielectrics) have reached $1\,cm^2\,V^{-1}\,s^{-1}$ for alkyl-substituted oligothiophenes [19] and $5\,cm^2\,V^{-1}\,s^{-1}$ for pentacene [20–22].

An interesting alternative to solution-processed polymers and vacuum-grown small-molecules was developed by Herwig and Müllen during the early 1990s in the form of solution-deposited pentacene [23]. The initial rationale was to combine the best of two worlds – that is, the simplicity of solution-processing with the large carrier mobility of pentacene. Herwig and Müllen (and later Afzali and coworkers [24]; see Figure 6.11) synthesized a soluble pentacene precursor that was

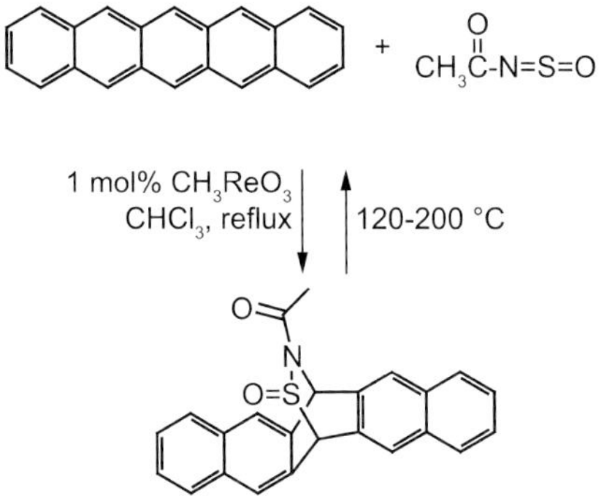

Figure 6.11 Synthesis of a soluble pentacene precursor and thermally induced conversion of the precursor to pentacene. (Reproduced with permission from Ref. [24].)

spin-coated and subsequently converted to pentacene at elevated temperature. Carrier mobilities for thermally converted pentacene are between 0.1 and $1\,\mathrm{cm^2\,V^{-1}\,s^{-1}}$, depending on the conversion temperature (130 to 200 °C).

The concept of solution-processable, high-mobility, small-molecule organic semiconductors was further developed by Anthony and colleagues, who designed and synthesized a number of soluble pentacene and anthradithiophene derivatives that do not require chemical conversion after deposition. Two examples – triisopropylsilyl (TIPS) pentacene and triethylsilyl (TES) anthradithiophene – are shown in Figure 6.12 [25]. In addition to providing solubility in common organic solvents, the functionalization of pentacene and anthradithiophene at the center rings can be utilized to strategically tune the molecular packing in the solid state in order to induce π-stacking with reduced intermolecular distances. With optimized solution-deposition, carrier mobilities as large as $3\,\mathrm{cm^2\,V^{-1}\,s^{-1}}$ have been achieved with these materials [26–28].

Organic transistors prepared with any of the semiconductors discussed so far operate efficiently only as p-channel transistors; that is, the drain currents in these

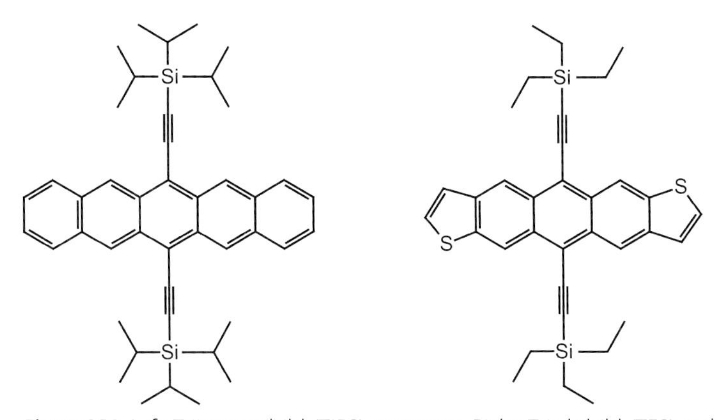

Figure 6.12 Left: Triisopropylsilyl (TIPS) pentacene. Right: Triethylsilyl (TES) anthradithiophene.

transistors are almost exclusively due to positively charged carriers. Currents due to negatively charged carriers are almost always extremely small in these materials, and often too small to be measurable, even when a large positive gate potential is applied to induce a channel of negatively charged carriers. Several explanations for the highly unbalanced currents have been suggested. One explanation is that the difference in mobility between the two carrier types is very large, due either to different scattering probabilities or perhaps to different probabilities for charge trapping, either at grain boundaries within the film or at defects at the dielectric interface [29, 30]. Another explanation is that charge injection from the contacts is highly unbalanced due to different energy barriers for positively and negatively charged carriers.

Interestingly, there are a number of organic semiconductors that show usefully large mobilities for negatively charged carriers. These materials include perfluorinated copper phthalocyanine ($F_{16}CuPc$), a variety of naphthalene and perylene tetracarboxylic diimide derivates, fluoroalkylated oligothiophenes, and the fullerene C_{60}. The carrier mobilities measured in n-channel FETs based on these materials are in the range of 0.01 to 1 $cm^2 V^{-1} s^{-1}$. Some of these materials are very susceptible to redox reactions and thus have poor environmental stability. For example, C_{60} shows mobilities as large as 5 $cm^2 V^{-1} s^{-1}$ when measured in ultra-high vacuum [31, 32], but when exposed to air the mobility drops rapidly by as many as four or five orders of magnitude. Similar degradation has been reported for some naphthalene and perylene tetracarboxylic diimide derivatives [33, 34]. Other materials, in particular $F_{16}CuPc$ [35], have been found to be very stable in air, although the exact mechanisms that determine the degree of air stability are still unclear. Air-stable n-channel organic FETs with carrier mobilities as large as 0.6 $cm^2 V^{-1} s^{-1}$ [36] have been reported.

6.3
Device Structures and Manufacturing

From a technological perspective, the most useful organic transistor implementation is the thin-film transistor (TFT). The TFT concept was initially proposed and developed by Weimer during the 1960s for transistors based on polycrystalline inorganic semiconductors, such as evaporated cadmium sulfide [37]. The idea was later extended to TFTs based on plasma-enhanced chemical-vapor deposited (PECVD) hydrogenated amorphous silicon (a-Si:H) TFTs [38]. Today, a-Si:H TFTs are widely employed as the pixel drive devices in active-matrix liquid-crystal displays (AMLCDs) which accounted for $50 billion in global sales in 2005. Organic TFTs were first reported during the 1980s [4, 39]. To produce an organic TFT, the organic semiconductor and other materials required (gate electrode, gate dielectric, source and drain contacts) are deposited as thin layers on the surface of an electrically insulating substrate, such as glass or plastic foil. Depending on the sequence in which the materials are deposited, three organic TFT architectures can be distinguished, namely inverted staggered, inverted coplanar, and top-gate (see Figure 6.13).

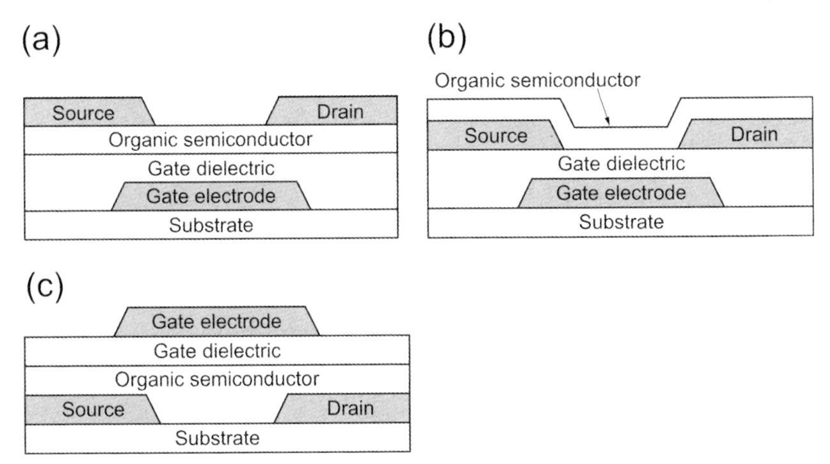

Figure 6.13 Schematic cross-sections of the (a) inverted staggered, (b) inverted coplanar, and (c) top-gate organic TFT structures.

Each of these structures has certain advantages and disadvantages:

- The inverted coplanar architecture allows for the use of photolithography or solution-based printing to pattern source and drain contacts with high resolution and short contact spacing, without exposing the organic semiconductor to potentially harmful process chemicals such as organic solvents, photoresists, and etchants [40]. However, as the contacts in organic transistors are not easily doped, the coplanar structure is often associated with larger contact resistance compared with the staggered architecture [41].

- The inverted staggered TFT structure is usually implemented by evaporating the source/drain metal through a shadow mask (also called an aperture or stencil mask). In this way, the organic semiconductor is not exposed to process chemicals, and the contact resistance can be quite low, as the entire area underneath the contacts is available for charge injection, though with shadow masks it is more difficult to reduce the channel length reliably below about 10 μm.

- For the top-gate structure the source and drain contacts can be patterned with high resolution prior to the deposition of the semiconductor, and the contact resistance can be as low as for the inverted staggered architecture. However, the deposition of a gate dielectric and a gate electrode on top of the semiconductor layer means that great care must be exercised to avoid process-induced degradation of the organic semiconductor, and the possibility of material mixing at the semiconductor/dielectric interface must be taken into account.

A variety of methods exists for the deposition and patterning of the individual layers of the TFT. For example, gate electrodes and source and drain contacts are often made using vacuum-deposited inorganic metals. Non-noble metals, such as aluminum or chromium, are suitable for the gate electrodes in the inverted device

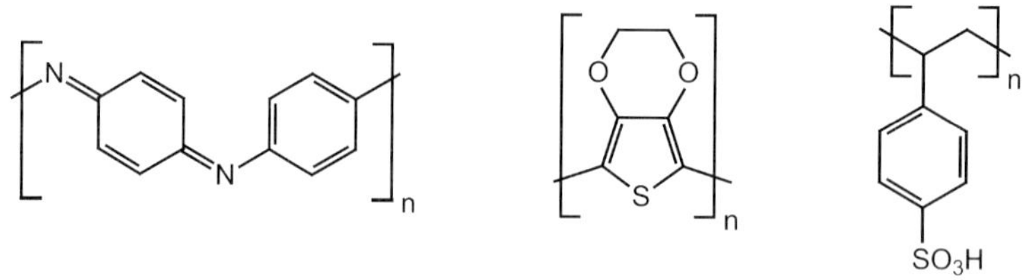

Figure 6.14 Left: Polyaniline. Right: Poly(3,4-ethylenedioxythiophene) (PEDOT) and poly (styrene sulfonic acid) PSS.

structures, as these metals have excellent adhesion on glass substrates. Noble metals (most notably gold) are a popular choice for the source and drain contacts, as they tend to give lower contact resistance than other metals. The metals are conveniently deposited by evaporation in vacuum and can be patterned either by photolithography and etching or lift-off, by lithography using an inkjet-printed wax-based etch resist [42], or simply by deposition through a shadow mask [43].

An alternative to inorganic metals are conducting polymers, such as polyaniline and poly(3,4-ethylenedioxythiophene):poly(styrene sulfonic acid) (PEDOT:PSS; see Figure 6.14). These are chemically doped conjugated polymers that have electrical conductance in the range between 0.1 and 1000 S cm^{-1}. The way in which continuous advances in synthesis and material processing have improved the conductivity of Baytron® PEDOT:PSS over the past decade is shown in Figure 6.15. Unlike inorganic metals, conducting polymers can be processed either from organic solutions or from aqueous dispersions, so gate electrodes and source and drain contacts for organic TFTs can be prepared by spin-coating and photolithography [44], or by inkjet-printing [45].

One important aspect in organic TFT manufacturing is the choice of the gate dielectric. Depending on the device architecture (inverted or top-gate), the dielectric material and the processing conditions (temperature, plasma, organic solvents, etc.)

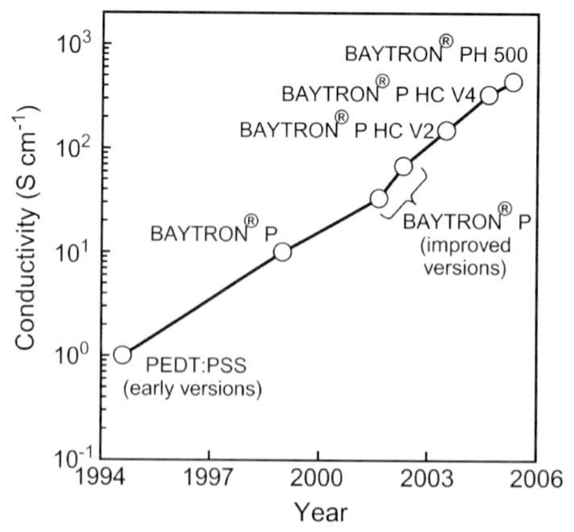

Figure 6.15 Improvements in the electrical conductivity of the conducting polymer PEDOT:PSS. (Adapted from a graph kindly provided by Stephan Kirchmeyer, H. C. Starck, Leverkusen, Germany.)

must be compatible with the previously deposited device layers and with the substrate. For example, chemical-vapor-deposited (CVD) silicon oxide and silicon nitride – which are popular gate dielectric materials for inorganic (amorphous or polycrystalline silicon) TFTs – may not be suitable for use on flexible polymeric substrates, as the high-quality growth of these films often requires temperatures that exceed the glass transition temperature of many polymeric substrate materials.

The thickness of the gate dielectric layer is usually a compromise between the requirements for large gate coupling, low operating voltages, and small leakage currents. Large gate coupling (i.e. a large dielectric capacitance) means that the transistors can be operated with low voltages, which is important when the TFTs are used in portable or handheld devices that are powered by small batteries or by near-field radio-frequency coupling, for example. Also, a large dielectric capacitance ensures that the carrier density in the channel is controlled by the gate and not by the drain potential, which is especially critical for short-channel TFTs. One way to increase the gate dielectric capacitance is to employ a dielectric with larger permittivity ε ($C = \varepsilon/t$). However, as Veres [46], Stassen [47], and Hulea [48] have pointed out, the carrier mobility in organic field effect transistors is systematically reduced as the permittivity of the gate dielectric is increased, presumably due to enhanced localization of carriers by local polarization effects (see Figure 6.16).

Alternatively, low-permittivity dielectrics with reduced thickness or thin multilayer dielectrics with specifically tailored properties may be employed. The greatest concern with thin dielectrics is the inevitable increase in gate leakage due to defects and quantum-mechanical tunneling as the dielectric thickness is reduced. A number of promising paths towards high-quality thin dielectrics with low gate leakage for low-voltage organic TFTs have recently emerged. One such approach is the anodization of aluminum, which has resulted in high-quality aluminum oxide films thinner than 10 nm providing a capacitance around 0.4 μF cm^{-2}. Combined with an ultra-thin molecular SAM, such dielectrics can provide sufficiently low leakage currents to allow the fabrication of functional low-voltage organic TFTs with large carrier

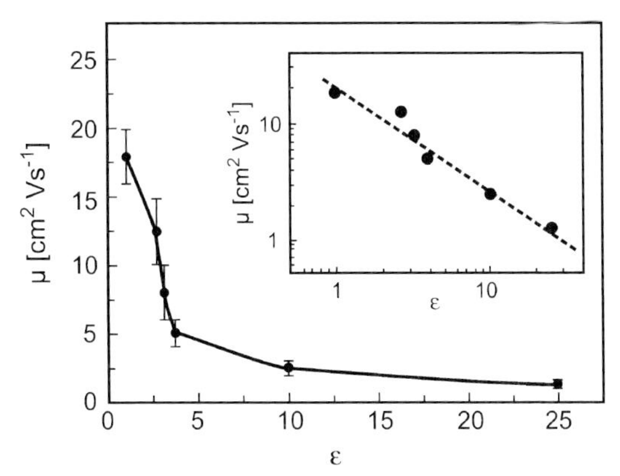

Figure 6.16 Relationship between the permittivity of the gate dielectric and the carrier mobility in the channel. (Reproduced with permission from Ref. [47].)

mobility [49]. Another path is the use of very thin crosslinked polymer films prepared by spin-coating [50, 51]. With a thickness of about 10 nm, these dielectrics provide capacitances as large as 0.3 μF cm^{-2} and excellent low-voltage TFT characteristics. Finally, the use of high-quality insulating organic SAMs or multilayers provides a promising alternative [52–54]. Such molecular dielectrics typically have a thickness of 2 to 5 nm and a capacitance between 0.3 and 0.7 μF cm^{-2}, depending on the number and structure of the molecular layers employed, and they allow organic TFTs to operate with voltages between 1 and 3 V.

6.4
Electrical Characteristics

Like silicon metal-oxide-semiconductor-field-effect-transistors (MOSFETs), organic TFTs are metal-insulator-semiconductor FETs in which a sheet of mobile charge carriers is induced in the semiconductor by applying an electric field across the gate dielectric. Silicon MOSFETs normally operate in inversion – that is, the drain current is due to minority carriers generated by inverting the conductivity at the semiconductor/dielectric interface from p-type to n-type (for n-channel MOSFETs) or from n-type to p-type (for p-channel MOSFETs). The contact regions of silicon MOSFETs are heavily doped so that minority carriers are easily injected at the source and extracted at the drain, while the undesirable flow of majority carriers from drain to source is efficiently blocked by a space charge region.

Organic TFTs typically utilize intrinsic semiconductors. Positively charged carriers are accumulated near the dielectric interface when a negative gate-source voltage is applied, or negative charges are accumulated with a positive gate bias. Source and drain are usually implemented by directly contacting the intrinsic semiconductor with a metal. Depending on the choice of materials for the semiconductor and the metal, organic TFTs may operate as p-channel, n-channel, or ambipolar transistors. In p-channel and n-channel TFTs, the transport of one type of carrier is far more efficient than that of the other carrier type, either because the semiconductor/metal contacts greatly favor the injection or extraction of one carrier type over the other, or because the mobilities in the semiconductor are very different (perhaps due to different trapping rates), or because the electronic properties of the semiconductor/dielectric interface allow the accumulation of only one type of carrier, but not the other. In ambipolar organic TFTs the injection and transport of both positive and negative carriers is possible and both carrier types contribute to the drain current.

Despite the fact that the transport physics in organic transistors is different from that in silicon MOSFETs, the current–voltage characteristics can often be described with the same formalism:

$$I_{\mathrm{D}} = \frac{\mu C_{\mathrm{diel}} W}{L}\left((V_{\mathrm{GS}} - V_{\mathrm{th}})V_{\mathrm{DS}} - \frac{V_{\mathrm{DS}}^2}{2}\right) \quad \text{for } |V_{\mathrm{GS}} - V_{\mathrm{th}}| > |V_{\mathrm{DS}}| \tag{6.1}$$

$$I_{\mathrm{D}} = \frac{\mu C_{\mathrm{diel}} W}{2L}(V_{\mathrm{GS}} - V_{\mathrm{th}})^2 \quad \text{for } |V_{\mathrm{DS}}| > |V_{\mathrm{GS}} - V_{\mathrm{th}}| > 0 \tag{6.2}$$

Equation (6.1) describes the relationship between the drain current I_D, the gate-source voltage V_{GS} and the drain-source voltage V_{DS} in the linear regime, while Eq. (6.2) relates I_D, V_{GS} and V_{DS} in the saturation regime. C_{diel} is the gate dielectric capacitance per unit area, μ is the carrier mobility, W is the channel width, and L is the channel length of the transistor. For silicon MOSFETs, the threshold voltage V_{th} is defined as the minimum gate-source voltage required to induce strong inversion. Although this definition cannot strictly be applied to organic TFTs, the concept is nonetheless useful, as the threshold voltage conveniently marks the transition between the different regions of operation.

Figure 6.17 shows the current–voltage characteristics of an organic TFT that was manufactured on a glass substrate using the inverted staggered device structure (see Figure 6.13a) with a thin layer of vacuum-evaporated pentacene as the semiconductor, a self-assembled monolayer gate dielectric, and source/drain contacts prepared by evaporating gold through a shadow mask [54]. The device operates as a p-channel transistor with a threshold voltage of -1.2 V. By fitting the current–voltage characteristics to Eqs. (6.1) or (6.2), the carrier mobility μ can be estimated; for this particular device it is about $0.6\,\mathrm{cm^2\,V^{-1}\,s^{-1}}$ ($C_{diel} = 0.7\,\mu\mathrm{F\,cm^{-2}}$, $W = 100\,\mu\mathrm{m}$, $L = 30\,\mu\mathrm{m}$).

Equations (6.1) and (6.2) describe the drain current for gate-source voltages above the threshold voltage. Below the threshold voltage there is a region in which the drain current depends exponentially on the gate-source voltage. This is the *subthreshold region*; for the TFT in Figure 6.17 it extends between about -0.5 V and about -1 V. Within this voltage range the drain current is due to carriers that have sufficient thermal energy to overcome the gate-controlled barrier near the source and mainly diffuse, rather than drift, to the drain:

$$I_D = I_0 \exp\left(\frac{q|V_{GS} - V_{th}|}{nkT}\right) \quad \text{for } V_{GS} \text{ between } V_{th} \text{ and } V_{SO} \tag{6.3}$$

The slope of the $\log(I_D)$ versus V_{GS} curve in the subthreshold region is determined by the ideality factor n and the temperature T (q is the electronic charge, k is

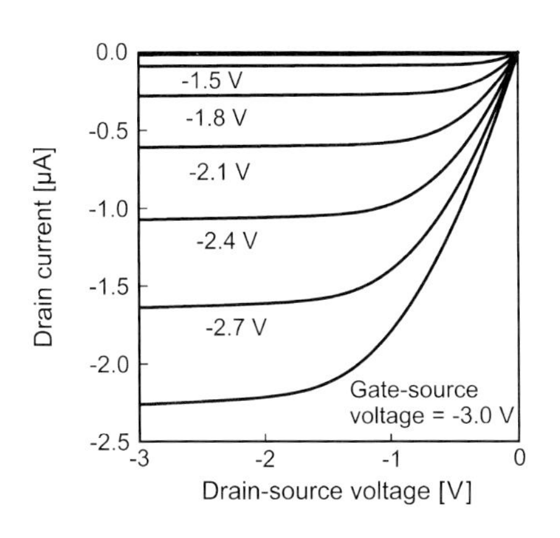

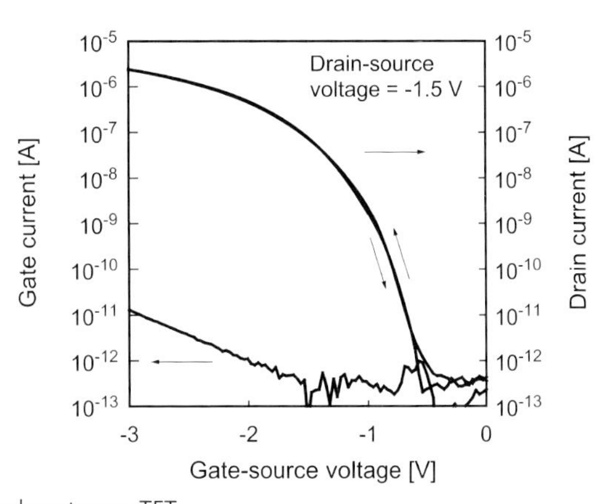

Figure 6.17 The electrical characteristics of a p-channel pentacene TFT.

Boltzmann's constant, and V_{SO} is the switch-on voltage which marks the gate-source voltage at which the drain current reaches a minimum [55]). It is usually quantified as the inverse subthreshold slope S (also called subthreshold swing):

$$S = \frac{\partial V_{GS}}{\partial(\log_{10} I_D)} = \frac{nkT}{q} \ln 10 \tag{6.4}$$

The ideality factor n is determined by the density of trap states at the semiconductor/dielectric interface, N_{it}, and the gate dielectric capacitance, C_{diel}:

$$n = 1 + \frac{qN_{it}}{C_{diel}} \tag{6.5}$$

$$S = \frac{kT}{q} \ln 10 \left(1 + \frac{qN_{it}}{C_{diel}}\right) \tag{6.6}$$

When N_{it}/C_{diel} is small, the ideality factor n approaches unity. Silicon MOSFETs often come close to the ideal room-temperature subthreshold swing of 60 mV per decade, as the quality of the Si/SiO_2 interface is very high. In organic TFTs the semiconductor/dielectric interface is typically of lower quality, and thus the subthreshold swing is usually larger. The TFT in Figure 6.17 has a subthreshold swing of 100 mV dec^{-1}, from which an interface trap density of 3×10^{12} cm^{-2} V^{-1} is calculated.

The subthreshold region extends between the threshold voltage V_{th} and the switch-on voltage V_{SO}. Below the switch-on voltage (-0.5 V for the TFT in Figure 6.17) the drain current is limited by leakage through the semiconductor, through the gate dielectric, or across the substrate surface. This off-state current should be as small as possible. The TFT in Figure 6.17 has an off-state current of 0.5 pA, which corresponds to a on off-state resistance of 3 TΩ.

To predict an upper limit for the dynamic performance of the transistor, it is useful to calculate the cut-off frequency [56]. This is the frequency at which the current gain is unity, and is determined by the transconductance and the gate capacitance:

$$f_T = \frac{g_m}{2\pi C_{gate}} \tag{6.7}$$

The transconductance g_m is defined as the change in drain current with respect to the corresponding change in gate-source voltage:

$$g_m = \frac{\partial I_D}{\partial V_{GS}} \tag{6.8}$$

Thus, the transconductance can be extracted from the current–voltage characteristics; for the pentacene TFT in Figure 6.17 the transconductance is about 2 µS at $V_{GS} = -2.5$ V. The transconductance is related to the other transistor parameters as follows:

$$g_m = \frac{\mu C_{diel} W}{L} V_{DS} \quad \text{in the linear regime} \tag{6.9}$$

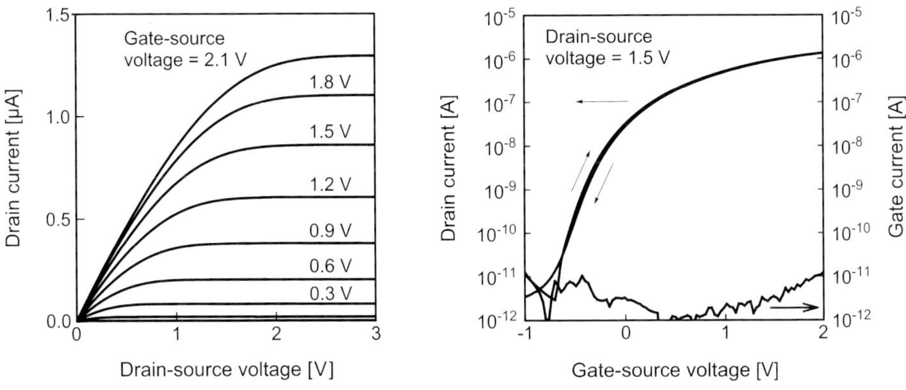

Figure 6.18 The electrical characteristics of an n-channel $F_{16}CuPc$ TFT.

$$g_{m} = \frac{\mu C_{diel} W}{L}(V_{GS} - V_{th}) \quad \text{in the saturation regime} \tag{6.10}$$

The gate capacitance C_{gate} is the sum of the intrinsic gate capacitance (representing the interaction between the gate and the channel charge) and the parasitic gate capacitances (including the gate/source overlap capacitances and any fringing capacitances). For the transistor in Figure 6.17 the total gate capacitance is estimated to be about 30 pF, so the cut-off frequency will be on the order of 10 kHz.

The current–voltage characteristics of an n-channel TFT based on perfluorinated copper phthalocyanine ($F_{16}CuPc$) are shown in Figure 6.18 [54]. It has a threshold voltage of -0.2 V, a switch-on voltage of -0.8 V, an off-state current of about 5 pA, and a carrier mobility of about 0.02 $cm^2 V^{-1} s^{-1}$ ($C_{diel} = 0.7\,\mu F\,cm^{-2}$, $W = 1000$ μm, $L = 30$ μm). The transconductance is about 0.8 μS at $V_{GS} = 1.5$ V and the gate capacitance is about 300 pF, so the TFT will have a cut-off frequency of about 500 Hz.

The availability of both p-channel and n-channel organic TFTs makes the implementation of organic complementary circuits possible. From a circuit design perspective, complementary circuits are more desirable than circuits based only on one type of transistor, as complementary circuits have smaller static power dissipation and greater noise margin [54]. The schematic and electrical characteristics of an organic complementary inverter with a p-channel pentacene TFT and an n-channel $F_{16}CuPc$ TFT are shown in Figure 6.19.

In a complementary inverter the p-channel FET is conducting only when the input is low ($V_{in} = 0$ V), while the n-channel FET is conducting only when the input is high ($V_{in} = V_{DD}$). Consequently, the static current in a complementary circuit is essentially determined by the leakage currents of the transistors and can be very small (less than 100 pA for the inverter in Figure 6.19). As a result, the output signals in the steady states are essentially equal to the rail voltages V_{DD} and ground. During switching there is a brief period when both transistors are simultaneously in the low-resistance on-state and a significant current flows between the V_{DD} and ground rails. Thus, most of the power consumption of a complementary circuit is due to switching, while the static power dissipation is very small.

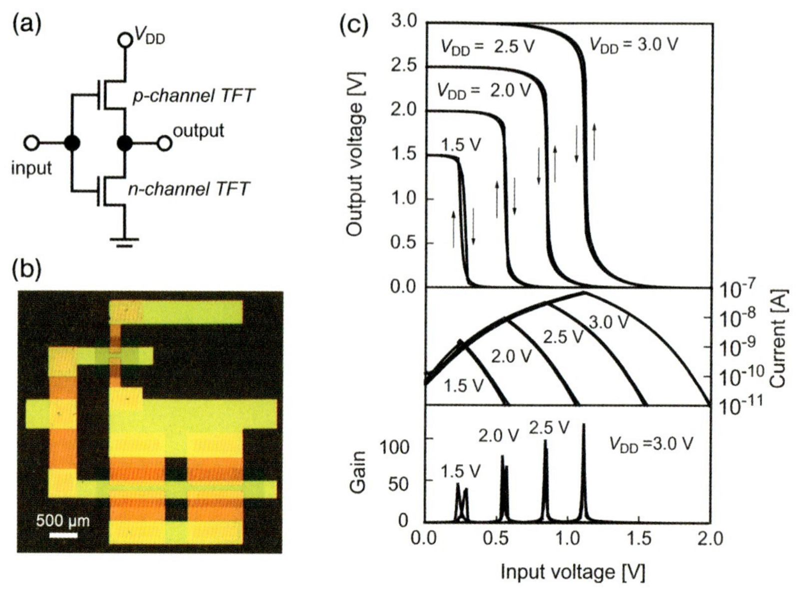

Figure 6.19 (a) Schematic, (b) actual photographic image and (c) transfer characteristics of an organic complementary inverter based on a p-channel pentacene TFT and an n-channel $F_{16}CuPc$ TFT.

The dynamic performance of the inverter is limited by the slower of the two transistors, in this case the n-channel $F_{16}CuPc$ TFT. Figure 6.20 shows, in graphical from, the inverter's response to a square-wave input signal with an amplitude of 2 V and a frequency of 500 Hz – that is, the cut-off frequency of the $F_{16}CuPc$ TFT.

To allow organic circuits to operate at higher frequencies, it is necessary to increase the transconductance and reduce the parasitic capacitances. From a materials point of view, this can be done by developing new organic semiconductors that provide larger carrier mobilities [36]. Ideally, the carrier mobilities of the p- and n-channel TFTs should be similar. From a manufacturing point of view, the critical dimensions of the devices must be reduced – that is, the channel length and overlap capacitances must be made smaller. However, as the channel length of organic TFTs is reduced, the

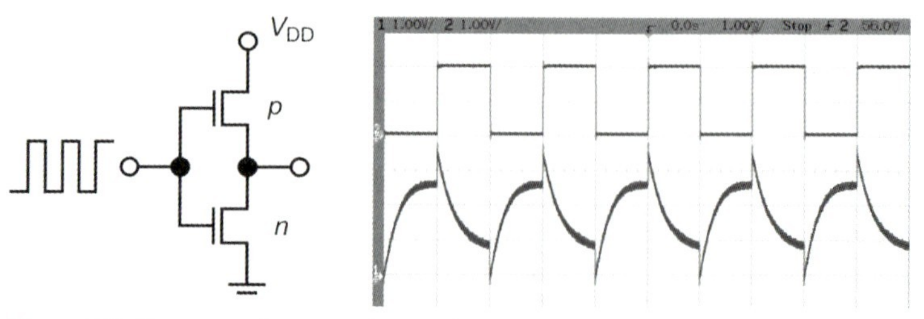

Figure 6.20 Response of an organic complementary inverter to a square-wave input signal with an amplitude of 2 V and a frequency of 500 Hz. Both TFTs have a channel length of 30 µm.

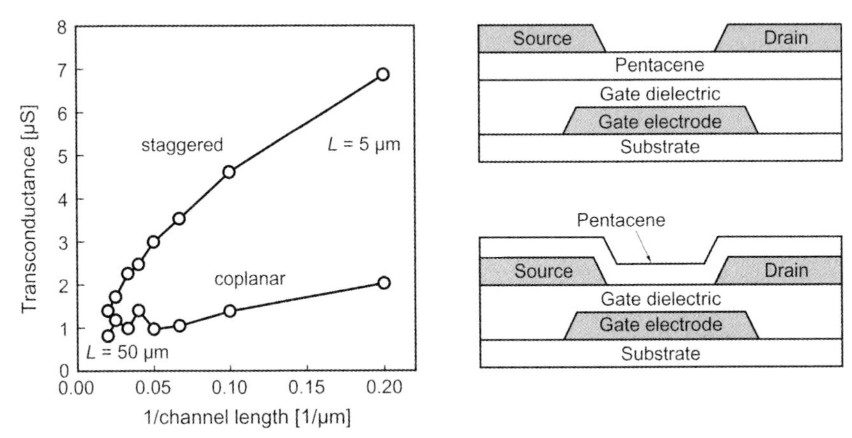

Figure 6.21 Transconductance as a function of channel length for two series of pentacene TFTs (top: inverted staggered configuration; bottom: inverted coplanar configuration). The channel width is 100 μm.

transconductance does not necessarily scale as predicted by Eqs. (6.9) and (6.10). The main reason for this is that the contact resistance in organic TFTs can be very large, as the contacts are typically not doped. Consequently, as the channel length is reduced, the drain current becomes increasingly limited by the contact resistance (which is independent of channel length), rather than by the channel resistance. This is shown in Figure 6.21 for pentacene TFTs in the inverted staggered configuration and in the inverted coplanar configuration, both with channel length ranging from 50 μm to 5 μm and all with a channel width of 100 μm.

For long channels ($L = 50\,\mu m$), where the effect of the contact resistance on the TFT characteristics is small, the transconductance is similar for both technologies ($g_m \sim 1\,\mu S$; $\mu \sim 0.5\,cm^2\,V^{-1}\,s^{-1}$). The difference between the two devices configurations becomes evident when the channel length is reduced. For the coplanar TFTs the potential benefit of channel length scaling is largely lost due to the significant contact resistance ($\sim 5 \times 10^4\,\Omega\cdot cm$). The staggered configuration offers significantly smaller contact resistance ($\sim 10^3\,\Omega\cdot cm$), as the area available for charge injection from the metal into the carrier channel is larger (given by the gate/contact overlap area), and as a result the transconductance for short channels (5 μm) is significantly larger in the case of the staggered TFTs (7 μS versus 2 μS). The staggered TFT with a channel length of 5 μm and a transconductance of 7 μS has a total gate capacitance of about 5 pF, so the cut-off frequency is estimated to be on the order of 200 kHz (at an operating voltage in the range of 2–3 V).

6.5 Applications

Unlike single-crystal silicon transistors, organic TFTs can be readily fabricated on glass or flexible plastic substrates, and this makes them useful for a variety of

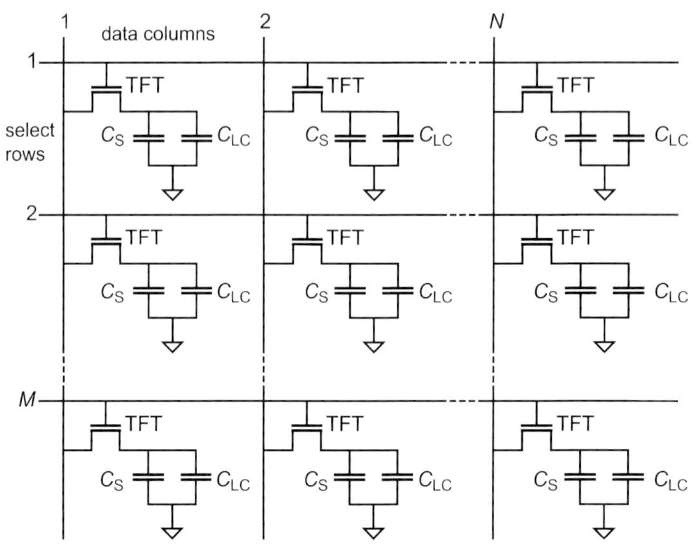

Figure 6.22 Schematic of an active-matrix liquid-crystal display.

large-area electronic applications, such as active-matrix flat-panel displays. In an active-matrix display, each of the pixels is individually controlled (and electrically isolated from the rest of the display) by a TFT circuit in order to reduce undesirable cross-talk and to increase fidelity and color depth. In active-matrix displays that utilize voltage-controlled display elements, such as a liquid crystal or an electrophoretic cell, each pixel circuit consists simply of a single TFT; in this case the display matrix has as many TFTs as it has pixels. If the display employs a current-controlled electro-optical device, such as a light-emitting diode, a more complex TFT circuit with two or more TFTs must be implemented in each pixel.

Figure 6.22 shows the circuit schematic of an active-matrix liquid-crystal display (AMLCD). The display is operated by applying a select voltage to one of the rows in order to switch all TFTs in that row to the low-resistance on-state. (All other rows are held at a lower potential that keeps the TFTs in these rows in the high-resistance off-state.) Data voltages that correspond to the desired brightness levels for each of the pixels in the selected row are then applied to each of the N columns. This charges the capacitors in the selected row to the applied data voltage. The time required to charge the capacitors (t_{select}) is determined by the on-state resistance of the TFTs, by the capacitances of the storage capacitor (C_S), the liquid crystal cell (C_{LC}) and the data lines, and by the maximum allowed deviation from the target voltage. Once the select voltage is removed, the capacitors are isolated and the charge is retained in the pixels. In this manner all rows are addressed one by one, and all pixel capacitors are charged to the desired voltage. The time required to sequentially address all M rows, and thus update the entire display, is the frame time, $t_{\text{frame}} = M \cdot t_{\text{select}}$.

In order to avoid visible flicker, the display information must be updated at least 50 times per second – that is, t_{frame} must be about 20 ms, or less. If a pixel capacitance ($C_S + C_{LC}$) of 1 pF is assumed, and if no more than 1% of the stored charge is allowed to leak from the pixel during the t_{frame}, then the minimum required TFT off-state resistance can be estimated:

$$R_{off} \geq \frac{t_{frame}}{0.01 C_{pixel}} \tag{6.11}$$

Thus, for a t_{frame} of 20 ms and a pixel capacitance of 1 pF, the TFTs must have an off-state resistance of 2 TΩ, or greater.

For an extended graphics array (XGA) display with 768 rows and a t_{frame} of 20 ms the time available to charge the capacitors in one row is $t_{select} = t_{frame}/M = 26\,\mu s$. If a combined (pixel plus data line) capacitance of 2 pF is assumed, and it is specified that the capacitors be charged to within 1% of the target data voltage, then the maximum allowed TFT on-state resistance can be estimated:

$$R_{on} \leq \frac{t_{select}}{4.6 C_{pixel}} \tag{6.12}$$

For a t_{select} of 26 μs and a capacitance of 2 pF this sets an upper limit of about 2 MΩ for the on-state resistance of the TFTs. In order to create a small on-state resistance the TFTs are operated in the linear regime by applying a select voltage that is larger than the largest data voltage (plus the threshold voltage). In the linear regime (when the gate-source voltage is much larger than the drain-source voltage) the channel resistance is approximately given by:

$$R_{on} \sim \frac{L}{\mu C_{diel} W (V_{GS} - V_{th})} \tag{6.13}$$

Assuming a carrier mobility of $0.5\,cm^2\,V^{-1}\,s^{-1}$, a gate dielectric capacitance of $0.1\,\mu F\,cm^{-2}$, and an overdrive voltage $|V_{GS} - V_{th}|$ of 10 V, the on-state resistance requirement (2 MΩ) can be met with a TFT geometry of $W/L = 1$ (where W and L are the channel width and channel length of the transistor, respectively). A W/L ratio near or equal to unity is desirable, as this means that the transistor occupies a relatively small fraction of the total pixel area. Taking into account both the off-state and on-state resistance requirements ($R_{off} > 2$ TΩ, $R_{on} < 2$ MΩ), the TFTs must have an on/off ratio of at least 10^6.

These requirements can be met by state-of-the-art organic TFTs. An early demonstration of an active-matrix polymer-dispersed liquid-crystal (PDLC) display with solution-processed polythienylenevinylene (PTV) TFTs was developed by Philips Research in 2001 [57]. In 2005, Sony reported an active-matrix twisted-nematic liquid-crystal (TN-LC) display with vacuum-deposited pentacene TFTs [58]. Also in 2005, Polymer Vision demonstrated a flexible roll-up display based on electrophoretic microcapsules (electronic ink) and solution-processed pentacene TFTs (see Figure 6.23).

Unlike liquid-crystal valves and electrophoretic microcapsules, light-emitting diodes (OLEDs) are current-controlled display elements and thus require a more complex pixel circuit. The simplest implementation of an active-matrix organic light-emitting diode (AMOLED) pixel is shown in Figure 6.24. When a select voltage is

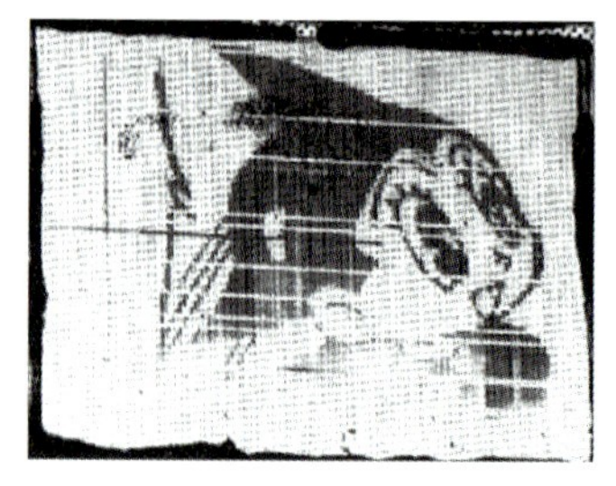

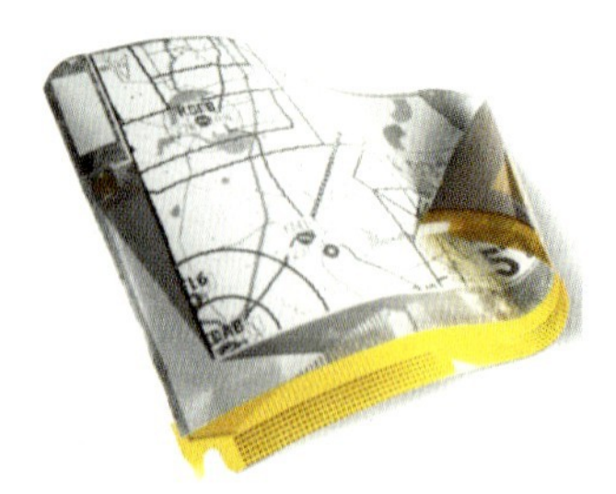

Figure 6.23 Left: A 64 × 64 pixel active-matrix polymer-dispersed liquid-crystal display with solution-processed polymer TFTs developed by Philips. (Reproduced with permission from Ref. [57].) Center: A 160 × 120 pixel active-matrix twisted-nematic liquid-crystal display with vacuum-deposited pentacene TFTs developed by Sony. (Reproduced with permission from Ref. [58].) Right: A 320 × 240 pixel active-matrix electronic-ink display with solution-processed pentacene TFTs developed by Polymer Vision. (Reproduced from: H. E. A. Huitema et al., Roll-up Active-matrix Displays, in: *Organic Electronics*, Wiley-VCH, 2006.)

applied, transistor T_1 switches to the low-resistance on-state so that capacitor C_S can be charged through the data line to a voltage corresponding to the desired luminous intensity. The voltage across C_S is the gate-source voltage of transistor T_2, and thus determines the drain current of T_2 and thereby the luminance of the OLED. When the select voltage is removed, T_1 switches off and the charge is retained on C_S, so T_2 remains active and drives a constant OLED current for the remainder of the frame time.

The on-state and off-state resistance requirements for the select transistor T_1 in an OLED pixel are similar to those for the TFT in a liquid-crystal or electrophoretic pixel – that is, they can be met by a TFT with $W/L \sim 1$. The drive transistor T_2 in an OLED pixel is usually operated in saturation, and must have a sufficiently large drain current to drive the OLED to the desired brightness. State-of-the-art small-molecule OLEDs have luminous efficiencies on the order of 2 to 50 cd A^{-1}, depending on emission color, material selection, and process technology [59]. For a typical display brightness of 100 cd m^{-2} and a pixel size of 6×10^{-4} cm^2 (which corresponds to a resolution of 100 dpi), this requires a maximum drive current up to about 3 μA. In the saturation regime the drain current of the transistor is given by Eq. (6.2). Assuming a carrier mobility of 0.5 $cm^2 V^{-1} s^{-1}$, a gate dielectric capacitance of 0.1 μF cm^{-2}, and an overdrive voltage $|V_{GS} - V_{th}|$ of 5 V, the drive current requirement (3 μA) can be met

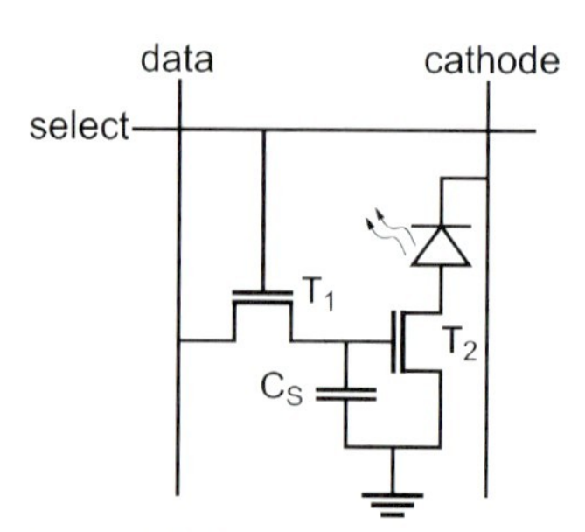

Figure 6.24 Schematic of a two-transistor active-matrix OLED pixel.

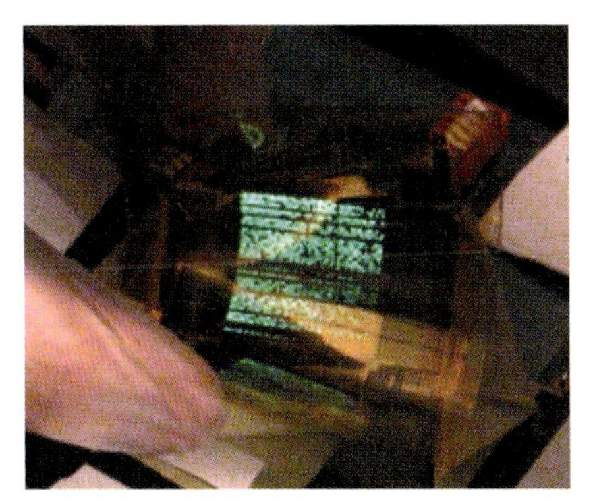

Figure 6.25 A flexible 48 × 48 pixel active-matrix organic light-emitting diode (OLED) display with pentacene organic TFTs developed at Penn State University. (Reproduced with permission from Ref. [60].)

with a TFT geometry of $W/L = 5$. Thus, the static transistor performance requirements for active-matrix OLED displays can be met by organic TFTs T_1 and T_2 occupying only a small fraction of the pixel area. A photograph of an active-matrix OLED display with two pentacene TFTs and a bottom-emitting OLED in each pixel [60] is shown in Figure 6.25.

Compared with liquid-crystal and electrophoretic displays, active-matrix OLED displays are far more demanding as far as the uniformity and stability of the TFT parameters are concerned. For example, if the TFT threshold voltage in a liquid-crystal display changes over time, or is not uniform across the display, the image quality is not immediately affected as the select voltage is usually large. In an OLED display, however, the threshold voltage of transistor T_2 directly determines the drive current and thus the OLED brightness. Consequently, even small differences in threshold voltage have a dramatic impact on image quality and color fidelity. In order to reduce or eliminate the effects of non-uniformities or time-dependent changes of the TFT parameters, more complex pixel circuit designs have been proposed [61]. In these designs, additional TFTs are implemented to make the OLED current independent of the threshold voltages of the TFTs. A pixel circuit with a larger number of TFTs is likely to occupy a greater portion of the total pixel area, but may significantly improve the performance of the display.

A second potential application for organic TFTs is in large-area sensors for the spatially resolved detection of physical or chemical quantities, such as temperature, pressure, radiation, or pH. As an example, Figure 6.26 shows the schematic of an active-matrix pressure sensor array. Mechanical pressure exerted on a sensor element leads to a reversible and reproducible change in the resistance of the sensor element. To allow external circuitry to access the resistance of each individual sensor it is necessary to integrate a transistor with each sensor element. During operation, the rows of the array are selected one by one to switch the TFTs in the selected row to the low-resistance state (similar to the row-select procedure in an active-matrix display) and the resistance of the each sensor element in the selected row is measured

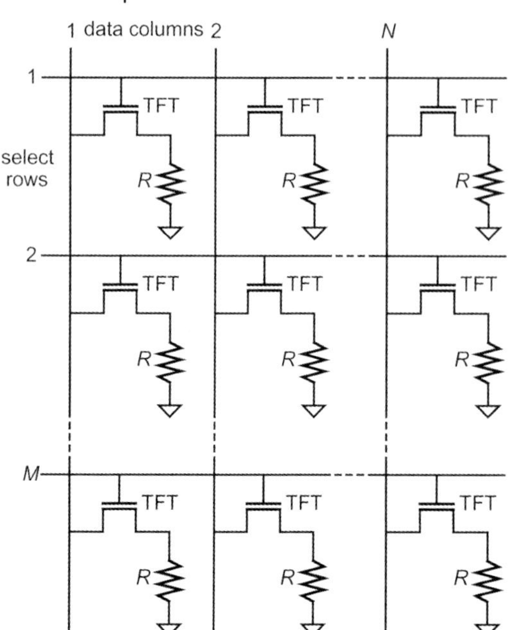

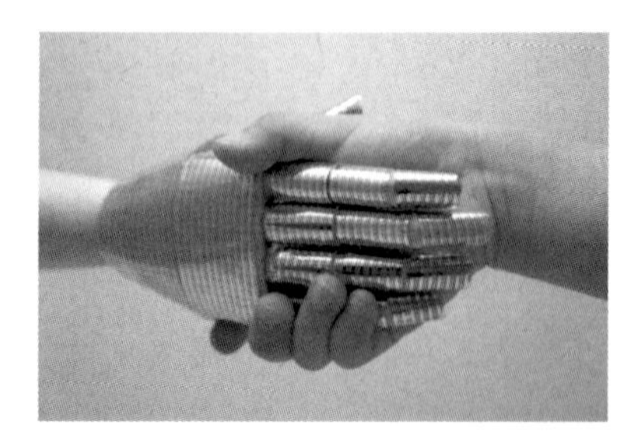

Figure 6.26 Left: Schematic of an active-matrix array with resistive sensor elements. Right: Demonstration of an artificial skin device with organic TFTs. (Reproduced from T. Someya et al., Large-area detectors and sensors, in: *Organic Electronics*, Wiley-VCH, 2006.)

through the data lines by external circuitry. This is repeated for each row until the entire array has been read out. The result is a map of the 2-D distribution of the desired physical quantity (in this case, the pressure) over the array. By reading the array continuously a dynamic image can be created (again, similar to an active-matrix display). One application of a 2-D pressure sensor array is a fingerprint sensor for personal identification purposes. Another interesting application is the combination of spatially resolved pressure and temperature sensing over large conformable surfaces to create the equivalent of sensitive skin for human-like robots capable of navigating in unstructured environments [62].

6.6 Outlook

Organic transistors are potentially useful for applications that require electronic functionality with low or medium complexity distributed over large areas on unconventional substrates, such as glass or flexible plastic film. Generally, these are applications in which the use of single-crystal silicon devices and circuits is either technically or economically not feasible. Examples include flexible displays and sensors. However, organic transistors are unlikely to replace silicon in applications characterized by large transistor counts, small chip size, large integration densities, or high-frequency operation. The reason is that, in these applications, the use of silicon MOSFETs is very economical. For example, the manufacturing cost of a silicon MOSFET in a 1-Gbit memory chip is on the order of $\$10^{-9}$, which is less than the cost of printing a single letter in a newspaper.

The static and dynamic performance of state-of-the-art organic TFTs is already sufficient for certain applications, most notably small or medium-sized flexible displays in which the TFTs operate with critical frequencies in the range of a few tens of kilohertz. Strategies for increasing the performance of organic TFTs include further improvements in the carrier mobility of the organic semiconductor (either through the synthesis of new materials, through improved purification, or by enhancing the molecular order in the semiconductor layer) and more aggressive scaling of the lateral transistor dimensions (channel length and contact overlap). For example, an increase in cut-off frequency from 200 kHz to about 2 MHz can be achieved either by improving the mobility from 0.5 $cm^2 V^{-1} s^{-1}$ to about 5 $cm^2 V^{-1} s^{-1}$ (assuming critical dimensions of 5 μm and an operating voltage of 3 V), or by reducing the critical dimensions from 5 μm to about 1.6 μm (assuming a mobility of 0.5 $cm^2 V^{-1} s^{-1}$ and an operating voltage of 3 V). A cut-off frequency of about 20 MHz is projected for TFTs with a mobility of 5 $cm^2 V^{-1} s^{-1}$ and critical dimensions of 1.6 μm (again assuming an operating voltage of 3 V).

However, these improvements in performance must be implemented without sacrificing the general manufacturability of the devices, circuits, and systems. This important requirement has fueled the development of a whole range of large-area, high-resolution printing methods for organic electronics. Functional printed organic devices and circuits have indeed been demonstrated using various printing techniques, but further studies are required to address issues such as process yield and parameter uniformity.

One of the most critical problems that must be solved before organic electronics can begin to find use in commercial applications is the stability of the devices and circuits during continuous operation, and while exposed to ambient oxygen and humidity. Early product demonstrators have often suffered from short lifetimes due to a rapid degradation of the organic semiconductor layers. However, recent advances in synthesis, purification, processing, in addition to economically viable encapsulation techniques, have raised the hope that the degradation of organic semiconductors is not an insurmountable problem and that organic thin-film transistors may soon be commercially utilized.

References

1 W. Warta, N. Karl, Hot holes in naphthalene: High, electric-field dependent mobilities, *Phys. Rev. B* 1985, **32**, 1172.

2 M. C. J. M. Vissenberg, M. Matters, Theory of field-effect mobility in amorphous organic transistors, *Phys. Rev. B* 1998, **57**, 12964.

3 G. Horowitz, R. Hajlaoui, P. Delannoy, Temperature dependence of the field-effect mobility of sexithiophene. Determination of the density of

traps, *J. Phys. III France* 1995, **5**, 355.

4 A. Tsumura, H. Koezuka, T. Ando, Macromolecular electronic device: Field-effect transistor with a polythiophene thin film, *Appl. Phys. Lett.* 1986, **49**, 1210.

5 A. Assadi, C. Svensson, M. Willander, O. Inganas, Field-effect mobility of poly(3-hexylthiophene), *Appl. Phys. Lett.* 1988, **53**, 195.

6 R. D. McCullough, R. D. Lowe, M. Jayaraman, D. L. Anderson, Design, synthesis, and control of conducting polymer architectures: Structurally homogeneous poly(3-alkylthiophenes), *J. Org. Chem.* 1993, **58**, 904.

7 Z. Bao, A. Dodabalapur, A. Lovinger, Soluble and processable regioregular poly (3-hexylthiophene) for thin film field-effect transistor applications with high mobility, *Appl. Phys. Lett.* 1996, **69**, 4108.

8 H. Sirringhaus, P. J. Brown, R. H. Friend, M. M. Nielsen, K. Bechgaard, B. M. W. Langeveld-Voss, A. J. H. Spiering, R. A. J. Janssen, E. W. Meijer, P. Herwig, D. M. de Leeuw, Two-dimensional charge transport in self-organized, high-mobility conjugated polymers, *Nature* 1999, **401**, 685.

9 B. S. Ong, Y. Wu, P. Liu, S. Gardner, High-performance semiconducting polythiophenes for organic thin-film transistors, *J. Am. Chem. Soc.* 2004, **126**, 3378.

10 I. McCulloch, M. Heeney, C. Bailey, K. Genevicius, I. MacDonald, M. Shkunov, D. Sparrowe, S. Tierney, R. Wagner, W. Zhang, M. L. Chabinyc, R. J. Kline, M. D. McGehee, M. F. Toney, Liquid-crystalline semiconducting polymers with high charge-carrier mobility, *Nature Mater.* 2006, **5**, 328.

11 M. Madru, G. Guillaud, M. Al Sadoun, M. Maitrot, C. Clarisse, M. Le Contellec, J. J. Andre, J. Simon, The first field effect transistor based on an intrinsic molecular semiconductor, *Chem. Phys. Lett.* 1987, **142**, 103.

12 C. Clarisse, M. T. Riou, M. Gauneau, M. Le Contellec, Field-effect transistor with diphthalocyanine thin film, *Electronics Lett.* 1988, **24**, 674.

13 G. Horowitz, D. Fichou, X. Peng, Z. Xu, F. Garnier, A field-effect transistor based on conjugated alpha-sexithienyl, *Solid State Commun.* 1989, **72**, 381.

14 F. Garnier, G. Horowitz, X. Z. Peng, D. Fichou, An all-organic 'soft' thin film transistor with very high carrier mobility, *Adv. Mater.* 1990, **2**, 592.

15 Y. Y. Lin, D. J. Gundlach, S. F. Nelson, T. N. Jackson, Pentacene-based organic thin-film transistors, *IEEE Trans. Electron. Dev.* 1997, **44**, 1325.

16 Y. Y. Lin, D. J. Gundlach, S. F. Nelson, T. N. Jackson, Stacked pentacene layer organic thin film transistors with improved characteristics, *IEEE Electr. Dev. Lett.* 1997, **18**, 606.

17 W. Kalb, P. Lang, M. Mottaghi, H. Aubin, G. Horowitz, M. Wuttig, Structure-performance relationship in pentacene/ Al_2O_3 thin-film transistors, *Synth. Metals* 2004, **146**, 279.

18 S. Y. Yang, K. Shin, C. E. Park, The effect of gate dielectric surface energy on pentacene morphology and organic field-effect transistor characteristics, *Adv. Funct. Mater.* 2005, **15**, 1806.

19 M. Halik, H. Klauk, U. Zschieschang, G. Schmid, S. Ponomarenko, S. Kirchmeyer, W. Weber, Relationship between molecular structure and electrical performance of oligothiophene organic thin film transistors, *Adv. Mater.* 2003, **15**, 917.

20 T. W. Kelley, L. D. Boardman, T. D. Dunbar, D. V. Muyres, M. J. Pellerite, T. P. Smith, High-performance OTFTs using surface-modified alumina dielectrics, *J. Phys. Chem. B* 2003, **107**, 5877.

21 S. Z. Weng, W. S. Hu, C. H. Kuo, Y. T. Tao, L. J. Fan, Y. W. Yang, Anisotropic field-effect mobility of pentacene thin-film transistor: Effect of rubbed self-assembled monolayer, *Appl. Phys. Lett.* 2006, **89**, 172103.

22 S. Lee, B. Koo, J. Shin, E. Lee, H. Park, H. Kim, Effects of hydroxyl groups in polymeric dielectrics on organic transistor performance all-organic active matrix flexible display, *Appl. Phys. Lett.* 2006, **88**, 162109.

23 P. Herwig, K. Müllen, A soluble pentacene precursor: Synthesis, solid-state conversion into pentacene and application in a field-effect transistor, *Adv. Mater.* 1999, **11**, 480.

24 A. Afzali, C. D. Dimitrakopoulos, T. L. Breen, High-performance, solution-processed organic thin film transistors from a novel pentacene precursor, *J. Am. Chem. Soc.* 2002, **124**, 8812.

25 M. M. Payne, S. R. Parkin, J. E. Anthony, C. C. Kuo, T. N. Jackson, Organic field-effect transistors from solution-deposited functionalized acenes with mobilities as high as 1 cm^2/Vs, *J. Am. Chem. Soc.* 2005, **127**, 4986.

26 C. C. Kuo, M. M. Payne, J. E. Anthony, T. N. Jackson, TES anthradithiophene solution-processed OTFTs with 1 cm^2/V-s mobility, 2004 International Electron Devices Meeting Technical Digest, 2004, p. 373.

27 S. K. Park, C. C. Kuo, J. E. Anthony, T. N. Jackson, High mobility solution-processed OTFTs, 2005 International Electron Devices Meeting Technical Digest, 2005, p. 113.

28 K. C. Dickey, J. E. Anthony, Y. L. Loo, Improving organic thin-film transistor performance through solvent-vapor annealing of solution-processable triethylsilylethynyl anthradithiophene, *Adv. Mater.* 2006, **18**, 1721.

29 L. L. Chua, J. Zaumseil, J. F. Chang, E. C. W. Ou, P. K. H. Ho, H. Sirringhaus, R. H. Friend, General observation of n-type field-effect behaviour in organic semiconductors, *Nature* 2005, **343**, 194.

30 R. Schmechel, M. Ahles, H. von Seggern, A pentacene ambipolar transistor: Experiment and theory, *J. Appl. Phys.* 2005, **98**, 084511.

31 J. Yamaguchi, S. Yaginuma, M. Haemori, K. Itaka, H. Koinuma, An in-situ fabrication and characterization system developed for high performance organic semiconductor devices, *Jpn. J. Appl. Phys.* 2005, **44**, 3757.

32 K. Itaka, M. Yamashiro, J. Yamaguchi, M. Haemori, S. Yaginuma, Y. Matsumoto, M. Kondo, H. Koinuma, High-mobility C_{60} field-effect transistors fabricated on molecular-wetting controlled substrates, *Adv. Mater.* 2006, **18**, 1713.

33 H. E. Katz, J. Johnson, A. J. Lovinger, W. Li, Naphthalenetetracarboxylic diimide-based n-channel transistor semiconductors: Structural variation and thiol-enhanced gold contacts, *J. Am. Chem. Soc.* 2000, **122**, 7787.

34 P. P. L. Malenfant, C. D. Dimitrakopoulos, J. D. Gelorme, L. L. Kosbar, T. O. Graham, A. Curioni, W. Andreoni, N-type organic thin-film transistor with high field-effect mobility based on a N,N′-dialkyl-3,4,9,10-perylene tetracarboxylic diimide derivative, *Appl. Phys. Lett.* 2002, **80**, 2517.

35 M. M. Ling, Z. Bao, Copper hexafluorophthalocyanine field-effect transistors with enhanced mobility by soft contact lamination, *Org. Electronics* 2006, **7**, 568.

36 B. A. Jones, M. J. Ahrens, M. H. Yoon, A. Facchetti, T. J. Marks, M. R. Wasielewski, High-mobility air-stable n-type semiconductors with processing versatility: Dicyanoperylene-3,4:9,10-bis (dicarboximides), *Angew. Chem. Int. Ed.* 2004, **43**, 6363.

37 P. K. Weimer, The TFT – A new thin film transistor, *Proc. IRE* 1962, **50**, 1462.

38 P. G. LeComber, W. E. Spear, A. Ghaith, Amorphous silicon field-effect device and possible application, *Electron. Lett.* 1979, **15**, 179.

39 F. Ebisawa, T. Kurokawa, S. Nara, Electrical properties of polyacetylene/polysiloxane interface, *J. Appl. Phys.* 1983, **54**, 3255.

40 D. J. Gundlach, T. N. Jackson, D. G. Schlom, S. F. Nelson, Solvent-induced phase transition in thermally evaporated

pentacene films, *Appl. Phys. Lett.* 1999, **74**, 3302.

41 D. J. Gundlach, L. Zhou, J. A. Nichols, T. N. Jackson, P. V. Necliudov, M. S. Shur, An experimental study of contact effects in organic thin film transistors, *J. Appl. Phys.* 2006, **100**, 024509.

42 A. C. Arias, S. E. Ready, R. Lujan, W. S. Wong, K. E. Paul, A. Salleo, M. L. Chabinyc, R. Apte, R. A. Street, Y. Wu, P. Liu, B. Ong, All jet-printed polymer thin-film transistor active-matrix backplanes, *Appl. Phys. Lett.* 2004, **85**, 3304.

43 D. V. Muyres, P. F. Baude, S. Theiss, M. Haase, T. W. Kelley, P. Fleming, Polymeric aperture masks for high performance organic integrated circuits, *J. Vac. Sci. Technol. A* 2004, **22**, 1892.

44 G. H. Gelinck, T. C. T. Geuns, D. M. de Leeuw, High-performance all-polymer integrated circuits, *Appl. Phys. Lett.* 2000, **77**, 1487.

45 C. W. Sele, T. von Werne, R. H. Friend, H. Sirringhaus, Lithography-free, self-aligned inkjet printing with sub-hundred-nanometer resolution, *Adv. Mater.* 2005, **17**, 997.

46 J. Veres, S. D. Ogier, S. W. Leeming, D. C. Cupertino, S. M. Khaffaf, Low-k insulators as the choice of dielectrics in organic field-effect transistors, *Adv. Funct. Mater.* 2003, **13**, 199.

47 A. F. Stassen, R. W. I. de Boer, N. N. Iosad, A. F. Morpurgo, Influence of the gate dielectric on the mobility of rubrene single-crystal field-effect transistors, *Appl. Phys. Lett.* 2004, **85**, 3899.

48 I. N. Hulea, S. Fratini, H. Xie, C. L. Mulder, N. N. Iossad, G. Rastelli, S. Ciuchi, A. F. Morpurgo, Tunable Fröhlich polarons in organic single-crystal transistors, *Nature Mater.* 2006, **5**, 982.

49 L. A. Majewski, R. Schroeder, M. Voigt, M. Grell, High performance organic transistors on cheap, commercial substrates, *J. Phys. D* 2004, **37**, 3367.

50 M. H. Yoon, H. Yan, A. Facchetti, T. J. Marks, Low-voltage organic field-effect transistors and inverters enabled by ultrathin cross-linked polymers as gate dielectrics, *J. Am. Chem. Soc.* 2005, **127**, 10388.

51 S. Y. Yang, S. H. Kim, K. Shin, H. Jeon, C. E. Park, Low-voltage pentacene field-effect transistors with ultrathin polymer gate dielectrics, *Appl. Phys. Lett.* 2006, **88**, 173507.

52 M. Halik, H. Klauk, U. Zschieschang, G. Schmid, C. Dehm, M. Schütz, S. Maisch, F. Effenberger, M. Brunnbauer, F. Stellacci, Low-voltage organic transistors with an amorphous molecular gate dielectric, *Nature* 2004, **431**, 963.

53 M. H. Yoon, A. Facchetti, T. J. Marks, σ-π molecular dielectric multilayers for low-voltage organic thin-film transistors, *Proc. Natl. Acad. Sci. USA* 2005, **102**, 4678.

54 H. Klauk, U. Zschieschang, J. Pflaum, M. Halik, Ultralow-power organic complementary circuits, *Nature* 2007, **445**, 745.

55 E. J. Meijer, C. Tanase, P. W. M. Blom, E. van Veenendaal, B. H. Huisman, D. M. de Leeuw, T. M. Klapwijk, Switch-on voltage in disordered organic field-effect transistors, *Appl. Phys. Lett.* 2002, **80**, 3838.

56 H. Klauk, U. Zschieschang, M. Halik, Low-voltage organic thin-film transistors with large transconductance, *J. Appl. Phys.* 2007, **102**, 074514.

57 H. E. A. Huitema, G. H. Gelinck, J. B. P. H. van der Putten, K. E. Kuijk, K. M. Hart, E. Cantatore, D. M. de Leeuw, Active-matrix displays driven by solution processed polymeric transistors, *Adv. Mater.* 2002, **14**, 1201.

58 K. Nomoto, N. Hirai, N. Yoneya, N. Kawashima, M. Noda, M. Wada, J. Kasahara, A high-performance short-channel bottom-contact OTFT and its application to AM-TN-LCD, *IEEE Trans. Electr. Dev.* 2005, **52**, 1519.

59 P. Wellmann, M. Hofmann, O. Zeika, A. Werner, J. Birnstock, R. Meerheim, G. He, K. Walzer, M. Pfeiffer, K. Leo, High-efficiency p-i-n organic light-emitting diodes with long lifetime, *J. Soc. Information Display* 2005, **13**, 393.

60 L. Zhou, A. Wanga, S. C. Wu, J. Sun, S. Park, T. N. Jackson, All-organic active matrix flexible display, *Appl. Phys. Lett.* 2006, **88**, 083502.

61 A. Kumar, A. Nathan, G. E. Jabbour, Does TFT mobility impact pixel size in AMOLED backplanes? *IEEE Trans. Electr. Dev.* 2005, **52**, 2386.

62 T. Someya, Y. Kato, T. Sekitani, S. Iba, Y. Noguchi, Y. Murase, H. Kawaguchi, T. Sakurai, Conformable, flexible, large-area networks of pressure and thermal sensors with organic transistor active matrixes, *Proc. Natl. Acad. Sci. USA* 2005, **102**, 12321.

7
Carbon Nanotubes in Electronics

M. Meyyappan

7.1
Introduction

Since the discovery of carbon nanotubes (CNTs) in 1991 [1] by Sumio Iijima of the NEC Corporation, research activities exploring their structure, properties and applications have exploded across the world. This interesting nanostructure exhibits unique electronic properties and extraordinary mechanical properties, and this has prompted the research community to investigate the potential of CNTs in numerous areas including, among others, nanoelectronics, sensors, actuators, field emission devices, and high-strength composites [2]. Although recent progress in all of these areas has been significant, the routine commercial production of CNT-based products is still years away. This chapter focuses on one specific application field of CNTs, namely electronics, and describes the current status of developments in this area. This description is complemented with a brief discussion of the properties and growth methods of CNTs, further details of which are available in Ref. [2].

7.2
Structure and Properties

A carbon nanotube is, configurationally, a graphene sheet rolled up into a tube (see Figure 7.1). If it is a single layer of a graphene sheet, the resultant structure is a single-walled carbon nanotube (SWNT), but with a stack of multiple layers a multi-walled carbon nanotube (MWNT) emerges. The SWNT is a tubular shell made from hexagonal rings of carbon atoms, with the ends of the shells capped by a dome-like, half-fullerene molecules [3]. They are classified using a nomenclature (n, m) where n and m are integer indices of two graphene unit lattice vectors ($\mathbf{a_1}, \mathbf{a_2}$) corresponding to the chiral vector of a nanotube, $c_a = n\mathbf{a_1} + m\mathbf{a_2}$. Based on the geometry, when $n = m$, the resulting structure is commonly known as an "arm chair" nanotube, as shown in Figure 7.1. The (n, 0) structure is called the "zig zag nanotube", while all other

Nanotechnology. Volume 4: Information Technology II. Edited by Rainer Waser
Copyright © 2008 WILEY-VCH Verlag GmbH & Co. KGaA, Weinheim
ISBN: 978-3-527-31737-0

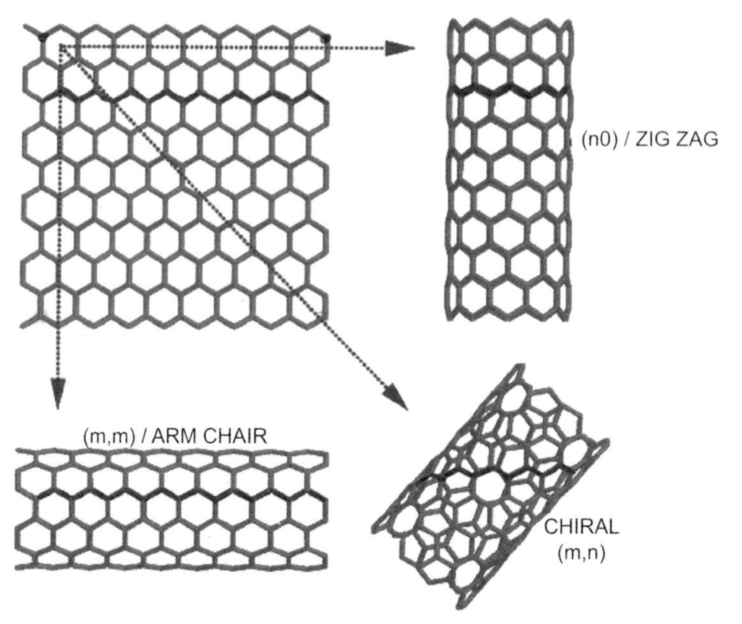

Figure 7.1 A strip of graphene sheet rolled into a carbon nanotube; m and n are chiral vectors.

structures are simply known as "chiral nanotubes". It is important to note that, at the time of this writing, exquisite control over the values of m and n is not possible. Transmission electron microscopy (TEM) images of a SWNT and a MWNT are shown in Figure 7.2, where the individual SWNTs are seen to have a diameter of about 1 nm. The MWNT has a central core with several walls, and a spacing close to 0.34 nm between the two neighboring walls (Figure 7.2b).

A SWNT can be either metallic or semiconducting, depending on its chirality – that is, the values of n and m. When (n − m)/3 is an integer, the nanotube is metallic, otherwise it is semiconducting. The diameter of the nanotube is given by $d = (a_g/\pi)(n^2 + mn + m^2)^{0.5}$, where a_g is the lattice constant of graphite. The strain energy

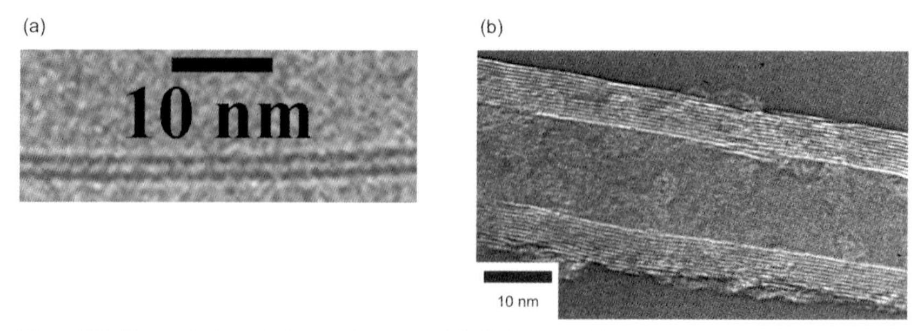

Figure 7.2 Transmission electron microscopy (TEM) images of (a) single-walled carbon nanotube and (b) a multi-walled carbon nanotube. (Image courtesy of Lance Delzeit.)

caused in the SWNT formation from the graphene sheet is inversely proportional to its diameter. There is a minimum diameter that can afford this strain energy, which is about 0.4 nm. On the other hand, the maximum diameter is about 3 nm, beyond which the SWNT may not retain its tubular structure and ultimately will collapse [3].

In the case of MWNTs, the smallest inner diameter found experimentally is about 0.4 nm, but typically is around 2 nm. The outer diameter of MWNT can be as large as 100 nm. Both, SWNTs and MWNTs, while preferentially being defect-free, have been observed experimentally in various defective forms such as bent, branched, helical, and even toroidal nanotubes.

The bandgap of a semiconducting nanotube is given by $E_g = 2d_{cc}\gamma/d$, where d_{cc} is the carbon–carbon bond length (0.142 nm), and γ is the nearest neighbor-hopping parameter (2.5 eV). Thus, the bandgap of semiconducting nanotubes of diameters between 0.5 and 1.5 nm may be in the range of 1.5 to 0.5 eV. The resistance of a metallic SWNT is $h/(4e^2) \approx 6.5\ \mathrm{K\Omega}$, where h is Planck's constant. However, experimental measurements typically show higher resistance due to the presence of defects, impurities, structural distortions and the effects of coupling to the substrate and/or contacts.

In addition to their interesting electronic properties, SWNTs exhibit extraordinary mechanical properties. For example, the Young's modulus of a (10,10) SWNT is over 1 TPa, with a tensile strength of 75 GPa. The corresponding values for graphite (in-plane) are 350 and 2.5 GPa, whereas the values for steel are 208 and 0.4 GPa [3]. Nanotubes can also sustain a tensile strain of 10% before fracturing, which is remarkably higher than other materials. The thermal conductivity of the nanotubes is substantially high [3, 4], with measured values being in the range of 1800 to 6000 $\mathrm{W\,mK^{-1}}$ [5].

7.3 Growth

The oldest process for preparing SWNTs and MWNTs is that of arc synthesis [6], with laser ablation subsequently being introduced during the 1990s to produce CNTs [7]. These bulk production techniques and large quantities are necessary when using CNTs in composites, gas storage. and similar applications. For electronics applications, it may be difficult to adopt "pick and place" strategies using bulk-produced material. Assuming that the need for *in-situ* growth approaches that currently used to produce devices for silicon-based electronics, it is important at this point to describe the techniques of chemical vapor deposition (CVD) and plasma-enhanced chemical vapor deposition (PECVD), both of which allow CNT growth on patterned substrates [8].

Chemical vapor deposition is a frequently used technique in silicon integrated circuit manufacture when depositing thin films of metals, semiconductors and dielectric materials. The CVD of CNTs typically involves a carbon-bearing feedstock such as CO or hydrocarbons including methane, ethylene, and acetylene. It is important to maintain the growth temperature below that of the pyrolysis temperature of a particular hydrocarbon in order to avoid the production of amorphous carbon. The CNT growth is facilitated by the use of a transition metal catalyst, the choice comprising iron, nickel, palladium, or cobalt. These metals can be thermally evaporated as a thin film on the substrate, or sputtered using ion beam sputtering or magnetron sputtering. Alternatively, the catalyst metal can be applied to the substrate, starting from the metal-containing salt solution and passing through a number of steps such as precipitation, mixing, evaporation, drying, and annealing. It should be noted that solution-based techniques are more cumbersome and much slower than physical techniques such as sputtering. In addition, they may not be amenable for working with patterned substrates. Regardless of the approach used, the key here is to deposit the catalyst in the form of particles in order to facilitate nanotube growth. The characterization of as-deposited catalysts using TEM and atomic force microscopy [9] reveals that the particles are in the range of 1 to 10 nm in size. The catalyst deposition may be restricted to selected locations of the wafer through lithographic patterning. The type of lithography (optical, electron-beam, etc.) needed would be dictated primarily by the feature size of the patterns.

Typically, CNT-CVD is performed at atmospheric pressure and temperatures of 550 to 1000 °C. Low-pressure processes at several torr have also been reported. SWNTs require higher growth temperatures (above 800 °C) than MWNTs, whereas the latter can be grown at temperatures as low as 550 °C. Lower temperatures (<500 °C) may not be possible if the catalytic activation and realistic reaction/growth rates occur only at such elevated temperatures. At present, this restriction poses a serious problem for the adoption of CVD as an *in-situ* process in the device fabrication sequence, as common masking materials cannot withstand such high temperatures. Figure 7.3 shows bundles of SWNTs grown using methane with an iron catalyst prepared by ion beam sputtering. Typically, the SWNTs tend to bunch together to form bundles or ropes. Figure 7.4 shows a patterned MWNT growth on a silicon substrate using an iron catalyst, which appears to yield a vertical array of nanotubes. Although the ensemble appears vertical, a closer inspection would reveal – in all cases of thermal CVD – that the individual MWNT itself is actually not well aligned but is wavy.

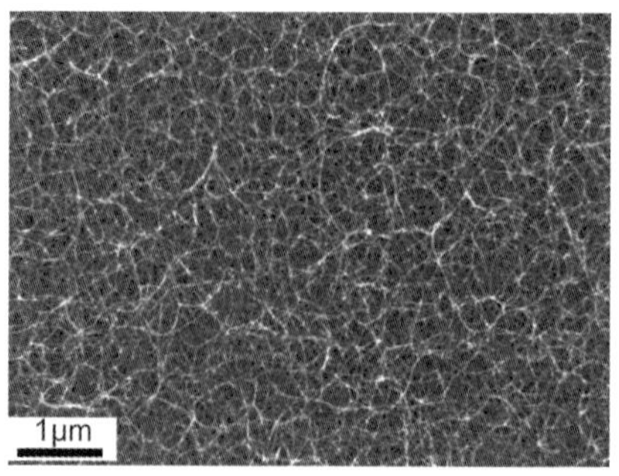

Figure 7.3 Bundles of SWNTs grown by chemical vapor deposition. (Image courtesy of Lance Delzeit.)

Figure 7.4 Patterned growth of multiwalled carbon nanotubes by CVD. (Image courtesy of H. T. Ng.)

In silicon integrated circuit manufacture, PECVD has emerged as a lower-temperature alternative to thermal CVD for the deposition of thin films of silicon, or its nitride or oxide. This strategy is not entirely successful in CNTgrowth, primarily because the growth temperature is tied to catalyst effectiveness as opposed to precursor dissociation [10]. Nevertheless, some reports have been made concerning nanotube growth at low temperature, or even at room temperature. However, these results are not reliable as they do not explicitly measure the growth temperature (i.e. the wafer temperature), but instead report only the temperatures on the bottom side of the substrate holder. Neither did any of these studies appreciate the fact that the plasma – and particularly the dc plasma used in most studies – heats the wafer substantially, particularly at the very high bias voltages commonly used. In such a case, even external heating via a heater may not be needed, and in most cases the temperature difference between the wafer and the bottom of the substrate holder may be several hundred degrees or more, depending on the input power [11]. Even if any degree of growth temperature reduction is achieved using PECVD, the material quality is relatively poor. Most of these structures are often conical in terms of configuration, with a continuously tapering diameter from the bottom to the top. Regardless of such issues, PECVD has one clear advantage over CVD, in that it enables the production of individual, freestanding, vertically aligned MWNT structures as opposed to individual, wavy nanotubes. These freestanding structures are invariably disordered with a bamboo-like inner core and, for that reason, are referred to as multi-walled carbon nanofibers (MWNFs) or simply carbon nanofibers (CNFs) [10]. PECVD is also capable of producing wavy MWNTs which are very similar to the thermally grown MWNTs.

To date, a variety of plasma sources have been used in CNT growth, including dc [12, 13], microwave [14], and inductive power sources [15]. The plasma efficiently breaks down the hydrocarbon feedstock, thus creating a variety of reactive radicals which are also the source for amorphous carbon. For this reason the feedstock is typically diluted with hydrogen, ammonia, argon or nitrogen to maintain the hydrocarbon fraction at less than about 20%. PECVD is performed at low pressures, typically in the range of 1 to 20 Torr. A scanning electron microscopy (SEM) image of PECVD-grown MWNFs is shown in Figure 7.5, wherein the individual structures are

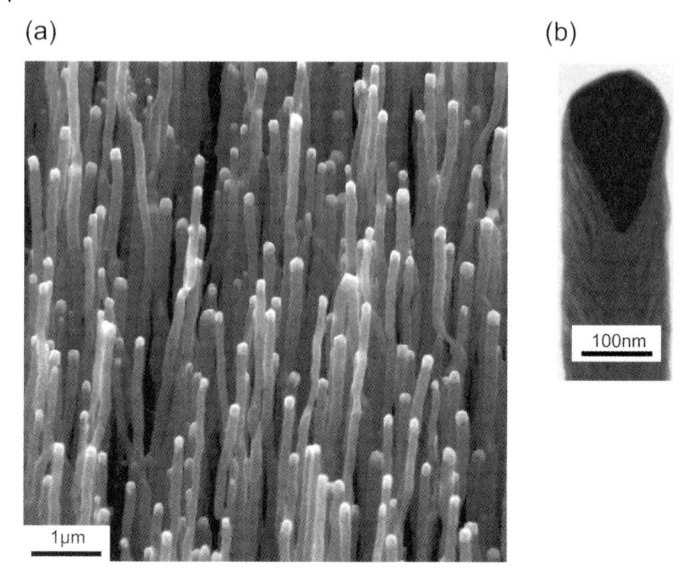

Figure 7.5 (a) SEM image showing vertical, freestanding carbon nanofibers grown by plasma-enhanced CVD. (Image courtesy of Alan Cassell.) (b) TEM image showing bamboo-like morphology and the catalyst particle at the head. (Image courtesy of Quoc Ngo and Alan Cassell.)

well separated and vertical. However, the TEM image reveals a disordered inner core and also the catalyst particle at the top. In contrast, in most cases of MWNT growth by thermal and plasma CVD, the catalyst particle is typically at the base of the nanotubes.

7.4 Nanoelectronics

Silicon complementary metal oxide semiconductor (CMOS) -based electronics has been moving forward impressively according to Moore's law, with 90-nm feature scale devices currently in production and 65-nm devices in the development stage. Research investigations are also well under way on the lower nodes and, as further miniaturization continues, a range of technological difficulties is anticipated, according to the Semiconductor Industry Association Roadmap [16]. These issues include lithography, novel dielectric materials, heat dissipation and efficient chip cooling to name a few. It was thought a few years ago that Si CMOS scaling may end at around 50 nm, beyond which alternatives such as CNT electronics or molecular electronics may be needed. However, this is no longer true as the current evidence suggests that scaling beyond 50 nm is possible, though with increased challenges. Regardless of when the need for transition to alternatives emerges, there are a few expectations from a viable alternative:

- The new technology must be easier and cheaper to manufacture than Si CMOS.
- A high current drive is needed with the ability to drive capacitances of interconnects of any length.
- A reliability factor enjoyed to date must be available (i.e. operating time > 10 years).
- A high level of integration must be possible ($>10^{10}$ transistors per circuit).
- A very high reproducibility is expected.
- The technology should not be handicapped with high heat dissipation problems currently forecast for the future-generation silicon devices, or attractive solutions must be available to tackle the anticipated heat loads.

Of course, the present status of CNT electronics has not yet reached the point where its performance in terms of the above goals can be evaluated. This is due to the fact that most efforts to date relate to the fabrication of single devices such as diodes and transistors, and little has been targeted at circuits and integration (as will be seen in the next section). In summary, the present status of CNT electronics evolution is similar to that of silicon technology between the invention of the transistor (during the late 1940s) and the development of integrated circuit in the 1960s. It would take at least a decade or more to demonstrate the technological progress required to meet the above-listed expectations.

7.4.1 Field Effect Transistors

The early attempts to investigate CNTs in electronics consisted of fabricating field effect transistors (FETs) with a SWNT as the conducting channel [17]. Tans *et al.* [18] reported first a CNT-FET where a semiconducting SWNT from a bulk-grown sample was transplanted to bridge the source and drain contacts separated by about a micron or more (see Figure 7.6). The contact electrodes were defined on a thick 300-nm SiO_2 layer grown on a silicon wafer acting as the back gate. The 1.4-nm tube with a corresponding bandgap of about 0.6 eV showed *I–V* characteristics indicating gate control of current flow through the nanotube. In the FET, the holes were the majority carriers and the conductance was shown to vary by at least six orders of magnitude. The device gain of this early device was below 1, due primarily to the thick gate oxide and high contact resistance.

At almost the same time, Martel *et al.* [19] presented their CNT-FET results using a similar back-gated structure. The oxide thickness was 140 nm, and 30 nm-thick

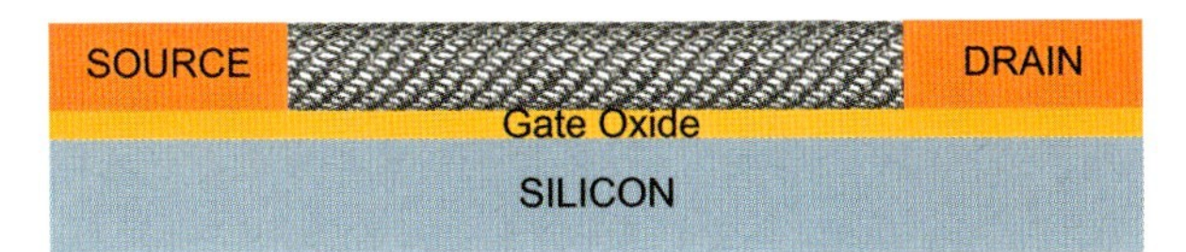

Figure 7.6 Schematic of an early CNT-FET with a back gate. A semiconducting nanotube bridges the source and drain, thus creating the conducting channel.

gold electrodes were defined using electron-beam lithography. The room-temperature I–V_{SD} characteristics showed that the drain current decreased strongly with increasing gate voltage, thus demonstrating CNT-FET operation through hole transport. The conductance modulation in this case spanned five orders of magnitude. The transconductance of this device was 1.7 nS at $V_{SD} = 10$ mV, with a corresponding hole mobility of 20 $cm^2 V^{-1} s^{-1}$. The authors concluded that the transport was diffusive rather than ballistic and, in addition, the high hole concentration was inherent to the nanotubes as a result of processing. This unipolar p-type device behavior suggested a Schottky barrier at the tube–contact metal interface. Later, the same IBM group [20] showed that n-type transistors could be produced simply by annealing the above p-type device in a vacuum, or by intentionally doping the nanotube with an alkali metal such as potassium.

All of the early CNT-FETs used the silicon substrate as the back gate. This unorthodox approach has several disadvantages. First, the resulting thick gate oxide required high gate voltages to turn the device on. Second, the use of the substrate for gating led to influencing all devices simultaneously. For integrated circuit applications, each CNT-FET needs its own gate control. Wind *et al.* [21] reported the first top gate CNT-FET which also featured embedding the SWNT within the insulator rather than exposing it to ambient, as had been done in the early devices. It was considered that such ambient exposure would lead to p-type characteristics and, as expected, the top gate device showed significantly better performance. A p-type CNT-FET with a gate length of 300 nm and gate oxide thickness of 15 nm showed a threshold voltage of –0.5 V and a transconductance of 2321 $\mu S \mu m^{-1}$. These results were better than those of a silicon p-MOSFET [22] with a much smaller gate length of 15 nm and an oxide thickness of 1.4 nm performing at a transconductance of 975 $\mu S \mu m^{-1}$. The CNT-FET also showed a three- to four-fold higher current drive per unit width compared to the above silicon device. Nihey *et al.* [23] also reported a top-gated device albeit with a thinner (6 nm) gate oxide TiO_2 with a higher dielectric constant. This device showed a 320 nS transconductance at a 100 mV drain voltage.

Most recently, Seidel *et al.* [24] reported CNT-FETs with short channels (18 nm), in contrast to all previous studies with micron-long channels. This group used nanotubes with diameters of 0.7 to 1.1 nm, and bandgaps in the range of 0.8 to 1.3 eV. The impressive performance of these devices included an on/off current ratio of 10^6 and a transconductance of 12 000 $\mu S \mu m^{-1}$. The current-carrying capacity was also very high, with a maximum current of 15 μA, corresponding to $10^9 A cm^{-2}$. Another recent innovation involved a nanotube-on-insulator (NOI) approach [25], similar to the adoption of silicon-on-insulator (SOI) by the semiconductor industry, which minimizes parasitic capacitance.

As noted above, many of the CNT-FET studies conducted to date have used SWNTs, this being due to their superior properties compared to other types of nanotubes such as MWNTs and CNFs. As the bandgap is inversely proportional to the diameter, large-diameter MWNTs are invariably metallic. Martel *et al.* [19] fabricated the first MWNT FETs showing a significant gate effect. The real advantage of MWNTs is that they can be grown vertically up to reasonable lengths for a given diameter. Choi *et al.* [26, 27] took advantage of this point to fabricate vertical transistors using MWNTs grown using an anodic alumina template which essentially contained nanopores of various diameter

Figure 7.7 Vertically aligned MWNT grown using a nanoporous template. (Image courtesy of W. B. Choi.)

such that they were able to control both the diameter and pore density. The vertically aligned MWNTs grown using nanopores are shown in Figure 7.7. Following this, the device fabrication consisted of depositing SiO_2 on top of the aligned nanotubes. The electrode was then attached to the nanotubes through electron-beam-patterned holes, and finally the top metal electrode was attached. A schematic of the CNT-FET array and a SEM image of a 10×10 array is shown in Figure 7.8. For these devices, the authors claimed a tera-level transistor density of $2 \times 10^{11}\,cm^{-2}$.

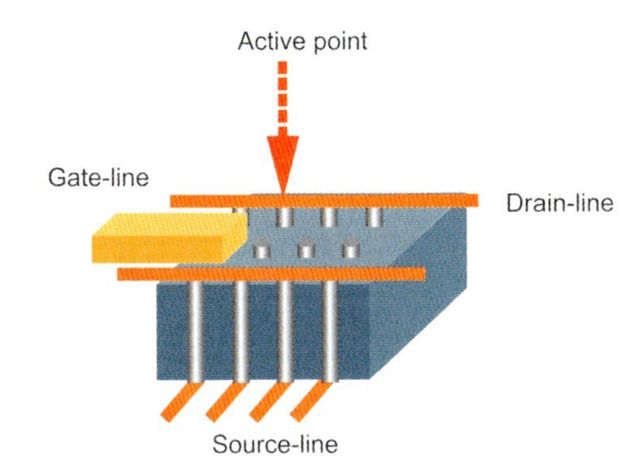

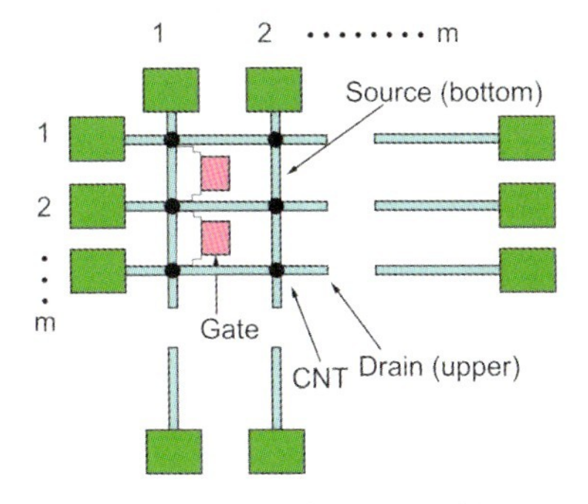

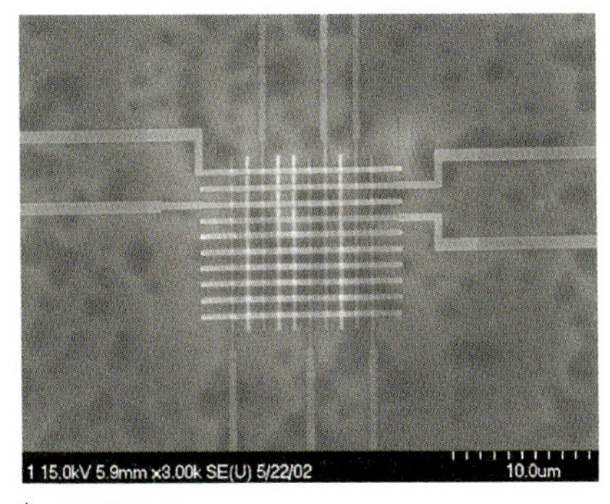

Figure 7.8 An array of CNT-FETs fabricated using the MWNTs in Figure 7.7. A schematic and an SEM image of an n × m array are shown. (Image courtesy of W. B. Choi.)

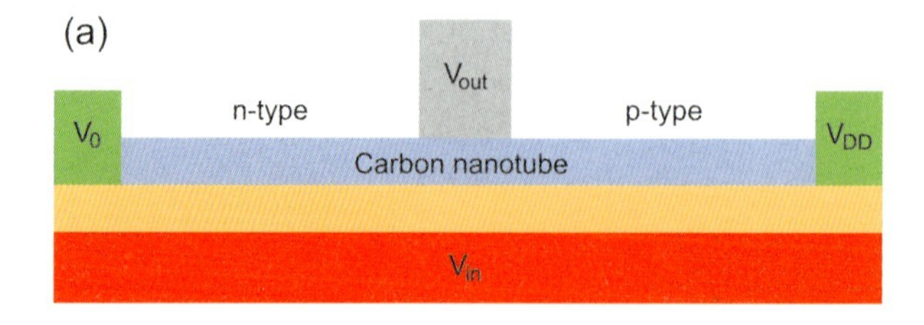

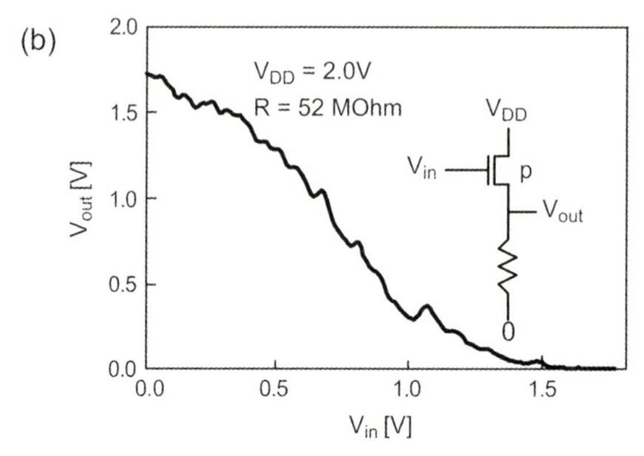

Figure 7.9 (a) n-type and p-type CNT-FETs. (b) Characteristics of an inverter circuit using the devices in (a). The inset shows the inverter circuit. (Reproduced from Ref. [28].)

Beyond single CNT-FETs, early attempts to fabricate circuit components have also been reported [28–30]. Liu *et al.* [28] fabricated both PMOS and CMOS inverters based on CNT-FETs. Their CMOS inverter connected two CNT-FETs in series using an electric lead of 2 mm length. One of the FETs was a p-type device, while the other – an n-type device – was obtained using potassium doping (see Figure 7.9). The devices used the silicon substrate as a back gate. The inverter was constructed and biased in the configuration shown in Figure 7.9b, after which a drain bias of 2.9 V was applied and the gate electrode was swept from 0 to 2.5 V, defining logics 0 and 1, respectively. As seen in the transfer curve in Figure 7.9b, when the input voltage is low (logic 0) the p-type and n-type devices were on and off, respectively (corresponding to their respective high- and low-conductance states). Then, the output is close to V_{DD}, producing a logic output of 1. When the input voltage is high (logic 1), the reverse is the case: the p-type transistor is off and the n-type device is on, with a combined output close to 0, producing a logic 0. The transfer curve in Figure 7.9b should not have a slope if this inverter were functioning ideally (which would correspond to a stepwise V_{out} versus V_{in} behavior). However, this first demonstration had a leaky p-device and so control of the threshold voltage of both devices was not perfect, thus leading to the slope seen in Figure 7.9b. Derycke *et al.* [29] also demonstrated an inverter using p- and n-type CNT-FETs. Beyond inverters, Bachtold *et al.* [30] fabricated circuits to perform logic operations such as NOR, and also constructed an ac ring oscillator.

As efforts continue in this direction, it is also important to consider issues such as $1/f$ noise, shot noise, and other similar concerns arising during operation. An analysis by Lin *et al.* [31] showed that the $1/f$ noise level in semiconducting SWNTs is correlated to the total number of charge carriers in the system. However, the noise level per carrier itself is not larger than that seen in silicon devices. Beyond the conventional binary logic approach, Raychowdry and Kaushik [32] discussed extensively implementation schemes for voltage-mode multiple-value logic (MVL) design. The MVL circuits reduce the number of operations per function and reduce the parasitics associated with routing and the overall power dissipation.

To date, the CNT-FET fabrication has essentially followed the silicon CMOS scheme by simply replacing the silicon conducting channel with a SWNT. This requires the presence of straight, aligned nanotubes controllably bridging a pair of electrodes laid out horizontally. As-grown SWNTs using any of the growth techniques, in contrast, exhibit a spaghetti-like morphology and occasionally consist of Y- and other types of junction. Menon and Srivastava [33, 34] postulated that such Y- and T-junctions are structurally stable and form the basis for three-terminal devices. However, such junctions in as-grown samples are, of course, neither controllable nor really amenable for further device processing. Satishkumer *et al.* [35] then set out to create these Y-junctions in a controllable manner using the anodic alumina template approach. Keeping two of the terminals at the same voltage, the two-terminal operation showed rectifying behavior when the voltage on the third terminal was varied. While their devices were constructed from MWNTs, SWNT Y-junctions have also been reported [36, 37], though FET operation using Y-junctions has not yet been demonstrated. Srivastava *et al.* [38] also proposed a radical neural tree architecture consisting of numerous Y-junctions (see Figure 7.10), wherein the concept is that the switching and processing of signals by these junctions in the tree would be similar to that of dentritic neurons in biological systems. In addition, acoustic, chemical or other signals may also be used instead of electrical signals.

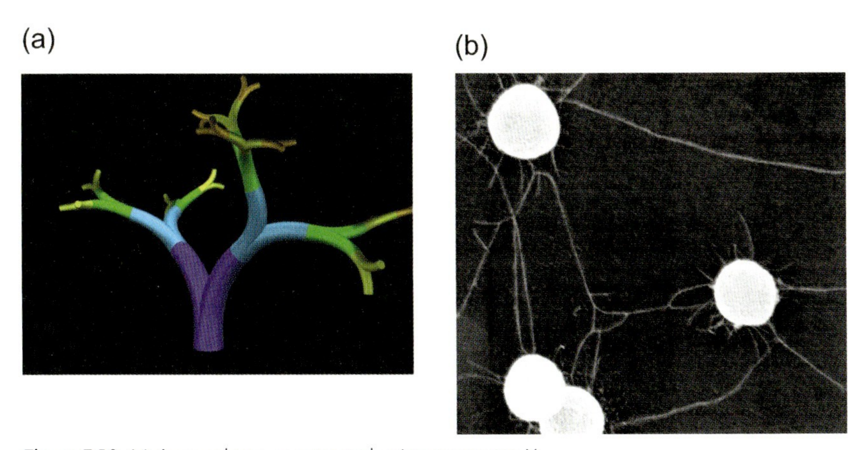

Figure 7.10 (a) A neural tree constructed using numerous Y-junctions of carbon nanotubes. (b) A network of interconnected SWNTs showing a few Y-junctions [36].

Figure 7.10b shows a very rudimentary attempt [36] to create such a CNT tree by utilizing self-assembled porous, collapsible polystrene/divinyl benzene microspheres to hold the catalyst. A controlled collapse of the spheres leads to the creation and release of the catalyst on the substrate for the CVD of SWNTs. Few Y-junctions showing an interconnected three-dimensional network of nanotubes are visible in Figure 7.10b.

7.4.2
Device Physics

The physics describing the operation of the CNT-FETs has been described in several theory papers [39–42] and summarized and reviewed elsewhere [17]. Here, the available information would be used to predict upper limits on CNT-FET performance. Yamada argues [17] that in nanoFETs the properties of the bulk material do not influence the device performance. In micro and macro FETs, the drain current is proportional to carrier mobility, which varies from material to material through its dependence on material-related properties such as effective mass and phonon scattering. In nanoFETs with ideal contacts, the drain current is determined by the transmission coefficient of an electron flux from the source to drain. When the carrier transport is ballistic, this coefficient is 1 and the material properties do not enter the picture directly. However, the material properties in practice enter indirectly as practical contacts (and hence the transmission coefficient at the contact) depend on the channel material and the interaction between the metal–channel semiconductor interface. The same applies to the preparation of the insulation material that determines the gate voltage characteristics. One-dimensional nanomaterials such as SWNTs inherently can suppress the short-channel effects arising from a deeper, broader distribution of carriers away from the gate, which occurs in reduced-size, two-dimensional silicon devices.

By using such an ideal device under ballistic transport and ideal contacts, Guo *et al.* [43] evaluated the performance of CNT-FETs. These authors considered a 1-nm SWNT with an insulator thickness of 1 nm and dielectric constant of 4. The geometry also was idealized to be a coaxial structure with contacts at either end of the nanotubes, and the gate wrapped around the nanotube. The computed on-off current ratio of 1120 was far higher than planar silicon CMOS devices with the same insulator parameters and power supply. The transconductance of this structure was also very high at 63 μS at 0.4 V, which was two orders of magnitude higher than any CNT-FET device discussed in Section 7.4.1. When a planar CNT-FET is compared with a planar Si-MOSFET with similar insulator parameters, the CNT-FET shows an on-current of 790 $\mu A\,\mu m^{-1}$ at $V_{DD} = 0.4$ V, in contrast to the 1100 $\mu A\,\mu m^{-1}$ value for silicon. In a following study, the same authors [44] computed the high-frequency performance of this ideal device and projected a unity gain cut-off frequency (f_T) of 1.8 THz. Their analysis also showed that the parasitic capacitance dominates the intrinsic gate capacitance by three orders of magnitude. In a similar investigation, Hasan *et al.* [45] computed f_T to be a maximum of 130 L^{-1} GHz, where L is the channel length in microns. As there is a desire to increase the current drive and reduce parasitic capacitance per tube, parallel array of nanotubes as the channel warrants consideration [44]. However, crosstalk becomes a serious issue in multiple tube systems. Leonard [46] analyzed this point, and established a length scale for tube separation below which inter-tube interaction becomes significant. For small channel lengths, this critical separation distance depends on the channel length; for long channel devices, the critical inter-tube separation distance is independent of channel length but depends on gate oxide thickness and dielectric constant.

7.4.3 Memory Devices

In relative terms, very few studies have been conducted on the use of nanotubes as memory devices. Rueckes *et al.* [47] proposed a crossbar architecture for constructing non-volatile random access memory with a density of 10^{12} element per cm^2 and an operation frequency of over 100 GHz. In this architecture, nanotubes suspended in a $n \times m$ array act like electromechanical switches with distinct on and off states. A carbon nanotube-based flash memory was fabricated by Choi *et al.* [48], in which the source-drain gap was bridged with a SWNT as a conducting channel and the structure had a floating gate and a control gate. By grounding the source and applying 5 V and 12 V at the drain and control gate, respectively, a writing of 1 was achieved. This corresponds to the charging of the floating gate. To write 0, the source was biased at 12 V, the control gate fixed at 0 V, and the drain allowed to float. Now, the electrons on the floating gate were tunneled to the source and the floating gate was discharged. In order to read, a voltage V_R was applied to the control gate and, depending on the state of the floating gate (1 or 0), the drain current was either negligible or finite, respectively. Choi *et al.* [48] reported an appreciable threshold modulation for their SWNT flash memory operation.

7.5 Carbon Nanotubes in Silicon CMOS Fabrication

Whilst the active role of CNTs in nanoelectronics (i.e. as a conducting channel in a transistor device) may be far away, it may play an important role in extending silicon nanoelectronics. Several areas exist in the Semiconductor Industry Association Roadmap [16] where CNTs may be useful, such as interconnects, heat dissipation and metrology.

7.5.1 Interconnects

One of the anticipated problems in the next few generations of silicon devices is that the copper interconnect would suffer from electromigration at current densities of 10^6 A cm^{-2} and above. The resistivity of copper increases significantly for wiring line widths lower than 0.5 μm. In addition, the etching of deep vias and trenches and void-free filling of copper in high-aspect ratio structures may pose technological challenges as progress continues along the Moore's law curve. All of these issues together demand investigation of alternatives to the current copper damascene process. In this regard, it is important to note that CNTs do not break down – even at current densities of 10^8 to 10^9 A cm^{-2} [49] – and hence can be a viable alternative to copper. Kreupl *et al.* [50] were the first to explore CVD-produced MWNTs in vias and contact holes. They measured a resistance of about 1 Ω for a 150 μm^2 via which contained about 10 000 MWNTs, thus yielding a resistance of 10 KΩ per nanotube. Further studies by this group demonstrated current densities of 5×10^8 A cm^{-2}, which exceeds the best results for metals, although the individual resistance of the MWNTs was still high at 7.8 KΩ. Srivastava *et al.* [51] provided a systematic evaluation of CNT and Cu interconnects and showed that, for local interconnects, nanotubes may not offer any advantages, partly due to the fact that practical implementations of nanotube interconnects have an unacceptably high contact resistance. On the other hand, their studies showed an 80% performance improvement with CNTs for long global interconnects. It is important to note that, even in the case of local interconnects, very few studies have been conducted on contact and interface engineering; however, with further investigation the situation may well improve beyond these early expectations.

Given the potential of CNTs as interconnects, it is necessary to devise a processing scheme that is compatible with the silicon integrated circuit fabrication scheme. In the via and contact hole schemes, Kreupl *et al.* [50] followed a traditional approach by simply replacing the copper filling step with a MWNT CVD step. If this proves to be reliable – and specifically if the MWNTs do not become unraveled during the chemical mechanical polishing (CMP) step – then it would be a viable approach, provided that a dense filling of vertical nanotubes can be achieved. Even then, the conventional challenges in deep aspect ratio etching and void-free filling of features, which arise due to shrinking feature sizes, remain. The dry etching of high-aspect ratio vias with vertical sidewalls will increasingly become a problem, and further processing studies must be performed to establish the viability of this approach. In the meantime, Li *et al.* [52] described an alternative bottom-up scheme (see Figure 7.11) wherein the CNT interconnect is first deposited using PECVD at prespecified locations. This is followed by tetraethylorthosilicate (TEOS) CVD of SiO_2 in the space between CNTs, and then by CMP to yield a smooth top surface of SiO_2 with embedded CNT interconnects. Top metallization completes the fabrication. The interconnects grown using PECVD in Ref. [52] are CNFs with a bamboo-like morphology. Whilst they are really vertical and freestanding compared to MWNTs, thus allowing ease of fabrication, their resistance is higher. This, combined with high contact resistance, resulted in a value of about 6 KΩ for a single 50-nm CNF. Further annealing to obtain higher quality CNFs and, more importantly, interface engineering to reduce contact resistance, can prove this approach valuable. It would also be useful for future three-dimensional architectures. A detailed theoretical study conducted by Svizhenko *et al.* [53] showed that almost 90% of the voltage drop occurs at the metal–nanotube interface, while only 10% is due to transport in the nanotube, thus emphasizing the need for contact interface engineering.

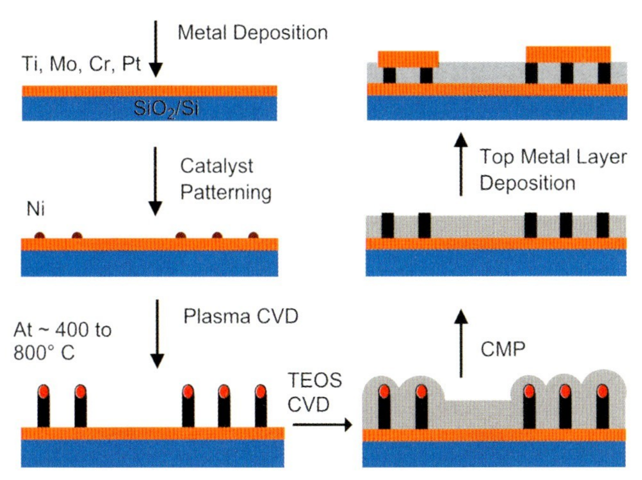

Figure 7.11 Carbon nanotube interconnect processing scheme for DRAM applications. TEOS = tetraethylorthosilicate.

7.5.2 Thermal Interface Material for Chip Cooling

The current trend in microprocessors is increasing operating frequency, decreasing dimensions, high packing density, and increasing power density. Together, these make thermal management in chip design a critical function to maintain the operating temperature at a prescribed, acceptable level. Otherwise, device reliability is severely compromised, and the speed of the microprocessor would also decrease with increasing operating temperatures [54]. The key figure-of-merit in thermal design and packaging is the thermal resistance, which is ΔT/input power. Here, ΔT is the temperature difference between the transistor junction and the ambient, which is fixed by the desirable operating junction temperature. As the power densities are on the rise, Shelling *et al.* [54] point out that the challenge is to develop high-conductivity structures which will accommodate the fixed ΔT, even with increased power densities.

Typically, the thermal packaging of a microprocessor consists of a heat spreader (primarily copper) and a heat sink. A variety of engineering designs is considered in all of the above to increase the heat-transfer efficiency [55] (which is beyond the scope of this chapter and not relevant at this point). However, one aspect which is relevant is a thermal interface material (TIM) commonly used to improve heat transfer between the chip and heat spreader, as well as between the heat spreader and the heat sink. Typically, a thermal grease has been used as TIM in the past, but more recent research on phase-change materials and polymers filled with high-conductivity particles has advanced knowledge of this subject. Carbon nanotubes exhibit very high axial thermal conductivity (see Section 7.2) which can be exploited in creating a TIM to address future thermal management needs.

Ngo *et al.* [56] reported a CNT-Cu composite for this purpose, wherein a PECVD-produced vertical CNF array is intercalated with copper using electrodeposition. As CNF surface coverage on the wafer from PECVD is only about 30–40% and air is a poor conductor, it becomes necessary to undertake a "gap-filling" effort with copper to cover the space between the nanotubes. This structure maintained its structural integrity at the 60 psi pressure normally used in packaging. Ngo *et al.* [56] reported a thermal resistance of about 0.1 $cm^2\,K\,W^{-1}$ for this structure, which makes it desirable for laptop, desktop, and workstation processor chips, although further improvements and reliability testing are required in this area.

7.5.3 CNT Probes in Metrology

Atomic force microscopy (AFM) is a versatile technique for imaging a wide variety of materials with high resolution. In addition to imaging metallic, semiconducting and dielectric thin films in integrated circuit manufacture, AFM has been advocated for critical dimension metrology. Currently, the conventional probes of either silicon or silicon nitride which are sited at the end of an AFM cantilever have a tip radius of curvature about 20–30 nm, which is obtained by micromachining or reactive ion etching. These probes exhibit significant wear during continuous use, and the worn probes can also break during tapping mode or contact mode operation. Carbon nanotube probes can overcome the above limitations due to their small size, high aspect ratio and the ability to buckle reversibly. Their use in AFM was first demonstrated by Dai *et al.* [57], while a detailed discussion of CNT probes and their construction and applications is also available [58].

A SWNT tip, attached to the end of an AFM cantilever, is capable of functioning as an AFM probe and provides better resolution than conventional probes. This SWNT probe can be grown directly using thermal CVD at the end of a cantilever [59]. An image of an iridium thin film collected using a SWNT probe is shown in Figure 7.12.

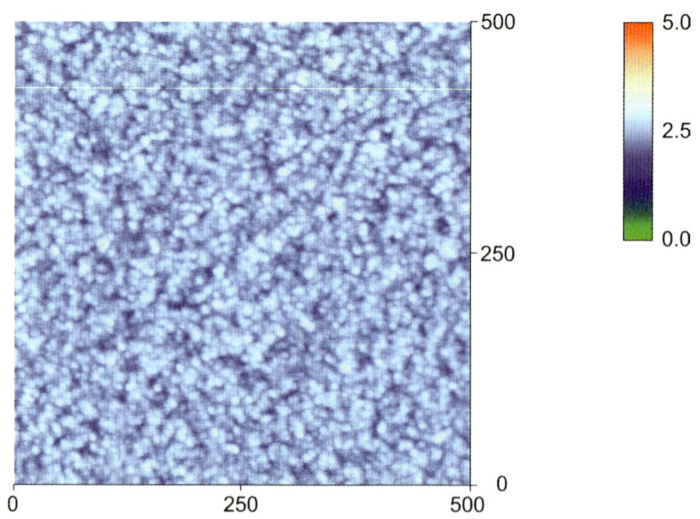

Figure 7.12 Atomic force microscopy image of an iridium thin film collected using a SWNT probe.

The nanoscale resolution is remarkable, but more importantly the tip has been shown to be very robust and significantly slow-wearing compared to conventional probes [59]. Due to thermal vibration problems, the SWNTs with a typical diameter of 1 to 1.5 nm cannot be longer than about 75 nm for probe construction. In contrast, however, the MWNTs – with their larger diameter – can form 2- to 3-μm-long probes. It is also possible to sharpen the tip of MWNTs to reach the same size as SWNTs, thus allowing the construction of long probes without thermal stability issues, but with the resolution of SWNTs [60]. Both, SWNT and sharpened MWNT probes have been used to image the semiconductor, metallic, and dielectric thin films commonly encountered in integrated circuit manufacture [58–60].

In addition to imaging, MWNT probes find another important application in the profilometry associated with integrated circuit manufacture. As via and other feature sizes continue to decrease, it will become increasingly difficult to use conventional profilometers to obtain sidewall profiles and monitor the depth of features. Although AFM is advocated as a replacement in this respect, the pyramidal nature of standard AFM probes would lead to artifacts when constructing the sidewall profiles of trenches. Hence, a 7- to 10-nm MWNT probe might be a natural choice for this task. An image of a MWNT probe for this purpose, and the results of profiling a photoresist pattern generated by interferometric lithography, are shown in Figure 7.13. While early attempts consisted of manually attaching a SWNT to a cantilever [57], followed by direct CVD of a nanoprobe on a cantilever [58, 59], Ye *et al.* [61] reported the first batch fabrication of CNT probes on a 100-mm wafer

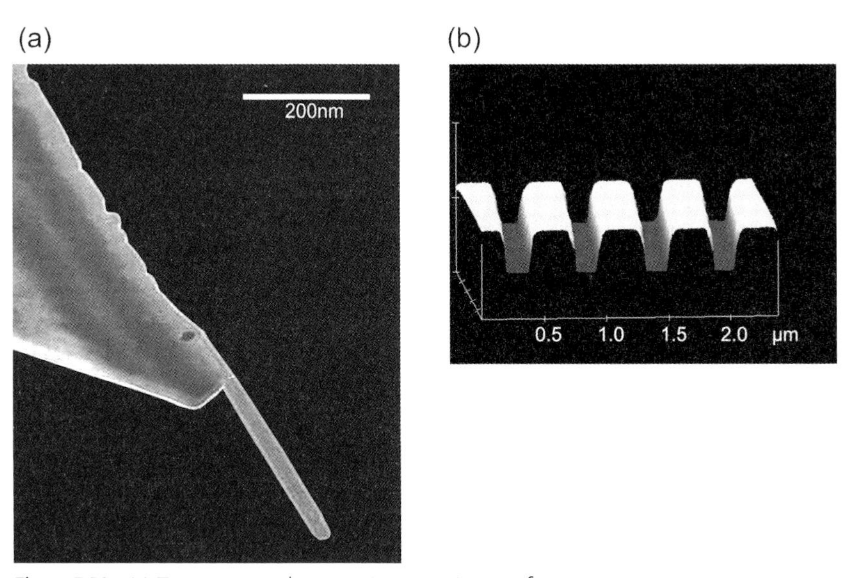

Figure 7.13 (a) Transmission electron microscopy image of a MWNT at the tip of an atomic force microscope cantilever. (b) Profile of a deep-UV photoresist pattern generated by interferometric lithography. The array has a pitch of 500 nm.

using PECVD. Unfortunately, the yield obtained was only modest, due mainly to difficulties encountered in controlling the angle of the nanotube to the plane.

7.6 Summary

In this chapter, the current status of CNT-based electronics for logic and memory devices has been discussed. Single-walled CNTs exhibit intriguing electronic properties that make them very attractive for future nanoelectronics devices, and early studies have confirmed this potential. Even with substantially longer channel lengths and thicker gate oxides, the performance of CNT-FETs is better than that of current silicon devices, although of course the design and performance of the former are far from being optimized. While all of this is impressive, the real challenge is in the integration of a large number of devices at reasonable cost to compete with and exceed the performance status quo of silicon technology at the end of the Moore's law paradigm. In addition, all of the studies conducted to date have been along the lines of following silicon processing schemes, with one-to-one replacement of a silicon channel with a CNT channel while maintaining the circuit and architectural schemes. Thus, aside from changing the channel material, there is no novelty in this approach. The structure and unique properties of SWNTs may be ideal for bold, novel architectures and processing schemes, for example in neural or biomimetic architecture, although very few investigations have been carried out in such non-traditional directions. Clearly, CNTs in active devices are a long-term prospect, at least a decade or more away. In the meantime, opportunities exist to include this extraordinary material into silicon CMOS fabrication not only as a high-current-carrying, robust interconnect but also as an effective heat-dissipating, thermal interface material.

Acknowledgments

The author is grateful to his colleagues at NASA Ames Center for Nanotechnology for providing much of the material described in this chapter.

References

1 S. Iijima, *Nature* 1991, **354**, 56.
2 M. Meyyappan (Ed.), *Carbon Nanotubes: Science and Applications*, CRC Press, Boca Raton, FL, 2004.
3 J. Han,in: M. Meyyappan (Ed.), *Carbon Nanotubes: Science and Applications*, CRC Press, Boca Raton, FL, 2004, Chapter 1.
4 M. A. Osman, D. Srivastava, *Nanotechnology* 2001, **12**, 21.
5 J. Hone, M. Whitney, C. Piskoti, A. Zetti, *Phys. Rev. B* 1999, **59**, R2514.
6 T. W. Ebbesen, P. M. Ajayan, *Nature* 1992, **358**, 220.
7 T. Guo, P. Nikolaev, A. Thess, D. T. Colbert, R. E. Smalley, *Chem. Phys. Lett.* 1995, **243**, 49.
8 M. Meyyappan,in: M. Meyyappan (Ed.), *Carbon Nanotubes: Science and Applications*,

CRC Press, Boca Raton, FL, 2004, Chapter 4 (and references therein).
9 L. Delzeit, B. Chen, A. M. Cassell, R. M. D. Stevens, C. Nguyen, M. Meyyappan, *Chem. Phys. Lett.* 2001, **348**, 368.
10 M. Meyyappan, L. Delzeit, A. Cassell, D. Hash, *Plasma Sources Sci. Technol.* 2003, **12**, 205.
11 K. Teo, D. Hash, R. Lacerda, N. L. Rupesinghe, M. B. Sell, S. H. Dalal, D. Bose, T. R. Govindan, B. A. Cruden, M. Chhowala, G. A. J. Amaratunga, M. Meyyappan, W. L. Milnes, *Nano Lett.* 2004, **4**, 921.
12 V. I. Merkulov, D. H. Lowndes, Y. Y. Wei, G. Eres, E. Voelkl, *Appl. Phys. Lett.* 2000, **76**, 3555.
13 M. Chhowalla, K. B. K. Teo, C. Ducati, N. L. Rupesinghe, G. A. J. Amaratunga, A. C. Ferrari, D. Roy, J. Robertson, W. I. Milne, *J. Appl. Phys.* 2001, **90**, 5308.
14 C. Bower, W. Zhu, S. Jin, O. Zhou, *Appl. Phys. Lett.* 2000, **77**, 830.
15 K. Matthews, B. A. Cruden, B. Chen, M. Meyyappan, L. Delzeit, *J. Nanosci. Nanotech.* 2002, **2**, 475.
16 International Technology Roadmap for Semiconductors (Semiconductor Industry Association, San Jose, CA 2001); http://public.itrs.net/.
17 T. Yamada,in: M. Meyyappan (Ed.), *Carbon Nanotubes: Science and Applications,* CRC Press, Boca Raton, FL, 2004, Chapter 7.
18 S. J. Tans, A. R. M. Verschueren, C. Dekker, *Nature* 1998, **393**, 49.
19 R. Martel, T. Schmidt, H. R. Shen, T. Hertel, Ph. Avouris, *Appl. Phys. Lett.* 1998, **76**, 2447.
20 V. Derycke, R. Martel, J. Appenzeller, Ph. Avouris, *Appl. Phys. Lett.* 2002, **80**, 2447.
21 S. J. Wind, J. Appenzeller, R. Martel, V. Derycke, Ph. Avouris, *Appl. Phys. Lett.* 2002, **80**, 3817.
22 B. Yu, *Proc. IEDM* 2001, 937.
23 F. Nihey, H. Hongo, M. Yudasaka, S. Iijima, *J. Appl. Phys.* 2002, **41**, L1049.
24 R. V. Seidel, A. P. Graham, J. Kretz, B. Rajasekharan, G. S. Duesberg, M. Liebau, E. Unger, F. Kreupl, W. Hoenlein, *Nano Lett.* 2005, **5**, 147.
25 X. Liu, S. Han, C. Zhou, *Nano Lett.* 2006, **6**, 34.
26 W. B. Choi, B. H. Cheong, J. J. Kim, J. Chu, E. Bae, *Adv. Funct. Mater.* 2003, **13**, 80.
27 W. B. Choi, E. Bae, D. Kang, S. Chae, B. H. Cheong, J. H. Ko, E. Lee, W. Park, *Nanotechnology* 2004, **15**, S512.
28 X. Liu, C. Lee, C. Zhou, J. Han, *Appl. Phys. Lett.* 2001, **79**, 3329.
29 V. Derycke, R. Martel, J. Appenzeller, Ph. Avouris, *Nano Lett.* 2001, **1**, 453.
30 A. Bachtold, P. Hadley, T. Nakanishi, C. Dekker, *Science* 2001, **294**, 1317.
31 Y. M. Lin, J. Appenzellar, J. Knoch, Z. Chen, Ph. Avouris, *Nano Lett.* 2006, **6**, 930.
32 A. Raychowdhury, K. Roy, *IEEE Trans. Nanotechnol.* 2005, **4**, 168.
33 M. Menon, D. Srivastava, *Phys. Rev. Lett.* 1997, **79**, 4453.
34 M. Menon, D. Srivastava, *J. Mater. Res.* 1998, **13**, 2357.
35 B. C. Satishkumar, P. J. Thomas, A. Govindaraj, C. N. R. Rao, *Appl. Phys. Lett.* 2000, **77**, 2530.
36 A. M. Cassell, G. C. McCool, H. T. Ng, J. E. Koehne, B. Chin, J. Li, J. Han, M. Meyyappan, *Appl. Phys. Lett.* 2003, **82**, 817.
37 W. B. Choi, unpublished results.
38 D. Srivastava, M. Menon, K. J. Cho, *Comput. Sci. Eng.* 2001, **3**, 42.
39 T. Yamada, *Appl. Phys. Lett.* 2000, **76**, 628.
40 T. Yamada, *Appl. Phys. Lett.* 2001, **78**, 1739.
41 T. Yamada, *Appl. Phys. Lett.* 2002, **80**, 4027.
42 T. Yamada, *Phys. Rev. B* 2004, **69**, 123408.
43 J. Guo, M. Lundstrom, S. Datta, *Appl. Phys. Lett.* 2002, **80**, 3192.
44 J. Guo, S. Hasan, A. Javey, G. Bosman, M. Lundstrom, *IEEE Trans. Nanotechnol.* 2005, **4**, 715.
45 S. Hasan, S. Salahuddin, M. Vaidyanathan, M. A. Alan, *IEEE Trans. Nanotechnol.* 2006, **5**, 14.
46 F. Leonard, *Nanotechnology* 2006, **17**, 2381.
47 T. Rueckes, K. Kim, E. Joselevich, G. Y. Tseng, C. L. Cheung, C. M. Lieber, *Science* 2000, **289**, 94.

48 W. B. Choi, S. Chae, E. Bae, J. W. Lee, B. Cheung, J. R. Kim, J. J. Kim, *Appl. Phys. Lett.* 2003, **82**, 275.

49 B. Q. Wei, R. Vajtai, P. M. Ajayan, *Appl. Phys. Lett.* 2001, **79**, 1172.

50 F. Kreupl, A. P. Graham, G. S. Duesberg, W. Steinhogl, M. Liebau, E. Unger, W. Honlein, *Microelec. Eng.* 2002, **64**, 399.

51 N. Srivastava, R. V. Joshi, K. Banerjee, *IEDM Proc.* 2005, 257.

52 J. Li, Q. Ye, A. Cassell, H. T. Ng, R. Stevens, J. Han, M. Meyyappan, *Appl. Phys. Lett.* 2003, **82**, 2491.

53 A. Svizhenko, M. P. Anantram, T. R. Govindan, *IEEE Trans. Nanotechnol.* 2005, **4**, 557.

54 P. Schelling, L. Shi, K. E. Goodson, *Mater. Today* 2005, 30.

55 R. Viswanatha, V. Wakharkar, A. Watwe, V. Lebonheur, *Intel Tech. J.* 2000, **Q3**, 1.

56 Q. Ngo, B. A. Cruden, A. M. Cassell, G. Sims, M. Meyyappan, J. Li, C. Yang, *Nano Lett.* 2004, **4**, 2403.

57 H. Dai, J. H. Hafner, A. G. Rinzler, D. T. Colbert, R. E. Smalley, *Nature* 1996, **384**, 147.

58 C. V. Nguyen,in: M. Meyyappan (Ed.), *Carbon Nanotubes: Science and Applications*, CRC Press, Boca Raton, FL, 2004, Chapter 6.

59 C. V. Nguyen, K. J. Chao, R. M. D. Stevens, L. Delzeit, A. M. Cassell, J. Han, M. Meyyappan, *Nanotechnology* 2001, **12**, 363.

60 C. V. Nguyen, C. So, R. M. D. Stevens, Y. Li, L. Delzeit, P. Sarrazin, M. Meyyappan, *J. Phys. Chem. B* 2004, **108**, 2816.

61 Q. Ye, A. M. Cassell, H. Liu, K. J. Chao, J. Han, M. Meyyappan, *Nano Lett.* 2004, **4**, 1301.

8
Concepts in Single-Molecule Electronics

Björn Lüssem and Thomas Bjørnholm

8.1
Introduction

Molecular electronics is a wide field of research, which consists of such diverging topics as organic light-emitting diodes (OLEDs), organic field effect transistors (OFETs; see Chapter 9) or, the topic of this chapter, single-molecule devices. Whereas, OLEDs and OFETs exploit the properties of a large number of molecules, in the field of single-molecule electronics an attempt is made to condense the entire functionality of an electronic device into a single molecule.

The field of (single) molecular electronics owes its significance to the tremendous downscaling that microelectronics has experienced during the past decades. In the ITRS roadmap [1], it is expected that, by the year 2013, the physical gate length of a transistor will scale down to 13 nm – that is, the transistor channel will consist of only a couple of atoms in a row. In order to obtain reliable devices, the composition of the devices must be controlled to only a few atoms – a demand that seems not to be feasible for conventional lithographic methods.

Chemistry – and especially organic chemistry – learned long ago how to control precisely the composition of a molecule to the last atom. Thus, the utilization of single organic molecules can be regarded as the ultimate miniaturization of electronic devices.

The concept of single-molecule electronics was first suggested in 1974 by Aviram and Ratner [2], who proposed that a single molecule consisting of a donor and an acceptor group could function as a diode. Unfortunately, however, at that time it was experimentally not feasible to test these predictions.

Molecular electronics gained impetus during the 1990s and early 2000s, when several molecular devices were proposed, including a single molecular switch [3] or a diode showing a negative differential resistance [4]. These early results raised great expectations, as evidenced by the election of molecular electronics as the "breakthrough of the year 2001" [5]. However, only two years later, the fledgling field

Nanotechnology. Volume 4: Information Technology II. Edited by Rainer Waser
Copyright © 2008 WILEY-VCH Verlag GmbH & Co. KGaA, Weinheim
ISBN: 978-3-527-31737-0

of molecular electronics experienced its first drawback when it was reported that some of the early results might be due to artifacts [6].

These reports on possible artifacts helped to settle the expectations laid on molecular electronics to a reasonable level, and in the current phase of development more emphasis has been placed on proving that the observed effects are in fact "molecular", and on identifying experimental set-ups that avoid the possible introduction of artifacts.

In this chapter, a brief overview is provided of the field of single-molecule electronics, beginning with a short theoretical introduction that aims to define the concepts and terminology used. (A more extensive explanation of the theory can be found in Part A of Volume III of this series.) In the following sections the text is more factual, and relates to how single molecules can actually be contacted and which functionalities they can provide. The means by which these molecules may be assembled to implement complex logical functions are then described, followed by a brief summary highlighting the main challenges of molecular electronics.

8.2
The General Set-Up of a Molecular Device

In this section, the basic concepts used in subsequent sections will be explained, and the presence of two domains of current transport – *strong coupling* and *weak coupling* – will be outlined.

Electrical transport across single molecules is remarkably different from conduction in macroscopic wires. In a large conductor, charge carriers move with a mean drift velocity v_d, which is proportional to the electric field, E. Together with the density of free charge carriers, this proportionality gives rise to Ohm's law.

For a single molecule this model is not applicable. Instead of considering drift velocities and resistances, which are only defined as average over a large number of charge carriers, concern is centered on the transmission of electrons across the molecule.

The general set-up of a molecular device is shown in Figure 8.1. The molecule is connected by two electrodes, labeled "source" and "drain", while the electrostatic potential of the molecule can also be varied by using a gate electrode.

If a voltage is applied between the source and drain (i.e. a negative voltage with respect to the drain), the electrochemical potential of the source, μ_S, shifts up and the potential of the drain, μ_D, moves down. An energy window is opened between these two potentials; in this energy window filled states in the source oppose empty states at the same energy in the drain.

However, as the two electrodes are isolated from each other, electrons cannot easily flow from the source to the drain. Only if a molecular level enters the energy window between μ_S and μ_D, can electron transport be mediated by these molecular levels (see Figure 8.1b). Therefore, each time the electrochemical potential of the source aligns with a molecular level the current rises sharply. In Figure 8.1b, for example, the source potential exceeds the lowest unoccupied molecular orbital (LUMO), and

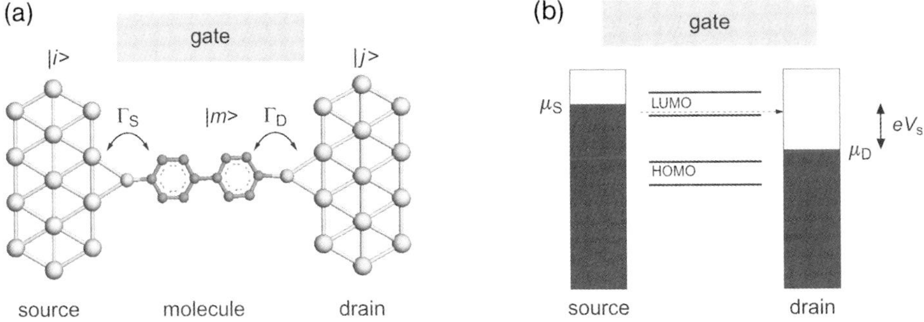

Figure 8.1 General set-up of a molecular device. In (a) a molecule is coupled to the source and drain (coupling strengths Γ_S and Γ_D). In (b) the molecule is replaced by its molecular levels. The electrodes are filled with electrons up to their electrochemical potential (indicated by the hatched area).

electrons can be transmitted across this level. Similarly, the drain potential can drop below the highest occupied molecular orbital (HOMO), which would also initiate electron flow.

8.2.1
The Strong Coupling Regime

Depending on the strength of the coupling of the electrodes with the molecule, there are two domains of the electron transport: the *weak coupling limit* and the *strong coupling limit*. To distinguish between these two limits, coupling strengths Γ_S and Γ_D can be defined that describe how strongly the electronic states of the source or drain $|i\rangle$, $|j\rangle$ interact with the molecular eigenstates $|m\rangle$. Γ_S and Γ_D have the dimension of energy; a high energy means that the electrode states can strongly couple with the molecular states.

High coupling energies therefore result in electronic wavefunctions of the electrodes that can extend into the molecule, so that charge can be easily transmitted from the source, across the molecule towards the drain. Thus, the current across the molecule can be expressed in terms of a transmission coefficient $T(E)$

$$I = \frac{2e}{h}\int_{\mu_D}^{\mu_S} T(E)dE \tag{8.1}$$

where e is the elementary charge and h is Planck's constant.

The transmission coefficient represents the transmission probability of electrons with a certain energy E to be transmitted across the molecule. This probability peaks at the molecular levels. It can be shown that the maximum conductance per molecular level $G_0 = \frac{\Delta I}{\Delta E \cdot e}$ becomes [7, 8]

$$G_0 = \frac{2e^2}{h} \tag{8.2}$$

This maximum conductance of a single electronic level is known as *quantum of conductance*, and corresponds to a resistance of 12.5 kΩ. Most interestingly, this conductance is not dependent on the length of the molecule as long as the ideal molecular level extends between the source and the drain electrodes.

8.2.2
The Weak Coupling Regime

In comparison to the strong coupling limit, current transport is remarkably different, if the coupling strengths Γ_S and Γ_D are weak. Here, the wavefunction of the electrodes cannot extend into the molecule and charge cannot be easily transmitted across the molecule. Rather, electrons must hop or tunnel *sequentially* from the source onto the molecule, and finally from the molecule to the drain. The lowest amount of charge that can be transferred onto the molecule is the elementary charge, e. This has an interesting consequence.

The molecule is electrostatically connected to the source, drain and gate electrode by the capacitances C_S, C_D, and C_G, respectively (see Figure 8.2a). Therefore, it is no longer sufficient that the electrochemical potential of the source aligns with the molecular level. Additionally, it must supply enough energy ($E_C(N)$, where N is the number of electrons) to charge the capacitances with an additional electron.

$$E_C(N+1) = \frac{1}{2}\frac{(N+1)^2e^2 - N^2e^2}{\underbrace{C_S + C_D + C_G}_{C_\Sigma}} = \frac{\left(N+\frac{1}{2}\right)e^2}{C_\Sigma} \tag{8.3}$$

To include this energy, the energy diagram shown in Figure 8.1b may be refined (see Figure 8.2b). Two levels are included in this diagram, which correspond to the HOMO (lower) and LUMO states shown in Figure 8.1b. The LUMO is floated upwards by $E_C(1) = \frac{e^2}{2C_\Sigma}$, while the HOMO is moved down by the same amount.

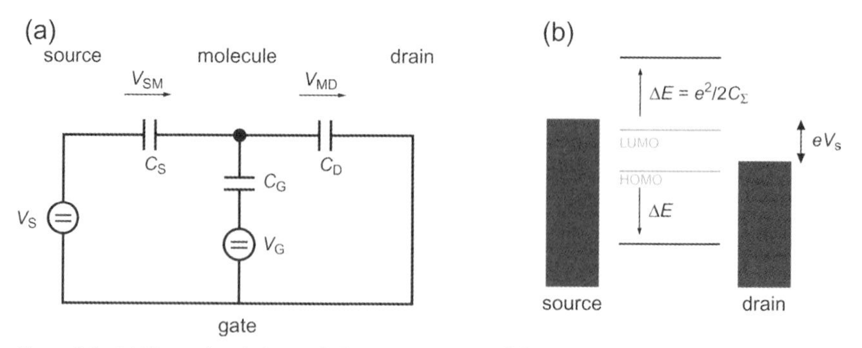

Figure 8.2 (a) The molecule is coupled to source, gate and drain by capacitances. (b) The energy level in the weak coupling limit. Additional charging energy must be provided by the source voltage.

Thereby, a large energy gap is opened within which no electrons can flow and hence the current is blocked; this effect is known as *coulomb blockade.*[1)]

8.3 Realizations of Molecular Devices

In the preceding section, the coupling of the molecule to the electrodes was described by the coupling strengths Γ_S and Γ_D. However, this theoretical description of contact between molecule and electrodes hides the complexity and difficulties that must be overcome in order to contact a single molecule. The main strategy that is followed to contact single molecules is to use specifically designed molecular anchoring groups that bind and self-organize on the contacts. In the following section, some key examples are provided of anchoring groups and self-organization strategies.

8.3.1 Molecular Contacts

Molecular end groups must provide a chemical bond to the contacting metal – that is, they must offer a self-organizing functionality. Furthermore, the nature of the contact determines the coupling strength Γ and, therefore, how strongly the molecular states couple with the electronic states of the electrodes. In the case of strong coupling, electrons can be easily transmitted across the molecule and the resistance of the molecule should be low; conversely, new effects such as coulomb blockade can occur for weak coupling, and this may be exploited for new devices. Thus, the suitable choice of molecular contact is one of the main issues in the design of a molecular device.

Various molecular contacts have been proposed. Besides the most common gold–sulfur bond, sulfur also binds to other metal such as silver [9] or palladium [10]. Sulfur may be replaced by selenium [11], which yields higher electronic coupling. A further increase in coupling strength is provided by dithiocarbamates [12, 13], which is explained by resonant coupling of the binding group to gold. Other binding groups include –CN [14], silanes [15], and molecules directly bound to either carbon [16] or silicon [17].

Using these binding groups, it is possible to contact single or at least a low number of molecules. In the past, several experimental set-ups have been developed which differ in the numbers of molecules contacted. Whereas some set-ups allow contacting single molecules (i.e. the method of *mechanically controlled break junctions, nanogaps* or *scanning probe methods*), in other arrangements the demand of a single molecule is relaxed and a small number of molecules is contacted (e.g. in the *crossed wire set-up* or in a *crossbar structure*). One further distinction between these set-ups is the number of electrodes that can contact each molecule – that is, if besides the source and drain a gate electrode is also present.

1) More details on single-electron effects can be found in Chapter 2 of Volume III in this series.

8.3.2
Mechanically Controlled Break Junctions

The concept of mechanically controlled break junctions dates back to 1985, when the method was used to obtain superconducting tunnel junctions [18]. In 1997, it was applied to contact a single molecule between two gold electrodes [19]. By comparing the current versus voltage characteristic of symmetric and asymmetric molecules, it was shown that only single molecules are contacted [20]. Since then, a variety of different molecules have been studied using this technique, and in particular the use of a low-temperature set-up was seen to provide a significant improvement in data quality [21–29].

The general set-up of a mechanically controlled break junction is shown in Figure 8.3. A metallic wire, which is thinned in the middle, is glued onto a flexible

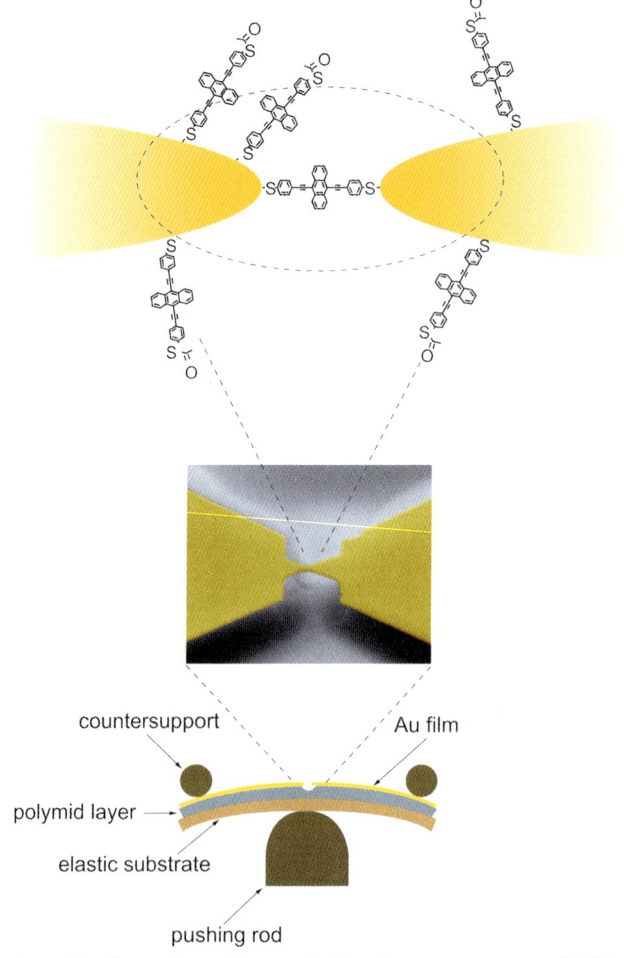

Figure 8.3 The mechanically controlled break junction. (From Ref. [30].)

substrate. Often, the wire is under etched so that a freestanding bridge is formed. Underneath the substrate, a piezo element can press the sample against two countersupports, which causes the substrate to bend upwards such that a strain is induced in the wire. If the strain becomes too large, the wire breaks and a small tunneling gap opens between the two parts of the wire. The length of the tunneling gap can be precisely controlled by the position of the piezo element.

To contact a single molecule, either a solution of the molecule is applied to the broken wire, or the molecules have already been preassembled onto the wire before it is broken. As described above, these molecules have chemical binding groups at both ends that easily bind to the material of the wire. As the molecule has binding groups at both ends it can bridge the tunneling gap if the length of the latter is properly adjusted. In this way a single-molecule device is formed.

Mechanically controlled break junctions represent a stable and reliable method for contacting single molecules. Most importantly, the correlation between molecular structure and current versus voltage characteristic can be studied, which will stimulate the understanding of the conduction through single molecules. At present, however, there is no way of integrating these devices – that is, it is not possible to contact a larger number of molecules in parallel and to combine these molecules into a logic device.

8.3.3 Scanning Probe Set-Ups

Due to its high spatial resolution, scanning probe methods (SPM) are well suited to contact single molecules, and several strategies have evolved during recent years.

One strategy is to contact a so-called self-assembled monolayer (SAM) of the molecule of interest with an atomic force microscope, using a conductive tip [31–34] (see Figure 8.4b). SAMs are formed by immersing a metallic bottom electrode into a solution of molecules (see Figure 8.4a) which must possess an end group that covalently binds to the metal layer. In the first layer, the molecules attach covalently to the metal; all following layers are then physisorbed onto this first chemisorbed layer. The physisorbed layers can easily be washed off using an additional rinsing step, such that only the first, chemisorbed, layer remains on the metal. A famous example of such a molecular end group/metal surface combination is sulfur on gold. Thiolates, and especially alkanethiols, are known to perfectly organize on a gold surface and to build a SAM that covers the gold electrode [35].[2)]

These SAMs can be contacted by a conductive atomic force microscopy (AFM) tip (see Figure 8.4b). Depending on the tip geometry, a low number of molecules (ca. 75) can be contacted [31]. Using this method, it has been shown that the current through alkanethiolates and oligophenylene thiolates decreases exponentially with the length of the molecule, and that the resistance of a molecule is dependent on the metal used to contact the molecule [36].

2) See also Chapter 9, which provides a broader introduction into self-organization and SAMs.

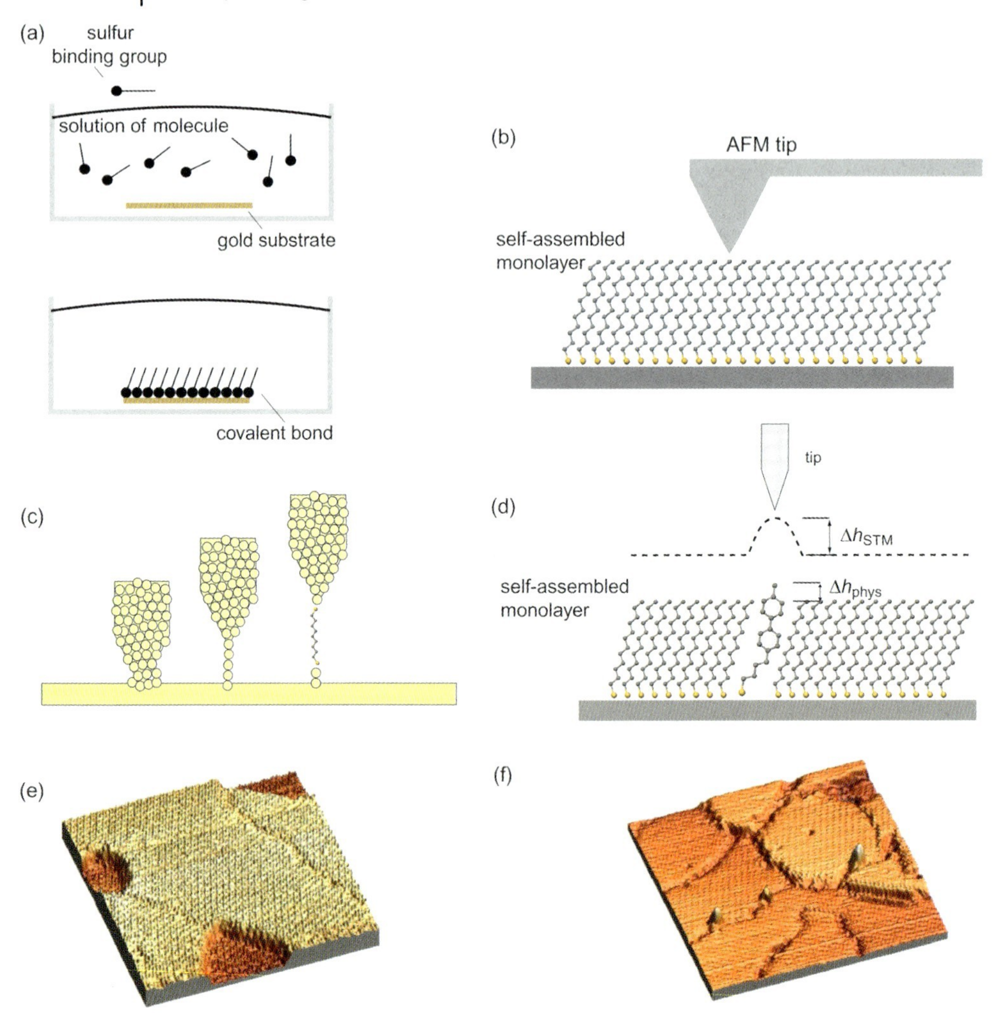

Figure 8.4 The different methods used to contact single molecules with SPM techniques. (a) The basic principle of self-assembled monolayer (SAM) formation. (b) A SAM of molecules is contacted by a conductive AFM tip. (c) The "tip crash" method, which forms a small tunneling gap. (d) Embedding conductive molecules in a SAM of insulating alkanethiols (molecules not drawn to scale). The height of the molecules can be used to deduce their conductivity. (e, f) Scanning tunneling microscopy (STM) image of a SAM of alkanethiols (e) and of oligo-phenylenevinylene molecules embedded into a SAM of alkanethiols. The molecules can be seen protruding from the SAM (f) [37].

An alternative method of contacting molecules using AFM or scanning tunneling microscopy (STM) is very similar to the break junction technique [38, 39]. A gold AFM or STM tip is moved into a gold substrate and subsequently slowly retracted (the "tip-crash" method shown in Figure 8.4c). Thereby, a thin gold filament is formed between the tip and the substrate. If the tip is moved too far away from the substrate, the filament will break and a small tunneling gap is opened between the substrate and the tip. The whole set-up is immersed in a solution of molecules that have functional binding groups at both ends. If the tunneling gap is approximately the size of the molecule, there is a probability that one molecule will bind to the tip and the substrate and will therefore bridge the gap.

As a third strategy of contacting single molecules, the molecules of interest can be embedded into a SAM of insulating alkanethiols (see Figure 8.4d–f). In this way it is possible to obtain single, isolated molecules which "protrude" from the surrounding alkanethiol SAM (see Figure 8.4f). The conductance of the molecule can be measured either by placing the STM tip above the molecule [40, 41] or by measuring the height difference between the embedded molecule and the surrounding alkanethiol SAM [37, 42, 43]. This height difference is not only dependent on the differences in length of the alkanethiol and the molecule but also reflects differences in the conductivities of the molecules. Therefore, the conductivity of the embedded molecule can be calculated from the height difference.

8.3.4 Crossed Wire Set-Up

This set-up consists of two crossed wires, which almost touch at their crossing point. One of these wires is modified with a SAM of the molecule of interest. A magnetic field is applied perpendicular to one wire, and a dc current is passed through this wire. This causes the wire to be deflected due to the Lorentz force, and consequently the separation of the two wires can be adjusted by setting the dc current [44–47].

It has been shown that the number of contacted molecules is dependent on the wire separation, and that the current versus voltage characteristics measured at different separations are all integer multiplies of a fundamental characteristic. Thus, it is proposed that this fundamental curve represents the characteristic of only a single molecule [45].

These measurements can also be carried out at cryogenic temperatures [46]. In this case, the vibronic states of the molecule can be identified which provide a "molecular fingerprint" and prove that the molecule is actively involved in the conduction process (see also Section 8.4.6) [46].

8.3.5 Nanogaps

Similar to the break junction method, in the nanogap set-up a small gap is formed in a thin metal wire. However, this gap is not formed by bending a flexible substrate and mechanically breaking the wire, but it is prepared on a rigid substrate by using various methods.

One such method is *electromigration*. The preparation of the nanogap starts with the definition of a thin metallic wire on an insulating substrate. A SAM is then deposited on top of this wire by immersing the sample into a molecular solution.

Subsequently, a voltage ramp is applied to the wire. If the current that flows through the wire becomes too large, the wire breaks due to electromigration, which is reflected by a drop in current. Such gaps are approximately 1 nm in width [48] and, with a certain degree of probability, are bridged by a molecule that was deposited onto the wire in advance; thus, a single molecular device has been built.

Alternatively, nanogaps can be formed by preparing an electrode pair with a gap of ~30 nm [49, 50] using electron-beam lithography. This gap can be shrunk to molecular dimensions by electrochemically depositing metal atoms [51, 52]. Combined with electrochemical etching, this method allows a precise control over the wire separation. A combination of lithography and low-temperature evaporation has also been used to fabricate 1- to 2-nm gaps directly on a gate oxide [53].

Although all of these techniques define the gaps laterally, the precise control over the vertical thickness of thin films can be exploited to define a vertical nanogap [54]. The preparation of these gaps starts with the deposition of a thin SiO_2 layer on top of a highly p-doped silicon bottom electrode (see Figure 8.5). A gold electrode is then deposited on top of the SiO_2 layer, and subsequently the oxide can be etched in hydrofluoric acid, thus yielding a thin gap between the Si bottom and Au top electrode.

A rich variety of molecules has been measured using nanogaps, including a coordination complex containing a Co atom [55], a divanadium molecule [56], C60 [57, 58], C140 [59], and phenylenevinylene oligomers [53, 60].

8.3.6
Crossbar Structure

In terms of integration, the crossbar structure is a very interesting device set-up where the demand of single molecules is relaxed, and a rather low number of molecules are contacted.

In order to obtain a crossbar structure, parallel metallic wires are deposited onto an insulating substrate. A SAM of the molecule of interest is then deposited on top of these wires. Orthogonally to the bottom electrodes, metallic wires are deposited onto the SAM. Thus, a single crossbar structure consists of many possible devices (e.g. see Figure 8.21 in Section 8.5.1).

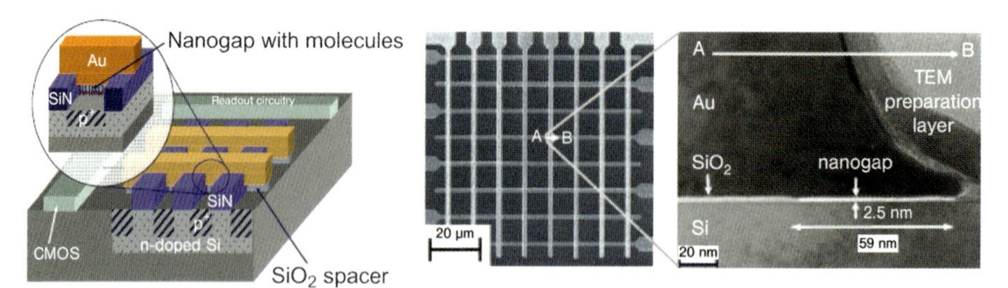

Figure 8.5 Vertical nanogaps integrated in a crossbar structure. (a) A schematic of the set-up. (b) Left: Scanning electron microscopy image of the crossbar, and (right) a transmission electron image of the nanogap. (From Ref. [54].)

The major technological problem in the crossbar set-up lies in the deposition of the top electrode. The metal/molecule interface may be unstable and metal ions can migrate through the molecular layer [61–63], thus shorting the device. The probability of metal ions penetrating the molecular layer is dependent on the molecular top group. A group which binds the metal at the top is more resistant, and many metal/molecular end groups have been examined, including Al on CO_2H [64], OH and OCH_3 [65], Cu, Au, Ag, Ca and Ti on OCH_3 [66, 67] or Au, Al and Ti on disulfides [68]. Ti is shown to be critical for metallization, because it reacts strongly with the molecule and partially destroys the SAM [69].

An alternative to these molecular end groups is to use aromatic end groups and to crosslink them with electron irradiation [70, 71]. This method yields stable Ni films on top of a molecular layer. Similarly, the molecular layer can be protected by a spun-on film of a highly conducting polymer film (e.g. PEDOT) [72].

In most devices the top electrode is deposited by evaporation techniques, so that the metal atoms arrive at the molecular layer with a high energy, and thc probability of the atoms punching through the layer is high. Attempts have been made to reduce the energy of the atoms by indirect evaporation and cooling the substrate [73], or by so-called "printing methods" in which the metal film is gently deposited on the molecular layer from a polymeric stamp [74, 75].

8.3.7 Three-Terminal Devices

The incorporation of a third electrode (the gate) opens up new possibilities. First, the molecular levels can be shifted upwards and downwards by the gate relative to the levels of the electrodes, which can be used to analyze the electronic structure of the molecule. The gate is also necessary for building a molecular transistor. As will be shown later, these transistors can be used to build logic circuits.

The basic working principle of a molecular single-electron transistor is illustrated in Figure 8.6. Without a gate voltage applied, no molecular states lie in the energy window between the source and drain potential and thus, the current is blocked. However, the molecular level can be moved down into the energy window, if the gate voltage is increased. Therefore, by applying a gate voltage it is possible to switch the transistor on.

In order to understand, how the gate can shift the molecular levels, the capacitive network shown in Figure 8.2a should be considered. The voltage between the source and the molecule, V_{SM}, and between the molecule and the drain, V_{MD}, is related to the source V_S and source and gate voltage V_G as follows:

$$V_{SM} = \frac{C_D + C_G}{C_\Sigma} V_S - \frac{C_G}{C_\Sigma} V_G \tag{8.4}$$

$$V_{MD} = \frac{C_S}{C_\Sigma} V_S + \frac{C_G}{C_\Sigma} V_G \tag{8.5}$$

with $C_\Sigma = C_S + C_D + C_G$

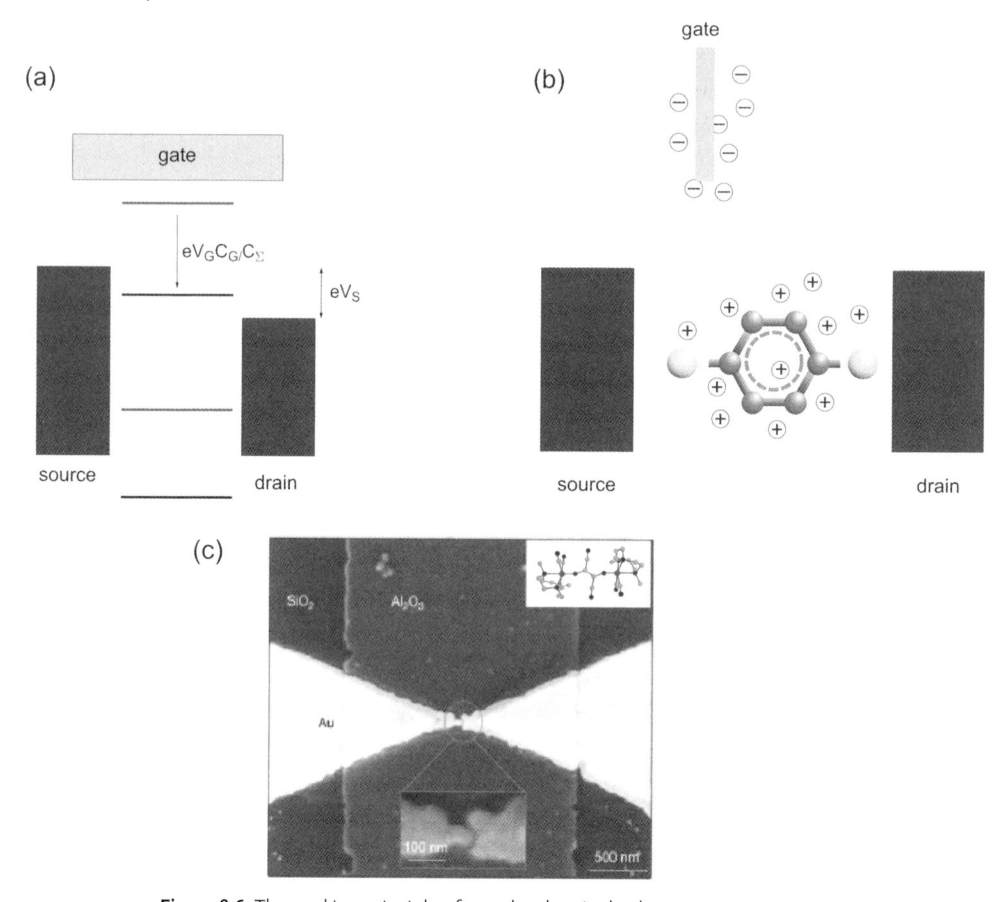

Figure 8.6 The working principle of a molecular single-electron transistor (a) and an electrochemical gate (b). In (c) a scanning electron image of a nanogap fabricated by the electromigration method is shown. The nanogap is prepared on an aluminum strip, which is covered by a thin Al_2O_3 layer and forms the gate. (From Ref. [56].)

Therefore, the molecular level shifts up or down relative to the source and drain energies. The amount of the shift is proportional to the term $\frac{C_G}{C_\Sigma} V_G$. In order to obtain good gate control, $\frac{C_G}{C_\Sigma}$ (the "gate-coupling parameter") must be large and, ideally, close to unity; hence, the gate must be placed very close to the molecule.

One elegant method of obtaining a high gate control is to use an electrochemical gate (c.f. Figure 8.6b). The molecular device (e.g. the nanogap or the STM set-up) is immersed in an electrolyte, and the source and drain voltages are varied relative to a reference electrode which is also immersed in the solution [38, 76, 78] and takes on the function of the gate. The effective gate distance is given by the thickness of the double layer of ions at the electrodes [38], which allows the application of high electric fields. Several molecules, including peralene tetracarboxylic diimide [76], a molecule containing a viologen group [79], oligo(phenylene ethynylene)s [80] or different transition metal complexes [77], have been studied using this type of gate.

Champagne *et al.* succeeded in including a gate in a break junction set-up [81] which consists of a freestanding, under-etched gold bridge deposited onto a silicon wafer. Underneath the bridge, the silicon is degenerately doped and serves as the gate electrode. The bridge is broken by the electromigration technique, and the size of the so-formed gap is adjusted by bending the silicon substrate. A C_{60} molecule is immobilized in this gap, so that a molecular transistor with a gate-molecule spacing of about 40 nm is realized.

The most straightforward method of including a gate is provided by the nanogap set-up. Here, the source and drain are formed on an insulating substrate; however, the insulating layer (e.g. SiO_2 or Al_2O_3) can be very thin, and the underlying (conductive) substrate may be used as gate [55–58, 82] (cf. Figure 8.6c). Compared to an electrochemical gate, this set-up has the advantage that the measurements can be conducted at cryogenic temperatures, which makes the observance of coulomb blockade effects easier and also allows the use of inelastic electron tunneling spectroscopy (see Section 8.4.6) to study the molecules.

8.3.8
Nanogaps Prepared by Chemical "Bottom-Up" Methods

Several strategies for nanogap preparation have been based on the pioneering studies of Brust *et al.* [83, 84], where the chemical preparation of metal nanoparticles was protected by a ligand shell. Two-dimensional arrays of such particles constitute a test bed that may be used to interconnect metal particles separated by a few nanometers by various organic molecules (see Figure 8.7) [85, 86].

By mixing hydrophobic nanoparticles with surfactants, more one-dimensional structures may be formed where molecules interconnect segments of gold nanowires separated by a few nanometers (see Figure 8.8a) [87, 88].

By using a single metal particle inserted in a metal gap prepared "top-down", two well-defined nanogaps may also be realized at the gap–particle interface (cf. Figure 8.8b) [90, 91].

Although all of these systems are easily prepared and are stable at room temperature, as yet it has not been possible to control the gap formation accurately enough to prepare a single gap bridged by a single molecule. Neither have individual gates been reported.

8.3.9
Conclusion

At present, the list of measurement set-ups used in molecular electronics is by far incomplete, and an ever-growing number of techniques is available, including the mercury drop method [92], nanogap preparation by the deposition of gold electrodes

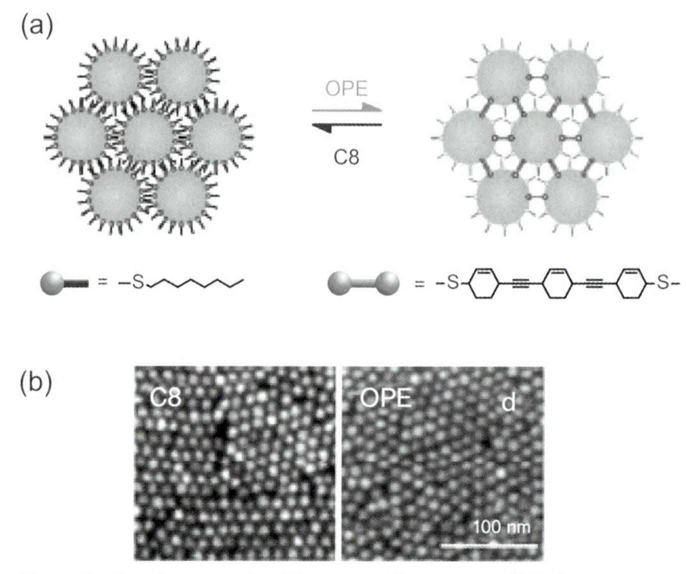

Figure 8.7 Two-dimensional gold nanoparticle arrays interlinked with octanethiol (left) and thiolated oligo(phenylene ethynylene) (right). (a) Schematic representation. (b) SEM image of the nanoparticle arrays. (From Ref. [85].)

through a shadow mask at shallow angles [53], the magnetic bead junction [90, 93], or the nanopore concept [94–97]. Each set-up has its own strengths and weaknesses: some allow the characterization of single molecules (e.g. break junction experiments and SPM set-ups), whereas others are interesting in terms of later applications (e.g. crossbar set-ups) or allow the inclusion of a gate electrode (nanogaps). It appears, however, that there is no ideal set-up, and often the intrinsic molecular behavior may be determined only by a combination of different experimental methods.

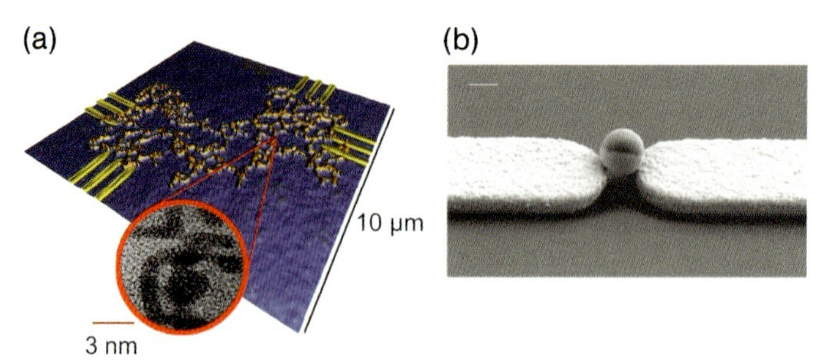

Figure 8.8 (a) Network of gold nanowires. The gaps can be bridged by molecules. (From ref. [89].) (b) A single nanoparticle immobilized between SAM-functionalized electrodes. (From Ref. [90].)

8.4
Molecular Functions

In this section, it is shown which molecules have been measured and which functionalities these molecules can provide. As the ultimate goal of molecular electronics is to provide a universal logic, each technology which aims to achieve this must fulfill several basic requirements [30].

The most basic requirement is that *a complete set of Boolean operators* can be built out of the molecular devices, such that every Boolean function can be obtained. One complete set of operators is for example a disjunction (OR) and an inversion (NOT), or alternatively, a conjunction (AND) and inversion. However, all complete sets have to include inverting gates – that is, an inversion.

Disjunction and conjunction can be relatively easily built out of a resistor and a diode (for a description, see Section 8.4.2.4), and so it is vital to identify molecules that conduct current only in one direction.

Similarly, complete fields of disjunctions and conjunctions can be implemented in crossbar structures in the form of so-called *programmable logic arrays* (PLA) (see Section 8.5.1). At the crossing points of these PLAs, molecules that can be switched on or off are needed; that is, the molecules must possess two conduction states, one isolating and one conducting. Molecules which demonstrate this behaviour include *hysteretic switches*, examples of which are described in Section 8.4.4.

These set-ups do not provide inverting logic, as negation is still missing. One set-up which provides inversion is the *molecular single electron transistor* (see Section 8.4.5). Even with only two terminal devices it is pfossible to construct an inversion using a so-called *crossbar latch* (see Section 8.4.4.1). Again, hysteretic switches are required for this. Inverting logic (e.g. an exclusive disjunction, XOR)[3] can also be built from a variant of a simple diode which displays a *negative differential resistance* (NDR) region – that is, a region in which the current drops with increasing voltage. These gates are described in Section 8.4.3.1.

A second requirement for the implementation of logic is that the gates must provide a means of *signal restoration*. At each stage of the circuit the signal voltage, which represents a logical 0 or 1, will be degraded. To be able to concatenate several logic gates, a means of restoring the original levels must be found, and this requirement can only be relaxed for small circuits, as long as the degradation of the signal voltage is tolerable.

In conventional CMOS logic, such restoration is provided by a non-linear transfer characteristic of the gates [30]. This strategy can also be followed with molecular single-electron transistors. Similarly, signal restoration can also be obtained by two terminal devices using hysteretic switches in the form of the crossbar latch or using NDR diodes in the form of the *molecular latch*, as proposed by Goldstein *et al.* [98].

Another requirement for the technology is that there must be elements that transmit signals across longer distances – that is, a type of *molecular wire* (see Section 8.4.1) must

3) The XOR gate can be converted into a negation if one input is fixed to "1".

be found. A molecular wire alone is insufficient, however, and a defined flow of information must be established, with feedback signals being prevented. This, again, can be achieved by using molecular diodes.

8.4.1
Molecular Wires

The most basic electronic device is a simple wire. However, in molecular electronics it is less easy than might be thought to construct a suitable molecule which transmits current with a low conductance across longer distances. So, what makes a molecule a good conductor? Starting from the short theoretical instruction provided in Section 8.2, certain conclusions can be drawn regarding the properties of an ideal molecular wire.

First, in order to obtain a low resistance a strong electronic coupling of the molecule and the electrodes is preferable. As discussed in Section 8.3.1, such a coupling can be obtained by choosing a suitable molecular binding group, for example a group that provides resonant coupling to the electrodes, such as dithiocarbamates.

Due to this choice of molecular binding group, the limit of strong coupling is valid and electrons are transmitted across the molecular levels. However, a prerequisite for such transmission is that this molecular level, extending from source to drain, actually exists. Extended molecular levels can be formed by delocalized π-systems, where in the tight binding approximation p_z orbitals of isolated carbon atoms add and form an extended, delocalized orbital. Therefore, aromatic groups (e.g. polyphenylene) are often used as the building blocks for molecular wires.

Another important property of a molecular wire is its ability to conduct current at a low bias, and therefore the molecular level used for transport should be close to the Fermi level of the contacts. Often, this requirement is expressed in terms of a low HOMO–LUMO gap.

Many molecules have been proposed as molecular wires, including polyene, polythiophene, polyphenylenevinylene, polyphenylene-ethynylene [99] (see Figure 8.9), oligomeric linear porphyrin arrays [100] or carbon nanotubes (CNTs) [[101] and references therein]. CNTs constitute a special category among molecular wire candidates as they may be either metallic or semiconducting, depending on their chirality. However, it is difficult to selectively prepare or isolate only one type of CNT, namely the metallic form. Furthermore, it is still challenging to organize and orient CNTs, although some techniques are available to arrange CNTs in a crossbar structure [e.g. see [102]]. A more detailed discussion on CNTs is provided in Chapter 10.

8.4.2
Molecular Diodes

Molecular diodes are the next step towards a higher complexity. Indeed, when combined with resistors, diodes are already sufficient to build AND and OR gates.

The first molecular electronic device to be proposed by Aviram and Ratner was just such a molecular diode [103], and consisted of a donor and an acceptor group

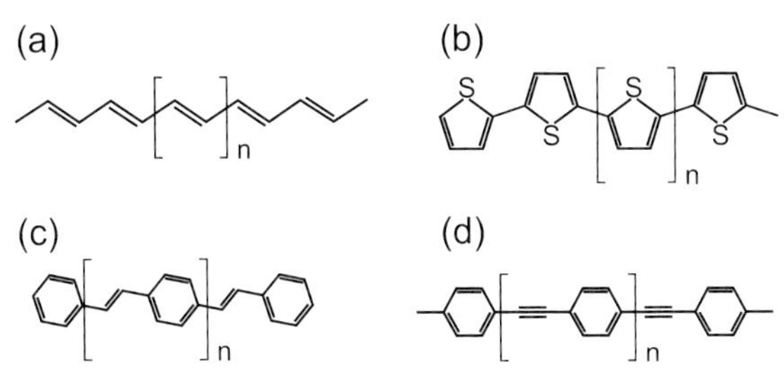

Figure 8.9 Building blocks for molecular wires: (a) polyene; (b) polythiophene; (c) polyphenylenevinylene; and (d) polyphenylene-ethynylene.

separated by a tunneling barrier. This set-up is often compared to the p- and n-layers in a conventional diode. As alternative to the Aviram–Ratner approach, a molecular diode can also be formed by asymmetric tunnel barriers at the source and drain electrodes [104]. This concept is based on the different electrostatic coupling of the electrodes. However, as will be seen later, it is very difficult to couple the molecule *symmetrically* to both electrodes, which in turn makes it difficult to distinguish between rectification due to the Aviram–Ratner mechanism and rectification due to asymmetric coupling.

8.4.2.1 **The Aviram–Ratner Concept**

The Aviram–Ratner concept is illustrated schematically in Figure 8.10. A molecule, which consists of an acceptor and a donor group, is connected to the source and drain. The acceptor and donor are isolated by a tunneling barrier, which ensures that the

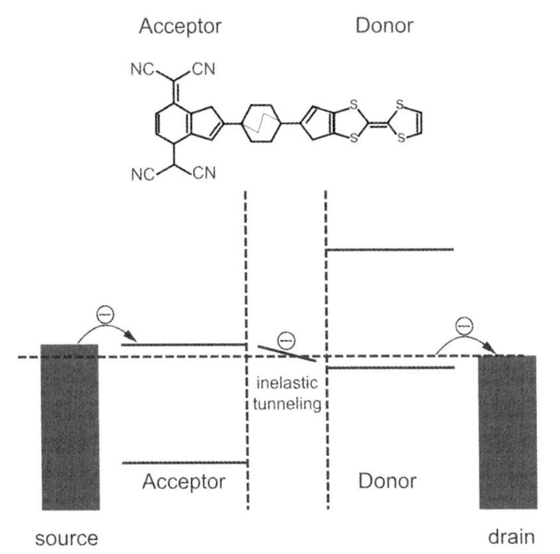

Figure 8.10 The diode proposed by Aviram and Ratner, and the suggested rectifying mechanism. For further details, refer to Section 8.4.2.1.

molecular levels of the two parts do not couple. The HOMO of the donor lies close to the Fermi level, in contrast to the acceptor, where the LUMO is adjacent to the Fermi level.

If a negative voltage V_S is applied to the diode (see Figure 8.2a for polarity), the potential of the source is raised with respect to the drain. Electrons can flow relatively easily from the source, across the acceptor and donor, towards the drain. However, at the opposite polarity a much higher voltage is needed to allow electrons to flow from drain to source. Thus, the molecule is considered to rectify the current.

8.4.2.2 **Rectification Due to Asymmetric Tunneling Barriers**

In contrast to the Aviram–Ratner mechanism, rectification due to asymmetric tunneling barriers is based on a difference in the source and drain capacitances. This difference can be obtained by attaching two insulating alkane chains to a conjugated part (see Figure 8.11). The alkane chains are functionalized with an end group, which provides binding functionality (e.g. sulfur for gold electrodes). The capacitance between the conjugated part of the molecule and the electrode is inversely proportional to the length of the alkane chains; varying these lengths is therefore a suitable way of adjusting the source and drain capacitances.

The rectifying mechanism can be explained by the energy diagram shown in Figure 8.11. The HOMO and the LUMO levels of the conjugated part of the molecule are included in the figure. These energy levels correspond to the molecular level of the (unbound) molecule plus the charging energy, as explained in Section 8.2. As can be seen in Figure 8.11, the levels lie asymmetrical with respect to the Fermi level of the electrodes.

Current can only flow when the electrochemical potential of the source or the drain aligns with, or even exceeds, the LUMO – that is, when $-eV_{SM}=\Delta$ for electrons flowing from source to drain, or $eV_{MD}=\Delta$ for the reverse bias. Here, Δ is the difference between the Fermi level of source and drain at zero bias and the LUMO.

V_{SM} and V_{MD} are given by Eqs. (8.4) and (8.5) (the gate capacitance must be set to zero). It follows for the voltage $V_{D\to S}$, at which electrons start to flow from drain to

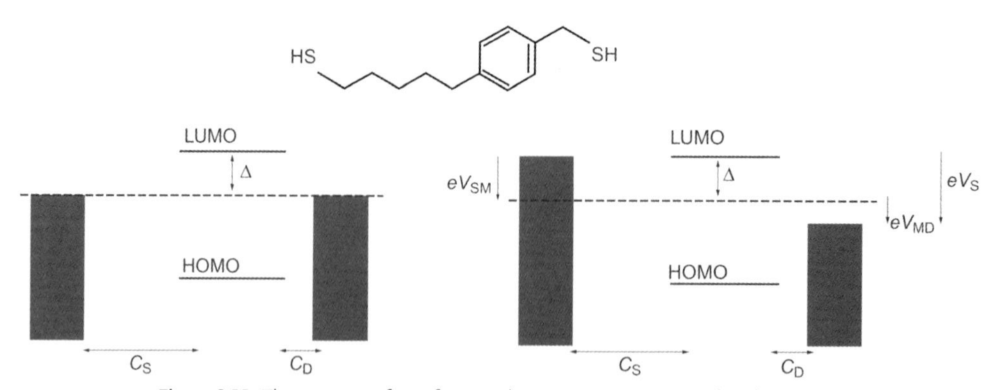

Figure 8.11 The concept of rectification due to asymmetric tunneling barriers.

source (which corresponds to a positive current flow from source to drain), and $V_{S \to D}$, at which electrons flow from source to drain [104]

$$V_{D \to S} = \frac{1 + C_S/C_D}{\underbrace{C_S/C_D}_{\equiv \eta}} \frac{\Delta}{e} = \frac{1+\eta}{\eta} \frac{\Delta}{e} \tag{8.6}$$

$$V_{S \to D} = -(1+\eta)\frac{\Delta}{e} \tag{8.7}$$

If the source capacitance is smaller than the drain capacitance ($\eta < 1$), $|V_{S \to D}|$ is smaller than $|V_{D \to S}|$ and electrons can flow from source to drain at a lower absolute bias than in the opposite direction; that is, the molecule shows a rectification behavior.

8.4.2.3 Examples

Starting from the molecule proposed by Aviram and Ratner (see Figure 8.10) [105], many other molecules containing acceptor and donor groups have been proposed [106]. One of the most extensively studied is γ-hexadecylquinolinium tricyannoquinodimethanide ($C_{16}H_{33}$ Q-3CNQ; c.f. Figure 8.12) [107–110]. Although this molecule consists of a donor and an acceptor group, it deviates from the normal Avriam–Ratner diode in that the two parts are not coupled by an insulating σ-group but rather by a delocalized π-group, which makes an analysis of the rectification behavior more difficult [109]. Furthermore, due to the alkane chain at one side of the molecule, it is often coupled asymmetrically to the electrodes. Thus, it is difficult to distinguish between the Aviram–Ratner mechanism and rectification due to asymmetric tunneling barriers. To circumvent this problem, decanethiol-coated gold STM tips are used as a second contact to a SAM of $C_{10}H_{21}$ Q-3CNQ, which places the same length of alkane groups at both ends of the donor and acceptor groups [111].

8.4.2.4 Diode–Diode Logic

Already with these rather simple diodes it is possible to build AND and OR gates in the form of so-called diode–diode logic [112]. In the AND gate (see Figure 8.13a) both

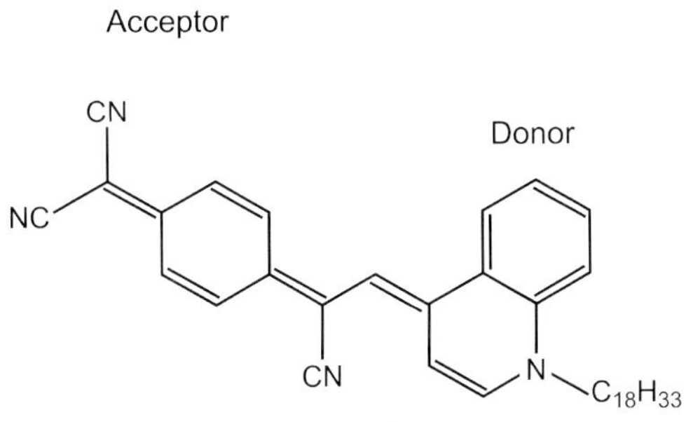

Figure 8.12 The molecular diode $C_{16}H_{33}$ Q-3CNQ in its neutral state.

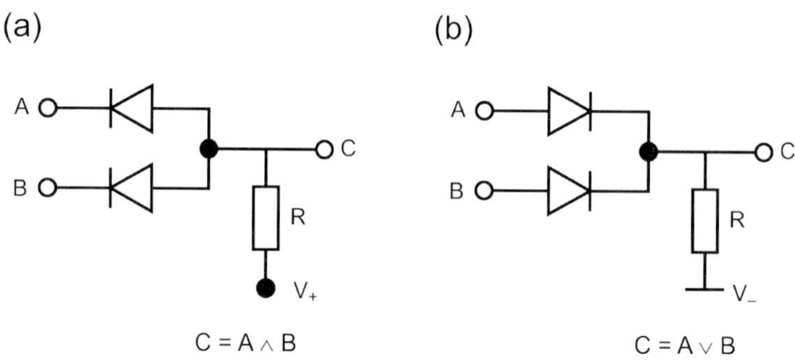

Figure 8.13 AND and OR gate using diode–diode logic.

inputs, A and B, are connected via reversely biased diodes and a resistor to the operating voltage. Only if both inputs are high (i.e. 1) is the output C high, and this results in an AND function.

The OR gate is shown in Figure 8.13b. In contrast to the AND gate, one input is already sufficient to push the output to a high voltage, and thus this gate implements a disjunction.

8.4.3
Negative Differential Resistance Diodes

Diode–diode logic yields AND and OR gates and, in order to obtain a complete set of Boolean variables, these gates can be combined with diodes that show a negative differential resistance (NDR). By using these modified gates it is possible to obtain inversion.

The NDR effect is illustrated schematically in Figure 8.14a. The current of the diode rises with increasing voltage up to a certain threshold voltage, above which the current drops. This rather odd behavior is already known from conventional semiconductors in the form of resonant tunneling diodes.

The concept of resonant tunneling can also be used for molecular NDR devices, as shown in Figure 8.14b and c [113]. The molecule consists of two conjugated molecular leads and an isolated benzene ring in the middle. In the absence of any inelastic processes, current can only flow if the molecular levels of the left lead of the isolated benzene ring and of the right lead, align. Such alignment occurs only at certain voltages, at which the levels are said to be in “resonance”. Such resonance is illustrated graphically in Figure 8.14c. If the voltage is detuned from this resonant value – for example, if it is further increased – then the current will drop and the device will show an NDR effect.

Another molecule, which shows a prominent NDR effect, is shown in Figure 8.15b [4, 94]. It exhibits peak-to-valley ratios as high as 1030:1, although the exact nature of the NDR effect observed in this molecule is currently a matter of intense research [114].

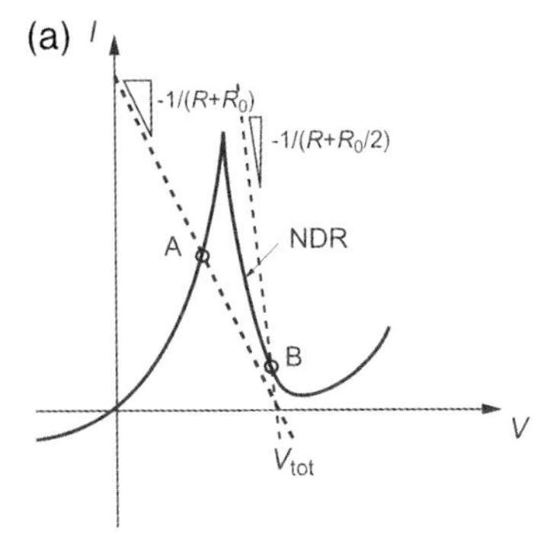

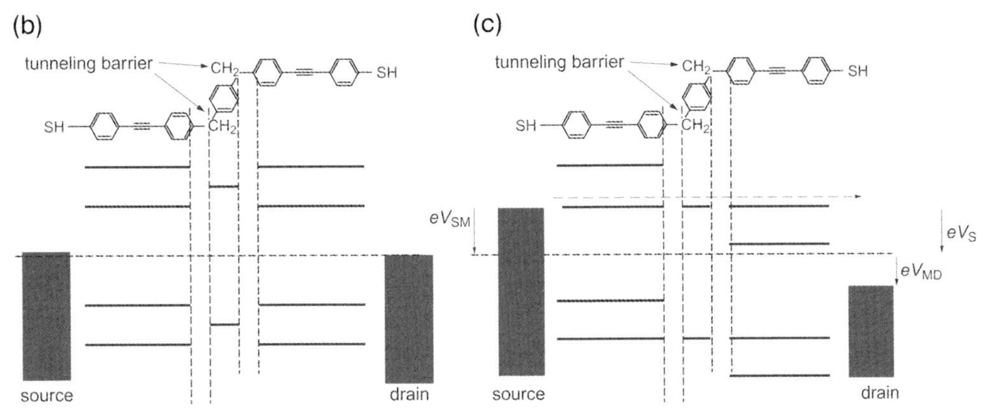

Figure 8.14 (a) The NDR effect. (b, c) The concept of a resonant tunneling diode consisting of a single molecule.

8.4.3.1 **Inverting Logic Using NDR Devices**

NDR devices can be used to obtain inverting logic. One example is shown in Figure 8.15a [112]; this gate implements an exclusive OR functionality – that is, the output C is high if either A or B is high, and low if both inputs are high or both inputs are low.

The XOR gate shown in Figure 8.15a resembles the normal OR gate shown in Figure 8.13b, the only difference being a NDR diode which is connected to the two diodes of inputs A and B. The voltage drop across the entire XOR circuit (V_{tot}, measured from input A and B to V–) is divided between a voltage drop across the NDR diode (V_{NDR}) and a voltage drop across the resistances R_0 and R (V_{R}). Assuming ohmic behavior for the resistances, it follows for the current flowing through the resistances:

$$I = \frac{V_{\text{R}}}{R + R_0} = \frac{V_{\text{tot}} - V_{\text{NDR}}}{R + R_0} \quad \text{if either A or B is high} \tag{8.8}$$

$$I = \frac{V_{\text{R}}}{R + R_0/2} = \frac{V_{\text{tot}} - V_{\text{NDR}}}{R + R_0/2} \quad \text{if both, A and B, are high} \tag{8.9}$$

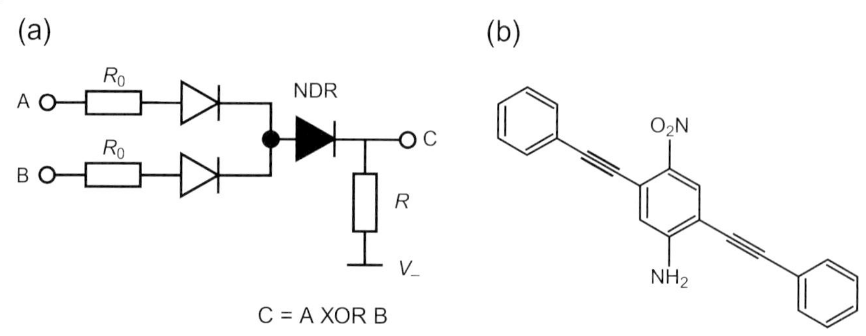

Figure 8.15 (a) XOR gate using NDR-diodes. (b) A well-known molecule that shows NDR.

These two characteristics, called "load lines", are included in Figure 8.14a. The crossing points of the NDR characteristic with these load lines are the operating points for only one input high (point A), or for both inputs high (point B). It transpires that, if both inputs are high, the NDR diode is forced into its valley region, and therefore only a low current flows and only a low voltage drops across *R*. Thereby, the output signal C goes low. From this XOR gate it is easy to obtain inversion; the only change to be made is to set one input (e.g. A) fixed to "1". The output C is then simply the negation of B.

8.4.4
Hysteretic switches

As noted above, hysteretic switches can be used to build PLAs and to yield signal restoration and inversion. The general current versus voltage characteristic of a hysteretic switch is shown in Figure 8.16a.

A hysteretic switch displays two conduction states: one insulating, and one conducting. It is possible to toggle between the two by applying a voltage which

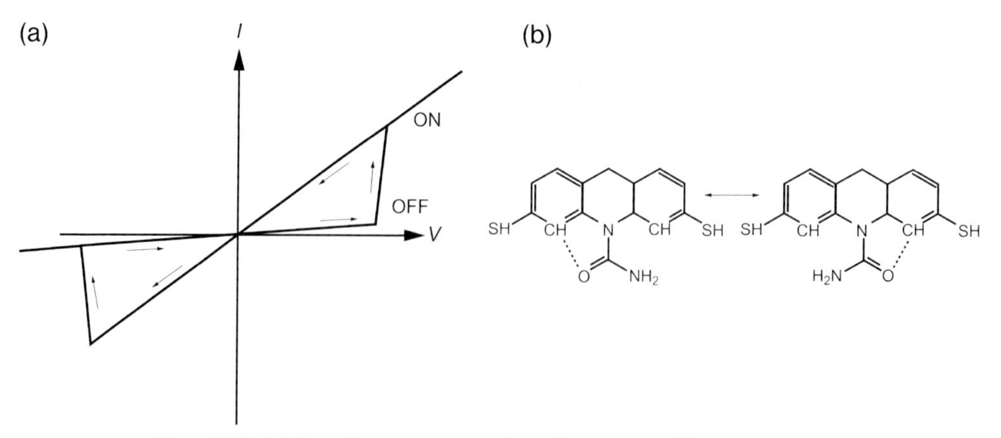

Figure 8.16 (a) General current versus voltage characteristic of a hysteretic switch. (b) A proposed molecule that would be expected to show switching effects.

exceeds a certain threshold value. For example, in Figure 8.16 a positive voltage is needed to switch the device on (i.e. from the insulating to the conducting state), and a negative voltage to switch it off.

Such bistability can be obtained when the molecule possesses two different states that are almost equal in energy, that are separated by an energy barrier, and that show different conduction behaviors. Different origins of these two states are conceivable [99]; for example, they may result from redox processes, from a change in configuration of the molecule, a change in conformation, or a change in electronic excitation.

The molecule shown in Figure 8.16b is an example of a molecule which has been proposed, in theory, to show hysteretic switching [115]. It consists of a fixed molecular backbone (the *stator*) and a side group with a high dipole moment (the *rotor*). By the application of an electric field, the rotor orients its dipole moment in the direction of the field. Bistability is obtained by the formation of hydrogen bonds between the stator and rotor that fix the latter in one of two stable positions relatively to the stator. The two conduction states are due to different conformations of the molecule (stabilized by hydrogen bonds). Switching is initiated by interaction of the dipole moment of the rotor with the electric field.

Bipyridyl-dinitro oligophenylene-ethynylene dithiols (BPDN-DT) are other examples of switching molecules. These are a variation of the molecule shown in Figure 8.15b, and their bistability has recently been confirmed by using various measurement techniques [28, 93].

Rotaxanes and catenanes are, even if controversially discussed, additional candidates for molecular switches. These molecules consist of two interlocked rings such that, by reducing or oxidizing the molecule, one ring rotates within the other. Two stable, neutral states which differ in the position of the inner ring are thus realized. The switching is therefore initiated by a redox process. The two different states are provided by the different conformations of the molecule, similar to the molecule shown in Figure 8.16b.

Although the preliminary results showing a switching effect of this molecule were questioned (see Section 8.4.6), recent results have confirmed the original proposed switching mechanism and indicates that, in the case of earlier results, this mechanism was occasionally hidden by artifacts [75, 116, 117].

8.4.4.1 **The Crossbar Latch: Signal Restoration and Inversion**

As noted in Section 8.4.2.4, AND and OR gates can be built using simple diodes. However, for a complete set of Boolean variables, negation is also required. Signal inversion can be obtained by NDR diodes (c.f. Section 8.4.3.1) or alternatively by the so-called crossbar latch, which also provides a means of signal restoration [118].

The crossbar latch (see Figure 8.17a) consists of two hysteretic switches which are connected to the signal line (at voltage V_L) and one control line (at voltage V_{CA} or V_{CB}). The two switches are oppositely oriented.

The idealized current versus voltage characteristic of a hysteretic switch shown in Figure 8.16 is assumed. By application of a positive voltage, the switch opens; an opposite voltage is then needed to close the switch. These voltages are always those

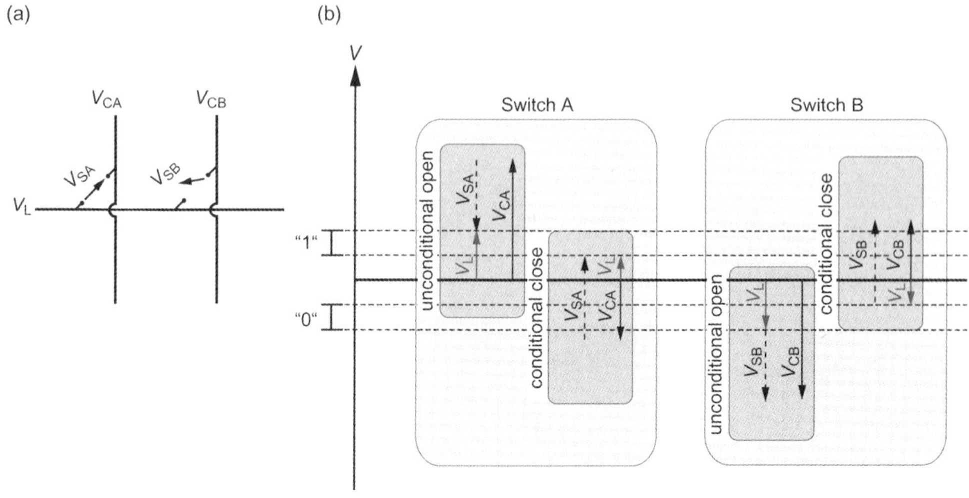

Figure 8.17 The crossbar latch as proposed by Kuekes *et al.* [118].

that are applied *across* the molecules, depicted in Figure 8.17a as V_{SA} and V_{SB}. However, in the circuit shown in Figure 8.17a, only the voltages of the control lines V_{CA} and V_{CB} are set. Therefore, the voltage that is applied *across* the molecule depends on the voltage on the signal line, V_L.

$$V_{SA} = V_L - V_{CA} \tag{8.10}$$

$$V_{SB} = -V_L + V_{CB} \quad \text{(note the opposite orientation of switch B)} \tag{8.11}$$

The voltage on the signal line in Figure 8.17 represents the logical state. Voltage intervals are defined that represent a logical "1" or "0". In general, the signal is degraded, so that the signal level will be at the lower end of the intervals.

To yield signal restoration – that is, to pull the signal level up to the upper end of the defined interval – the following procedure can be followed:

- First, a large positive voltage is applied to the control line A, and a large negative voltage to control line B. This voltage is large enough to always open switches A and B, regardless of the voltage level of the signal line. This is shown in Figure 8.17b (unconditional open).

- Second, a small negative voltage is applied to control line A, and a small positive voltage to control line B. These voltages are so small that they close switch A only if the signal line carries a "1", and they close switch B if there is a "0" on the signal line (see conditional close in Figure 8.17b).

- Therefore, depending on the voltage on the signal line, switch A or B is closed and the opposite switch is open. To yield signal restoration, V_{CA} is connected to a full "1" signal and V_{CB} to "0". Inversion can also be obtained; the only modification is that a logical "1" must be connected to V_{CB} and "0" to V_{CA}.

This scheme shows that it is possible to build a complete set of Boolean variables with only two terminal devices. Therefore, hysteretic switches represent a very valuable element, which explains the high activity in this field of research.

8.4.5 Single-Molecule Single-Electron Transistors

In Section 8.3.7 it was shown how a third electrode – the gate – could be included into the device set-up. The most convenient method to prepare such three-terminal devices is a nanogap that is deposited onto a gate, which is isolated from the device by a thin insulator. A three-terminal device essentially forms a transistor. If the molecule is only weakly coupled to the source and drain, the transistor is termed a *single-molecule single-electron transistor* [119–121]. Whilst the basic working principles of such a transistor were described in Section 8.3.7, and a more extensive explanation is provided in Figure 8.18.

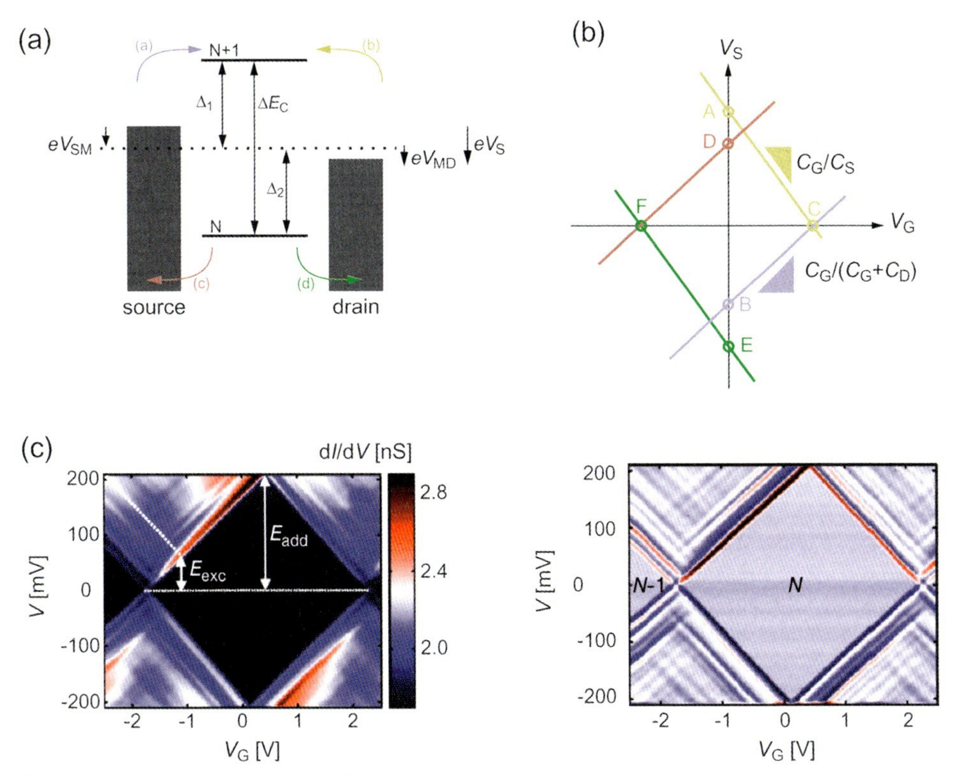

Figure 8.18 (a, b) The working principles of a single-molecule single-electron transistor. For details, see the text. (c) First (upper panel) and second (lower panel) derivatives of the current versus voltage characteristic of a single-molecule single-electron transistor containing an oligophenylenevinylene (OPV5) molecule. The black coulomb blockade diamond is clearly visible. (From Ref. [122].) The fine structure in the open state is due to vibrations of the molecule, and serves as a "fingerprint" of the molecular structure of the OPV5 molecule.

In Figure 8.18a, the Nth and $(N + 1)^{th}$ state of the molecule are shown. These two levels correspond, for example, to the HOMO and LUMO in Figure 8.2b. In Figure 8.18a, electrons can hop or tunnel onto or off the molecule by the processes (a) to (d) – that is, from the source or drain onto the molecule (processes a and b), or from the molecule to the source or drain (processes c and d).

Electrons can only flow from the source or the drain onto the molecule if the potential of the electrode aligns with (or exceeds) the molecular level. Below these potentials, no electrons can flow; rather, the current is blocked, which is known as coulomb blockade. This behavior is often visualized in a plot as in Figure 8.18b, in which the source voltage V_S is plotted against the gate voltage V_G.

In Figure 8.18b, four lines build a diamond, the color of these lines corresponding to processes (a) to (d) in Figure 8.18a. In the interior of the diamond, the source and gate voltages are too low to overcome the injection barriers, and the current is blocked. By increasing the source and gate voltage, the working point of the transistor can be moved to outside the diamond. As the working point crosses one line in the V_S/V_G plane, the current sets in (e.g. if it crosses the dark yellow line, electrons can hop from the drain onto the molecule).

Depending on the process, there are different barriers that must be surmounted by the electron. For electrons hopping from source onto the $(N + 1)^{th}$ level (process a), the voltage between source and molecule V_{SM} must exceed the barrier Δ_1 (see Figure 8.18 for a definition of Δ_1) – that is, $eV_{SM} \geq \Delta_1$.

V_{SM} and V_{MD} are governed by Eqs. (8.4) and (8.5), respectively. By combining Eqs. (8.4) and (8.5) with the conditions for current flow (e.g. $eV_{SM} \geq \Delta_1$ for process a), linear relationships are yielded between V_S and V_G that represent the equations of the four lines in Figure 8.18b.

Single-molecule single-electron transistors resemble in one important aspect conventional MOSFETs. The resistance between source and drain can be controlled by the gate voltage – that is, these transistors represent electronic switches and can be used to build logical circuits. As an example, an inverter based on a single-electron transistor is shown in Figure 8.19. In fact, logic circuits consisting of conventional single-electron transistors have already been presented [121, 123]. Single-electron transistors consisting of single molecules have also been realized

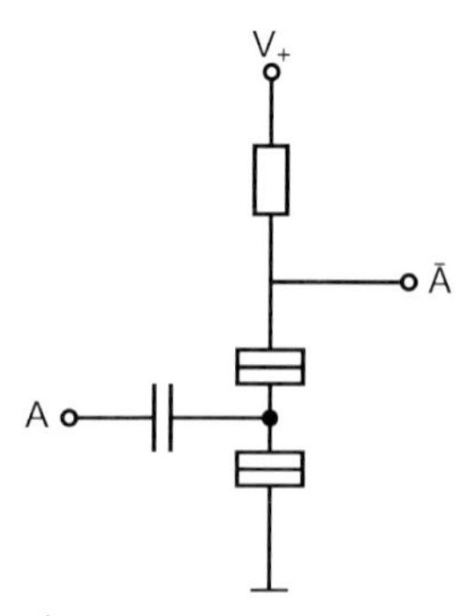

Figure 8.19 Inverting gate using a single-electron transistor.

[53, 55, 57, 59, 60, 82, 122], and an inverter consisting of a multiwall carbon nanotube has also been built [124]. For further information on single-electron devices, the reader is referred to Chapter 2 in Volume I of this Handbook, and to Chapter 6 in the present volume.

8.4.6 Artifacts in Molecular Electronic Devices

As noted above, molecules can provide a rich variety of functions. However, contacting single molecules reaches the limits (or even extends the limits) of current technology, and only recently has it been reported that some results in the field of molecular electronics were due to artifacts [6]. The most prominent example of this was the rotaxanes (see Section 8.4.4). In fact, it has been reported that the observed switching effect is independent of the molecule, and is thus an effect of the entire set-up, including the contacts and interfaces, and is not purely "molecular" [125]. However, this does not necessarily mean that the proposed switching mechanism is incorrect. It is rather covered by artifacts and may in effect exist in other experimental set-ups [75, 116, 126].

8.4.6.1 Sources of Artifacts

Several sources of artifacts are conceivable. One is concerned with the high electric fields that are applied across the molecular junction. Although only low voltages are applied ($\sim$1–2 V), the junctions are very thin, which generates high field strengths. These field strengths can cause metal atoms from the electrodes to migrate into the molecular layer and finally to shorten it by a metal filament [6, 128]. These filaments are thin, so that they can be easily broken by high currents flowing through them, for example due to resistive heating or to electromigration. Thus, the filaments can form and break and therefore, they can switch a device into a low- and high-impedance state, which can mitigate molecular switching.

One recent experiment concerning the formation of filaments was conducted by Lau *et al.* [127]. A Langmuir–Blodgett film of steric acid ($C_{18}H_{36}OH$) was sandwiched between a titanium and platinum electrode (see Figure 8.20). It was possible to switch this device between a high and a low conduction state, and a current versus voltage characteristic similar to that shown in Figure 8.16a was obtained. To examine the switching effect, Lau and colleagues scanned the device area using AFM, while simultaneously applying a bias voltage through external leads. In that way, the conductance of the whole device could be measured and correlated to the actual position of the AFM tip. Two-dimensional maps of conductance versus tip position were obtained for the on and off states (see Figure 8.20b).

The AFM tip exerts a certain pressure on the top electrode, which locally compresses the monolayer by $\sim$0.2 Å. This compression did not alter the conductance in the off state. Independently of the position of the AFM tip, the conductance stayed almost constant (see bottom image in Figure 8.20b). However, in the on state, sharp peaks in conductance were observed, which appeared when the tip was scanning across a localized spot on the top electrode (see red spike in the middle

(a)

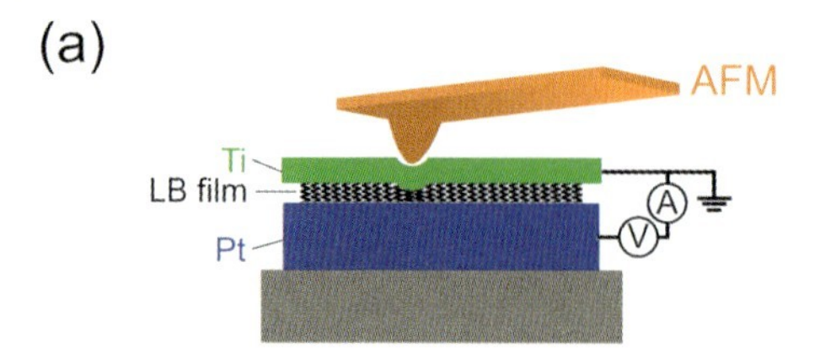

(b)

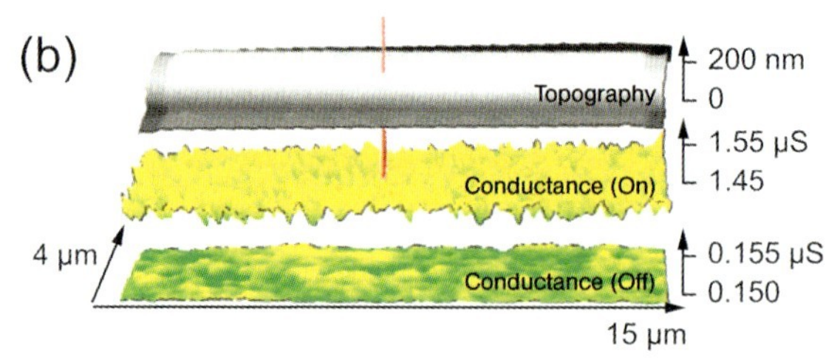

Figure 8.20 An experiment conducted by Lau *et al.* to examine the switching effect observed in their devices. Combining current versus voltage and atomic force measurements, these authors observed nanoscale switching centers, which are interpreted in terms of conductive filaments that almost bridge the two electrodes. (From Ref. [127].)

image in Figure 8.20b). These peaks were interpreted as "nanoasperities", in which the effective distance between the top and bottom electrode is reduced. These nanoasperities might be due to conducting filaments that almost bridge the electrodes.

Besides filaments, there are other sources of artifacts, such as oxidation of the metal electrode (i.e. titanium), which can also induce hysteretic current–voltage responses [129] or the formation of charge traps at the electrode/molecule interface [130]. Similarly, nanogaps produced by the electromigration technique (see Section 8.3.5) can, even in the absence of molecules, show "molecular" features, for example coulomb blockade effects with addition energies that are in the range as would expected for single molecules [82]. These effects are ascribed to small metallic grains within the junction.

To rule out these artifacts and to verify that the observed effect is truly molecular, several strategies have been followed [126]. One straightforward approach is systematically to vary the composition of the molecule (e.g. its length) and to study the influence of this variation on the current versus voltage characteristic [96]. Other strategies are to use a variety of test set-ups to rule out any systematic errors due to the measurement set-up [93], to completely avoid metallic electrodes and thereby eliminate the possibility of metallic filaments [131].

An alternative approach is to study the vibrational states of the molecule by using *inelastic electron tunneling spectroscopy* (IETS). This technique is highly sensitive to the molecular vibrations which open additional inelastic channels through which electrons can tunnel. However, these vibration states smear out in energy at room temperature, and can only be observed at cryogenic temperatures. In the current versus voltage characteristic, they are visible as peaks in the second derivative [46, 132]. For three-terminal devices, the second derivative can be plotted in the source voltage (V_S) versus gate voltage (V_G) plane, and the vibrational modes are then visible as lines running parallel to the diamond edges (cf. Figure 8.18c) [122].

The vibrational states are an intrinsic property of the molecule, and thus IETS provides a "molecular fingerprint" to prove that the molecule is actively involved in current transport.

8.4.7 Conclusions

Molecular electronic devices provide a wealth of functions. The appeal of molecular electronics is, amongst other things, based on the variety of these functions. The target is to make use of new effects that appear at these small dimensions (e.g. coulomb blockade or resonant tunneling effects, conformational changes of the molecule) for novel electronic devices.

Whilst it has been shown how these molecular devices can be combined to form small logical gates (e.g. in the form of diode–diode logic or the crossbar latch), this raises one important question: How can molecules be assembled so that these gates are formed?

8.5 Building Logical Circuits: Assembly of a Large Number of Molecular Devices

In the previous sections it has been shown how single (or at least a low number of) molecules can be contacted, and which functionalities these molecules can provide. Furthermore, it has been described, how these single molecules can be combined to small logic gates, for example as AND, OR, XOR gates, or as the crossbar latch. In this section, the discussion proceeds one step further to determine how a large number of devices can be assembled. And what implications does the use of single molecules have for the architecture of future logic circuits? As already discussed (in Section 8.3), for single-molecule devices there is at present no method available to deterministically place a single molecule on a chip. Thus, reliance must be placed on statistical processes and the ability of the molecules to self-organize, for example in the form of SAMs.

This dependence on self-organization bears the first implication for the architecture of molecular devices. Self-organization will always result in very regular structures, which have only a low information content [133]. In comparison to current CMOS circuits, in which a huge number of transistors is connected statically to other transistors, and which include therefore a high information content, this lack of information must be fed into the molecular circuit by an additional, post-fabrication training step. In other words, the technology and architecture has to be re-configurable.

A second implication of the self-organization process is given by the fact that these circuits will always contain defective parts, and the high yields necessary for CMOS architectures are not feasible. Again, the defective sites on the molecular chip must be identified and isolated in a post-fabrication step.

Several architectures have been proposed for future molecular circuits, and these differ in how strongly they rely on self-organization. PLAs based on crossbars, for example, use self-organization only for the preparation of the SAMs. The electrodes that contact the SAM are commonly defined by lithography (although techniques are available to prepare crossbars completely by self-organization, e.g. [102]). In contrast to PLAs, the *Nano-Cell* architecture (see Section 8.5.2) relies completely on self-organization.

8.5.1
Programmable Logic Arrays Based on Crossbars

In Section 8.4.2.4 it was described, how simple AND and OR gates can be implemented using diodes and resistances only. In the following it will be shown, how large arrays of these gates can be implemented using crossbars, as described in Section 8.3.6.

The equivalent circuit of a crossbar is shown in Figure 8.21a and b. A SAM of rectifying diodes is contacted by orthogonal top and bottom electrodes. These arrays can easily be converted into AND and OR gates. For an AND circuit, the vertical electrodes must be connected to the high voltage, and the horizontal lines to the input variables (see Figure 8.21a). Similarly, for OR, the horizontal lines are connected to the low voltage and the vertical lines are the signal lines.

However, these circuits have one drawback, in that only a single disjunction or conjunction can be implemented. If, for example, a simple AND connection of two input variables is built, all other horizontal input lines must be set to the high voltage. Therefore, at the end of each vertical line, only the conjunction of A and B is computed.

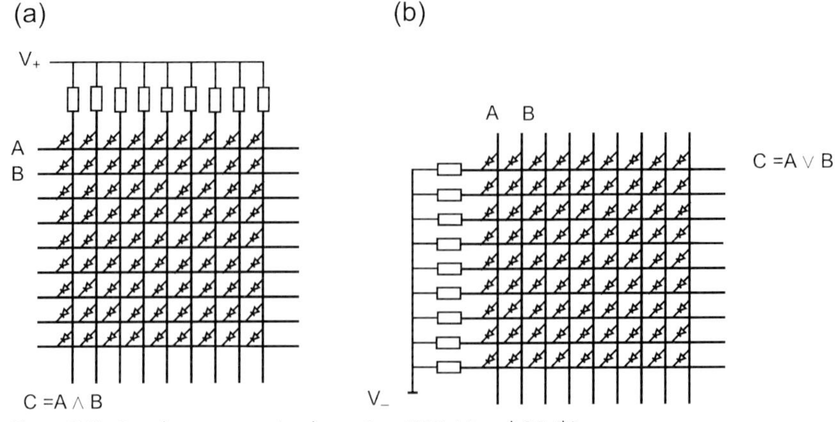

Figure 8.21 Crossbar structure implementing AND (a) and OR (b).

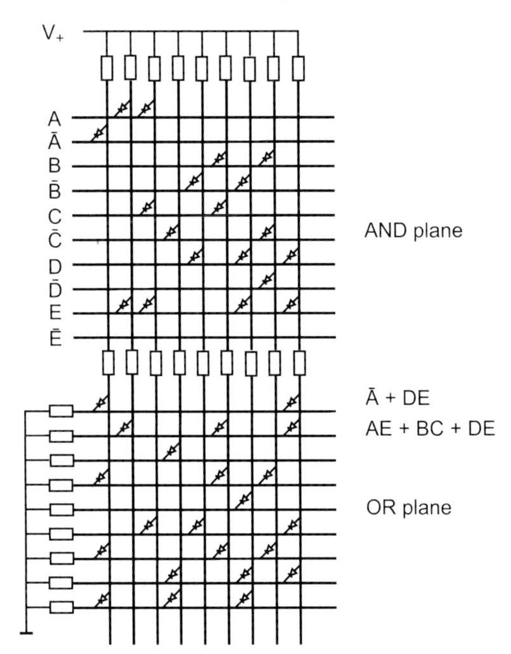

Figure 8.22 A programmable logic array consisting of an AND and an OR plane.

This problem can be circumvented if some diodes can be switched off – that is, if some input lines can be isolated from the output lines. Thus, a combination of a hysteretic switch (as described in Section 8.4.4) and a diode, for example by asymmetrically coupling a bistable molecule to the electrodes, would be highly beneficial. By using these switches, individual crossing points could be switched "off" and "on" by the application of a high-voltage pulse. The state of the molecules at the crossing points will therefore determine the logical function that is computed.

The AND and the OR gate can be combined to a PLA, as shown in Figure 8.22. The output of the AND plane is fed into the OR plane. Most diodes at the crossing points are switched off, so that certain Boolean functions are realized. The Boolean functions of the first two horizontal lines in the OR plane are given in Figure 8.22.

Such a PLA can compute every Boolean function if not only the input variables but also the negation of them is supplied to the PLA. Therefore, the negation of each variable must be computed, which can be done by using NDR-diodes as described in Section 8.4.3.1. Again, these NDR diodes can be implemented in a crossbar, so that all negations of all input signals can be realized simultaneously. As an alternative, the negation can be supplied by the surrounding CMOS circuitry, which would result in hybrid molecular/CMOS circuits.

Based on hysteretic switching diodes, the PLA is re-configurable. Therefore, the logic functions are programmed in a post-fabrication step; defective junctions can be

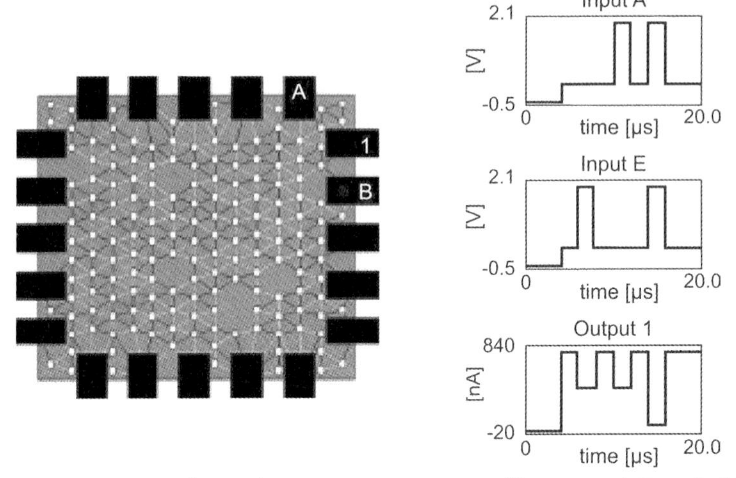

Figure 8.23 NanoCell trained as NAND gate, as proposed by Tour *et al.* (From Ref. [134].)

identified and disregarded in the circuit. This makes the PLA architecture a promising architecture for molecular electronics. A more detailed explanation of architectures based on crossbars is provided in Chapter 11 of this volume.

8.5.2 NanoCell

Although the crossbar set-up relies on self-organization, a "quasi-regular" structure is imposed by the orthogonal top and bottom electrode [133]. An architecture, which consists completely of random patterns, is the so-called NanoCell.

The structure of a NanoCell is shown in Figure 8.23 [134, 135]. It consists of a self-assembled, two-dimensional network of metallic particles which are randomly interlinked with molecules (cf. Section 8.3.8). In order to provide inverting logic, molecules exhibiting an NDR effect are used (see Section 8.4.3.1). The network is contacted by large metallic leads at its sides.

Due to the random arrangement of metal particles and molecules, the NanoCell must be trained or programmed to fulfill a certain task. To train this circuit, the molecules must be bistable and, similar to the PLAs, the molecules can be switched off and on by large voltage pulses. Therefore, the molecules must exhibit a combination of NDR effect and hysteretic switching. In Figure 8.23, the molecules are represented by lines connecting two metal particles; a white line represents an open switch, and a black line a closed switch.

In real applications, the molecular network is completely random – that is, neither are the positions of the individual molecules known, nor can a certain single molecule be addressed. The only knowledge about the circuit can be obtained through the contact pins, and most probably only bundles of molecules and not individual molecules can be switched by the application of voltage pulses to the contacts.

However, for a proof of concept, Tour *et al.* assumed the case of omnipotent programming [134], which means that the position of each molecule is known and that it can be individually programmed to its low or high state.

Based on this assumption, the network can be trained for a certain task; for example, to perform a NAND operation, as shown in Figure 8.23. The state of the network can be described by a list of the switching states of all molecules, typically, which molecule is open or which is closed. A function can be defined, which evaluates by how much the output ("1" in Figure 8.23) resembles a NAND combination of the inputs A and B. The task of training is now reduced to find a network state, which sufficiently minimizes (or, depending on the definition, maximizes) the evaluation function.

Tour and coworkers have used a genetic algorithm to identify such a network state – that is, to determine which molecules must be switched on, and which off. The output signal is determined by a SPICE simulation and compared to the desired functionality. Using this genetic algorithm, Tour and colleagues were able to train NanoCells as inverters, NAND gates, or complete 1-bit adders [134].

8.6 Challenges and Perspectives

Molecular electronics represents an exciting and promising field of research, but it imposes huge demands on the technology and is at the border of what is currently feasible. Many challenges remain for further research, as well in the design of molecules and in the assembly and architecture of future devices. For example, molecular interconnects must be found that can conduct current across larger distances, the rectification ratio of molecular diodes must be increased and, as a key element, bistable switching molecules must be identified and optimized. Future architectures of molecular devices will have to incorporate the statistical nature of the assembly of molecules. The optimum architecture will start from a random arrangement of molecules and will be trained the molecular network, as has been attempted with the NanoCell set-up. However, this training step is mathematically highly complex and has only been solved for the simplifying case of omnipotent programming. It seems that there is a price to pay for the low information content of these random structures with a complex training algorithm.

In CMOS, molecular electronics has a very strong competitor. CMOS has been so successful in the past, because it combines high integration densities, a high switching speed, and a low power consumption [133]. If molecular electronics is to at least complement CMOS in the future, it must outperform CMOS in one or more of these key characteristics. It is proposed that molecular electronics will be small in size and will exhibit low switching energies [1, 133], but the switching speed will be low. However, these predictions are very uncertain and will need to be substantiated by further research.

References

1 International Technology Roadmap for Semiconductors; 2005 Edition. public.itrs.net. 2006.
2 A. Aviram, M. Ratner, *Chem. Phys. Lett.* 1974, **29**, 277–283.
3 C. P. Collier, G. Mattersteig, E. W. Wong, Y. Luo, K. Beverly, J. Sampaio, F. M. Raymo, J. F. Stoddart, J. R. Heath, *Science* 2000, **289**, 1172–1175.
4 J. Chen, M. A. Reed, A. M. Rawlett, J. M. Tour, *Science* 1999, **286**, 1550–1552.
5 R. F. Service, *Science* 2001, **294**, 2442–2443.
6 R. F. Service, *Science* 2003, **302**, 556.
7 S. Datta, *Nanotechnology* 2004, **15**, S433–S451.
8 S. Datta, *Quantum Transport – Atom to Transistor*, Cambridge University Press, 2005.
9 M. Zharnikov, S. Frey, H. Rong, Y. J. Yang, K. Heister, M. Buck, M. Grunze, *Phys. Chem. Chem. Phys.* 2000, **2**, 3359–3362.
10 J. C. Love, D. B. Wolfe, R. Haasch, M. L. Chabinyc, K. E. Paul, G. M. Whitesides, R. G. Nuzzo, *J. Am. Chem. Soc.* 2003, **125**, 2597–2609.
11 L. Patrone, S. Palacin, J. P. Bourgoin, J. Lagoute, T. Zambelli, S. Gauthier, *Chem. Phys.* 2002, **281**, 325–332.
12 P. Morf, F. Raimondi, H. G. Nothofer, B. Schnyder, A. Yasuda, J. M. Wessels, T. A. Jung, *Langmuir* 2006, **22**, 658–663.
13 J. M. Wessels, H. G. Nothofer, W. E. Ford, F. von Wrochem, F. Scholz, T. Vossmeyer, A. Schroedter, H. Weller, A. Yasuda, *J. Am. Chem. Soc.* 2004, **126**, 3349–3356.
14 J. Chen, L. C. Calvet, M. A. Reed, D. W. Carr, D. S. Grubisha, D. W. Bennett, *Chem. Phys. Lett.* 1999, **313**, 741–748.
15 A. Ulman, *Chem. Rev.* 1996, **96**, 1533–1554.
16 S. Ranganathan, I. Steidel, F. Anariba, R. L. McCreery, *Nano Lett.* 2001, **1**, 491–494.
17 M. R. Kosuri, H. Gerung, Q. M. Li, S. M. Han, P. E. Herrera-Morales, J. F. Weaver, *Surface Sci.* 2005, **596**, 21–38.
18 J. Moreland, J. W. Ekin, *J. Appl. Physics* 1985, **58**, 3888–3895.
19 M. A. Reed, C. Zhou, C. J. Muller, T. P. Burgin, J. M. Tour, *Science* 1997, **278**, 252–254.
20 J. Reichert, R. Ochs, D. Beckmann, H. B. Weber, M. Mayor, H. von Lohneysen, *Phys. Rev. Lett.* 2002 88.
21 H. B. Weber, J. Reichert, F. Weigend, R. Ochs, D. Beckmann, M. Mayor, R. Ahlrichs, H. von Lohneysen, *Chem. Phys.* 2002, **281**, 113–125.
22 H. B. Weber, J. Reichert, R. Ochs, D. Beckmann, M. Mayor, H. von Lohneysen, *Physica E: Low-Dimensional Systems Nanostructures* 2003, **18**, 231–232.
23 J. Reichert, H. B. Weber, M. Mayor, H. von Lohneysen, *Appl. Phys. Lett.* 2003, **82**, 4137–4139.
24 M. Mayor, C. von Hanisch, H. B. Weber, J. Reichert, D. Beckmann, *Angew. Chem. Int. Ed.* 2002, **41**, 1183–1186.
25 M. Mayor, H. B. Weber, J. Reichert, M. Elbing, C. von Hanisch, D. Beckmann, M. Fischer, *Angew. Chem. Int. Ed.* 2003, **42**, 5834–5838.
26 M. Elbing, R. Ochs, M. Koentopp, M. Fischer, C. von Hanisch, F. Weigend, F. Evers, H. B. Weber, M. Mayor, *Proc. Natl. Acad. Sci. USA* 2005, **102**, 8815–8820.
27 D. Dulic, S. J. van der Molen, T. Kudernac, H. T. Jonkman, J. J. D. de Jong, T. N. Bowden, J. van Esch, B. L. Feringa, B. J. van Wees, *Phys. Rev. Lett.* 2003 91.
28 E. Lortscher, J. W. Ciszek, J. Tour, H. Riel, *Small* 2006, **2**, 973–977.
29 R. H. M. Smit, Y. Noat, C. Untiedt, N. D. Lang, M. C. van Hemert, J. M. van Ruitenbeek, *Nature* 2002, **419**, 906–909.
30 R. Waser (Ed.), *Nanoelectronics and Information Technology*, 2nd edition. Wiley-VCH, 2005.
31 D. J. Wold, C. D. Frisbie, *J. Am. Chem. Soc.* 2001, **123**, 5549–5556.
32 D. J. Wold, C. D. Frisbie, *J. Am. Chem. Soc.* 2000, **122**, 2970–2971.

33 D. J. Wold, R. Haag, M. A. Rampi, C. D. Frisbie, *J. Phys. Chem. B* 2002, **106**, 2813–2816.

34 A. M. Rawlett, T. J. Hopson, L. A. Nagahara, R. K. Tsui, G. K. Ramachandran, S. M. Lindsay, *Appl. Phys. Lett.* 2002, **81**, 3043–3045.

35 F. Schreiber, *Prog. Surface Sci.* 2000, **65**, 151–256.

36 J. M. Beebe, V. B. Engelkes, L. L. Miller, C. D. Frisbie, *J. Am. Chem. Soc.* 2002, **124**, 11268–11269.

37 K. Moth-Poulsen, L. Patrone, N. Stuhr-Hansen, J. B. Christensen, J. P. Bourgoin, T. Bjornholm, *Nano Lett.* 2005, **5**, 783–785.

38 X. L. Li, B. Q. Xu, X. Y. Xiao, X. M. Yang, L. Zang, N. J. Tao, *Faraday Disc.* 2006, **131**, 111–120.

39 B. Q. Xu, X. Y. Xiao, N. J. Tao, *J. Am. Chem. Soc.* 2003, **125**, 16164–16165.

40 D. I. Gittins, D. Bethell, D. J. Schiffrin, R. J. Nichols, *Nature* 2000, **408**, 67–69.

41 Y. Yasutake, Z. J. Shi, T. Okazaki, H. Shinohara, Y. Majima, *Nano Lett.* 2005, **5**, 1057–1060.

42 L. A. Bumm, J. J. Arnold, M. T. Cygan, T. D. Dunbar, T. P. Burgin, L. Jones, D. L. Allara, J. M. Tour, P. S. Weiss, *Science* 1996, **271**, 1705–1707.

43 B. Lussem, L. Muller-Meskamp, S. Karthauser, R. Waser, M. Homberger, U. Simon, *Langmuir* 2006, **22**, 3021–3027.

44 A. S. Blum, J. G. Kushmerick, S. K. Pollack, J. C. Yang, M. Moore, J. Naciri, R. Shashidhar, B. R. Ratna, *J. Phys. Chem. B* 2004, **108**, 18124–18128.

45 J. G. Kushmerick, J. Naciri, J. C. Yang, R. Shashidhar, *Nano Lett.* 2003, **3**, 897–900.

46 J. G. Kushmerick, J. Lazorcik, C. H. Patterson, R. Shashidhar, D. S. Seferos, G. C. Bazan, *Nano Lett.* 2004, **4**, 639–642.

47 J. G. Kushmerick, D. B. Holt, J. C. Yang, J. Naciri, M. H. Moore, R. Shashidhar, *Phys. Rev. Lett.* 2002, 89.

48 H. Park, A. K. L. Lim, A. P. Alivisatos, J. Park, P. L. Mceuen, *Appl. Phys. Lett.* 1999, **75**, 301–303.

49 S. Kronholz, S. Karthauser, A. van der Hart, T. Wandlowski, R. Waser *Microelectronics J.* 2006, **37**, 591–594.

50 S. Kronholz, S. Karthauser, G. Meszaros, T. Wandlowski, A. van der Hart, R. Waser, *Microelectronic Eng.* 2006, **83**, 1702–1705.

51 C. Z. Li, H. X. He, N. J. Tao, *Appl. Phys. Lett.* 2000, **77**, 3995–3997.

52 S. Boussaad, N. J. Tao, *Appl. Phys. Lett.* 2002, **80**, 2398–2400.

53 S. Kubatkin, A. Danilov, M. Hjort, J. Cornil, J. L. Bredas, N. Stuhr-Hansen, P. Hedegard, T. Bjornholm, *Nature* 2003, **425**, 698–701.

54 E. Ruttkowski, R. J. Luyken, Y. Mustafa, M. Specht, F. Hofmann, M. Städele, W. Rösner, W. Weber, R. Waser, L. Risch, *Proceedings, 2005 5th IEEE Conference on Nanotechnology*, Volume 1, pp. 438–441, 2005.

55 J. Park, A. N. Pasupathy, J. I. Goldsmith, C. Chang, Y. Yaish, J. R. Petta, M. Rinkoski, J. P. Sethna, H. D. Abruna, P. L. Mceuen, D. C. Ralph, *Nature* 2002, **417**, 722–725.

56 W. J. Liang, M. P. Shores, M. Bockrath, J. R. Long, H. Park, *Nature* 2002, **417**, 725–729.

57 H. Park, J. Park, A. K. L. Lim, E. H. Anderson, A. P. Alivisatos, P. L. Mceuen, *Nature* 2000, **407**, 57–60.

58 A. V. Danilov, S. E. Kubatkin, S. G. Kafanov, T. Bjornholm, *Faraday Disc.* 2006, **131**, 337–345.

59 A. N. Pasupathy, J. Park, C. Chang, A. V. Soldatov, S. Lebedkin, R. C. Bialczak, J. E. Grose, L. A. K. Donev, J. P. Sethna, D. C. Ralph, P. L. Mceuen, *Nano Lett.* 2005, **5**, 203–207.

60 S. Kubatkin, A. Danilov, M. Hjort, J. Cornil, J. L. Bredas, N. Stuhr-Hansen, P. Hedegard, T. Bjornholm, *Curr. Appl. Physics* 2004, **4**, 554–558.

61 M. J. Tarlov, *Langmuir* 1992, **8**, 80–89.

62 G. C. Herdt, D. R. Jung, A. W. Czanderna, *Prog. Surface Sci.* 1995, **50**, 103–129.

63 T. Ohgi, H. Y. Sheng, H. Nejoh, *Appl. Surface Sci.* 1998, **132**, 919–924.

64 G. L. Fisher, A. E. Hooper, R. L. Opila, D. L. Allara, N. Winograd, *J. Phys. Chem. B* 2000, **104**, 3267–3273.

65 G. L. Fisher, A. V. Walker, A. E. Hooper, T. B. Tighe, K. B. Bahnck, H. T. Skriba, M. D. Reinard, B. C. Haynie, R. L. Opila, N. Winograd, D. L. Allara, *J. Am. Chem. Soc.* 2002, **124**, 5528–5541.

66 A. V. Walker, T. B. Tighe, B. C. Haynie, S. Uppili, N. Winograd, D. L. Allara, *J. Phys. Chem. B* 2005, **109**, 11263–11272.

67 A. V. Walker, T. B. Tighe, O. M. Cabarcos, M. D. Reinard, B. C. Haynie, S. Uppili, N. Winograd, D. L. Allara, *J. Am. Chem. Soc.* 2004, **126**, 3954–3963.

68 B. de Boer, M. M. Frank, Y. J. Chabal, W. R. Jiang, E. Garfunkel, Z. Bao, *Langmuir* 2004, **20**, 1539–1542.

69 T. B. Tighe, T. A. Daniel, Z. H. Zhu, S. Uppili, N. Winograd, D. L. Allara, *J. Phys. Chem. B* 2005, **109**, 21006–21014.

70 Y. Tai, A. Shaporenko, H. Noda, M. Grunze, M. Zharnikov, *Adv. Mater.* 2005, **17**, 1745–1749.

71 Y. Tai, A. Shaporenko, W. Eck, M. Grunze, M. Zharnikov, *Langmuir* 2004, **20**, 7166–7170.

72 H. B. Akkerman, P. W. M. Blom, D. M. de Leeuw, B. de Boer, *Nature* 2006, **441**, 69–72.

73 H. Haick, M. Ambrico, J. Ghabboun, T. Ligonzo, D. Cahen, *Phys. Chem. Chem. Phys.* 2004, **6**, 4538–4541.

74 Y. L. Loo, D. V. Lang, J. A. Rogers, J. W. P. Hsu, *Nano Lett.* 2003, **3**, 913–917.

75 K. T. Shimizu, J. D. Tabbri, J. J. Jelincic, N. A. Melosh, *Adv. Mater.* 2006, **18**, 1499–1504.

76 B. Q. Xu, X. Y. Xiao, X. M. Yang, L. Zang, N. J. Tao, *J. Am. Chem. Soc.* 2005, **127**, 2386–2387.

77 T. Albrecht, K. Moth-Poulsen, J. B. Christensen, A. Guckian, T. Bjornholm, J. G. Vos, J. Ulstrup, *Faraday Disc.* 2006, **131**, 265–279.

78 T. Albrecht, K. Moth-Poulsen, J. B. Christensen, J. Hjelm, T. Bjornholm, J. Ulstrup, *J. Am. Chem. Soc.* 2006, **128**, 6574–6575.

79 W. Haiss, H. van Zalinge, S. J. Higgins, D. Bethell, H. Hobenreich, D. J. Schiffrin, R. J. Nichols, *J. Am. Chem. Soc.* 2003, **125**, 15294–15295.

80 X. Y. Xiao, L. A. Nagahara, A. M. Rawlett, N. J. Tao, *J. Am. Chem. Soc.* 2005, **127**, 9235–9240.

81 A. R. Champagne, A. N. Pasupathy, D. C. Ralph, *Nano Lett.* 2005, **5**, 305–308.

82 H. S. J. van der Zant, Y. V. Kervennic, M. Poot, K. O'Neill, Z. de Groot, J. M. Thijssen, H. B. Heersche, N. Stuhr-Hansen, T. Bjornholm, D. Vanmaekelbergh, C. A. van Walree, L. W. Jenneskens, *Faraday Disc.* 2006, **131**, 347–356.

83 M. Brust, M. Walker, D. Bethell, D. J. Schiffrin, R. Whyman, *J. Chem. Soc. - Chem. Commun.* 1994, 801–802.

84 C. J. Kiely, J. Fink, M. Brust, D. Bethell, D. J. Schiffrin, *Nature* 1998, **396**, 444–446.

85 J. Liao, L. Bernard, M. Langer, C. Schonenberger, M. Calame, *Adv. Mater.* 2006, **18**, 2444–2447.

86 J. M. Tour, L. Cheng, D. P. Nackashi, Y. X. Yao, A. K. Flatt, S. K. St Angelo, T. E. Mallouk, P. D. Franzon, *J. Am. Chem. Soc.* 2003, **125**, 13279–13283.

87 T. Hassenkam, K. Moth-Poulsen, N. Stuhr-Hansen, K. Norgaard, M. S. Kabir, T. Bjornholm, *Nano Lett.* 2004, **4**, 19–22.

88 T. Hassenkam, K. Norgaard, L. Iversen, C. J. Kiely, M. Brust, T. Bjornholm, *Adv. Mater.* 2002, **14**, 1126–1130.

89 K. Norgaard, T. Bjornholm, *Chem. Commun.* 2005, 1812–1823.

90 D. P. Long, C. H. Patterson, M. H. Moore, D. S. Seferos, G. C. Bazan, J. G. Kushmerick, *Appl. Phys. Lett.* 2005, **86**, 153105.

91 T. Dadosh, Y. Gordin, R. Krahne, I. Khivrich, D. Mahalu, V. Frydman, J. Sperling, A. Yacoby, I. Bar-Joseph, *Nature* 2005, **436**, 677–680.

92 R. Haag, M. A. Rampi, R. E. Holmlin, G. M. Whitesides, *J. Am. Chem. Soc.* 1999, **121**, 7895–7906.

93 A. S. Blum, J. G. Kushmerick, D. P. Long, C. H. Patterson, J. C. Yang, J. C.

Henderson, Y. X. Yao, J. M. Tour, R. Shashidhar, B. R. Ratna, *Nature Mater.* 2005, **4**, 167–172.

94 J. Chen, M. A. Reed, *Chem. Phys.* 2002, **281**, 127–145.

95 M. A. Reed, J. Chen, A. M. Rawlett, D. W. Price, J. M. Tour, *Appl. Phys. Lett.* 2001, **78**, 3735–3737.

96 W. Y. Wang, T. Lee, M. A. Reed, *Phys. Rev. B* 2003, 68.

97 C. Zhou, M. R. Deshpande, M. A. Reed, L. Jones, J. M. Tour, *Appl. Phys. Lett.* 1997, **71**, 611–613.

98 S. C. Goldstein, D. Rosewater, Solid-State Circuits Conference 2002. Digest of Technical Papers, ISSCC 2002, Volume 1, p. 204.

99 M. Mayor, H. B. Weber, R. Waser,in: R. Waser (Ed.), *Nanoelectronics and Information Technology* Wiley-VCH Weinheim 2003, pp. 503–525.

100 M. J. Crossley, P. L. Burn, *J. Chem. Soc. - Chem. Commun.* 1991, 1569–1571.

101 R. L. Carroll, C. B. Gorman, *Angew. Chem. Int. Ed.* 2002, **41**, 4379–4400.

102 A. Ismach, E. Joselevich, *Nano Lett.* 2006, **6**, 1706–1710.

103 A. Aviram, M. Ratner, *Chem. Phys. Lett.* 1974, **29**, 277–283.

104 P. E. Kornilovitch, A. M. Bratkovsky, R. S. Williams, *Phys. Rev. B* 2002, **66**, 165436.

105 A. Aviram, M. Ratner, *Chem. Phys. Lett.* 1974, **29**, 277–283.

106 R. M. Metzger, *Chem. Phys.* 2006, **326**, 176–187.

107 G. J. Ashwell, J. R. Sambles, A. S. Martin, W. G. Parker, M. Szablewski, *J. Chem. Soc. - Chem. Commun.* 1990, 1374–1376.

108 A. S. Martin, J. R. Sambles, G. J. Ashwell, *Phys. Rev. Lett.* 1993, **70**, 218–221.

109 C. Krzeminski, C. Delerue, G. Allan, D. Vuillaume, R. M. Metzger, *Phys. Rev. B* 2001, 6408.

110 R. M. Metzger, B. Chen, U. Hopfner, M. V. Lakshmikantham, D. Vuillaume, T. Kawai, X. L. Wu, H. Tachibana, T. V. Hughes, H. Sakurai, J. W. Baldwin, C. Hosch, M. P. Cava, L. Brehmer, G. J. Ashwell, *J. Am. Chem. Soc.* 1997, **119**, 10455–10466.

111 G. J. Ashwell, R. Hamilton, L. R. H. High, *J. Mater. Chem.* 2003, **13**, 1501–1503.

112 J. C. Ellenbogen, J. C. Love, *Proc. IEEE* 2000, **88**, 386–426.

113 M. A. Reed, *Proc. IEEE* 1999, **87**, 652–658.

114 J. Taylor, M. Brandbyge, K. Stokbro, *Phys. Rev. B* 2003, 68.

115 P. E. Kornilovitch, A. M. Bratkovsky, R. S. Williams, *Phys. Rev. B* 2002, **66**, 245413.

116 K. Norgaard, B. W. Laursen, S. Nygaard, K. Kjaer, H.-R. Tseng, A. H. Flood, J. F. Stoddart, T. Bjornholm, *Angew. Chem. - Int. Ed.* 2005, **44**, 7035–7039.

117 J. W. Choi, A. H. Flood, D. W. Steuerman, S. Nygaard, A. B. Braunschweig, N. N. P. Moonen, B. W. Laursen, Y. Luo, E. DeIonno, A. J. Peters, J. O. Jeppesen, K. Xu, J. F. Stoddart, J. R. Heath, *Chemistry - A European Journal* 2005, **12**, 261–279.

118 P. J. Kuekes, D. R. Stewart, R. S. Williams, *J. Appl. Physics* 2005, **98**, 049901.

119 P. Hedegard, T. Bjornholm, *Chem. Phys.* 2005, **319**, 350–359.

120 K. K. Likharev, *Proc. IEEE* 1999, **87**, 606–632.

121 K. Uchida, in: R. Waser (Ed.), *Nanoelectronics and Information Technology*, Wiley-VCH, Weinheim, 2006, pp. 425–443.

122 E. A. Osorio, K. O'Neill, N. Stuhr-Hansen, O. F. Nielsen, T. Bjornholm, H. S. J. van der Zant *Adv. Mater.* 2007, **19**, 281–285.

123 K. Uchida, J. Koga, R. Ohba, A. Toriumi, *IEEE Trans. Electron Devices* 2003, **50**, 1623–1630.

124 K. Ishibashi, D. Tsuya, M. Suzuki, Y. Aoyagi, *Appl. Phys. Lett.* 2003, **82**, 3307–3309.

125 D. R. Stewart, D. A. A. Ohlberg, P. A. Beck, Y. Chen, R. S. Williams, J. O. Jeppesen, K. A. Nielsen, J. F. Stoddart, *Nano Lett.* 2004, **4**, 133–136.

126 A. H. Flood, J. F. Stoddart, D. W. Steuerman, J. R. Heath, *Science* 2004, **306**, 2055–2056.

127 C. N. Lau, D. R. Stewart, R. S. Williams, M. Bockrath, *Nano Lett.* 2004, **4**, 569–572.

128 V. V. Zhirnov, R. K. Cavin, *Nature Mater.* 2006, **5**, 11–12.

129 W. R. McGovern, F. Anariba, R. L. McCreery, *J. Electrochem. Soc.* 2005, **152**, E176–E183.

130 C. A. Richter, D. R. Stewart, D. A. A. Ohlberg, R. S. Williams, *Appl. Physics A - Mater. Sci. Process.* 2005, **80**, 1355–1362.

131 J. L. He, B. Chen, A. K. Flatt, J. J. Stephenson, C. D. Doyle, J. M. Tour, *Nature Mater.* 2006, **5**, 63–68.

132 W. Y. Wang, T. Lee, I. Kretzschmar, M. A. Reed, *Nano Lett.* 2004, **4**, 643–646.

133 M. R. Stan, P. D. Franzon, S. C. Goldstein, J. C. Lach, M. M. Ziegler, *Proc. IEEE* 2003, **91**, 1940–1957.

134 J. M. Tour, W. L. Van Zandt, C. P. Husband, S. M. Husband, L. S. Wilson, P. D. Franzon, D. P. Nackashi, *IEEE Trans. Nanotechnol.* 2002, **1**, 100–109.

135 C. P. Husband, S. M. Husband, J. S. Daniels, J. M. Tour, *IEEE Trans. Electron Devices* 2003, **50**, 1865–1875.

9
Intermolecular- and Intramolecular-Level Logic Devices

Françoise Remacle and Raphael D. Levine

9.1
Introduction and Background

Today, there is an intense research activity in the field of nanoscale logic devices towards miniaturization and qualitative improvement in the performance of logic circuits [1–16]. A radical and potentially very promising approach is the search for quantum computing [17–24], and this is reviewed as an emerging technology in Ref. [25]. Other alternatives are based on neural networks [26], on DNA-based computing [27–31], or on molecular quantum cellular automata [32–37]. Single-electron devices should also be mentioned because if they use chemically synthesized quantum dots (QDs) they are "molecular" in nature [38–40]. Devices that have been implemented rely on the ability to use molecules as switches and/or as wires, an approach known as "molecular electronics" [5, 12, 41–44]. This approach is currently being extended in several interesting directions, including the modification of the electronic response of the molecule through changing its Hamiltonian [45–47]. In this chapter, these topics are first reviewed, after which ongoing studies on an alternative computational model, where the molecule acts not as a switch but as an entire logic circuit, are discussed. Both, electrical and optical inputs and outputs are considered. Advantage is then taken of the discrete quantum states of molecules to endow the circuits with memory, such that a molecule acts as a finite state logic machine. Speculation is also made as to how such machines can be programmed. Finally, the potential concatenation of molecular logic circuits either by self-assembly or by directed synthesis so as to produce an entire logic array, is discussed. In this regard, directed deposition is also a possible option [48–52].

9.1.1
Quantum Computing

Quantum computing can be traced to Feynman, who advocated [53] the use of a quantum computer instead of a classical computer to simulate quantum systems. The rational is that quantum systems, when simulated classically, are very demanding in computing resources. A quantum state is described by two "numbers" – its amplitude

Nanotechnology. Volume 4: Information Technology II. Edited by Rainer Waser
Copyright © 2008 WILEY-VCH Verlag GmbH & Co. KGaA, Weinheim
ISBN: 978-3-527-31737-0

and its phase – and the number of quantum states, N, grows exponentially with the number of degrees of freedom of the system. The sizes of the matrices necessary to describe a system quantum mechanically scales as N^2, and for large systems this number becomes rapidly prohibitively large. A computer that operates quantum mechanically will require far less resources because the computations can be massively parallel due to the superposition principle of quantum mechanics. Conceptually, to compute quantum mechanically required the extension of classical Boolean logic to quantum logic [19, 54] and to set up quantum logic gates [55–61] that operate reversibly [62, 63].

In quantum computing, the logic is processed via the coupling structure between the levels of the Hamiltonian. Typically, this coupling is induced by external electrical and magnetic fields. Nuclear magnetic resonance (NMR) is a particularly promising direction for both pump and probe [20, 64]. Quantum implementations are very encouraging for search algorithms [65–68], where a power law reduction in the number of queries can be obtained. For operations where the answer is more complex than a YES or NO – such as Fourier transform operations – the read-out remains a key problem because of the collapse of the wave function when one of its component is read. One very successful outcome of quantum computing and quantum information is *cryptography* [69], where the very effective factorization algorithms (public key cryptography, Shor algorithm [70, 71]) show the potential that is available. Quantum computing has very much caught the popular imagination, and several excellent introductory books (e.g. Ref. [72]) are now available.

9.1.2
Quasiclassical Computing

The essential difference between quantum computing and the approach discussed here is that a quantum gate operates on both the amplitude and the phase of the quantum state. The phase is very sensitive to noise, and quantum computing theorists have devised various ways to protect the phase from external unwanted perturbations [73–79] or to seek to correct a corrupted phase. Because the authors' background is in molecular dynamics and coherent control, they are aware that the phase of quantum states is extremely difficult to protect, and in this chapter adopt a quasi-classical approach [80] where, while the time evolution of the molecular system is quantal, what matters in terms of inputs and outputs are the populations of the states – that is, the square modulus of the amplitudes. This approach relies on classical logic and does not require reversible gates. There are two special characteristics of quantum computing: parallelism and entanglement. Currently, the authors' investigations center on understanding the potential of the quasi-classical approach in terms of parallelism.

9.1.3
A Molecule as a Bistable Element

Another very successful approach is molecular electronics, which aims to provide molecular-based computing by using the molecule as a switch [5, 11, 12, 42, 43, 81, 82].

In the following sections, it is explained that the essential difference with what is advocated in this chapter is that the molecule can do – and has been shown to do – much more than act as a switch.

Quantum cellular automata (QCA) represents another promising approach to molecular computing where the molecules are not used as switches but as structured charge containers and information transmitted via coupled electric fields [32–34]. The charge configuration of a cell composed of a few QDs – and, more recently, of molecular complexes – is the support for encoding the binary information. Most studies on QCA consists of theoretical design, with very few experimental implementations. While the QD-based implementations [33, 34] operate at very low temperatures, theoretical modeling predicts that molecular quantum cellular automata could be operated at room temperature [36, 37, 83].

9.1.4 Chemical Logic Gates

Early proposed chemical implementations of logic gates were based on the response of molecules in solution to light or changes in chemical species concentrations [84–86]. Using photo-induced electron transfer where the emitted fluorescence is modulated by the concentrations of ions species in solution, different kinds of realizations of uni and binary gates (i.e. OR, NOT, AND, XOR, etc.) have been proposed [11, 14, 87, 88], and it has been shown that AND and XOR gates can be combined to lead to half adder and/or subtractor [4, 8, 30, 89–93] and full adder/subtractor [94–98] implementations. Other well-studied systems that lead to similar levels of logic complexity are those built on molecular motors (catenane, rotaxane), where the inputs can be communicated photochemically and electrochemically [11], or on DNA oligonucleotides [29–31]. These ways of providing the inputs and reading the outputs are rather slow, however, and do not exploit the complexity of the quantum level structure and intra- and inter-molecular couplings. Nor is it clear how to reduce the size of such devices. All of these approaches were largely limited to two bit operations (half adder or subtractors). In 2001 [94, 95], the level of the three bits operation with an optical full adder had already been reached, while more recently a cyclable full adder on a QD by electrical addressing was proposed [99]. This last example shows that QDs, and not only molecules proper, can be used to act as more than a switch.

The emphasis in this chapter is on demonstrating single-molecule rather than chemical computing. The proposal is that it is possible to use the complexity of molecules to integrate logic circuits of increasing sophistication on a single molecule or on supramolecular assemblies. Clearly, at the present time signal-to-noise considerations mean that the independent response of more than a single molecule is needed. Beyond the independent molecule, however, the proposal is to concatenate single molecules in the sense that the logic output from one molecule is the logic input to another. This is achieved by (rapid) intermolecular coupling.

9.1.5
Molecular Combinational Circuits

Combinational circuits are the simplest logic units, being built of logic gates (the most common are NOT, AND, OR, XOR, NOR, NAND) and providing a specific set of outputs, given a specific set of inputs. These circuits have no memory. In transistor-based computer circuits even the simpler such gates require to be built as a network of switches. Studies on molecular combinational circuits forms part of an intense research effort aimed at recasting into logic functions the fact that molecules can respond selectively to stimuli (inputs) of different forms (chemical, optical, electrical) and produce outputs (chemical, optical, electrical). The advantage is that a molecule which implements a combinational circuit acts not as a single switch but as an entire gate. However, most of molecular gates proposed until now have been based on chemistry in solution, and use at one stage or another a chemical stimulus (e.g. concentration of ions such as H^+, Na^+) coupled to optical or electrical stimuli. This leads rather slow rates of processing of the information and concatenation. In contrast, the authors' studies do not involve chemical inputs, and this allows faster rates to be reached. It has already been shown that, within less than 1 ps ($=10^{-12}$ s), it is possible to implement combinational circuits on a single molecule using selective photo-excitation for providing the inputs and the intra- and inter-molecular dynamics for processing the information. (See Refs. [100–102] for an example of sub-ps logic gates using femtosecond ($=10^{-15}$ s) pulses as input, and Refs. [94, 103] for an example of concatenation by intermolecular coupling.)

One advantage of using molecules in the gas phase is that far fewer molecules are required than in solution. In the preliminary gas-phase schemes, the reading of the outputs is achieved by detecting fragments of the molecule used to implement logic, which means that this particular molecule is not available for a new computation. This is not intrinsically a problem because about 10 molecules are needed to obtain a good signal-to-noise ratio; hence, the computation can be continued with other molecules present on the sample. This represents a problem for cycling, however, and is why the target is to explore the possibilities offered by non-destructive optical and electrical reading. Another route to explore is to increase the number of operations (more than 32 bits) performed on a single molecule by using the fast intramolecular dynamics for concatenation, which decreases the need for cycling and the need for I/O. As discussed above and further elaborated below, this will have major implications for the miniaturization of logic devices by implementing compound logic at the hardware level.

In this way it was possible to implement logic functions on a single molecule up to and including the ability to program – that is, to use the same physics to realize different computations.

In terms of technology transfer it is clear, however, that what the industry would very much prefer is some form of extension to the CMOS technology. So, the need is to combine the many advantages of working with molecules in the gas phase with the need to anchor molecules on the surface. In this connection, studies with self-assembled small arrays of QDs is of central interest.

9.1.6
Concatenation, Fan-Out and Other Aspects of Integration

It is an essential characteristic of modern logic circuits that the gates are integrated. Specifically, the output of one gate can be accepted as the input of the next gate; the two gates are thereby *concatenated*. Very simple examples of concatenation are the NotAND (=NAND) or NotOR, etc. gates. It should also be possible to deliver the output of one gate as input to several gates. In 2001, it was shown [94] how to concatenate the logic performed by two molecules using electronic energy transfer as the vehicle for the information forwarding. It is clear that electron transfer – including electron transfer between QDs, proton transfer, and vibration energy transfer – can all be used for this purpose. However, this remains an as-yet poorly traveled course and further studies are clearly called for as it is a possible key to high-density circuits.

9.1.7
Finite-State Machines

So far, only combinational logic has been discussed, where combinational gates combine the inputs to produce an output. A *finite-state machine* does more – and the "more" is very essential and is also something well suited to what a molecule can do. A finite-state machine accepts inputs to produce an output that is dependent both on the inputs *and* on the current state of the machine. In addition to producing an output, such a machine will update the state of the machine so that it is ready for the next operation. Technically, what the finite-state machine has is a memory, and the circuit has as many different states of memory as states of the machine (see Figure 9.1). It has been shown possible to build very simple finite-state and Turing machines at the molecular level [101, 104]. Molecules have a high capacity for memory because of their many quantum states; for example, in conformers, after radiationless transition, the molecule can undergo relaxation and be in a stable state for a long time. Retrieval of the information may then be effected by either optical or electrical pulses (see below).

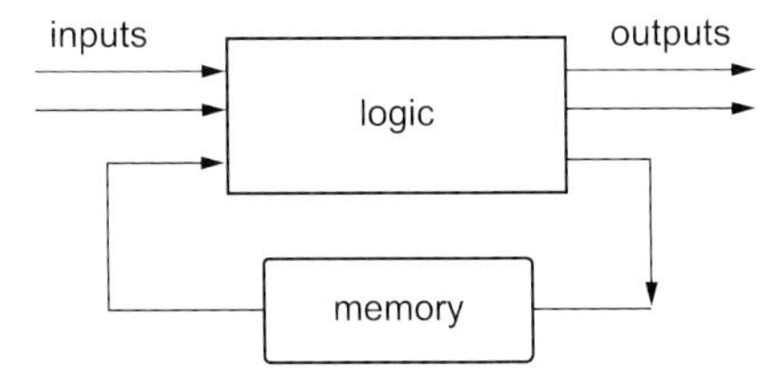

Figure 9.1 A schematic diagram of a finite-state machine. This consists of a combinational circuit ("logic") to which a memory element is connected to form a feedback path. The molecular implementation of this simple machine has been described in the authors' exploratory studies.

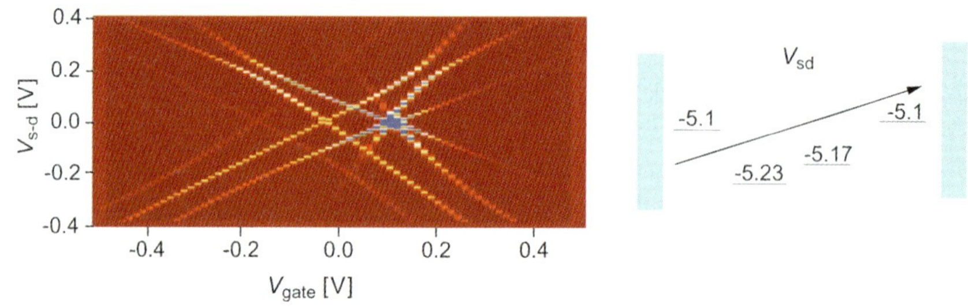

Figure 9.2 Differential conductance, dI/dV plotted as a function of the gate (V_{gate}) and the source-drain ($V_{s\text{-}d}$) voltages. The computation is for the four-level system shown where the coupling between the sites is 0.05 eV. The Stark shift of the site energies due to the source-drain bias is included in the Hamiltonian. The effect of the gate voltage is to shift the energies of system with respect to the Fermi energy of the electrodes.

A finite-state molecular machine has been proposed that can be cycled based on laser pulses [101, 104]. In recent investigations, the same optical finite-state machine has been used to implement a cyclable full adder and subtractor [96]. In a different direction, recently a molecular finite-state machine in the gas phase has been proposed by showing that a linear sequential machine can be designed using the vibrational relaxation process of diatomic molecules in a buffer gas. An alternative to optical pulses for providing inputs is to apply *voltage pulses*. In a recent study on QD assemblies (e.g. Refs. [105, 106]), it was shown that the application of a gate voltage allows the molecular orbitals to be tuned in resonance with the nanoelectrodes so that a current flows, whereas there is no current in the absence of the gate voltage (see Figures 9.2 and 9.3). A similar control is also possible on a single molecule [107–109] and for single QDs or several coupled dots.

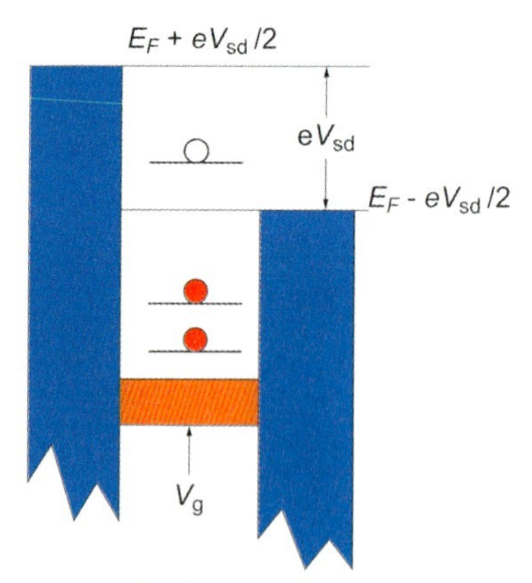

Figure 9.3 Schematic representation of a three-terminal device. The gating voltage is applied to a dieletric layer perpendicularly to the direction of the source-drain voltage. This scheme, which is equally useful for molecules and quantum dots, shows that either can be made to act as a finite-state device.

9.1.8
Multi-Valued Logic

So far, it has been taken as given that an input or an output can have one of two possible values. A laser beam can be on or off, a molecule can fluoresce or not, and a charge can transfer or not, and so on. Therefore, two-valued or Boolean logic is being implemented. Ever since Shannon showed, in 1938, the equivalence between a Boolean logic gate and a network of switches, computer circuits have been assembled from switches. First, the switches were electromechanical in nature, followed by vacuum tubes and then the transistor. Now that the discrete nature of the carriers of electrical charge has made it unclear as to how the size of the transistor might be further reduced, a major – and very serious – effort has been made to use a molecule as a switch.

Previously, several ways have been discussed of using a molecule as an entire logic gate (= a connected net of switches), or even as a finite-state machine rather than simply as a switch. There is, however, another possible generalization, which also is well-suited to what molecules are and what they can do – namely, to go to *multi-valued logic* [110–112]. What this means is that there are more than two allowed mutually exclusive answers to every question. This makes the numbers to be dealt with shorter (e.g. 1001 is the binary number that is written as 9 in base 10). In their studies, the present authors took three as a compromise between shorter numbers (e.g. 10 is ternary for 9) and the errors that can result in making a choice between too many alternatives. The molecular physics is straightforward, to take zero, one or two electrons in the valence orbital [113]. But clearly, there are other choices such as a Raman transition to several different final vibrational states. Molecules are willing and able to be pumped and to be probed to multiple states. It is not clear, however, if the industry is willing to learn to go beyond two.

9.2
Combinational Circuits by Molecular Photophysics

Combinational circuits are made of concatenated combinational gates. A logic gate accepts inputs and implements a particular function of the inputs to provide an output. In most implementations to date, the inputs and outputs are Boolean (binary) variables that can take one of two values, that is 0 or 1. Boolean logic gates can be one input–one output gates, like the NOT gate or two input–one output gates. There are in total 16 functions of two binary variables, and among these the AND, OR, XOR and INH gates are among the most commonly used in combinational circuits. The way in which half adders and full adders can be implemented are discussed in the following section, but initially attention is centered on elementary gates such as AND or OR to set the scene. The truth tables of these gates are provided in Table 9.1, and correspond respectively to taking the MIN and the MAX of the two Boolean inputs x and y. The AND gate can also be viewed as the product of the two inputs, $x \times y$. This gives an output 1 only if the two inputs are both 1. It can be implemented by two switches in series, while two switches in

Table 9.1 Truth tables of the binary AND, OR, XOR and INH gates.

INPUT		OUTPUT			
x	*y*	*x* AND *y*	*x* OR *y*	*x* XOR *y*	*x* INH *y*
0	0	0	0	0	0
1	0	0	1	1	1
0	1	0	1	1	0
1	1	1	1	0	0

parallel correspond to the OR gate. Another binary gate which is often used in combinational circuits is the XOR gate (see Table 9.1) which corresponds to the addition of the binary inputs modulo 2, $x \times y$.

The addition modulo 2 of two binary numbers works exactly as would be expected, in base 10. When the two binary inputs are both 1, the result of their sum modulo 2 is 0, but a 1 (the carry digit) must be reported in the next column of binary digits [114] (see Table 9.2).

When using the XOR and the AND gates, it is possible to build a half adder. This has two inputs, the two numbers to be added modulo 2, and two outputs, the sum modulo 2 of the two inputs, realized by a XOR gate and the carry, realized by an AND gate, the truth tables for which are shown in Table 9.1. A half adder is an important building block for molecular combinational circuits because, by concatenating two half adders, a full adder can be built. Usually, as shown by an example in Table 9.2, binary numbers of length more than one digit must be added. A full adder is more complex than a half adder, in that it takes the carry digit from the addition of the previous two digits into account (called the "carry in"). A full adder has therefore three inputs – the two digits to be added and the carry digit from the previous addition – and two outputs – the sum modulo 2 of the three inputs and the carry digit for the next addition (see Table 9.2) – called the "carry out".

In photophysics logic implementations, the inputs are provided by laser pulses, their physical action being to excite the molecule, typically to electronic excited states. The logical value 1 is encoded as the laser being "on", and the logical value 0 as the laser being "off". The molecules in the sample act independently of one another, and there is uncertainty as to whether every molecule absorbs light. It is therefore not the case that a single molecule suffices to provide an output. When working with an ensemble of molecules, there is no need to read a strict "yes" or a strict "no" from each molecule.

Table 9.2 Addition of the two binary numbers $x = 010$ and $y = 111$.

Carry	**1**	**1**		
x		0	1	0
y		1	1	1
Sum	1	0	0	1

What is needed is to excite enough molecules to be above the threshold for the detection of light absorption. The detection of the outputs is by fluorescence and/or by detecting ions. The detection of ions is relatively easy, so by monitoring the absorption by photoionization of the molecule, it is sufficient if only a small number of molecules respond to the input. An excited molecular ion typically fragments, and the detection of ionic fragments can also be used to encode outputs. Although ionic fragments can be detected very efficiently, the price is that the molecule self-destructs at the end of the computation and so the gate cannot be cycled. The details of an optically addressed half and full adder that can be cycled are provided in Section 9.1.3 [96].

First, however, the implementation of a half adder will be discussed and two approaches for doing this will be compared. It will then be shown how to implement a fault-tolerant full adder on a single molecule [115] using photophysics in the gas phase of the 2-phenylethyl-*N,N*-dimethylamine (PENNA) molecule [116–118]. This implementation follows the lines of the 2001 implementation of a full adder on the NO molecule [95]. Finally, the realization of a full adder by concatenation of two half adders is discussed, where the logic variables are transmitted between the two half adders by energy transfer between two aromatic molecules that are photoexcited [94] in solution.

9.2.1
Molecular Logic Implementations of a Half Adder by Photophysics

A discussed above, a half adder has two outputs: Addition modulo 2 is implemented by the XOR gate, and the carry digit is the result of an AND operation. A molecular realization of the logical XOR operation is challenging because the output must be 0 when both inputs are applied. For a photophysics implementation of the inputs this means that when both lasers are on, there is not the output that is observed when only one laser is on. The realization of the AND gate is comparatively easier because an output is produced only if both inputs are on, and so any reproducible experiment, which requires two inputs for generating an output can implement an AND gate [103]. The truth table for the implementation of a half-adder by optical excitation is given in Table 9.3.

Two-photon resonance-enhanced absorption by aromatic chromophores has been used as an effective way to implement an XOR operation on optical inputs at the molecular level [94, 95, 104]. The two photons (each of a somewhat different color and

Table 9.3 Truth table for an implementation of a half-adder by photoexcitation.

x(laser 1)	*y* (laser 2)	Carry (AND)	Sum (XOR)	(carry, sum)
0	0	0	0	(0,0)
1	0	0	1	(0,1)
0	1	0	1	(1,0)
1	1	1	0	(1,1)

therefore distinguishable) represent the possible inputs. For an XOR gate to be physically realizable, the following conditions must be satisfied. First, for aromatic molecules or for molecules with aromatic chromophores the resonant level, typically the first optically bright electronically excited state, S_1, has a fairly broad absorption band consisting of many vibronic transitions. So, two photons of different frequencies (and therefore distinguishable) can be absorbed with a similar cross-section. This provide the OR part of the XOR gate (second and third lines of Table 9.3): if laser light of either frequency is on, the output can be identified as the fluorescence. The exclusive part results from the following effect: it is often the case that, having absorbed one photon, the cross-section of an aromatic chromophore to absorb a second photon is higher. That the bottleneck for two-UV photon absorption is often the absorption of the first photon has been realized since the earliest days of visible/UV multiphoton ionization/dissociation [119, 120], and has been used extensively since then. Therefore, when two light pulses are applied the system need not remain in S_1 for a significant length of time because it can absorb a second photon. Whether it preferentially does so depends on the particular molecule. Either input laser can excite from the ground state to S_1. Therefore, the fluorescence signal will increase when both lasers are on. The input in this case is 1,1 and from Table 9.3 the need is to detect the output 1,0. In the presence of two photons the fluorescence from S_1 (a rather slow, >5 ns scale, process in PENNA) may fall, but the more secure detection of the presence of both input beams is the increase of the number of ions. There is an increase rather than pure onset because a single laser can also cause ionization, but this occurs with low intensity. A high ionization intensity corresponds to a simultaneous input of both lasers. In practice, the two events "high ionization efficiency" and "low ionization efficiency" can easily be distinguished by the use of discriminators and analog electronics. A simulation of the temporal response of the molecule to the laser pulses, so as to show that this is possible, is available in Ref. [115]. To conclude, the 1,1 input is identified as a high ionization signal.

The use of two lasers of different colors is dictated by the need to represent two distinct inputs, but there is a clear physical advantage if the two frequencies differ by more than a vibrational frequency in S_1: because the S_1 and S_0 vibrational frequencies are not exactly the same, the down-pumping by the other laser is not resonance enhanced and thus improbable, as shown schematically in Figure 9.4.

There are therefore two ways to implement a half adder (see Table 9.4). The direct way is to detect separately the sum and the carry and to assign them to a different experimental probe. In the case of the PENNA molecule (see Figure 9.5) the inputs are encoded as UV lasers being off or on. For the sum digit, the experimental probe is the detection of the fluorescence from the S_1 electronic state. The carry digit is the result of the absorption of two photons. In the case of PENNA, the absorption of two UV photons at the chromophore end causes local ionization, followed by charge migration to the N-end of the chain. The carry digit is therefore encoded as the detection of N-fragments. As discussed above, the absorption of a second photon decreases the intensity of the fluorescence from S_1, but typically does not quench it enough so that detection of the carry digit through N-end ions is preferable.

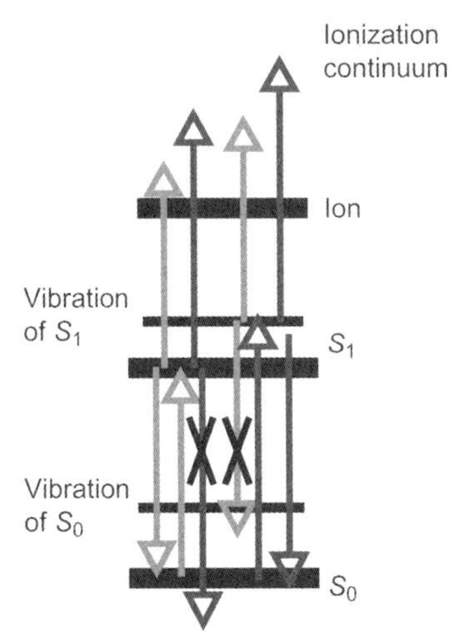

Figure 9.4 Two-photon ionization with the two inputs being lasers of unequal frequency can be made fault-tolerant to stimulated emission. This is particularly so if one laser operates on the 0,0 transition while the other pumps a vibrationally excited level of S_1. For details, see the text.

One way of implementing a fault-tolerant half adder is to combine the result of the carry and the sum digit into one word (carry, sum). This is shown in the fifth column of Table 9.3. As can be seen, the four distinct pairs of inputs of the half adder – that is, (0,0), (1,0), (0,1) and (1,1) – corresponds only to three distinct outputs of (0,0), (0,1) and (1,0). This is because in addition modulo 2, if the carry is 1, the value of the sum is necessarily 0. Therefore, instead of assigning a separate physical probe for the sum and the carry, it is possible to assign a physical probe for the three different logical values of the word (carry, sum). It should be noted that the binary meaning of the word (carry, sum) corresponds to the number of inputs with value 1. In the case of PENNA, the choice was made to assign the word (0,1) to the presence above threshold

Table 9.4 Truth table and experimental probes used for the two ways of implementing a half-adder on PENNA.

x (UV(1))	*y* (UV(2))	Carry (AND)	Sum (XOR)	Probe for carry (AND)	Probe for XOR	Probe for words (carry,sum)
0	0	0	0	no N-end fragment	no fluo from S_1	no output signal (0,0)
1	0	0	1	no N-end fragment	fluo from S_1	fluorescence from S_1 (0,1)
0	1	0	1	no N-end fragment	fluo from S_1	fluorescence from S_1 (0,1)
1	1	1	0	N-end fragment	no fluo from S_1	N-end fragment (1,0)

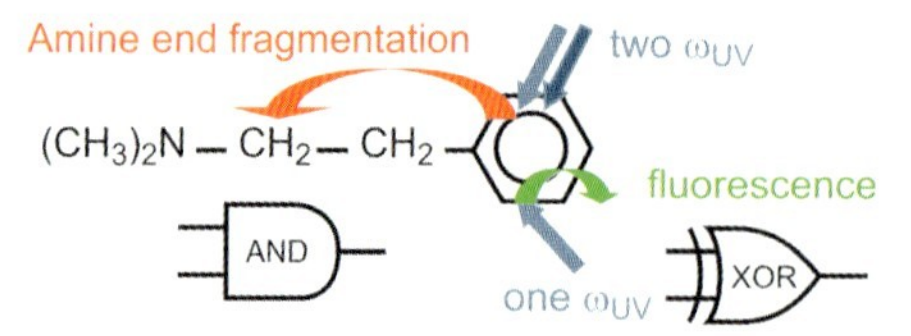

Figure 9.5 Schematic representation of the two-photon excitation scheme used to implement a half adder on PENNA.

of a fluorescence signal from S_1, and the word (1,0) to the detection above a threshold value of a N-end fragments. The fault tolerance of this scheme arises from the fact that in case of inputs (1,1), a N-end fragment ion detection will be reported, irrespective of the intensity of the fluorescence signal from S_1. The scheme is fault tolerant with respect to the extent of fluorescence from S_1 in case of two-photon excitation. Detecting the (0,0) output is straightforward as no excitation is provided. The two ways of implementing a half adder on PENNA are summarized in Table 9.4.

This half-adder self-destructs at the end of the computation because the local ionization at the chromophore end causes the PENNA ion to fragment. However, this scheme allows for a remarkable sensitivity in the detection of the outputs because very few ions can already be detected with a good signal-to-noise ratio. Although this involves an ensemble of molecules, not all of which provide an answer,. However, response is needed from only 100 molecules to obtain acceptable statistics, and this response occurs quite rapidly.

In a full adder, the word (1,1) as an output is allowed, so that a full adder has four distinct binary words as outputs, namely (0,0), (0,1), (1,0) and (1,1). As discussed above, it has also one more input than the half adder, the carry digit. It is shown in the next section how a full adder can be implemented on the PENNA molecule using the same fault-tolerant scheme for probing the outputs. This manner of implementation of a full adder is contrasted with the implementation based on the concatenation of two half adders, in which the sum and the carry are detected separately.

9.2.2
Two Manners of Optically Implementing a Full Adder

A full adder has three inputs and produces two outputs, the sum which is the addition modulo 2 of the two inputs and the carry in digit and the carry out, which for the next cycle of computation becomes the carry in:

$$\text{sum out} = x \oplus y \oplus \text{carry in} \tag{9.1}$$

$$\begin{aligned}\text{carry out} &= (x \otimes y) + ((x \oplus y) \otimes \text{carry in}) \\ &= (x \otimes y) + (x \otimes \text{carry in}) + (y \otimes \text{carry in})\end{aligned} \tag{9.2}$$

where $\oplus$ means addition modulo 2 (XOR), $\otimes$ means binary product (AND) and $+$ means OR. The carry out logic equation can be simplified to

$$\text{carry out} = x \otimes y + x \otimes \text{carry in} + y \otimes \text{carry in} \tag{9.3}$$

In the implementation on PENNA that was proposed above, the two inputs x and y are encoded as for the half adder, by two UV photons with slightly different wavelengths. The carry in digit is encoded as a laser pulse of green light which is intense enough that two photons can be absorbed to allow the transition to the S_1 state by a non-resonance-enhanced two-green photon transition. The four outputs words (carry out, sum): (0,0), (0,1), (1,0) and (1,1) are detected each by a distinct experimental probe. As in the half adder implementation of PENNA, the output word (0,1) is detected as fluorescence from the S_1 state while the output (1,0) as the presence of N-end fragment ions. The detection of the output (0,0) is straightforward as it corresponds to no inputs. The output (1,1) corresponds to the three inputs having the value 1; that is, the PENNA molecule is excited by two UV (x and y) and two green photons (carry in). Experimentally, this amount of energy is causing fragmentation at the C-end (instead of fragmentation at the N-end which occurs when only two inputs are 1, in which case it is only the equivalent in energy of two UV photons). The presence of C-end fragment ions above a given threshold is therefore used to detect the (1,1) output. The excitation scheme and experimental probes of the outputs are shown in Figure 9.6 and the corresponding truth table in Table 9.5. Note that, as in the case of the half adder, the output word (carry, sum) counts in binary how many inputs are 1.

Another way to implement a full adder is by concatenation of two half adders. The corresponding combination logic circuit is shown in Figure 9.7.

The physical implementation, by photoexcitation of a donor–acceptor complex in solution[94], is an example of intermolecular concatenation of two half adders by energy transfer. It demonstrates that one molecule is able to communicate its logical output to another molecule. The implementation is on a specific pair (rhodamine 6G–azulene), for which considerable data are available, but the scheme is general enough to allow a wide choice of donor and acceptor pairs. The first half-adder is realized on rhodamine 6G, and the second half adder on azulene. The midway sum is transmitted from the first to the second half adder by electronic energy transfer between rhodamine 6G and azulene.

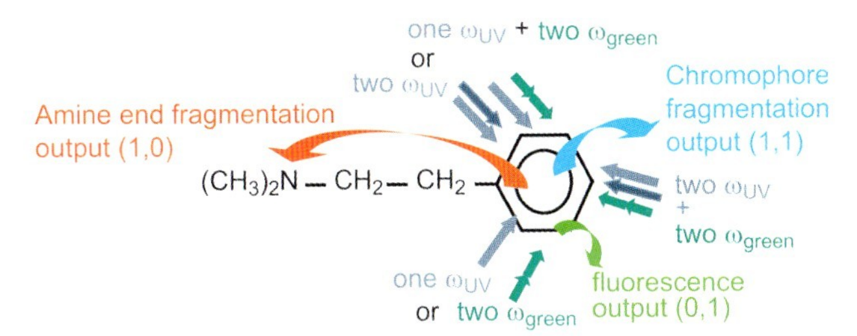

Figure 9.6 Excitation scheme of PENNA for implementing a full adder. Fluorescence from the S_1 state of the phenyl ring codes for carry-out = 0, sum-out = 1, fragmentation at the amine end codes for carry out = 1, sum out = 0 and fragmentation at the chromophore end for carry out = 1, sum out = 1 (see also Table 9.5). The two binary digits to be added are encoded as the two UV photons being on or off. The carry-in is encoded as excitation by two green photons.

Table 9.5 Truth table and detection scheme for the outputs for the optical implementation of a full adder on PENNA.

x (UV(1))	y (UV(2))	Carry in (vis, two-photon)	Carry out	Sum out	Output word (carry,sum)	Probe for output word
0	0	0	0	0	(0,0)	No signal
1	0	0	0	1	(0,1)	fluorescence from S_1
0	1	0	0	1	(0,1)	Fluorescence from S_1
1	1	0	1	0	(1,0)	N-end fragment
0	0	1	0	1	(0,1)	Fluorescence from S_1
1	0	1	1	0	(1,0)	N-end fragment
0	1	1	1	0	(1,0)	N-end fragment
1	1	1	1	1	(1,1)	C-end fragment

The physical realization of the two half adders relies (as in the case of the half adder implementation on PENNA discussed above) on the fact that the absorption of one or two UV photons (inputs x and y) by the donor (acceptor) molecule leads to distinct outputs. However here, unlike the case of PENNA, the absorption of the second photon does not lead to ionization, but rather to absorption by a second excited electronic state, S_2. In general, a molecular half adder will be available for molecules which have a detectable one-photon and a detectable two-photon absorption. This seems to go against Kasha's rule [121], but in fact there are enough exceptions. Azulene and many of its derivatives provide one class. The emission from the second electronically excited state, S_2, is often as strong or stronger as the fluorescence from S_1 [122, 123]. More in general, emission from S_2 is not forbidden; rather, due to competing non-radiative processes it often has a low quantum yield but it is definitely detectable, particularly so as it is much to the blue as compared to the emission from S_1. If necessary, the emission from S_2 can be detected by photon counting. There is, therefore, a case where the outputs of the XOR gate and the AND gate that constitute the half adder can be probed separately, with sufficient fidelity. The output of the XOR gate of the first half adder is encoded as populating the S_1 state of rhodamine 6G, while the output of the XOR gate of the second half adder (the sum out) is encoded as detecting fluorescence of the S_1 state of azulene. The output of the first AND gate

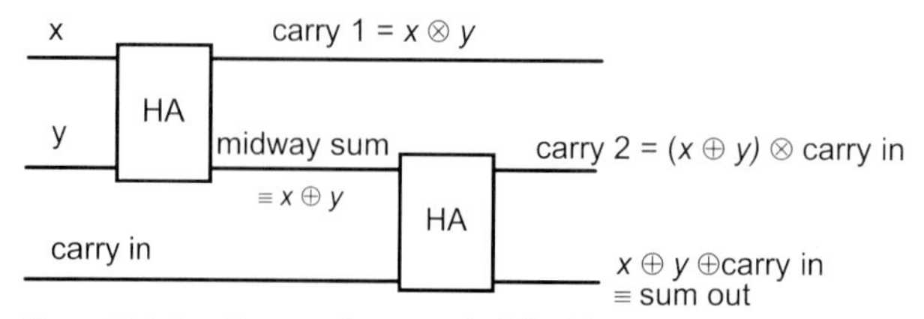

Figure 9.7 Combinational circuit of a full adder implemented by concatenation of two half adders.

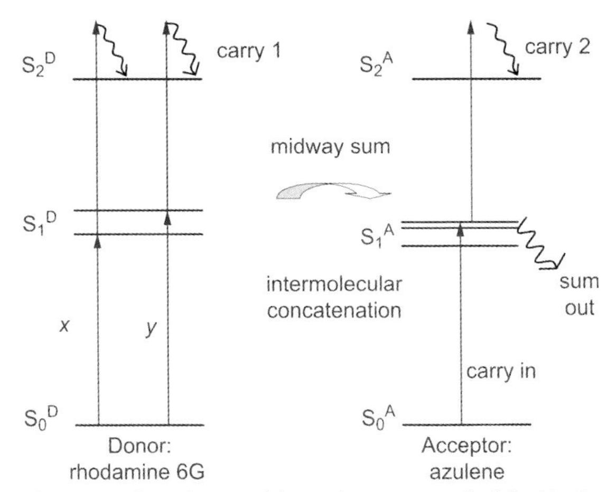

Figure 9.8 Photophysics of the implementation of a full adder by concatenation of two half adders on the donor–acceptor complex rhodamine 6G–azulene in solution.

(carry 1 in Figure 9.7) is probed by fluorescence from the S_2 of rhodamine 6G and correspondingly, the output of the AND of second half-adder (carry 2 in Figure 9.7) by fluorescence of the S_2 state of azulene. The concatenation between the two half adders is performed by the fairly rapid [124] intermolecular electronic energy transfer. Specifically, the well-characterized [125–127] transfer from the S_1 level of rhodamine 6G to the S_1 of azulene was proposed.

The photophysical scheme for the implementation of a full adder on the donor–acceptor complex rhodamine 6G–azulene is summarized in Figure 9.8. The S_1 level of rhodamine 6G can be readily pumped with photons absorbed within the $S_0 \rightarrow S_1$ band. The frequencies ω_1 ($\equiv$ input x) $= 18\,797\ \text{cm}^{-1}$ (the second harmonic of the Nd-YAG laser) and ω_2 ($\equiv$ input y) $= 18\,900\ \text{cm}^{-1}$ are taken. This is not needed for the full adder, but the absorption to S_1 can be detected through its emission at about $17\,500\ \text{cm}^{-1}$ [125, 126]. This emission is logically equivalent to $x \oplus y$ because if the intensity is high enough due to two photons being present, the donor will be pumped either directly to S_2 or to higher levels, followed by ultrafast non-radiative relaxation to S_2. The large absorption cross-section of $2.5 \times 10^{-18}\ \text{cm}^2\ \text{molecule}^{-1}$ [128] for the $S_1 \rightarrow S_n$ ($n \geq 2$) of rhodamine 6G ensures efficient pumping of S_2. The emission from S_2 is at about $23\,250\ \text{cm}^{-1}$, with a quantum yield of about 10^{-4} [126]. It is this emission which serves to logically implement the left AND gate, and it is equivalent to $x \otimes y$ (denoted as carry 1). The S_1 level of the donor transfers the energy, via the Forster mechanism [124] to the azulene acceptor, the S_1 level of which is at $14\,400\ \text{cm}^{-1}$, that emits in the 13 400 to $11\,000\ \text{cm}^{-1}$ range [129]. This emission provides the logical sum output [Eq. (9.1)]. The S_2 level of azulene has its absorption origin at $28\,300\ \text{cm}^{-1}$, and so it can be reached from S_1 by a third photon of frequency $14\,400\ \text{cm}^{-1}$. The same photon can also pump ground-state azulene to its S_1 level. Emission (or lack thereof) from S_2 of azulene at $26\,670\ \text{cm}^{-1}$ [127] provides the carry 2 bit. The carry 1 and the carry 2 cannot be equal to 1 together, because if the carry 1 is 0, the midway sum is 0, meaning that even if the carry in is 1, the carry 2 cannot be 1. In other words:

$$carry\ out = carry\ 1 + carry\ 2 \tag{9.4}$$

The carry out is therefore physically probed by monitoring the fluorescence from the S_2 states of rhodamine 6G and azulene, which logically corresponds to the first line of Eq. (9.2).

The advantage of an all-optical scheme for the full adder implementation compared to the implementation on PENNA as discussed above is that the adder does not self-destruct at the end of the computation. Another advantage is that it operates relatively rapidly. The energy transfer rate for a solution of 10^{-3} M azulene, estimated using the S_1 fluorescence spectrum of rhodamine 6G and the absorption spectrum of azulene, is about $10^{10}\,s^{-1}$. This rate is sufficient for present needs, but it can be increased [130] if the two chromophores are incorporated within a single molecular unit using a short bridge to connect them [124]. The increase in the rate will be particularly significant (five orders of magnitude) if the bridge is rigid [130, 131]. It should be emphasized that a rigid bridge is required to achieve a very high rate. Many other couples based on commonly used laser dyes as donors and azulene derivatives [128, 132] may also be utilized for implementation of the logic gate [133, 134].

9.3 Finite-State Machines

Finite-state (also called sequential) machines are combinational circuits with memory capability. The memory registers are the internal state(s) of the machine [135, 136]. As in a combinational circuit, the outputs of the machine depend on the inputs, but in addition the output also depends on the current state of the machine. It is this dependence of the output on the state of the machine that endows finite-state machines with a "memory". The memory of the machine corresponds to the state of the experimental system, and this state can be changed by applying suitable perturbations, such as optical or voltage pulses. As in the other logic schemes, the "logic" part is an encoding of the subsequent dynamics of the system.

The finite-state machine computational model takes advantage of two aspects that are natural for quantum systems:

- A physical quantum system has discrete internal states and its response to perturbation will in general depend on what state it is in.
- Perturbations can be applied sequentially, so that the machine can be cycled.

By taking advantage of the two points above, the implementation of several forms of finite-state machine was proposed: a simple set-reset that can be either optically [101] or electrically addressed [106]; an optical flip-flop [104] and full adder and subtractor [96]; an electrically addressed full adder [106] and a electro-optically addressed counter [137]. Beyond that it has been shown, using optical addressing, that a molecule can be programmed and behaves (almost) like a Turing machine [101]. The caveat "almost" is introduced because a molecule can have only a finite number of quantum states, whereas a Turing machine has an unlimited memory. Possibly this is not a true limitation since if indeed the number of quantum states of the universe is finite – sometimes known as the "holographic bound" – then no physical system can strictly act as a Turing machine.

In this section, a review is conducted of optically addressed finite-state machines, up to a full adder (Section 9.3.1) and an electrically addressed machine (Section 9.3.2). If molecules and/or supramolecular assemblies are to offer an inherent advantage over the paradigm of switching networks, it will likely be through each molecule acting as a finite-state unit.

9.3.1 Optically Addressed Finite-State Machines

Laser pulses are used to optically address atomic or molecular discrete quantum states. All of the schemes discussed here are based on the Stimulated Raman Adiabatic passage (STIRAP) pump-probe control scheme, that allows the population of the quantum states of atoms or molecules to be manipulated. The advantages of the STIRAP control scheme for implementing finite-state machines are that the external perturbation can induce a change of state with a very high efficiency (close to 100%), and that the residual noise which accumulates when the machine is cycled can be erased by resetting it. Moreover, the perturbation has a distinctly different effect on the system depending on the initial state. These advantages are supported by experimental results for atomic (i.e. Ne [138]) and molecular systems (i.e. SO_2 [139], NO [140]), and the dynamics is well-described by solving the quantum mechanical time-dependent Schrödinger equation [141–143].

Here, the operation of finite-state machines using quantum simulations on a three-level system with a Λ-level scheme is described (see Figure 9.9). The pump pulse, with photons of frequency ω_P, is nearly resonant (up to a detuning Δ_P) with the $1 \rightarrow 2$ transition, while the Stokes pulse, with photons of frequency ω_S, is nearly resonant (up to a detuning Δ_S) with the $2 \rightarrow 3$ transition. Levels 1 and 3 are long-lived, but level 2 is metastable because it can fluoresce. The spontaneous emission from level 2 provides a readable output. The important feature of this level structure is that there are two routes for going from level 1 to level 3. The first route is a kinetic or

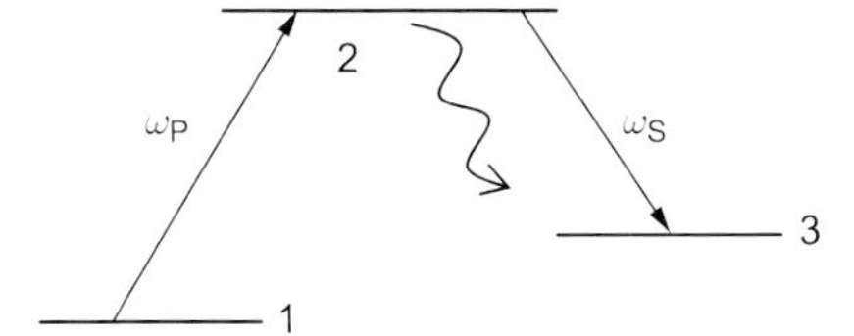

Figure 9.9 The Λ-level scheme is a STIRAP experiment. The pump transition driven by ω_P is between levels 1 and 2, while the Stokes or dump transition is between levels 2 and 3 induced by ω_S. The population from level 2 is detected by its fluorescence.

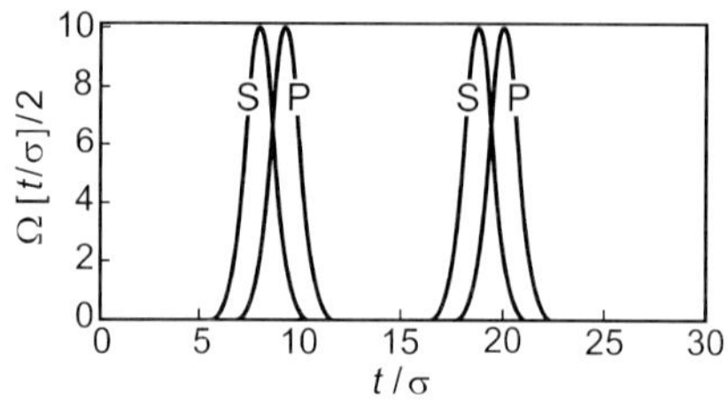

Figure 9.10 Resistance–time profile of an individual SP pulse and of the sequence of the two pulses, as used in the simulation reported in Figure 9.11. The plot is as a function of time in reduced units t/σ, where σ is the width of the S and the P pulses and the two widths are taken as equal.

"intuitive" pump scheme where first a pulse of frequency ω_P is applied so that population is transferred from level 1 to the excited level 2. From this level, the population can fluoresce or be transferred to level 3 by induced emission, using a Stokes pulse of frequency ω_S. The second route is the counter-intuitive or STIRAP route that takes population from level 1 to level 3 with almost no population in level 2, and therefore no spontaneous emission from level 2. In this counter-intuitive route, the Stokes (S) pulse of frequency ω_S is applied first and the Pump (P) pulse of frequency ω_P is somewhat delayed and applied subsequent to the dump pulse, preferably so that its front still overlaps the tail of the pulse of frequency ω_S (see also Figure 9.10).

For the purpose of constructing a finite machine, the following observation is used. Suppose that passage occurs from level 1 to level 3 by the counter-intuitive route; this means that the S pulse is on first and the P pulse is delayed. This promotes (almost) all molecules from level 1 to level 3. The same set of two pulses can now be applied, keeping their respective order in time, and this will drive almost all molecules back to level 1. However, since the system is in level 3 this order of pulses constitutes a kinetic route, and therefore state 2 is populated as an intermediate state. In the simulation of finite-state machines, the pulse conditions are chosen such that the kinetic route leaves a few percent of molecules in level 2. The level 2 will fluoresce, and this will serve as the signature of the kinetic route. In other words, the set of two pulses – Stokes followed immediately by pump – drives the molecule from level 1 to level 3, or vice-versa, depending on what state it is in. The fluorescence from level 2 is the signature of which transition was driven. It should be noted that, by using the S and the P pulses as two distinct inputs, with either one being on or off, it could be shown possible to implement a set–reset finite-state machine and also a programmable Turing machine on the Λ-level structure [94]. In order to keep the discussion as simple as possible, these schemes will not be discussed here. In the present discussion the input is the presence or absence of the superposition of both pulses.

For the purpose of implementing a flip-flop and a full adder and subtractor, the optical input is defined as a Stokes pulse at the frequency ω_S, followed in time by a pump pulse at frequency ω_P. The input is referred to as a SP pulse (see Figure 9.10). The overlap in time between the two pulses and their intensity is adjusted such that if

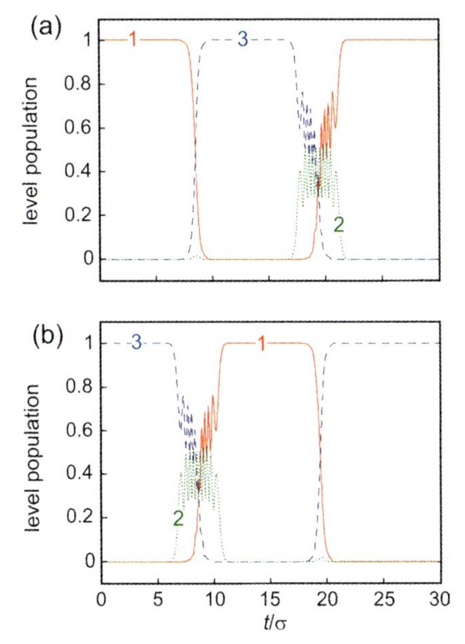

Figure 9.11 Population of the levels 1 (red), 2 (green) and 3 (blue) as a function of time when two SP pulses are applied (see Figure 9.10). The reason why such a sequence of two pulses is useful as a way of reading the state of the machine is explained in the text. The populations are computed as $|c_i(t/\sigma)|^2$ by numerical integration of the time-dependent Schrödinger equation. (a) The system is initially in level 1. The first SP pulse takes it to level 3 via the STIRAP route, with essentially no transient population in level 2. The second SP pulse takes the population in level 3 back to level 1. This second transition occurs via the kinetic route and significant transient population in level 2. (b) The system is initially in level 3. The population transfer from level 3 to level 1 goes via the kinetic route while the second SP pulse takes the system back to level 3 via the STIRAP route.

the system is initially in level 1, the SP pulse transfers to level 3 via the STIRAP route, without any significant population in level 2. On the other hand, if the system is initially in level 3, the SP pulse will pump it down to level 1 via the kinetic route, that can be detected by spontaneous emission from level 2. The time profile of the SP pulse is shown in Figure 9.10.

The quantum simulations illustrating the two routes that are possible using a SP pulse as an input are shown in Figure 9.11. The purpose of the simulation is to show that, by either route, the SP pulse achieves an essentially 100% population transfer between levels 1 and 3. The main source of noise is the spontaneous emission from level 2 that can end up either in level 1 or 3 or to yet another level, in which case the molecule is lost from the ensemble. To achieve a population transfer close to 100% between levels 1 and 3, rather intense pulses are needed, with the result that in the kinetic route the population in level 2 remains low (see Figure 9.10). Only a few photons are necessary to detect the output, so that detecting the output does not introduce too much noise. After a few cycles, the noise accumulation can be corrected for by resetting the machine (see Ref. [104]).

For the three-level structure shown in Figure 1.9, the Hamiltonian in the rotating wave approximation takes the form [141, 144, 145]:

$$\mathbf{H}=\frac{1}{2}\begin{pmatrix} 2\omega_1 & \Omega_P(t)\exp(i\omega_P t) & 0 \\ \Omega_P(t)\exp(-i\omega_P t) & 2\omega_2 & \Omega_S(t)\exp(-i\omega_S t) \\ 0 & \Omega_S(t)\exp(i\omega_S t) & 2\omega_3 \end{pmatrix} \tag{9.5}$$

where the two pairs of levels are coupled by nearly resonant transient laser pulses. The Rabi frequency [141, 146] is denoted as $\Omega(t)$. It is given by the product of the amplitude of the laser pulse, $E(t)$ and the transition dipole, μ: $\Omega(t)=\mu\ E(t)/h$. The central frequency of the Pump and Stokes lasers is almost resonant with the $1\rightarrow 2$ and the $2\rightarrow 3$ transitions; that is, $\omega_P=\omega_2-\omega_1-\Delta_P$ and $\omega_S=\omega_2-\omega_3-\Delta_S$ and the detunings are small and taken to be equal in the simulation, $\Delta_P=\Delta_S=\Delta$. Therefore, the two lasers are off resonance for the transitions for which they are not intended. In the rotating-wave approximation, the Hamiltonian couples between levels using only the component of the oscillating electrical field that is in resonance or nearly so for the two levels. The Hamiltonian [Eq. (9.5)] is that used in earlier studies of STIRAP [141, 147–149].

The Hamiltonian [Eq. (9.5)] can be recast in the interaction picture where it takes the form:

$$\tilde{\mathbf{H}}=\frac{1}{2}\begin{pmatrix} 0 & \Omega_P(t)\exp(-i\Delta_P t) & 0 \\ \Omega_P(t)\exp(i\Delta_P t) & 0 & \Omega_S(t)\exp(i\Delta_S t) \\ 0 & \Omega_S(t)\exp(-i\Delta_S t) & 0 \end{pmatrix} \tag{9.6}$$

The wavefunction of the system, $\psi(t)$, is a linear combination of the three levels, with time-dependent coefficients:

$$\psi(t)=\sum_{i=1}^{3}\tilde{c}_i(t)|i\rangle,\quad \tilde{c}_i(t)=c_i(t)\exp(-i\omega_i t) \tag{9.7}$$

where $\tilde{c}(t)$ are the coefficients in the interaction picture. These satisfy the matrix equation of the time-dependent Schrödinger equation, $i\,d\tilde{\mathbf{c}}/dt=\mathbf{H}\tilde{\mathbf{c}}$, which is solved numerically without invoking the adiabatic approximation [150]. The total probability, $\mathbf{c}^T\mathbf{c}=\tilde{\mathbf{c}}^T\tilde{\mathbf{c}}$, is conserved because the Hamiltonian is Hermitian.

Figure 1.11 shows the effect of acting with two SP pulses successively, for the system being initially in level 1 [panel a and in level 3 (panel b)]. The time profile of the sequence of two SP pulses in shown in Figure 9.10.

The simulations start with the molecule either in level 1 (Figure 9.11, panel a), or level 3 (panel b). The sequence of pulses as shown in Figure 9.10 returns the system to the level it started from. In a single cycle of the machine the SP pulse is applied only once. Parameters of the simulation given in reduced time units (t/σ) are: $\Omega_P(t/\sigma)=\Omega_S(t/\sigma)=20.05\ \exp\left(-((t/\sigma)-\tau_i)^2/2\right)$, with $\tau_{s1}=8$, $\tau_{p1}=9.25$, $\tau_{s2}=18.75$, $\tau_{p2}=20$. The detuning $\Delta=\Delta_S=\Delta_P=4(\sigma/t)$. The area of the pulse, $A(t)=\int\Omega(t/\sigma)d(t/\sigma)$ is $6.38\ \pi$. These details are quoted since the achievement of an essentially complete population transfer by the kinetic route (as shown in Figure 9.11) is sensitive to the intensity of the pulse and also to the detuning.

The physics shown in Figure 9.11 is all that is required to implement finite-state machines. The implementation of a full adder and a full subtractor are discussed below; these are implemented in a cyclable manner, with each full addition or subtraction requiring two steps. The inputs x and y are both encoded as an SP pulse. The duration of a computer time step is taken to be somewhat longer than the duration of the input SP pulse. Here (unlike Section 9.3.1, where the combinational circuits implement a full adder) there is no need for concatenation because the carry (borrow) is encoded in the state of the machine for the first step and the midway sum (the midway difference) is encoded in the state of the machine for the second step. This is a major advantage of finite-state machines. The state of the machine encodes intermediate values needed for the computation.

A logic value of 0 is encoded for the carry-in or of the borrow digit as the molecule being in level 1, and a logic value of 1 as the molecule being in level 3. During the course of the discussion, it will also be shown (see Table 9.8) how the first cycle of the operation can also be logically interpreted as a T-flip-flop [136] (T for toggle) machine.

In order to cycle an adder after two optical inputs, the machine should be in a state that corresponds to the carry for the next addition, so that it is ready for the next operation. At present, a scheme which does exactly that cannot be devised, as two more operations are required in order for the machine to be ready for the next cycle. The reason for this is that, as shown below, at the end of the two cycles, the sum out is encoded in the state of the machine. So, the first requirement is to read the state of the machine in order to obtain the sum out as an output. This can be readily done by applying a SP pulse (as explained in Ref. [104] and shown in Figure 9.11). If the machine is in logical state 1 (level 3), an output from level 2 will be obtained, whereas if it is in logical state 0 (level 1) there will be no input. The machine is then restored to state 0 (level 1) by applying a second SP pulse if needed. Next, the carry out must be encoded in the internal state of the machine. If fluorescence was observed either in the first step or in the second step of the addition, it means that the carry is 1 and a SP pulse must be input in order to bring the machine to internal state 1 (level 3). Depending on the value of the sum and the carry out, the preparation of the machine for the next cycle may be automatic, in the sense that reading the sum out can coincide with encoding the carry in.

In a full addition, the order into which the three inputs, x, y and carry in, are added does not matter. This is unlike the case of a full subtractor, where the order does matter – that is, $x - y$ and $y - x$ differ by a sign. In order that the first step is the same for the full addition and the full subtraction discussed below, the process is started by adding the carry in and the y input digit. The finite-state machine implementation of a full adder goes along lines similar to the combinational circuit implementation by concatenation of two half adders. It is simply the order of adding the three inputs that differs, in order to take advantage of the memory provided by the internal state of the machine. The first step can be summarized by the Boolean equations:

$$\begin{aligned} state(t+1) &= carry\ in \oplus y \\ carry\ 1 &= carry\ in \otimes y \end{aligned} \tag{9.8}$$

The corresponding truth table is given in Table 9.6.

The carry 1 is logically represented as the output of the machine after the first step (at time $t + 1$) and if its value is 1, fluorescence is detected from level 2. This only

Table 9.6 Truth table for the first half addition.

state(*t*) ≡ carry in	*y*(*t*) ≡ SP pulse	*state*(*t* + 1) (XOR) ≡ midway sum	*output*(*t* + 1) (AND) ≡ carry 1
0 (level 1)	0	0 (level 1)	0
0 (level 1)	1	1 (level 3)	0
1 (level 3)	0	1 (level 3)	0
1 (level 3)	1	0 (level 1)	1

occurs if the input is (1,1) – that is, the carry in was 1 (system in level 3) and the input SP pulse is 1, so that is induces a transition from level 3 to level 1 via the kinetic route. At the next interval the second digit, x, is input as a SP pulse. The truth table is given in Table 9.7, and corresponds to the following logic equations.

State $(t+2)$ is the XOR sum of the three inputs (x, y, and the *carry in*):

$$state(t+2) = state(t+1) \oplus x = (state(t) \oplus y) \oplus x = carry\ in \oplus y \oplus x \quad (9.9)$$

and corresponds the sum out given by Eq. (9.1) above. Using a bar to denote negation

$$\begin{aligned} carry2 &= state(t+1) \otimes x = (carry\ in \oplus y) \otimes x \\ &= (\overline{carry\ in} \otimes y + carry\ in \otimes \bar{y}) \otimes x \\ &= x \otimes y \otimes \overline{carry\ in} + x \otimes \bar{y} \otimes carry\ in \end{aligned} \quad (9.10)$$

The carry out is obtained by reading fluorescence from level 2, either at time $t+1$ or at $t+2$, $carry\ out = carry\ 1 + carry\ 2$, which corresponds to Eq. (9.4) above.

It can now be shown how encoding level 1 as the logical value 1 of the internal state of the machine and level 3 as the logical value 0 and still using a SP pulse as the input, known as y, leads to different state equations and different machines. With this convention, the following logic equations are obtained:

$$state(t+1) = \bar{y}(t) \otimes state(t) + y(t) \otimes \overline{state(t)} \quad (9.11)$$

$$Output(t) = y(t) \otimes \overline{state(t)} \quad (9.12)$$

Table 9.7 Truth table for the second half addition.

State(t + 1) ≡ midway sum	*x*(*t* + 1) ≡ SP pulse	State(*t* + 2) (XOR) ≡ sum	Output(*t* + 2) (AND) ≡ carry 2
0 (level 1)	0	0 (level 1)	0
0 (level 1)	1	1 (level 3)	0
1 (level 3)	0	1 (level 3)	0
1 (level 3)	1	0 (level 1)	1

Table 9.8 Truth table for the operation of the machine with logical encoding level 1 ≡ 1 and level 3 ≡ 0.

State (t)	*y(t)* (SP pulse)	*State (t + 1)*	*Output (t)*
0 (level 3)	0	0 (level 3)	0
0 (level 3)	1	1 (level 1)	1 (kinetic)
1 (level 1)	0	1 (level 1)	0
1 (level 1)	1	0 (level 3)	0 (STIRAP)

The equation for the next state corresponds to a XOR operation identical to Eq. (9.8), while the logical equation for the output corresponds to an INH gate (see Table 9.1). The truth table corresponding to the logic Eqs. (9.11) and (9.12) is given in Table 9.8.

This machine can be logically interpreted in two different ways. The first approach is to note that the machine's output monitors the direction of the change of state as induced by the input. The output is 1 if the pulse induces the logical change of state is $0 \rightarrow 1$. For the change $1 \rightarrow 0$ there is no output. Viewed in this manner [104], the machine is a flip-flop because it maintains a binary state until directed by the input to switch state. Specifically, the machine is similar to a T flip-flop [136] because a single input toggles the state. Flip-flops are key components as they provide a memory element for storing one bit. The data in Table 9.8 show that the state indeed flips, but the machine has no provision for knowing what is the present state of the machine. As discussed above, knowledge of the state of the machine can be readily implemented by applying two SP pulses – one to interrogate the state and one to restore the machine to its initial state. This is in the sense that the machine is endowed with memory.

Another way in which to view the machine represented by Eqs. [9.11–9.12] and Table 9.8 is to see it as a half subtractor, where the minuent digit x is encoded in the state of the machine and the subtrahend y as a SP pulse, so that the machine computes $x - y$. In a half subtractor, the difference is given by the XOR of the two digits (so that it is equivalent to the sum) but instead of a carry a borrow is needed, which is given by the INH function (see Table 9.1).

$$diff = x \oplus y \tag{9.13}$$

$$borrow = x \otimes \bar{y} \tag{9.14}$$

Therefore, the state at $t + 1$ [Eq. (9.11)] gives the difference while the output [Eq. (9.12)] gives the borrow. Another way of implementing a half subtractor is discussed in Ref. [96], where the initial convention of level $1 \equiv 0$ and level $3 \equiv 1$ is maintained but the input is "reverse" and is now a PS pulse. It can be readily checked that by encoding x in the state and y as a PS pulse, Eqs. [9.11–9.12] are obtained for the next state and for the output.

There are two ways to implement a full subtractor (see Ref. [96] for details). The first method is by combining two half subtractors, along the lines used for the full adder discussed above. The other method is more interesting because it closely mimics the implementation of the full adder, which means that the same logic device can be used, either to add or to subtract. This is what is meant by the ability to program a molecule: the same set of levels and of inputs can be used to implement different logic operations.

9.3.2
Finite-State Machines by Electrical Addressing

Until now, only the implementation of combinational circuits and finite-state machines in the gas phase or in solution have been discussed. Here, attention is focused on logic machine implementations on QD arrays. Because of the confinement induced by their nanometer size, QDs have also discrete quantum states but otherwise they are closer to solid-state devices. This is particularly the case for lithographic QDs where the confinement is induced by external voltages that confine electrons in a finite region embedded in a solid-state semi-conductor layer with a high dielectric constant. In this case, the electrons in the QD behave in good approximation as a 2-D electron gas[151]. At this point, interested is centered on a more "chemical" form of QDs – that is, metallic or semiconducting nanosize clusters passivated by organic ligands, for example thioalkane chains. The role of the ligands is to prevent aggregation of the colloidal nanoparticles and to ensure confinement of the electrons in the nanocluster. These QDs are prepared using a wet chemical method, and typically present a size dispersion of at best 5 to 10% in diameter of the cluster. When the size dispersion is narrow enough, they can self-assemble into ordered chains or arrays, and in that sense that they become closer to solid-state devices. They behave in many respects like artificial atoms [152–155] and can be used to make artificial solids [156–161]. When an ordered domain [105] or a chain of QDs [40]can be tethered between electrodes, they can be electrically addressed and probed. This is this type of arrangement used for implementing logic.

Although the details of the system matter a great deal, when discussing the principle of operation of the first logic implementation the only observation needed is that there can be one or more discrete level(s) that can be accessed by varying the electric potential across the dot. The second example discussed in this subsection is built on a system of coupled QDs, and such assemblies have been realized experimentally [162].

Initially, a three-terminal device is considered (see Figure 9.3) so that both a source-drain voltage, V_{sd}, can be applied across the system, and a gate voltage, V_g, in the perpendicular direction. Advantage is taken of the discrete level structure of the QD tethered between the three electrodes of the device to perform more complex operations at the hardware level than is usually done on a transistor.

The implementation of a set–reset finite-state machine is discussed in detail at this point. This is a machine with two logical states, that can accept two inputs, a set input and a reset input. The role of the set input is to bring the machine to logical state 1 if it was in logical state 0, and to do nothing if it is already in state 1. The role of the reset input is to bring the machine back to logical state 0 if it was in state 1 and to do nothing

Table 9.9 Operation of a set–reset machine.

Present state	Set input	Reset input	Name of action	Next State
0	0	0	No change	0
0	1	0	set	1
1	1	0	set	1
1	0	0	No change	1
1	0	1	reset	0
0	0	1	reset	0

if it is already in 0. The case where the two inputs are both 1 is not defined. The operation of the set–reset machine is summarized in Table 9.9, and a state diagram is shown in Figure 9.12.

Here, a single QD tethered in a three-terminal device is considered, that is submitted to a source-drain and to a gate bias. Its discrete level structure is described using the "orthodox" theory [163–165], which assumes that discrete level structure of the QD is due solely to quantization of charge on the dot. The one-electron level spacing of the dot is assumed to be continuous because it is much smaller than the change in electrostatic energy of the QD that occurs when an electron is added to or removed from it by varying the source-drain or the gate bias. The electrostatic energy of a QD with N electrons in a three-terminal device is given by

$$U_N = \frac{Q^2}{2C_T} = \frac{N^2e^2}{2C_T} + \frac{Ne}{C_T}\sum_i C_iV_i + \frac{1}{2C_T}\left(\sum_i C_iV_i\right)^2 \tag{9.15}$$

where Q the effective charge on the dot is given by

$$Q = C_T\Phi = Ne + \sum_{i=l,r,g} C_iV_i \tag{9.16}$$

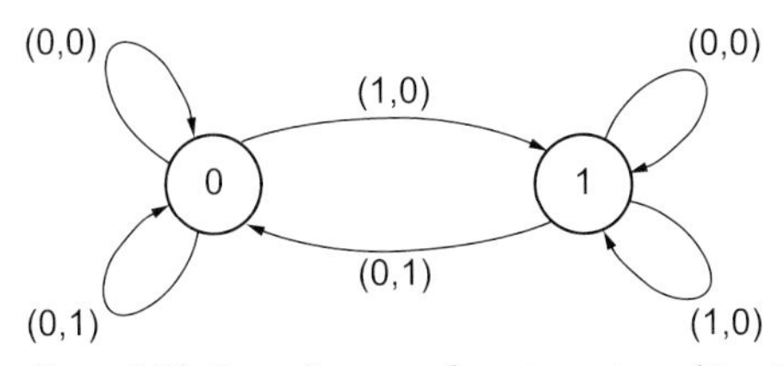

Figure 9.12 State diagram of a set–reset machine. The two possible values of the logical state of the machine are represented by the two circles denoted as 0 or 1. The arrows show the state changing transitions induced by the inputs. The inputs are given next to the arrows as (set,reset).

+1*e* from left
−1*e* to the left
Ne
+1*e* from right
−1*e* to the right

Figure 9.13 Electron transfer to/from the left and right electrode possible for a *N* QD in a three-terminal device.

In Eqs. (9.16) and (9.15), Φ is the electrostatic potential, C_T is the total capacitance of the system ($C_T = C_l + C_r + C_g$ where $C_{l,r}$ are the capacitances of the junctions to the left and right electrodes), and C_g is the capacitance of the gate electrode. $V_{l,r}$ are the source and the drain voltages. For a given gate voltage, an electron will be transferred to the dot or will leave the dot to the left or the right electrode when one of its discrete level falls within the energy window opened by the source-drain bias, V_{sd}, which is the difference between the bias of the right and on the left electrode. As shown in Figure 9.13, if the dot possesses initially *N* electrons, there are therefore four resonance conditions for electron transfer to/from the left and the right electrodes.

The four resonance conditions are:

$$\begin{aligned}
\Delta E_{l\rightarrow QD} &= \frac{e}{C_T}\left(\frac{e}{2} + (Ne + C_g V_g)\right) - e\frac{V}{2} \\
\Delta E_{QD\rightarrow l} &= \frac{e}{C_T}\left(\frac{e}{2} - (Ne + C_g V_g)\right) + e\frac{V}{2} \\
\Delta E_{r\rightarrow QD} &= \frac{e}{C_T}\left(\frac{e}{2} + (Ne + C_g V_g)\right) + e\frac{V}{2} \\
\Delta E_{QD\rightarrow R} &= \frac{e}{C_T}\left(\frac{e}{2} - (Ne + C_g V_g)\right) - e\frac{V}{2}
\end{aligned} \tag{9.17}$$

where $V = V_{sd}$ and a symmetric junction is assumed so that $C_l = C_r$ and $V_l = V_r = V/2$. $\Delta E = U_N - U_{N\pm1}$ must be ≤ 0 for the process to be allowed. It is the free energy difference for adding or removing an electron to the QD. Note that when only the charge on the dot is quantized, $U_N - U_{N\pm1}$ varies linearly with the applied source-drain and gate bias. The threshold for transferring an electron is given by the resonance condition, $\Delta E = 0$, which allows stability maps to be drawn of the charged QD as a function the gate and the source-drain bias. A stability map for $N = 0$, 1 electrons on the dot is shown in Figure 9.14. The areas in gray are the zones where the number of electrons on the QD is stable.

In the "orthodox" theory [164] the rates of transfer from the QD to the source and the drain electrodes are given by

$$\Gamma = \frac{2}{e^2 R}\,\frac{-\Delta E}{1 + \exp(\Delta E/kT)} \xrightarrow{T\rightarrow 0K} \frac{2}{e^2 R}\,|\Delta E|\theta(-\Delta E) \tag{9.18}$$

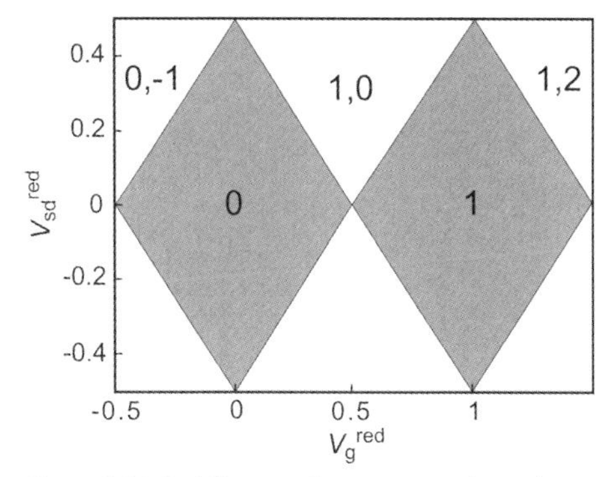

Figure 9.14 Stability map for a quantum dot with $N=0$ and $N=1$ electrons, plotted using Eq. (9.17) as a function of the V_g and V_{sd} in reduced units. $V_g^{red} = C_g V_g/e$ corresponds to the number of electrons on the dot, $V_g^{red} = V_{sd} C_T/2e$.

where R is the resistance of the junction through which the electron passes, and is inversely proportional to the coupling between the QD and the electrode.

For implementation of the set–reset, two charges states of the QD are used, namely $N-1$ and N, where N is the number of extra electrons on the QD. For the simulation shown below, $N=0$ and $N=1$ were utilized. The logical state 0 of the set–reset machine was encoded as the QD with $N=0$ extra electrons, and the logical state 1 of the machine was encoded as the QD with $N=1$ extra electrons. From Figure 9.13, it can be seen that there are two rates for adding an electron to a QD with $N=0$, $\Gamma_{r\to QD}$ and $\Gamma_{l\to QD}$, and two rates for removing an electron from a $N=1$ QD, $\Gamma_{QD\to r}$ and $\Gamma_{QD\to l}$. Their analytical forms at $T=0$ K are

$$
\begin{aligned}
\Gamma_{QD\to l, l\to QD} &\propto \pm\left(\frac{e}{2C_T}+\frac{C_g V_g}{C_T}\right)\mp\frac{V}{2} \\
\Gamma_{QD\to r, r\to QD} &\propto \pm\left(\frac{e}{2C_T}+\frac{C_g V_g}{C_T}\right)\pm\frac{V}{2}
\end{aligned}
\qquad (9.19)
$$

These four rates are plotted in Figure 9.15 for a fixed gate voltage as a function of the source-drain bias, V.

In order for the set–reset machine to operate properly, a set voltage must be chosen such that the rate of transfer of an electron from the left electrode to the QD with $N=0$ is much larger than the rate for leaving the dot with $N=1$ to the right electrode, so that an electron is added to the dot and stays on the dot for a finite time. For the reset voltage, it is sufficient that the rate of leaving the dot with $N=1$ to the left electrode is significant. The rate $\Gamma_{r\to QD}$ corresponds to adding an electron to the QD with $N=0$. The operation of the set–reset device is more robust if the resistance of the right junction is much larger than that of the left one.

To check that the set–reset machine operates properly, the probability of getting an extra electron on the QD is monitored as a function of time while applying a time-dependent source-drain bias. By defining Q as the probability for having $N=1$ extra

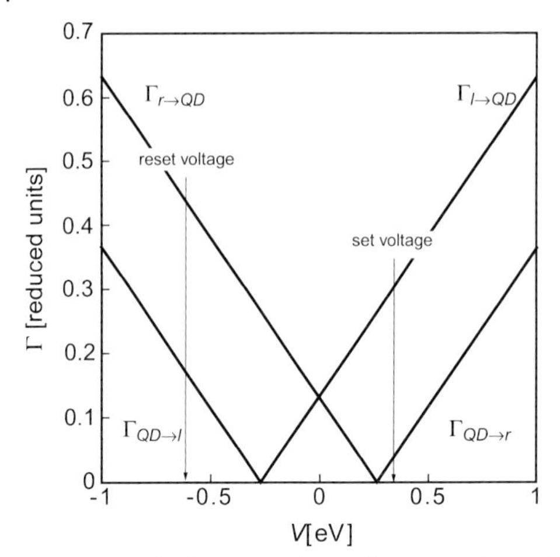

Figure 9.15 The four rates relevant for a QD with $N=0$ and $N=1$ electrons computed as a function of the source-drain bias V with a value of the gate voltage $V_g=-0.8$ V. $C_g=C_l=C_r=0.2$ aF. The reduced units for Γ are $2/e^2R$, where R is the resistance of the junction. A symmetric junction is considered so that $R=R_l=R_r$. R is inversely proportional to the coupling between the QD and the left (right) electrode.

electrons on the dot, and P_l and P_r as the probabilities for this extra electron to be on the left and on right electrode, respectively, the following kinetic scheme is obtained:

$$\begin{aligned}\frac{dP_l}{dt} &= \Gamma_{QD\to l}Q(t) - \Gamma_{l\to QD}P_l(t)\\ \frac{dQ}{dt} &= \Gamma_{l\to QD}P_l(t) + \Gamma_{r\to QD}P_r(t) - (\Gamma_{QD\to l} + \Gamma_{QD\to r})Q(t)\\ \frac{dP_r}{dt} &= \Gamma_{QD\to r}Q(t) - \Gamma_{r\to QD}P_r(t)\end{aligned} \tag{9.20}$$

The time profile of the applied source-drain bias (a) and result of the integration of the kinetic scheme (b) are shown in Figure 9.16. The logical state 0 of the device is defined as $P_l \gg Q$, while state 1 is defined as $Q \gg P_l$. It can be seen that the effect of the set voltage is to fill the dot with one extra electron, whilst applying the reset pulse empties the dot of that extra electron. This shows that a single QD with two electrically addressable discrete levels can operate as a set–reset machine. It has also been shown that varying not only the source-drain but also the gate bias allows a full adder to be implemented on this system [106].

This subsection is concluded with a description of the implementation of another form of finite-state machine, a counter, on an array of two QDs anchored on a surface. This implementation is based on a scheme for addressing and/or reading the states of the dots electrically or optically that has been experimentally realized and characterized [162] (see Figure 9.17).

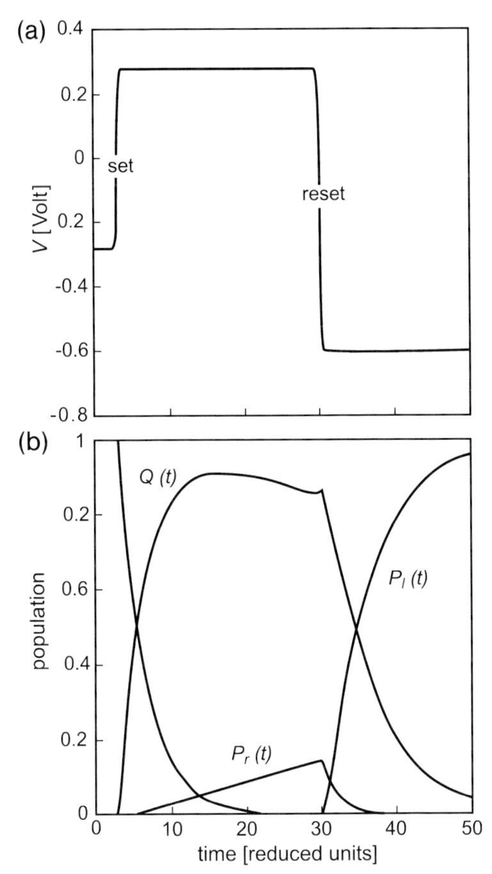

Figure 9.16 (a) Voltage–time profile and (b) time-dependent probability for the extra electron to be localized on the QD, $Q(t)$, on the left electrode, $P_l(t)$, and on the right electrode, $P_r(t)$. In Eq. (9.20), the rates are expressed in reduced units and the time is scaled correspondingly.

A counter [136] is a machine that is able to accept N inputs and to provide an output for every N inputs. The states of the counter are S_i, $i = 0, 1, 2, \ldots, N-1$ and transition can occur between successive states only when an input i, $i = 0, 1, 2, \ldots, N-1$, is received. After the count of N inputs, the state S_{N-1} is reached, an output is produced and the next input resets the counter to its initial state, S_0.

In the implementation based on the device shown in Figure 9.17 the index of the state is determined by the number of extra electrons on the Au QD. This number can be controlled optically or electrically (further details may be found in Refs. [137, 162]).

The counter functions as follows. First, the CdS QD is irradiated in a solution of TEA (10^{-2} M) so as to charge the Au QD. The irradiation is then stopped. The initial state, S_0, corresponds physically to the Au QD charged with four extra electrons. This number can be determined using surface plasmon resonance spectroscopy, by measuring the shift of the plasmon resonance of the gold surface due to the charging

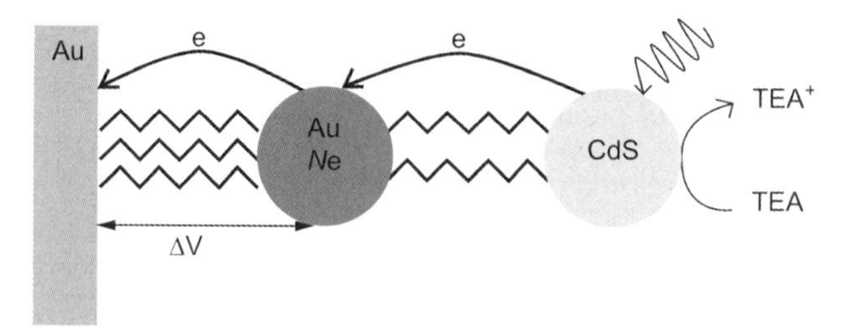

Figure 9.17 An experimentally realized [162] opto-electrically addressed array of two QDs anchored on a surface. The device can be used as a counter. The Au QDs are linked to the gold surface by a long-chain alkyl monolayer, and covalently to a semi-conducting CdS QD. The optical excitation of the CdS QD induces an electron transfer to the Au QD that is compensated by triethanolamine (TEA) present in the electrolyte solution. While the optical excitation is on, extra electrons accumulate on the Au QD and a potential drop is maintained across the junction between the Au QDs and the conductive gold surface. The charging of the Au QD can be optically monitored by changes in the resonance spectrum of the surface plasmon.

of the Au QD. On each occasion that an input is to be provided an input the index of the state must be incremented; this is done by decreasing the potential applied to the Au surface by a step sufficient to discharge an electron onto the surface. The magnitude of the required voltage drop is determined by the capacitance of the Au QD – that is, by the energy needed to charge or discharge the dot by one electron [137]. In the experiment the Au QD is passivated by a ligand, tiopronin, that has a high dielectric constant (16; see Ref. [162]), so that the charging energy is exceptionally low ($\approx$30 meV for Au QD of 2.3 ± 0.5 nm diameter). When the dot is fully discharged, after four voltage drops, S_4 is reached; this is the last state of the counter for which there are no extra electrons on the dot. At this point, the counter must be returned to the state S_0, so that it is available for the next counting cycle, and an output signal must then be provided. Unlike the usual system for counters, in this scheme the last input does not reset the counter to the state S_0. For this reason, even though for a dot with four extra electrons, there are five states – S_0, S_1, S_2, S_3, and S_4 – and modulo 4 is counted rather than modulo 5, the last step, $S_3 \rightarrow S_4$, is used to produce the output and reset the counter. The output is produced by monitoring the disappearance of the plasmon angle shift or by measuring the value of the surface potential. To reset the counter, the CdS dot is irradiated again. It should be noted that, in principle, the maximal number of four extra electrons on the dot is not a limitation, but up to 15 oxidation states of monolayer-protected Au QDs have been reported [166]. It is possible, therefore, to implement counters with a higher value of N.

9.4 Perspectives

The entire discussion in this chapter is based on the premise that there is a desire to design molecule-based logic circuits and not only switches. Results to date that have been validated by *proof of concept* experiments, include:

- the implementation of *combinational circuits on a single molecule;*
- the *concatenation of logic operations*, whether performed on different molecules (intermolecular) or performed within the same molecule by communicating results carried out on different functional groups (intramolecular);
- the implementation of a *finite-state logic machine on a single molecule* and beyond that, *programming* of a single molecule; and
- using both electrical and optical addressing and readout with the advantage that it is not necessary to be able to address many states, because with two states a full adder can already be performed.

Technical studies in progress exploit these results towards increasing the logical capacity and depth (= number of switches) that can be implemented on a single molecule, or on a supramolecular assembly by the application to multifunctional group molecules where the intramolecular dynamics are used to concatenate the logical operations carried separately by the different groups. Next, in the order of integration is the assembly and concatenation of an array of molecules or an array of quantum dots.

Further studies are also needed to take even greater advantage of the large number of quantum states available, in a hierarchical order (electronic, backbone vibrations, torsions, rotations), which allows the processing in one cycle of far more information than a binary (classical or quantum) gate and, in the same direction, the use of more sophisticated optical and electrical inputs and readouts.

The first results to reach the level of technological implementation will most likely be the use of a single molecule not as a switch but rather as a combinational circuit. This will likely happen in the context of the architecture of a 2-D array cross-bar, which is the favored device geometry as foreseen by Hewlett Packard and others. However, even this progress will take time before it becomes a technology. The essential difference to be advocated is that at each node is placed not a switch but a single molecule acting as the equivalent of an entire network of switches. The very fast logic is conducted within the node, but the slower, wire-mediated communication between the nodes will remain. In the second round, communication between the nodes will be carried out by concatenation through self-assembly of the array using molecular recognition. Part of this endeavor is to achieve realistic programming abilities with special reference to selective intramolecular dynamics.

The key further breakthroughs that are currently required include:

- The design of molecular logic circuits that can be cycled reliably many times, and to explore whether this can be done using all-optical schemes.
- Input/output operations that reduce dissipation and allow fan-out and macroscopic interface, with special reference to the use of pulse shaping, electrical read/write and integrate storage within the logic unit.
- Beyond what is already available, it will be necessary to improve concatenation in order to reduce not only the need for cycling but also for interfacing with the macroscopic world. This will in turn lead to a need for molecular systems with special reference to devices on surfaces and their application as logic units.

Acknowledgments

These studies were supported by the EC FET-Open project MOLDYNLOGIC, the US-Israel Binational Science Foundation, BSF, Jerusalem, Israel and the EC NoE FAME.

References

1 C. Joachim, J. K. Gimzewski, A. Aviram, *Nature* 2000, **408**, 541.
2 C. P. Collier, E. W. Wong, M. Belohradsk, F. M. Raymo, J. F. Stoddart, P. J. Kuekes, R. S. Williams, J. R. Heath, *Science* 1999, **285**, 391.
3 C. P. Collier, G. Mattersteig, E. W. Wong, Y. Luo, K. Beverly, J. Sampaio, F. M. Raymo, J. F. Stoddart, J. R. Heath, *Science* 2000, **289**, 1172.
4 P. R. Ashton, R. Ballardini, V. Balzani, A. Credi, K. R. Dress, E. Ishow, C. J. Kleverlaan, O. Kocian, J. A. Preece, N. Spencer, J. F. Stoddart, M. Venturi, S. Wenger, *Chem-Eur. J.* 2000, **6**, 3558.
5 M. A. Reed, J. M. Tour, *Sci. Am.* 2000, **282**, 86.
6 R. M. Metzger, *Acc. Chem. Res.* 1999, **32**, 950.
7 R. M. Metzger, *J. Mater. Chem.* 2000, **10**, 55.
8 A. P. de Silva, N. D. McClenaghan, *J. Am. Chem. Soc.* 2000, **122**, 3965.
9 A. P. de Silva, Y. Leydet, C. Lincheneau, N. D. McClenaghan, *J. Phys. - Cond. Mater.* 2006, **18**, S1847.
10 Y. Luo, C. P. Collier, J. O. Jeppesen, K. A. Nielsen, E. Delonno, G. Ho, J. Perkins, H.-R. Tseng, T. Yamamoto, J. F. Stoddart, J. R. Heath, *ChemPhysChem* 2002, **3**, 519.
11 V. Balzani, A. Credi, M. Venturi, *ChemPhysChem* 2003, **4**, 49.
12 J. M. Tour, *Molecular Electronics*, World Scientific, River Edge, USA, 2003.
13 T. Nakamura, *Chemistry of Nanomolecular Systems: Towards the Realization of Molecular Devices*, Volume 70, Springer, Berlin, 2003.
14 F. M. Raymo, *Adv. Mater.* 2002, **14**, 401.
15 F. M. Raymo, M. Tomasulo, *Chem.- Eur. J.* 2006, **12**, 3186.
16 R. Waser, *Nanoelectronics and Information Technology*, Wiley-VCH, Weinheim, 2003.
17 M. A. Nielsen, I. L. Chuang, *Quantum Computation and Quantum Information*, Cambridge University Press, Cambridge, 2000.
18 C. H. Bennett, D. P. DiVincenzo, *Nature* 2000, **404**, 247.
19 D. Deutsch, *Proc. R. Soc. Lond.* 1985, **A 400**, 97.
20 N. Gershenfeld, I. L. Chuang, *Sci. Am.* 1998, **278**, 66.
21 S. J. Glaser, T. Schulte-Herbruggen, M. Sieveking, O. Schedletzky, N. C. Nielsen, O. W. Sorensen, C. Griesinger, *Science* 1998, **280**, 421.
22 A. Steane, *Nature* 2003, **422**, 387.
23 D. R. Glenn, D. A. Lidar, V. A. Apkarian, *Mol. Phys.* 2006, **104**, 1249.
24 M. A. Nielsen, M. R. Dowling, M. Gu, A. C. Doherty, *Phys. Rev. A* 2006, 73.
25 D. Bouwmeester, A. Ekert, A. Zeilinger, *The Physics of Quantum Information*, Springer, Berlin, 2000.
26 S. Haykin, *Neural Networks*, Prentice-Hall, Upper Saddle River, 1999.
27 J. Chen, D. H. Wood, *Proc. Natl. Acad. Sci. USA* 2000, **97**, 1328.
28 Y. Benenson, R. Adar, T. Paz-Elizur, Z. Livneh, E. Shapiro, *Proc. Natl. Acad. Sci. USA* 2003, **100**, 2191.
29 M. N. Stojanovic, D. Stefanovic, *J. Am. Chem. Soc.* 2003, **125**, 6673.
30 D. Margulies, G. Melman, C. E. Felder, R. Arad-Yellin, A. Shanzer, *J. Am. Chem. Soc.* 2004, **126**, 15400.

31 A. Okamoto, K. Tanaka, I. Saito, *J. Am. Chem. Soc.* 2004, **126**, 9458.
32 P. D. Tougaw, C. S. Lent, *J. Appl. Phys.* 1994, **75**, 1818.
33 A. O. Orlov, I. Amlani, G. H. Bernstein, C. S. Lent, G. L. Snider, *Science* 1997, **277**, 928.
34 I. Amlani, A. O. Orlov, G. Toth, G. H. Bernstein, C. S. Lent, G. L. Snider, *Science* 1999, **284**, 289.
35 L. Boni, M. Gattobigio, G. Iannaccone, M. Macucci, *J. Appl. Phys.* 2002, **96**, 3169.
36 H. Qi, S. Sharma, Z. H. Li, G. L. Snider, A. O. Orlov, C. S. Lent, T. P. Fehlner, *J. Am. Chem. Soc.* 2003, **125**, 15250.
37 J. Twamley, *Phys. Rev. A* 2003, **67**, 052328.
38 D. L. Klein, R. Roth, A. K. L. Lim, A. P. Alivisatos, P. L. McEuen, *Nature* 1997, **389**, 699.
39 W. Liang, M. P. Shores, J. L. Long, M. Bockrath, H. Park, *Nature* 2002, **417**, 725.
40 D. N. Weiss, X. Brokmann, L. E. Calvet, M. A. Kastner, M. G. Bawendi, *Appl. Phys. Lett.* 2006, **88**, 143507.
41 A. Nitzan, M. A. Ratner, *Science* 2003, **300**, 1384.
42 J. R. Heath, M. A. Ratner, *Physics Today* 2003, **56**, 43.
43 A. H. Flood, J. F. Stoddart, D. W. Steuerman, J. R. Heath, *Science* 2004, **306**, 2055.
44 C. Joachim, M. A. Ratner, *Nanotechnology* 2004, **15**, 1065.
45 J. Fiurasek, N. J. Cerf, I. Duchemin, C. Joachim, *Physica E* 2004, **24**, 161.
46 S. Ami, M. Hliwa, C. Joachim, *Chem. Phys. Lett.* 2003, **367**, 662.
47 R. Stadler, S. Ami, M. Forshaw, C. Joachim, *Nanotechnology* 2002, **13**, 424.
48 A. Bezryadin, C. Dekker, *Appl. Phys. Lett.* 1997, **71**, 1273.
49 S. Karthauser, E. Vasco, R. Dittmann, R. Waser, *Nanotechnology* 2004, **15**, S122.
50 C. R. Barry, J. Gu, H. O. Jacobs, *Nano Lett.* 2005, **5**, 2078.
51 B. Lussem, L. Muller-Meskamp, S. Karthauser, R. Waser, *Langmuir* 2005, **21**, 5256.
52 J. J. Urban, D. V. Talapin, E. V. Shevchenko, C. B. Murray, *J. Am. Chem. Soc.* 2006, **128**, 3248.
53 R. P. Feynman, *Feynman Lectures on Computations, reprint with corrections*, Perseus Publishing, Cambridge, MA, 1999.
54 D. Deutsch, *Proc. R. Soc. Lond.* A 1989, **425**, 73.
55 D. P. DiVincenzo, *Proc. R. Soc. Lond. A* 1998, **454**, 261.
56 R. Cleve, A. Ekert, C. Macchiavello, M. Mosca, *Proc. R. Soc. Lond. A* 1998, **454**, 339.
57 A. Ekert, R. Jozsa, *Proc. R. Soc. Lond. A* 1998, **356**, 1769.
58 R. Jozsa, *Proc. R. Soc. Lond. A* 1998, **454**, 323.
59 D. Loss, D. P. DiVincenzo, *Phys. Rev. A* 1998, **57**, 120.
60 G. Burkard, D. Loss, D. P. DiVincenzo, *Phys. Rev. B* 1999, **59** 2070.
61 K. R. Brown, D. A. Lidar, K. B. Whaley, *Phys. Rev. A* 2001, **65**, 012307.
62 C. H. Bennett, *IBM J. Res.* 1973, **17**, 525.
63 C. H. Bennett, *Int. J. Theoret. Phys.* 1982, **21**, 905.
64 D. Cory, A. Fahmy, T. Havel, *Proc. Natl. Acad. Sci. USA* 1997, **94**, 1634.
65 L. K. Grover, in Proceedings 28th ACM Symposium on the Theory of Computing, 1996.
66 D. Deutsch, R. Jozsa, *Proc. R. Soc. Lond. A* 1992, **439**, 553.
67 L. K. Grover, *Phys. Rev. Lett.* 1997, **79**, 4709.
68 T. Tulsi, L. K. Grover, A. Patel, *Quant. Inf. Comp.* 2006, **6**, 483.
69 C. H. Bennett, F. Bessette, G. Brassard, L. Slavail, J. Smolin, *J. Crypt.* 1992, **5**, 3.
70 P. W. Shor, in: S. Goldwasser (Ed.), Proceedings, 35th Annual Symposium on the Foundations of Computer Science, IEEE Computer Society Press, Los Alamitos, CA, 1994.
71 A. Ekert, R. Jozsa, *Rev. Mod. Phys.* 1996, **68**, 733.

72 J. Brown, *The Quest for Quantum Computer*, Simon & Schuster, New York, 2001.
73 J. Preskill, *Proc. R. Soc. Lond. A* 1998, **454**, 385.
74 J. Preskill, *Physics Today* 1999, **52**, 24.
75 E. Knill, R. Laflamme, L. Viola, *Phys. Rev. Lett.* 2000, **84**, 2525.
76 D. P. DiVincenzo, D. Bacon, J. Kempe, G. Burkard, K. B. Whaley, *Nature* 2000, **408**, 339.
77 D. Bacon, K. R. Brown, K. B. Whaley, *Phys. Rev. Lett.* 2001, **87**, 247902.
78 E. S. Myrgren, K. B. Whaley, *Quant. Inf. Proc.* 2004, **2**, 309.
79 G. Schaller, S. Mostame, R. Schutzhold, *Phys. Rev. A* 2006, 73.
80 F. Remacle, R. D. Levine, *Proc. Natl. Acad. Sci. USA* 2004, **101**, 12091.
81 A. Aviram, M. A. Ratner, *Chem. Phys. Lett.* 1974, **29**, 277.
82 J. C. Ellenbogen, J. C. Love, *Proc. IEEE* 2000, **88**, 386.
83 C. S. Lent, B. Isaksen, M. Lieberman, *J. Am. Chem. Soc.* 2003, **125**, 1056.
84 A. Credi, V. Balzani, S. J. Langford, J. F. Stoddart, *J. Am Chem. Soc.* 1997, **119**, 2679.
85 A. P. de Silva, I. M. Dixon, H. Q. N. Gunaratne, T. Gunnlaugsson, P. R. S. Maxwell, T. E. Rice, *J. Am. Chem. Soc.* 1999, **121**, 1393.
86 V. Balzani, A. Credi, F. M. Raymo, J. F. Stoddart, *Angew. Chem. Int. Ed.* 2000, **39**, 3349.
87 A. P. de Silva, N. D. McClenaghan, *Chem. Eur. J.* 2004, **10**, 574.
88 F. M. Raymo, R. J. Alvarado, S. Giordani, M. A. Cejas, *J. Am. Chem. Soc.* 2003, **125**, 2361.
89 G. J. Brown, A. P. de Silva, S. Pagliari, *Chem. Commun.* 2002, 2461.
90 X. F. Guo, D. Q. Zhang, G. X. Zhang, D. B. Zhu, *J. Phys. Chem. B* 2004, **108**, 11942.
91 J. Andreasson, G. Kodis, Y. Terazono, P. A. Liddell, S. Bandyopadhyay, R. H. Mitchell, T. A. Moore, A. L. Moore, D. Gust, *J. Am. Chem. Soc.* 2004, **126**, 15926.
92 R. Baron, O. Lioubashevski, E. Katz, T. Niazov, I. Willner, *Angew. Chem.* 2006, **45**, 1572.
93 K. Szacilowski, W. Macyk, G. Stochel, *J. Am. Chem. Soc.* 2006, **128**, 4550.
94 F. Remacle, S. Speiser, R. D. Levine, *J. Phys. Chem. A* 2001, **105**, 5589.
95 F. Remacle, E. W. Schlag, H. Selzle, K. L. Kompa, U. Even, R. D. Levine, *Proc. Natl. Acad. Sci. USA* 2001, **98**, 2973.
96 F. Remacle, R. D. Levine, *Phys. Rev. A* 2006, **73**, 033820.
97 H. Lederman, J. Macdonald, D. Stefanovic, M. N. Stojanovic, *Biochemistry* 2006, **45**, 1194.
98 D. Margulies, G. Melman, A. Shanzer, *J. Am. Chem. Soc.* 2006, **128**, 4865.
99 F. Remacle, J. R. Heath, R. D. Levine, *Proc. Natl. Acad. Sci. USA* 2005, **102**, 5653.
100 K. L. Kompa, R. D. Levine, *Proc. Natl. Acad. Sci. USA* 2001, **98**, 410.
101 F. Remacle, R. D. Levine, *J. Chem. Phys.* 2001, **114**, 10239.
102 T. Witte, C. Bucher, F. Remacle, D. Proch, K. L. Kompa, R. D. Levine, *Angew. Chem.* 2001, **40**, 2512.
103 F. Remacle, R. Weinkauf, D. Steinitz, K. L. Kompa, R. D. Levine, *Chem. Phys.* 2002, **281**, 363.
104 D. Steinitz, F. Remacle, R. D. Levine, *ChemPhysChem.* 2002, **3**, 43.
105 F. Remacle, K. C. Beverly, J. R. Heath, R. D. Levine, *J. Phys. Chem. B* 2003, **107**, 13892.
106 F. Remacle, J. R. Heath, R. D. Levine, *Proc. Natl. Acad. Sci. USA* 2005, **102**, 5653.
107 F. Remacle, R. D. Levine, *Faraday Disc.* 2006, **131**, 46.
108 A. W. Ghosh, T. Rakshit, S. Datta, *Nano Lett.* 2004, **4**, 565.
109 X. Li, B. Xu, X. Xiao, X. Yang, L. Zang, N. Tao, *Faraday Disc.* 2006, **131**, 111.
110 S. L. Hurst, *IEEE Trans. Comp.* 1984, **C-33**, 1160.
111 D. C. Rine, *Computer Science and Multiple-Valued Logic*, North-Holland, Amsterdam, 1977.
112 Wikipedia, http://en.wikipedia.org/wiki/Multi-valued_logic 2006.

113 G. C. Schatz, M. A. Ratner, *Quantum Mechanics in Chemistry*, Prentice-Hall, New York, 1993.
114 M. M. Mano, C. R. Kime, *Logic and Computer Design Fundamentals*, Prentice-Hall, Upper Saddle River, NJ, 2000.
115 F. Remacle, R. Weinkauf, R. D. Levine, *J. Phys. Chem. A* 2006, **110**, 177.
116 W. Cheng, N. Kuthirummal, J. Gosselin, T. I. Solling, R. Weinkauf, P. Weber, *J. Phys. Chem. A* 2005, **109**, 1920.
117 L. Lehr, T. Horneff, R. Weinkauf, E. W. Schlag, *J. Phys. Chem. A* 2005, **109**, 8074.
118 R. Weinkauf, L. Lehr, A. Metsala, *J. Phys. Chem. A* 2003, **107**, 2787.
119 R. B. Bernstein, *J. Phys. Chem.* 1982, **86**, 1178.
120 R. B. Bernstein, *Chemical Dynamics via Molecular Beam and laser techniques*, Oxford University Press, New York, 1982.
121 G. Wiswanath, M. Kasha, *J. Chem. Phys.* 1956, **24**, 574.
122 M. Beer, H. C. Longuett-Higgins, *J. Chem. Phys.* 1955, **23**, 1390.
123 J. W. Sidman, D. S. McClure, *J. Chem. Phys.* 1955, **24**, 757.
124 S. Speiser, *Chem. Rev.* 1996, **96** 1953.
125 I. Kaplan, J. Jortner, *Chem. Phys. Lett.* 1977, **52**, 202.
126 I. Kaplan, J. Jortner, *Chem. Phys.* 1978, **32**, 381.
127 S. Speiser, *Appl. Phys. B* 1989, **49**, 109.
128 S. Speiser, N. Shakkour, *Appl. Phys. B* 1985, **38**, 191.
129 M. Orenstein, S. Kimel, S. Speiser, *Chem. Phys. Lett.* 1978, **58**, 582.
130 N. Lokan, M. N. Paddow-Row, T. A. Smith, M. LaRosa, K. P. Ghiggino, S. Speiser, *J. Am. Chem. Soc.* 1999, **121**, 2917.
131 S. Speiser, F. Schael, *J. Mol. Liq.* 2000, **86**, 25.
132 S. Speiser, *Opt. Commun.* 1983, **45**, 84.
133 U. Peskin, M. Abu-Hilu, S. Speiser, *Optical Mater.* 2003, **24**, 23.
134 S. Speiser, *J. Luminescence* 2003, **102**, 267.
135 T. L. Booth, *Sequential Machines and Automata Theory*, Wiley, New York, 1968.
136 Z. Kohavi, *Switching and Finite Automata Theory*, Tata McGraw-Hill, New Delhi, 1999.
137 F. Remacle, I. Willner, R. D. Levine, *ChemPhysChem.* 2005, **6**, 1.
138 J. Martin, B. W. Shore, K. Bergmann, *Phys. Rev. A* 1996, **54**, 1556.
139 T. Halfmann, K. Bergmann, *J. Chem. Phys.* 1996, **104**, 7068.
140 A. Kuhn, S. Steuerwald, K. Bergmann, *Eur. Phys. J. D* 1998, **1**, 57.
141 B. W. Shore, *The Theory of Coherent Atomic Excitation: Multilevel Atoms and Incoherence*, Wiley, New York, 1990.
142 B. W. Shore, K. Bergmann, J. Oreg, S. Rosenwaks, *Phys. Rev. A* 1991, **44**, 7442.
143 N. V. Vitanov, B. W. Shore, K. Bergmann, *Eur. Phys. J. D* 1998, **4**, 15.
144 V. S. Malinovsky, D. J. Tannor, *Phys. Rev. A* 1997, **56**, 4929.
145 S. A. Rice, M. Zhao, *Optical Control of Molecular Dynamics*, Wiley, New York, 2000.
146 C. Cohen-Tannoudji, J. Dupont-Roc, G. Grynberg, *Atom-Photon Interactions*, Wiley, New York, 1992.
147 K. Bergmann, H. Theuer, B. W. Shore, *Rev. Mod. Phys.* 1998, **70**, 1003.
148 D. J. Tannor, *Introduction to Quantum Mechanics: A Time Dependent Perspective*, University Science Books, Sausalito, CA, 2005.
149 N. V. Vitanov, M. Fleischhauer, B. W. Shore, K. Bergmann, *Adv. At. Mol. Opt. Phys.* 2001, **46**, 55.
150 M. P. Fewell, B. W. Shore, K. Bergmann, *Aust. J. Phys.* 1997, **50**, 281.
151 L. P. Kouwenhoven, D. G. Austing, S. Tarucha, *Rep. Prog. Phys.* 2001, **64**, 701.
152 R. C. Ashoori, *Nature* 1996, **379**, 413.
153 M. A. Kastner, *Physics Today* 1993, **46**, 24.
154 U. Banin, Y. W. Cao, D. Katz, O. Millo, *Nature* 1999, **400**, 542.
155 M. Reed, *Sci. Am.* 1993, **268**, 98.
156 C. P. Collier, R. J. Saykally, J. J. Shiang, S. E. Henrichs, J. R. Heath, *Science* 1997, **277** 1978.
157 C. P. Collier, T. Vossmeyer, J. R. Heath, *Annu. Rev. Phys. Chem.* 1998, **49**, 371.

158 C. J. Kiely, J. Fink, J. G. Zheng, M. Brust, D. Bethell, D. J. Schiffrin, *Adv. Mater.* 2000, **12**, 640.

159 G. Markovich, C. P. Collier, S. E. Henrichs, F. Remacle, R. D. Levine, J. R. Heath, *Acc. Chem. Res.* 1999, **32**, 415.

160 G. Schmid, U. Simon, *Chem. Commun.* 2005, 697.

161 A. Taleb, V. Russier, A. Courty, M. P. Pileni, *Phys. Rev. B* 1999, **59**, 13350.

162 M. Zayats, A. B. Kharitonov, S. P. Pogorelova, O. Lioubashevski, E. Katz, I. Willner, *J. Am. Chem. Soc.* 2003, **125**, 16006.

163 R. I. Shekhter, *Sov.-Phys. JETP* 1973, **36**, 747.

164 D. V. Averin, A. N. Korotkov, K. K. Likharev, *Phys. Rev. B* 1991, **44**, 6199.

165 I. O. Kulik, R. I. Shekhter, *Sov. Phys. JETP* 1975, **41**, 308.

166 B. M. Quinn, P. Liljeroth, V. Ruitz, T. Laaksonen, K. Kontturi, *J. Am. Chem. Soc.* 2003, **125**, 6644.

II
Architectures and Computational Concepts

Nanotechnology. Volume 4: Information Technology II. Edited by Rainer Waser
Copyright © 2008 WILEY-VCH Verlag GmbH & Co. KGaA, Weinheim
ISBN: 978-3-527-31737-0

10
A Survey of Bio-Inspired and Other Alternative Architectures

Dan Hammerstrom

10.1
Introduction

Since the earliest days of the electronic computer, there has always been a small group of people who have seen the computer as an extension of biology, and have endeavored to build computing models and even hardware that are inspired by, and in some cases are direct copies of, biological systems. Although biology spans a wide range of systems, the primary model for these early efforts has been neural circuits. Likewise, in this chapter the discussion will be limited to neural computation.

Several examples of these early investigations include McCulloch-Pitts *Logical Calculus of Nervous System Activity* [2], Steinbuch's *Die Lernmatrix* [3], and Rosenblatt's *Perceptron* [4]. At the same time, an alternate approach to intelligent computing, *Artificial Intelligence* (AI), that relied on higher-order symbolic functions, such as structured and rule-based representations of knowledge, began to demonstrate significantly greater success than the neural approach. In 1969, Minsky and Papert [5] of the Massachusetts Institute of Technology published a book that was critical of the then current "bio-inspired" algorithms, and which succeeded in eventually ending most research funding for that approach. Consequently, significant research funding was directed towards AI, and the field subsequently flourished. The AI approach, which relied on symbolic reasoning often represented by a first-order calculus and sets of rules, began to exhibit real intelligence, at least on toy problems. One reasonably successful application was the "expert" system, and there was even the development of a complete language, ProLog, dedicated to logical rule-based inference.

A few expert system successes were also enjoyed in actually fielded systems, such as Soar [6], the development of which was started by Alan Newell's group at Carnegie Mellon University. However, by the 1980s AI in general was beginning to lose its luster after 40 years of funding with ever-diminishing returns.

Since the 1960s, however, there have always been groups that continued to study biologically inspired algorithms, and two such projects – mostly as a result of their

Nanotechnology. Volume 4: Information Technology II. Edited by Rainer Waser
Copyright © 2008 WILEY-VCH Verlag GmbH & Co. KGaA, Weinheim
ISBN: 978-3-527-31737-0

being in the right place at the right time – had a huge impact which re-energized the field and led to an explosion of research and funding. The first project incorporated the investigations [7] of John Hopfield, a physicist at Caltech, who proposed a model of auto-associative memory based on physical principles such as the Ising theory of spin-glass. Although Hopfield nets were limited in capability and size, and others had proposed similar algorithms previously, Hopfield's formulation was both clean and elegant. It also succeeded in bringing many physicists, armed with sophisticated mathematical tools, into the field. The second project was the "invention" of the back-propagation algorithm by Rumelhart, Hinton, and Williams [8]. Although there too similar studies had been conducted previously [9], the difference with Rumelhart and colleagues was that they were cognitive scientists creating a set of techniques called parallel distributed processing (PDP) models of cognitive phenomena, where back-propagation was a part of a larger whole.

At this point, it would be useful to present some basic neuroscience, followed by details of some of the simpler algorithms inspired by this biology. This information will provide a strong foundation for discussing various biologically inspired hardware efforts.

10.1.1 Basic Neuroscience

In simplified terms, neural circuits consist of large numbers of parallel processing components, the *neurons*. These tend to be slow in operation, with typical switching times on the order of milliseconds, and consequently the brain uses significant parallelism rather than speed to perform its complex tasks. Adaptation comes in short- and long-term versions, and can result from a variety of complex interactions.

Although most neurons are exceptions to the canonical neuron shown in Figure 10.1, the neuron illustrated is sufficiently complex to demonstrate the basic principles. Via various ion channels, neurons maintain a constant negative voltage of approximately −70 mV relative to the ambient environment. This neuron consists of a *dendritic* tree for taking inputs from other neurons, a body or *soma*, which basically

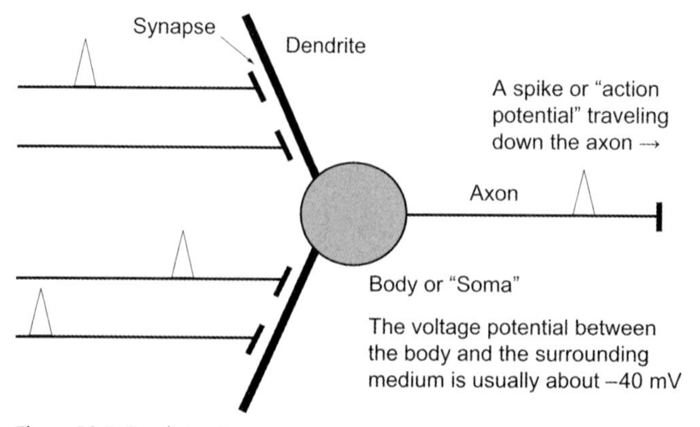

Figure 10.1 An abstract neuron.

performs charge summation, and an output, the *axon*. Inter-neuron communication is generally via pulses or spikes. Axons form synapses on dendrites and signal the dendrite by releasing small amounts of neurotransmitter, which is taken up by the dendrite.

Axons from other neurons connect via synapses onto the dendritic tree of each neuron. When an axon fires it releases a neurotransmitter into the junction between the *presynaptic* axon and the *postsynaptic* dendrite. The neurotransmitter causes the dendrite to depolarize slightly, and this charge differential eventually reaches the body or soma of the neuron, depolarizing the neuron.

When a neuron is sufficiently depolarized it passes a threshold which causes it to generate an *action potential*, or output spike, which moves down the axon to the next synapse. When an output spike is traveling down an axon, it is continuously regenerated allowing for arbitrary fan-out.

While the dendrites are depolarizing the neuron, the resting potential is slowly being restored, creating what is known as a "leaky integrator." Unless enough action potentials arrive within a certain time window of each other, the depolarization of the soma will not be sufficient to generate an output action potential.

In addition to accumulating signals and creating new signals, neurons also learn. When a spike arrives via an axon at a synapse, it "presynaptically" releases a neurotransmitter, which causes some depolarization of the postsynaptic dendrite. Under certain conditions, the effect of a single spike can be modified, typically increased or "facilitated", where it causes a greater depolarization at the synapse. When the effect that an action potential has on the postsynaptic neurons is enhanced, the synapse is said to be *potentiated*, and learning has occurred. One form of this is called long-term potentiation (LTP), as such potentiation has been shown to last for several weeks at a time and may possibly last much longer. LTP is one of the more common learning mechanisms, and has been shown to occur in several areas of the brain whenever the inputs to the neuron and the output of the neuron correlate within some time window. Learning correlated inputs and outputs is also called Hebb's law, named after Donald Hebb who proposed it in 1947. Synapses can also lose their facilitation; one example of this is a similar mechanism called long-term depression (LTD), which generally occurs when an output is generated and there is no input at a particular synapses.

Postsynaptic excitation can either be *excitatory* (an excitatory postsynaptic potential or EPSP), which leads to accumulating even more charge in the soma, or *inhibitory* (an inhibitory postsynaptic potential or IPSP), which tends to drain charge off of the soma, making it harder for a neuron to fire an action potential. Both capabilities are needed to control and balance the activation of groups of neurons. In one model, the first neuron that fires tends to inhibit the others in the group leading to what is called a "winner-take-all" function.

10.1.2 A Very Simple Neural Model: The Perceptron

One of the first biologically inspired models was the perceptron which, although very simple, was still based on biological neurons. The primary goal of a perceptron is to

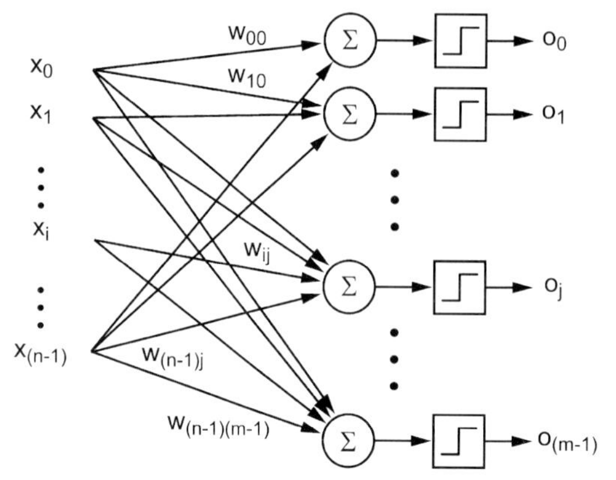

Figure 10.2 A single-layer perceptron.

do classification. Perceptron operation is very simple, as it has a one-dimensional synaptic weight vector and takes another, equal size, one-dimensional vector as input. Normally the input vector is binary and the weight vectors positive or negative integers. During training, a "desired" signal is also presented to the perceptron. If the output matches the training signal, then no action is taken; however, if the output is incorrect and does not match the training signal, then the weights in the weight vector are adjusted accordingly. The perceptron learning rule, which was one of the first instances of the "delta rule" used in most artificial neural models, incremented or decremented the individual weights depending on whether they were predictive of the output or not. A single-layer perceptron is shown in Figure 10.2. Basic perceptron operation is

$$o_j = f\left(\sum_{i=1}^{n} w_{ij}x_i\right) = f(W_j^T X)$$
$$O = f(W^T X)$$

where $f()$ is an activation function, and is a step function here (if $sum > 0$, then $f(sum) = 1$); however, $f()$ can also be a smooth function (see below). A "layer" has some number (two or more) of perceptrons, each with its own weight vector and individual output value, leading to a weight matrix and an output vector. In a single "layer" of perceptrons, each one sees the same input vector.

The Delta Rule, which is used during learning is,

$$\Delta w_{ij} = \alpha(d_j - o_j)x_i$$

where d_j is the desired output, and o_j is the actual output.

The delta rule is fundamental to most adaptive neural network algorithms. Rosenblatt proved that if a data set is linearly separable, the perceptron will eventually find a plane that separates the set. Figure 10.3 shows the two-dimensional (2-D) layout of data for, first, a linearly separable set of data and, second, for a non-linearly separable set. Unfortunately, if the data are not linearly separable the perceptron fails miserably, and this was the point of "...the book that killed neural networks",

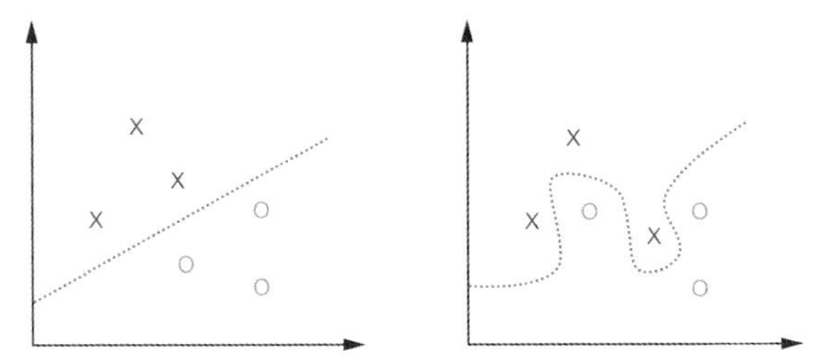

Figure 10.3 Linear (left) and non-linear (right) classification.

Perceptrons by Marvin Minsky and Seymour Papert (1968). Perceptrons cannot solve non-linearly separable problems; neither do they function in the kind of multiple layer structures that may be able to solve non-linear problems, as the algorithm is such that the output layer cannot tell the middle layer what its desired output should be. Attention is now turned to a description of the multi-layer perceptron.

10.1.3 A Slightly More Complex Neural Model: The Multiple Layer Perceptron

The invention of the multi-layer perceptron constituted, to some degree, a "victory" against the "evil empire" of symbolic computing. The Minsky and Papert book focused primarily, through numerous sophisticated examples, on how perceptrons, as envisioned at that time, could not solve non-linear problems, an excellent example being the XOR problem. Although there are a number of variations, back-propagation (BP) allowed the use of multiple, non-linear layers, and is sometimes referred to as a multi-layer perceptron (MLP).

Although the derivation of BP is beyond the scope of this chapter, some of the characteristics that allowed it to extend perceptron-like learning to multiple layers can be briefly summarized. First, instead of a discrete step function for output, a continuous activation function is used. As with the perceptron, there is also a training or "desired" signal, and by actually quantifying the error of the output as a function (generally least means square) an error surface is created. The gradient of the error surface can then be found and used to adjust the weights. By using the chain rule from calculus the error can then be back-propagated through the various levels.

In summary, the steps of the BP algorithm are:

- Present an input vector to the first layer of the network.
- Calculate the output for each layer, moving forward to network output.
- Calculate the error delta at the output (using actual output and the externally supplied target output, d).
- Use the computed error deltas at the output to compute the error deltas at the next layer, etc., moving backward through the net.
- After all error deltas have been computed for every node, use the delta rule to incrementally update all the weights in the network. (Note: the feedforward input activation values for each connection must be remembered.)

Even two-level networks can approximate complex, non-linear functions. Moreover, this technique generally finds good solutions, which are compact, leading to fast, feed-forward (non-learning) execution time. Although it has been shown to approximate Bayesian decisions (i.e. it results in a generally good estimate of where Baye's techniques would put the decision surface), it can have convergence problems due to many non-local minima. It is also computationally intensive, often taking days to train with complex large feature sets.

10.1.4 Auto-Association

Another important family of networks are associative networks, one example of which (as given above) is the Hopfield net. Here, details of a simple associative network developed by G. Palm [10] will be presented (this was in fact developed before the studies of Hopfield). One useful variation implements an *auto-associative network* that is an overly simplistic approximation of the circuits in the mammalian neocortex. Auto-associative networks have been studied extensively. In its simplest form, an associative memory maps an input vector to an output vector. When an input is supplied to the memory, its output is a "trained" vector with the closest match, assuming some metric, to the given input. Auto-association results when the input and output vectors are in the same space, with the input vector being a corrupted version of one of the training vectors. With *best-match association*, when a noisy or incomplete input vector is presented to the network, the "closest" training vector can usually be recalled reliably. In auto-association the output is fed back to the input, which may require several iterations to stabilize to a trained vector.

Here, it is assumed that the vectors are binary and the distance metric between two vectors is simply the number of bits that are different – that is, the Hamming distance.

The Palm associative network maps input vectors to output vectors, where the set of input vector to output vector mappings are noted as $\{(x^\mu, y^\mu), \mu = 1, 2, \ldots, M\}$. There are M mappings, and both x^μ and y^μ are binary vectors with size of m and n respectively. x^μ and y^μ are sparsely encoded, with $\sum_{i=1}^{m} x_i = l\ (l \ll m)$ and $\sum_{j=1}^{n} y_j = k$ $(k \ll n)$. Here, l and k are the numbers of active nodes (non-zero) in the input and output vectors, respectively. In training, the "clipped" Hebbian "outer-product," learning rule is generally used, and a binary weight matrix W is formed by $W = \vee_{\mu=1}^{M} [y^\mu \cdot (x^\mu)^T]$. Such batch computation has the weights computed off-line and then down-loaded into the network. It is also possible to learn the weights adaptively [11].

During recall, a noisy or incomplete input vector $\tilde{x}$ is applied to the network, and the network output is computed by $\tilde{y} = f(W \cdot \tilde{x} - \theta)$, θ is a global threshold, and $f()$ is the Heaviside step function, where an output node will be 1 (active) if its dendritic sum $x_i = \sum_{j=1}^{m} w_{ij}\tilde{x}_j$ is greater than the threshold θ; otherwise, it is 0. To set the threshold, the "k winners take all (k-WTA) rule" is used, where k is the number of active nodes in an output vector. The threshold, θ, is set so that only those nodes that have the k maximum dendritic sums are set to "1", and the remaining nodes are set to "0". The k-WTA threshold is very close to the minimum error threshold. The k-WTA operation plays the role of competitive lateral inhibition, which is a major component in all cortical circuits. In the BCPNN model of Lansner and his group [11], the nodes are divided into hypercolumns, typically $\sqrt{N}$ nodes in each of the $\sqrt{N}$ columns, with 1-WTA being performed in each column.

An auto-associative network starts with the associative model just presented and feeds the output back to the input, so that the x and y are in the same vector space and $l = k$. This auto-associative model is called an *attractor model* in that its state space creates an energy surface with most minima ("attractor basins") occurring when the state is equal to a training vector. Under certain conditions, given an input vector x', then the output vector y that has the largest conditional probability $P(x'|y)$ is the most likely training vector in a Bayesian sense. It is possible to define a more complex version with variable weights, as would be found during dynamic learning, which also allows the incorporation of prior probabilities [12].

10.1.5 The Development of Biologically Inspired Hardware

With BP and other non-linear techniques in hand, research groups began to solve more complex problems. Concurrent to this there was an explosion in neuroscience that was enabled by high-performance computing and sophisticated experimental technologies, coupled with an increasing willingness in the neuroscience community to begin to speculate about the function of the neural circuits being studied. As a result, research into artificial neural networks (ANNs) of all types gained considerable momentum during the late 1980s, continuing until the mid-1990s when the research results began to slow down. However, like AI before it – and fuzzy logic, which occurred concurrently – ANNs had trouble in scaling to solve the difficult problems in intelligent computing. Nevertheless, ANNs still constitute an important area of research, and ANN technologies play a key role in a number of real-world applications [13, 14]. In addition, they are responsible for a number of important breakthroughs.

During the heady years of the late 1980s and early 1990s, while many research groups were investigating theory, algorithms, and applications, others began to examine hardware implementation. As a consequence, there quickly evolved three schools of thought, though with imprecise dividing lines between them:

- The first concept was to build very specialized analog chips where, for the most part, the algorithms were hard-wired into silicon. Perhaps the best known was the aVLSI (low-power analog VLSI) technology developed by Carver Mead and his students at Caltech.
- The second concept was to build more general, highly parallel digital, but still fairly specialized chips. Many of the ANN algorithms were very computer-intensive, and it seemed that simply speeding up algorithm execution – and especially the learning phase – would be a big help in solving the more difficult problems and the commercialization of ANN technology. During the late 1980s and early 1990s these chips were also significantly faster than mainstream desktop technology; however, this second group of chips incorporated less biological realism than the analog chips.

- The third option was to use off-the-shelf hardware, digital signal processing (DSP) and media chips, and this ultimately was the winning strategy. This approach was successful because the chips were used in a broader set of applications and had manufacturing volume and software inertia in their favor. Their success was also assisted by Amdahl's law (see Section 10.2.1).

The aim of this chapter is to review examples of these biologically inspired chips in each of the main categories, and to provide detailed discussions of the motivation for these chips, the algorithms they were emulating, and architecture issues. Each of the general categories presented is discussed in greater detail as appropriate. Finally, with the realm of nano- and molecular-scale technology rapidly approaching, the chapter concludes with a preview of the future of biologically inspired hardware.

10.2
Early Studies in Biologically Inspired Hardware

The hardware discussed in this chapter is based on neural structures similar to those presented above, and, as such, is designed to solve a particular class of problems that are sometimes referred to as "intelligent computing". These problems generally involve the transformation of data across the boundary between the real world and the digital world, in essence from sensor readings to symbolic representations usable by a computer; indeed, this boundary has been called "the digital seashore".[1] Such transformations are found wherever a computer is sampling and/or acting on real-world data. Examples include the computer recognition of human speech, computer vision, textual and image content recognition, robot control, optical character recognition (OCR), automatic target recognition, and so on. These are difficult problems to solve on a computer, as they require the computer to find *complex structures and relationships* in massive quantities of low-precision, ambiguous, and noisy data. These problems are also very important, and an inability to solve them adequately constitutes a significant barrier to computer usage. Moreover, the list of ideas has been exhausted, as neither AI, ANNs, fuzzy logic, nor Bayesian networks[2] have yet enabled robust solutions.

At the risk of oversimplifying a complex family of problems, the solution to these problems will, somewhat arbitrarily, be partitioned into two domains: the "front end" and the "back end" (see Figure 10.4):

- *Front-end* operations involve more direct access to a signal, and include filtering and feature extraction.

1) Hiroshi Ishii, MIT Media Lab.

2) "A Bayesian network (or a belief network) is a probabilistic graphical model that represents a set of variables and their probabilistic independencies". Wikipedia, http://www.wikipedia.org/.

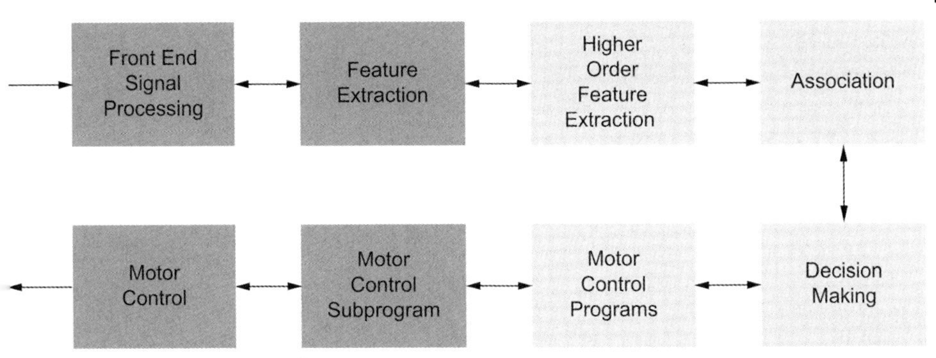

Figure 10.4 A canonical system.

- *Back-end* operations are more "intelligent", and include storing abstract views of objects or inter-word relationships.

In moving from front end to back end, the computation becomes increasingly interconnect driven, leveraging ever-larger amounts of diffuse data at the synapses for the connections. Much has been learned about the front end, where the data are input to the system and where there are developments in traditional as well as neural implementations. Whilst these studies have led to a useful set of tools and techniques, they have not solved the whole problem, and consequently more groups are beginning to examine the back-end – the realm of the cerebral cortex – as a source of inspiration for solving the remainder of the problem. Moreover, as difficult as the front-end problems are, the back-end problems are even more so. One manifestation of this difficulty is the "perception gap" discussed by Lazzaro and Wawrzynek [15], where the feature representations produced by more biologically inspired front-end processing are incompatible with existing back-end algorithms.

A number of research groups are beginning to refer to this "backend" as intelligent signal processing (ISP), which augments and enhances existing DSP by incorporating contextual and higher level knowledge of the application domain into the data transformation process. Simon Haykin (McMaster University) and Bart Kosko (USC) were editors of a special issue of the *Proceedings of the IEEE* [16] on ISP, and in their introduction stated:

> *"ISP uses learning and other 'smart' techniques to extract as much information as possible from signal and noise data."*

If you are classifying at Baye's optimal rates and you are still not solving the problem, what do you do next? The solution is to add more knowledge of the process being classified to your classification procedures, which is the goal of ISP. One way to do this is to increase the contextual information (e.g. higher-order relationships such as sentence structure and word meaning in a text-based application) available to the algorithm. It is these complex, "higher-order" relationships that are so difficult for us to communicate to existing computers and, subsequently, for them to utilize efficiently when processing signal data.

Humans make extensive use of contextual information. We are not particularly good classifiers of raw data where little or no context is provided, but we are masters of leveraging even the smallest amount of context to significantly improve our pattern-recognition capabilities.

One of the most common contextual analysis techniques in use today is the Hidden Markov Model (HMM) [17]. An HMM is a discrete Markov model with a finite number of states, that can be likened to a probabilistic finite-state machine. Transitions from one state to the next are determined probabilistically. Like a finite-state machine, a symbol is emitted during each state transition. In an HMM the selection of the symbol emission during each state transition is also probabilistic. If the HMM is being used to model a word, the symbols could be phonemes in a speech system. As the symbols are not necessarily unique to a particular state, it is difficult to determine the state that the HMM is in simply by observing the emitted symbols – hence the term "hidden." These probabilities can be determined from data of the real-world process that the HMM is being used to model. One variation is the study of Morgan and Bourlard [18], who used a large BP network to provide HMM emission probabilities.

In many speech-recognition systems, HMMs are used to find simple contextual structure in candidate phoneme streams. Most HMM implementations generate parallel hypotheses and then use a dynamic programming algorithm (such as the Viterbi algorithm) to find a match that is the most likely utterance (or most likely path through the model) based on the phonemes captured by preprocessing and capturing features from the speech stream, and the probabilities used in constructing the HMM. However, HMMs have several limitations:

- They grow very large if the probability space is complex, such as when pairs of symbols are modeled rather than single symbols; yet most human language has very high order structure.
- The "Markov horizon", which is a fundamental definition of Markov models and makes them tractable analytically, also contributes to the inability to capture higher-order knowledge. Many now believe that we have passed the point of diminishing returns for HMMs in speech recognition.

The key point here is that most neuro-inspired silicon – and in particular the analog-based components – is primarily focused on the front-end "DSP" part of the problem, since robust, generic back-end algorithms (and subsequently hardware) have eluded identification. It has been argued by some that if the front end was performed correctly, then the back-end would be easier, but whilst it is always easier to do better in front-end processing the room for improvement is smaller there. Without robust back-end capabilities, general solutions will be more limited.

10.2.1 Flexibility Trade-Offs and Amdhal's Law

During the 1980s and early 1990s, when most of the hardware surveyed in this chapter was created, there was a perception that the algorithms were large and

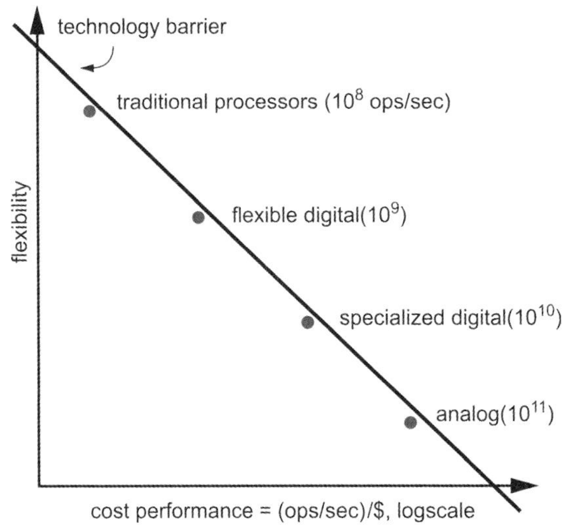

Figure 10.5 The flexibility–cost/performance trade-off.

complex and therefore slow to emulate on the computers of that time. Consequently, specialized hardware to accelerate these algorithms was required for successful applications. What was not fully appreciated by many was that the performance of general-purpose hardware was increasing faster than Moore's law, and that the existing neural algorithms did not scale well to the large sizes that would have fully benefited from special purpose hardware. The other problem was *Amdahl's law.*

As discussed above, these models have extensive concurrency which naturally leads to massively parallel implementations. The basic computation in these models is the multiply-accumulate operation that forms the core of almost all DSP and which can be performed with minimal, fixed point, precision. Also during the early 1990s, when many of the studies on neural inspired silicon were carried out, microprocessor technology was actually not fast enough for many applications.

The problem is that neural network silicon is highly specialized and there are specific risks involved in its development. One way to conceptualize the trade-offs involved in designing custom hardware is shown in Figure 10.5. Although *cost-performance*[3)] can be measured, flexibility cannot be assessed as easily, and so the graph in Figure 10.5 is more conceptual than quantitative. The general idea is that the more a designer hard-wires an algorithm into silicon, the better the cost-performance of the device, but the less flexible.

The line, which is moving slowly to the right according to Moore's law, shows these basic trade-offs and is, incidentally, not likely to be linear in the real world. Another

3) In this chapter, cost-performance is measured by (operations/second)/cost. The cost of producing a silicon chip is directly related (in a complex, non-linear manner) to the *area* of the chip. Larger chips are generally more expensive, as used here. More recently, however, other factors such as *power consumption* have become equally, if not more, important.

assumption is that the algorithms being emulated can be implemented at many different points on that scale, which is not always true either.

An important assumption in this analysis is that most applications require a wide range of computations, although there are exceptions, as in most real-world systems the "recognition" component is only a part of a larger system. So, when considering specialized silicon for such a system, the trade-off shown in Figure 10.5 must be factored into the analysis. More general-purpose chips tend to be useful on a larger part of the problem, but will lead to a sacrifice in cost-performance. On the other hand, the more specialized chips tend to be useful on a smaller part of the problem, but at a much higher cost-performance. Related to this trade-off then is Amdahl's law [19], which has always been a fundamental limitation to fully leveraging parallel computing,

> *Amdahl's law*: the speed-up due to parallelizing a computation is proportional to that portion of the computation that cannot be parallelized.

Imagine, for example, that there is a speech-recognition application, and 20% of the problem can be cast into a simplified parallel form. If a special-purpose chip was available that speeded up that 20% portion by 1000-fold, then the total system performance increase would be about 25%:

$$1/(0.8 + (0.2/1000)) = 1.25$$

Of course, depending on the cost of the 1000-fold chip, the 25% may still be worthwhile. However, if a comparably priced chip was available that was slower but more flexible and could parallelize 80% of the application, albeit with a more moderate speed increase (say 20-fold), then the total system performance would be over 400%:

$$1/(0.2 + (0.8/20)) = 4.17$$

Almost all computationally intensive pattern recognition problems have portions of sequential computation, even if it is just moving data into and out of the system, data reformatting, feature extraction, post-recognition tasks, or computing a final result. Amdahl's law shows that these sequential components have a significant impact on total system performance. As a result, the biggest problem encountered by many early neural network chips was that they tended to speed up a small portion of a large problem by moving to the right in Figure 10.5. For many commercial applications, after all was said and done, the cost of a specialized neural chip did not always justify the resulting modest increase in total system performance.

During the mid-1990s, desktop chips were doubling their performance every 18 to 24 months. Then, during the mid-1990s both Intel and AMD added on-chip SIMD coprocessing in the form of MMX which, for the Intel chips, has eventually evolved to SSE3 [20]. These developments, for the most part, spelled the death of most commercial neural net chips. However, in spite of limited commercial success most neural network chips were very interesting implementations, often with elegant engineering. A representative sample of some of these chips will be examined briefly in the remainder of this chapter.

It should not be concluded from the discussions so far that specialized chips are never economically viable. Rather, the continued success of graphics processors and DSPs are examples of specialized high-volume chips, and some neural networks chips[4] have found very successful niches. Nonetheless, it does illustrate some of the problems involved in architecting a successful niche chip. An example is the commercial DSP chips used for signal processing and related applications, these provide unique cost-performance, efficient power utilization, and just the right amount of specialization in their niche to hold their own in volume applications against general-purpose processors. In addition, they have enough volume and history to justify a significant software infrastructure.

In light of what is now known about Amdahl's law and ISP, the history and state of the art of neuro-inspired silicon can now be surveyed.

10.2.2
Analog Very-Large-Scale Integration (VLSI)

There is no question that the most elegant implementation technique developed for neural emulation is the sub-threshold CMOS technology pioneered by Carver Mead and his students at CalTech [21]. Most MOS field effect transistors (MOSFETs) used in digital logic are operated in two modes, either off (0) and on (1). For the *off* state the gate voltage is more or less zero and the channel is completely closed. For the *on* state, the gate voltage is significantly above the transistor threshold and the channel is saturated. A saturated *on* state works fine for digital, and is generally desired to maximize current drive. However, the limited gain in that regime restricts the effectiveness of the device in analog computation. This is due to the fact that the more gain the device has, the easier it is to leverage this gain to create circuits that perform useful computation and which also are insensitive to temperature and device variability.

However, if the gate voltage is positive (for the nMOS gate) but below the point where the channel saturates, the FET is still on, though with a much lower current. In this mode, which sometimes is referred to as "weak inversion", there is useful gain and the small currents significantly lower the power requirements, though FETs operating in this mode tend to be slower. Carver Mead's great insight was that when modeling biologically inspired circuits, significant computation could be carried out using simple FETs operating in the sub-threshold regime where, like real neurons, performance resulted from parallelism and not the speed of the switching devices. Moreover, as Carver and colleagues have shown, these circuits do a very good job approximating a number of neuroscience functions.

By using analog voltage and currents to represent signals, the considerable expense of converting signal data into digital, computing the various functions in digital, and then converting the signal data back to analog, was eliminated. Neurons operate slowly and are not particularly precise, yet when combined appropriately they perform complex and remarkably precise computations. The goal of the aVLSI

4) One example is General Vision; http://www.general-vision.com/.

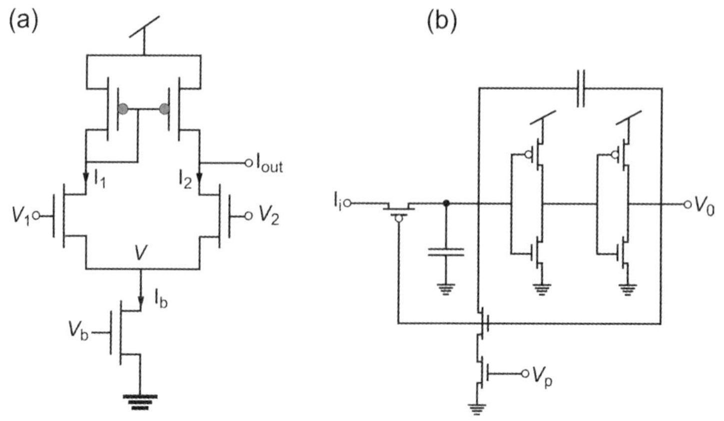

Figure 10.6 Basic aVLSI building blocks. (a) A transconductance amplifier; (b) an integrate and fire neuron.

research community has been to create elegant VLSI sub-threshold circuits that approximate biological computation.

One of the first chips developed by Carver *et al.* was the silicon retina [22]. This was an image sensor that performed localized adaptive gain control and temporal/spatial edge detection using simple "local neighborhood" functional extensions to the basic photosensitive cell. There subsequently followed a silicon cochlea and numerous other simulations of biological circuits.

Two examples of these circuits are shown in Figure 10.6. A transconductance amplifier (voltage input, current output) and an "integrate and fire" neuron are two of the most basic building blocks for this technology. The current state of aVLSI research is very well described by Douglas [23], of the Neuroinformatics Institute, ETH-Zurich:

> *Fifteen years of Neuromorphic Engineering: progress, problems, and prospects.* Neuromorphic engineers currently design and fabricate artificial neural systems: from adaptive single chip sensors, through reflexive sensorimotor systems, to behaving mobile robots. Typically, knowledge of biological architecture and principles of operation are used to construct a physical emulation of the target neuronal system in an electronic medium such as CMOS analog very large scale integrated (aVLSI) technology.
>
> Initial successes of neuromorphic engineering have included smart sensors for vision and audition; circuits for non-linear adaptive control; non-volatile analog memory; circuits that provide rapid solutions of constraint-satisfaction problems such as coherent motion and stereo-correspondence; and methods for asynchronous event-based communication between analog computational nodes distributed across multiple chips.

> *These working chips and systems have provided insights into the general principles by which large arrays of imprecise processing elements could cooperate to provide robust real-time computation of sophisticated problems. However, progress is retarded by the small size of the development community, a lack of appropriate high-level configuration languages, and a lack of practical concepts of neuronal computation.*

Although still a modest-sized community, research continues in this area, the largest group being that at ETH in Zurich. The commercialization of this technology has been limited, however, with the most notable success to date being that of Synaptics, Inc. This company created several products which used the basic aVLSI technology, the most successful being the first laptop touch pads.

10.2.3 Intel's Analog Neural Network Chip and Digital Neural Network Chip

During the "heyday" of neural network silicon, between 1986 and 1996, a major semiconductor vendor, Intel, produced two neural network chips. The first, the ETANN [24] (Intel part number 80170NX), was completely analog, but it was designed as a general-purpose chip for non-linear feed-forward ANN operation. There were two grids of analog "inner product" networks, each with 80 inputs and 64 outputs, and a total of 10 K (5 K for each grid) weights. The chip computed the two inner products simultaneously, taking about 5 μs for the entire operation. This resulted in a total performance (feed-forward only, no learning) of over two billion connections computed per second, where a connection is a single multiply-accumulate of an input-weight pair. All inputs and outputs were in analog. The weights were analog voltages stored on floating gates – with the chip being developed and manufactured by the flash memory group at Intel. Complementary signals for each input provided positive and negative inputs. An analog multiplier was used to multiply each input by a weight, current summation of multiplier outputs provided the accumulation, with the output being sent through a non-linear amplifier (giving roughly a sigmoid function) to the output pins.

Although not designed specifically to do real-time learning, it was possible to carry out "chip in the loop" learning where incremental modification of the weights was performed in an approximately stochastic fashion. Learning could also be done off-line and the weights then downloaded to the chip.

The ETANN chip had very impressive computational density, although the awkward learning and total analog design made it somewhat difficult to use. The multipliers were non-linear, which made the computation sensitive to temperature and voltage fluctuations. Ultimately, Intel retired the chip and moved to a significantly more powerful and robust all digital chip, the Ni1000.

The Ni1000 [25, 26] implemented a family of algorithms based on radial basis function networks (RBF [26]). This family included a variation of a proprietary algorithm created by Nestor, Inc., a neural network algorithm and software company.

Rather than doing incremental gradient descent learning, as can be seen with the BP algorithm, the Ni1000 used more of a template approach where each node represented a "basis" vector in the input space. The width of these regions, which was controlled by varying the node threshold, was reduced incrementally when errors were made, allowing the chip to start with crude over-generalizations of an input to output space mapping, and then fine-tune the mapping to capture more complex variations as more data are input. An input vector would then be compared to all the basis vectors, with the closest basis vector being the winner. The chip performed a number of concurrent basis computations simultaneously, and then, also concurrently, determined the classification of the winning output, both functions were performed by specialized hardware.

The Ni1000 was a two-layer architecture. All arithmetic was digital and the network parameters/weights were stored in Flash EEPROM. The first or hidden layer had 256 inputs of 16 bits each with 16 bit weights. The hidden layer had 1024 nodes and the second or output layer 64 nodes (classes). The hidden layer precision was 5 to 8 bits for input and weight precision. The output layer used a special 16-bit floating point format. One usage model was that of Bayesian classification, where the hidden layer learns an estimate of a probability density function (PDF) and the output layer classifies certain regions of that PDF into up to 64 different classes. At 40 MHz the chip was capable of over 10 billion connection computations per second, evaluating the entire network 40 K times per second with roughly 4 W peak power dissipation.

The Ni1000 used a powerful, compact, specialized architecture (unfortunately, space limitations prevent a more detailed description here, but the interested reader is referred to Refs. [25, 26]). The Ni1000 was much easier to use than the ETANN and provided very fast dedicated functionality. However, referring back to Figure 10.5, this chip was a specific family of algorithms wired into silicon. Having a narrower functionality it was much more at risk from Amdahl's law, as it was speeding up an even smaller part of the total problem. Like CNAPS (the Connected Network of Adapted Processors), it too was ultimately "run over by the silicon steam roller".

10.2.4
Cellular Neural Networks

Cellular neural networks (CNN) constitute another family of analog VLSI neural networks. This was proposed by Leon Chua in 1988 [27], who called it the "Cellular Neural Network", although now it is known as the "Cellular Non-Linear Network". Like aVLSI, CNN has a dedicated following, the most well-known being the Analogic and Neural Computing Laboratory of the Computer and Automation Research Institute of the Hungarian Academy of Sciences under the leadership of Tamas Roska. CNN-based chips have been used to implement vision systems, and complex image processing similar to that of the retina has been investigated by a number of groups [28].

Although there are variations, the basic architecture is a 2-D rectangular grid of processing nodes. Although the model allows arbitrary inter-node connectivity, most CNN implementations have only nearest-neighbor connections. Each cell computes its state based on the values of its four immediate neighbors, where the neighbor's

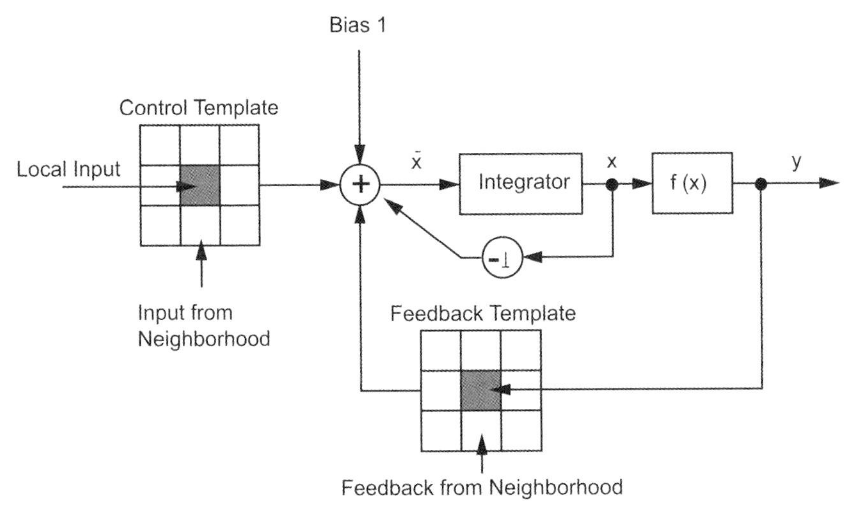

Figure 10.7 Basic cellular neural network (CNN) operation [72].

voltage and the derivative of this voltage are each multiplied by constants and summed. Each node then takes its new value and the process continues for another clock. This computation is generally specified as a type of filter, and is done entirely in the analog domain. However, as the algorithm steps are programmable one of the real strengths of CNN is that the inter-node functions and data transfer is programmable, with the entire array appearing as a digitally programmed array of analog-based processing elements. This is an example of a Single Instruction, Multiple Data (SIMD) architecture, which consists of an array of computation units, where each unit performs the same operation, but each on its own data. CNN programming can be complex and requires an intimate understanding of the basic analog circuits involved. The limited inter-node connectivity also restricts the chip to mostly "front-end" types of processing, primarily of images. A schematic of the basic CNN cell is shown in Figure 10.7.

Whereas, research and development continue, the technology has had only limited commercial success. As with aVLSI, it is a fascinating and technically challenging system, but in real applications it tends to be used for front-end problems and consequently is subject to Amdahl's law.

10.2.5 Other Analog/Mixed Signal Work

It is difficult to do justice to the large and rich area of biologically inspired analog design that has developed over the years. Other investigations include those of Murray [29], the former neural networks group at AT&T Bell Labs [30], Ettienne-Cummings [31], Principe [32], and many more that cannot be mentioned due to limited space. And today, some workers, such as Boahan, are beginning to move the processing further into the back end [33] by looking at cortical structures for early vision.

On returning to Figure 10.4, it can be seen that the first few boxes of processing require the type of massively parallel, locally connected feature extraction that CNN, aVLSI and other analog techniques provide. With regards to sensors, these can perform enhanced signal processing, and demonstrate better signal-to-noise ratios than more traditional implementations, providing such capabilities in compact, low-power implementations.

Although further studies are needed, there is a concern that the limited connectivity and computational flexibility make it difficult to apply these technologies to the back end. Although not a universally held opinion, the author feels that these higher-level association areas require a different approach to implementation. This general idea will be presented in more detail below, but first, it is important to examine another major family of neural network chips, the massively parallel digital processors.

10.2.6 Digital SIMD Parallel Processing

Concurrent with the development of analog neural chips, parallel effort was devoted to architecting and building digital neural chips. Although these could have dealt with a larger subset of pattern-recognition solutions they, like the analog chips, were mostly focused on neural network solutions to simple classification. A common design style that was well matched to the basic ANN algorithms was that of SIMD processor arrays. One chip that embodied that architecture was CNAPS, developed by Adaptive Solutions [34, 35].

The world of digital silicon has always flirted with specialized processors. During the early days of microprocessors, silicon limitations restricted the available functionality and as a result many specialized computations were provided by coprocessor chips. Early examples of this were specialized floating point chips, as well as graphics and signal processing. Following Moore's law, the chip vendors found that they could add increasing amounts of function and so began to pull some of these capabilities into the processor.

Interestingly, graphics and signal processing have managed to maintain some independence, and remain as external coprocessors in many systems. Some of the reasons for this were the significant complexity in the tasks performed, the software inertia that had built up around these functions, and the potential for very low power dissipation which is required for embedded signal processing applications, such as cell phones, PDAs, and MP3 players.

During the early 1990s it was clear that there was an opportunity to provide a significant speed-up of basic neural network algorithms because of their natural parallelism. This was particularly true in situations involving complex, incremental, gradient descent adaptation, as can be seen in many learning models. As a result, a number of digital chips were produced that aimed squarely at supporting both learning and non-learning network emulation.

It was clear from Moore's law that performance improvements and enhanced functionality continued for mainline microprocessors. This relentless march of the desktop processors was referred to as the "silicon steam roller" where, as the 1990s continued, it became increasingly difficult for the developers of specialized silicon to stay ahead of. At Adaptive Solutions, the goal was to avoid the steam roller by steering between having enough flexibility to solve most of the problem, to avoid Amdahl's law, and yet to have a sufficiently specialized function that allowed enough performance to make the chip cost-effective – essentially sitting somewhere in the middle of the line in Figure 10.5. This balancing act became increasingly difficult until eventually the chip did not offer enough cost-performance improvement in its target applications to justify the expense of a specialized coprocessor chip and board.

The CNAPS architecture consisted of a one-dimensional (1-D) processor node (PN) array in an SIMD parallel architecture [36]. To allow as much performance-price as possible, modest fixed point precision was used to keep the PNs simple. With small PNs the chip could leverage a specialized redundancy technique developed by Adaptive Solution's silicon partner, Inova Corporation. During chip testing, each PN could be added to the 1-D processor chain, or bypassed. In addition, each PN had a large power transistor (with a width of 20 000 λ) connecting the PN to ground. Laser fuses on the 1-D interconnect and the power transistor were used to disconnect and power down defective PNs. The testing of the individual PNs was done at wafer sort, after which an additional lasing stage (before packaging and assembly) would configure the dice, fusing in the good PNs and fusing out and powering down the bad PNs. The first CNAPS chip had an array of 8×10 (80) PNs fabricated, of which only 64 needed to work to form a fully functional die. The system architecture and internal PN architecture is shown in Figure 10.8.

Simulation and analysis was used to determine approximately the optimal PN size (the "unit of redundancy") and the optimal number of PNs. Ultimately, the die was almost exactly 2.5 cm (1 inch) on a side with 14 million transistors fabricated; this led to 12 dice per 15-cm (6-inch) wafer which, until recently, made it physically the largest processor chip ever made (see Figure 10.9).

The large version of the chip, called the CNAPS-1064, had 64 operational PNs and operated at 25 MHz with 6 W worst-case power consumption. Each PN was a complete 16-bit DSP with its own memory. Neural network algorithms tend to be vector/matrix-based and map fairly cleanly to a 1-D grid, so it was easy to have all PNs performing useful work simultaneously. The maximum compute rate then was 1.2 billion multiply-accumulates per second per chip, which was about 1000-fold faster than the fastest workstation at that time. Part of this speed up was due to the fact that each PN did several things in one clock to realize a single clock multiply-accumulate: input data to the PN, perform a multiply, perform an accumulate, perform a memory fetch, and compute the next memory address. During the late 1980s, DSP chips were able to perform such a single multiply-accumulate in one clock, but it was not until the Pentium Pro that desktop microprocessors reached a point where they performed most of these operations simultaneously.

When developing the CNAPS architecture, a number of key decisions were made, including the use of limited precision and local memories, architecture support for the C programming language, and I/O bandwidth.

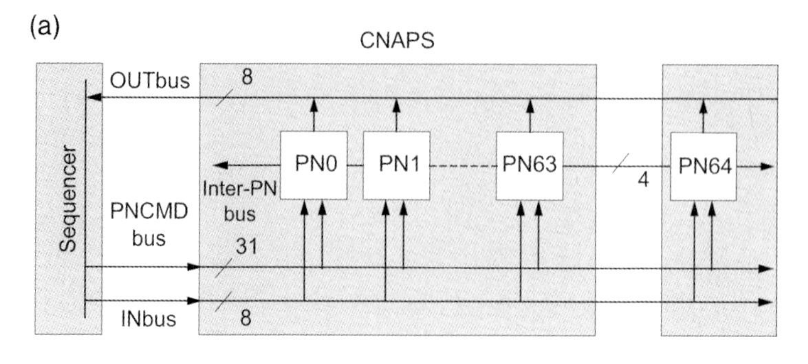

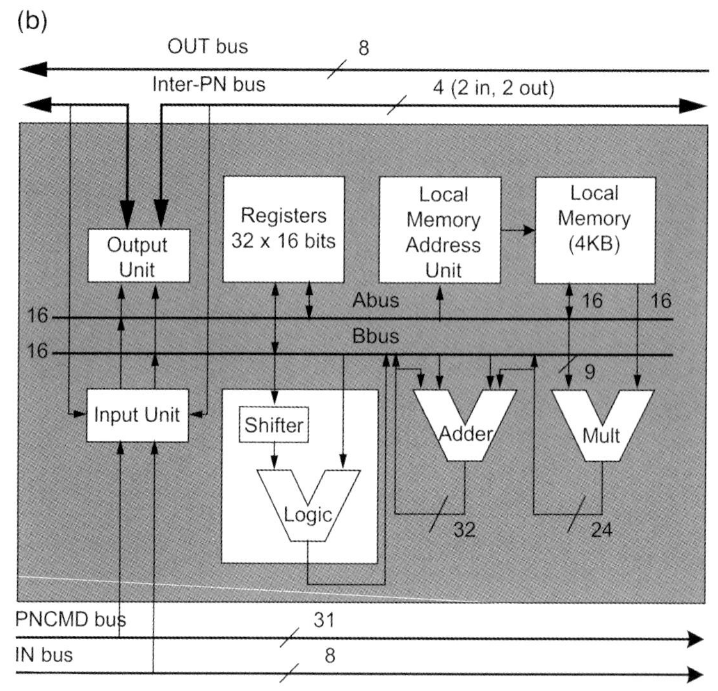

Figure 10.8 Connected Network of Adaptive Processors (CNAPS) architecture. (a) System architecture; (b) PN architecture.

At a time when the computing community was moving to floating-point computation, and the microprocessor vendors were pulling floating processing onto the processor chips and optimizing its performance, the CNAPS used limited precision, fixed point arithmetic. The primary reason for this decision was based on yield calculations, which indicated that a floating-point PN was too large to take advantage of PN redundancy. This redundancy bought an approximately twofold cost-performance improvement. Since a floating-point PN would have been two to three times larger than a fixed-point PN, the use of modest precision fixed-point arithmetic meant an almost sixfold difference in cost-performance. Likewise, simulation showed that

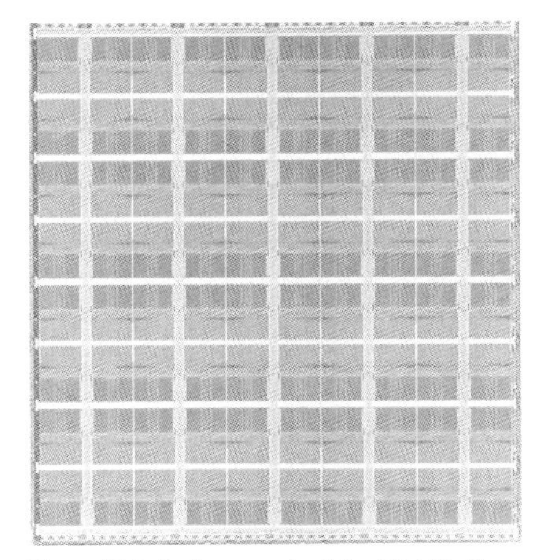

Figure 10.9 A photograph of the CNAPS die.

most of the intended algorithms could get by with limited precision fixed-point arithmetic, and this proved to be, in general, a good decision, as problems were rarely encountered with the limited precision. In fact, the major disadvantage was that it made programming more difficult, although DSP programmers had been effectively using fixed-point precision for many years.

The second decision was to use local, per PN, memory (4 KB of SRAM per PN). Although this significantly constrained the set of applications that could leverage the chip, it was absolutely necessary in achieving the performance goals. The reality was that it was unlikely that the CNAPS chip would have ever been built had performance been reduced enough to allow the use of off-chip memory. As with DSP applications, almost all memory access was in the form of long arrays that can benefit from some pre-fetching, but not much from caching.

The last two decisions – architecture and I/O bandwidth limitations – were driven by performance-price and design time limitations. One objective of the architecture was that it be possible for two integrated circuit (IC) engineers to create the circuits, logic schematics and layout (with some additional layout help) in one year. As a result, the architecture was very simple, which in turn made the design simpler and the PNs smaller, but programming was more difficult. One result of this strategy was that the architecture did not support the C language efficiently. Although there were some fabrication delays, the architecture, board, and software rolled out simultaneously and worked very well, and the first system was shipped in December 1991 and quickly ramped to a modest volume. One of the biggest selling products was an accelerator card for Adobe Photoshop which, in spite of Amdahl problems (poor I/O bandwidth), offered unprecedented performance.

By 1996, desktop processors had increased their performance significantly, and Intel was on the verge of providing the MMX SIMD coprocessor instructions. Although this first version of an SIMD coprocessor was not complete and was not particularly easy to use, the performance of a desktop processor with MMX reduced the advantages of the CNAPS chipset even further in the eyes of their customers, and people stopped buying.

Everybody knew that the silicon steam roller was coming, but it was moving much faster (and perhaps even accelerating, as some had suggested) than expected. In addition, Intel quickly enhanced the MMx coprocessor to the now current SSE3, which is a complete and highly functional capability. DSPs were also vulnerable to the microprocessor steam roller, but managed, primarily through software inertia and very low power dissipation, to hold their own.

Although there were other digital neural network processors, none of them achieved any significant level of success, and basically for the same reasons. Although at this point the discussion of all but a few of these others is limited by space, two in particular deserve mention.

10.2.7
Other Digital Architectures

One important digital neural network architecture was the Siemens SYNAPSE-1 processor developed by Ramacher and colleagues at Siemens Research in Munich [37]. The chip was similar to CNAPS in terms of precision, fixed-point arithmetic, and basic processor node architecture, but differed by using off-chip memory to store the weight matrices.

A SYNAPSE-1 chip contained a 2×4 array of MA16 "neural signal processors," each with a 16-bit multiplier and 48-bit accumulator. The chip frequency was 40 MHz, and one chip could compute about five billion connections per second with feedforward (non-learning) execution.

Recall that, in architecting the CNAPS, one of the most important decisions was whether to use on-chip per PN memory, or off-chip shared memory for storing the primary weight matrices. For a number of reasons, including the targeted problems space and the availability of a state-of-the-art SRAM process, Adaptive Solutions chose to use on-chip memory for CNAPS. However, for performance reasons this decision limited the algorithm and application space to those whose parameters fit into the on-chip memories. Although optimized for matrix-vector operations, CNAPS was designed to perform efficiently over a fairly wide range of computations.

The SYNAPSE-1 processor was much more of a matrix–matrix multiplication algorithm mapped into silicon. In particular, Ramacher and colleagues were able to take advantage of a very clever insight – the fact that in any matrix multiplication, the individual elements of the matrix are used multiple times. The SYNAPSE-1 broke all matrices into 4×4 chunks. Then, while the elements of one matrix were broadcast to the array, 4×4 chunks of the other matrix would be read from external memory into the array. In a 4×4-matrix by 4×4-matrix multiplication, each element in the matrix was actually used four times, which allowed the processor chip to support four times as many processor units for a given memory bandwidth than a processor not using this optimization.

On returning to Figure 10.5, it can be seen that the SYNAPSE-1 architecture increased performance by specializing the architecture to matrix–matrix multiplications.

Fortunately, most neural network computations can be cast in a matrix form, though it did restrict maximum machine performance to algorithms that performed matrix–matrix multiples. However, like the other digital neural network chips, the SYNAPSE-1 eventually lost out to high-performance microprocessor and DSP hardware.

10.2.8
General Vision

A similar chip to the Ni1000 was the ZISC (Zero Instruction Set Computer) developed by Paillet and colleagues at IBM in Paris. The ZISC chip was digital, employed basically a "vector template" approach, and was simpler and cheaper than the Ni1000 but implemented approximately the same algorithms. Today, the ZISC chip survives as the primary product of General Vision, Petaluma, California.

In addition to the CNAPS, SYNAPSE-1, ZISC, and Ni1000, several other digital chips have been developed either specifically or in part to emulate neural networks. HNC developed the SNAP, a floating-point SIMD standard cell-based architecture [38]. One excellent architecture is the SPERT [39], which was developed by groups at the University of California Berkeley and the International Computer Science Institute (ICSI) in Berkeley. SPERT was designed to perform efficient integer vector arithmetic and to be configured into large parallel arrays. A similar parallel processor array that was created from field-programmable gate arrays (FPGAs) and suited to neural network emulation was REMAP [40].

10.3
Current Directions in Neuro-Inspired Hardware

One limitation of traditional ANN algorithms is that they did not scale particularly well to very large configurations. As a result, commercial silicon was generally fast enough to emulate these models, thus reducing the need for specialized hardware. Consequently, with the exception of on-going studies in aVLSI and CNN, general research in neural inspired hardware has languished.

Today, however, activity in this area is picking up again, for two main reasons. The first reason is that computational neuroscience is beginning to yield algorithms that can scale to large configurations and have the potential for solving large, very complex problems. The second reason is the excitement of using molecular-scale electronics, which makes possible comparably scalable hardware. As will be seen, at least one of the projected nanoelectronic technologies is a complementary match to biologically inspired algorithms.

Today, a number of challenges face the semiconductor industry, including power density, interconnect reverse scaling, device defects and variability, memory bandwidth limitations, performance overkill, density overkill, and increasing design complexity. *Performance overkill* is where the highest-volume segments of the market are no longer performance/clock frequency-driven. *Density overkill* is where it is difficult for a design team to effectively design and verify all the transistors available to them on a single die. Although neither of these is a potential show-stopper, taken together they do create some significant challenges.

Another challenge is the growing reliance on parallelism for performance improvements. In general purpose applications, the primary source of parallelism has been within a single instruction stream, where many instructions can be executed simultaneously, sometimes even out of order. However, this instruction level parallelism (ILP) has its limits and becomes exponentially expensive to capture. Microprocessor manufacturers are now developing "multiple core" architectures, the goal of which is to execute multiple threads efficiently. As multiple core machines become more commonplace, software and application vendors will struggle to create parallel variations of their software.

Due to very small, high-resistance wires, many nano-scale circuits will be slow, and power density will be a problem because of high electric fields. Consequently, performance improvements at the nano-scale will also need to come almost exclusively from parallelism and to an even greater extent than traditional architectures.

When considering these various challenges, it is unclear which ones are addressed by nanoelectronics. In fact, nanoelectronics only addresses the end of Moore's law, and perhaps also the memory bandwidth problem. However, it also aggravates most other existing problems, notably signal/clock delay, device variability, manufacturing defects, and design complexity.

In proceeding down the path of creating nanoscale electronics, by far the biggest question is, how exactly will this technology be used? Can it be assumed that computation, algorithms, and applications will continue more or less as they have in the past? What should the research agenda be? Will the nanoscale processor of the future consist of thousands of $\times$ 86 cores with a handful of application-specific coprocessors? The effective use of nanoelectronics will require solutions to more than just an increased density; rather, total system solutions will need to be considered. Today, computing structures cannot be created in the absence of some sense of how they will be used and what applications they will enable. Any paradigm shift in applications and architecture will have a profound impact on the entire design process and the tools required, as well as the requirements placed on the circuits and devices themselves.

As discussed above, algorithms inspired by neuroscience have a number of interesting and potentially useful properties, including fine-grain and massive parallelism. These are constructed from slow, low-power, unreliable components, are tolerant of manufacturing defects, and are robust in the presence of faulty and failing hardware. They adapt rather than be programmed, they are asynchronous, compute with low precision, and degrade gracefully in the presence of faults. Most importantly, they are adaptive, *self-organizing* structures which promise some degree of design error tolerance, and solve problems dealing with the interaction of an organism/system with the real world. The functional characteristics of neurons, such such as analog operation, fault tolerance, slow, massive parallelism, are radically different from those of typical digital electronics. Yet, some of these characteristics match very well the basic characteristics such as large numbers of faults and defects, low speed, and massive parallelism that many research groups feel will characterize nanoelectronics systems.

Self-organization involves a system adapting (usually increasing in complexity) in response to an external stimulus. In this context, a system will learn about its environment and adjust itself accordingly, without any additional intervention. In order to achieve some level of self-organization, a few fundamental operating principles are required. Self-organizing systems are those that have been built with these principles in mind.

Recently, Professor Christoph von der Malsburg has defined a new form of computing science – "organic computing" – which deals with a variety of computations that are performed by biology. Organic computations are massively parallel, low precision, distributed, adaptive, and self-organizing. The neural algorithms discussed in this chapter form an important subset of this area (the interested reader is referred to the web site: www.organic-computing.org).

Several very important points should be made about biologically inspired models. The first point concerns the computational models and the applications they support. Biologically inspired computing uses a very different set of computational models than have traditionally been used. And subsequently they are aimed at a fairly specialized set of applications. Consequently, for the most part biological models are not a replacement for existing computation, but rather they are an enhancement to what is available now. Specialized hardware for implementing these models needs to be evaluated accordingly, and in the next few sections some of these models will be explored at different levels.

10.3.1
Moving to a More Sophisticated Neuro-Inspired Hardware

As mentioned above, it is the back end where the struggle with algorithms and implementation continues, and it is also the back end where potential strategic inflection points lie. Hence, the remainder of the chapter will focus on back-end algorithms and hardware.

The ultimate cognitive processor is the *cerebral cortex*. The cortex is remarkably uniform, not only across its different parts, but also across almost all mammalian species. Although current knowledge is far from providing an understanding of how the cerebral cortex does what it does, some of the basic computations are beginning to take shape. Nature has, it appears, produced a general-purpose computational device that is a fundamental component of higher level intelligence in mammals.

Some generally accepted notions about the cerebral cortex are that it represents knowledge in a sparse, distributed, hierarchical manner, and that it performs a type of Bayesian inference over this knowledge base, which it does with remarkable efficiency. This knowledge is added to the cortical data base by a complex process of adaptive learning.

One of the fundamental requirements of intelligent computing, the need to capture higher-order relationships. The problem with Bayesian inference is that it is an exponentially increasing computation in the number of variables (it has been shown to be NP-Hard, which means that the number of computational steps increases exponentially with the size of the problem); in other words, as order increases the computational overhead increases even more rapidly. Consequently, to use Bayesian inference in real problems, order is reduced to make them computationally tractable.

One common way to do this is to create a Bayesian network, which is a graph structure where the nodes represent variables and the arcs connecting the nodes represent dependencies. If there is reasonable independence between many of the variables then the network itself will be sparsely connected. Bayesian networks "factorize" the inference computation by taking advantage of the independence between different variables. Factorization does reduce the computational load, but at the cost of limiting the knowledge represented in the network. A custom network is also required for each problem.

Cortical networks appear to use sparse distributed data representations, where each neuron participates in a number of specific data representations. Distributed representations also diffuse information, topologically localizing it to the areas where it is needed and reducing global connectivity. Computing with distributed representations can be thought of as the hardware equivalent of spread spectrum communication, where pseudo-random sequences of bits are used to spread a signal across time and frequency. In addition to spreading inter-node communication, distributed representations also spread the computation itself. One hypothesis of cortex operation is that distributed representations of information are a form of extreme factorization, allowing efficient, massively parallel Bayesian inference.

Mountcastle [41, 42] conducted many pioneering studies in understanding the structural architecture of the cortex, including proposing the columnar organization. The fundamental unit of computation appears to be the *cortical minicolumn*, a vertically organized group of about 80 to 100 neurons which traverses the thickness of the gray matter ($\sim$3 mm) and is about 50 μm in diameter. Neurons in a column tend to communicate vertically with other neurons on different layers in the same column. These are subsequently organized into larger columns variously called just "columns", "cortical columns", "hypercolumns", or "modules." Braitenberg [43] postulates two general connectivity systems in cortex: "metric" (high-density connections to physically local cells, based on actual 2-D layout); and "ametric" (low-density point-to-point connections to all large groups of densely connected cells). Connectivity is significantly denser in the metric system, but to a limited extent.

One approach to creating cortical-like algorithms is to model each column as an auto-associative network, such as the Palm model discussed above. The columns are then sparsely connected to each other, but the specifics of the inter-column connections are still not certain and different research groups have expressed different ideas about this [44]. In addition, the neocortex has a definite hierarchical organization where there are as many feedback paths as feed-forward paths.

Among other things, the massive scale is probably one of the more important advantages of biological computation. Consequently, it is likely that useful versions of these algorithms will require networks with a million or more nodes. Back-end processing, because of a need to store large amounts of unique synaptic information, will most likely have simpler processing than is seen at the front end, albeit on a much larger scale.

Hecht-Nielsen [45] bases the inter-column (which he calls "regions") connections on conditional probabilities, which capture higher-order relationships. He also uses abstraction columns to represent groups of lower-level columns. He has demonstrated networks that perform a remarkable job of capturing aspects of English, as these networks consist of several billion connections and require a large computer cluster to execute.

Granger [46, 47] leverages nested distributed representations in a way that adds the temporal dimension, creating hierarchical networks that learn sequences of sequences. George and Hawkins [48] use model likelihood information ascending a hierarchy with model confidence information being fed back. Other researchers are also contributing to these ideas include Grossberg [49], Lansner [50, 51], Arbib [52], Roudi and Treves [53], Renart *et al.* [54], Levy *et al.* [55], and Anderson [56]. Clearly, this remains a dynamic area of research, and at this point there is no clear "winning" approach.

Another key feature of some of these algorithms is that there is an oscillatory sliding threshold that causes the more "confident" columns to converge to their attractors more quickly, the less-confident more slowly, while those of low confidence do not converge at all, taking a "NULL" state. This process is remarkably similar to the electromagnetic waves that flow through the cortex when it is processing data.

Connectivity remains one of the most important problems when first considering scaling to very large models. Axons are very fine and can be packed very densely in a three-dimensional (3-D) mesh. Interconnect in silicon generally operates in a 2-D plane, although with several levels, nine or more with today's semiconductor technologies. Most importantly, silicon communication is not continuously amplifying, as can be seen in axonal and some dendritic processes. The following result [57–59] demonstrates this particular problem.

> *Theorem*: Assume an unbounded or large rectangular array of silicon neurons where each neuron receives input from its N nearest neighbors; that is, the fan-out (divergence) and fan-in (convergence) is N. Each such connection consists of a single metal line, and the number of 2-D metal layers is much less than N. Then, the area required by the metal interconnect is $O(N^3)$.

So, if the fan-in per node is doubled from 100 to 200, the silicon area required for the metal interconnect increases by a factor of 8. This result means that, even for modest local connectivity, that portion of silicon area devoted to the metal interconnect will dominate. It has been shown that for some models even moderate multiplexing of interconnect would significantly decrease the silicon area requirements, without any real loss in performance [60]. Carver Mead's group at Caltech, and others, developed the address event representation (AER), a technique for multiplexing a number of pulse streams on a single metal wire [61, 62]. When analog computation is used, signals can be represented by the time between action-potential-like "spikes". These signal "packets" or "pulses" are transmitted asynchronously the moment they occur, with the originating unit's address, on a single shared bus. This "pseudo-digital" representation allows multiplexing of the bus and the retention of analog and temporal information, without expensive conversions.

In studying the potential implementations of cortical structures, an efficient connection multiplexing architecture was developed where data transfer occurs via overlapping, hierarchical buses [63]. This structure, *The Broadcast Hierarchy* (TBH), allows simultaneous high-bandwidth local connectivity and long-range connectivity, thereby providing a reasonable match to many biological connectivity patterns.

The details of a relevant proposed hybrid CMOS/nanoelectronic technology, CMOL, are presented in the next section.

10.3.2 **CMOL**

Likharev has proposed CMOL (Cmos/MOLecular hybrid) as an implementation strategy for charge-based[5] nanoelectronic devices. Likharev's group has analyzed a number of examples of CMOL configurations, including memory, reconfigurable logic, and neuromorphic CrossNets [64–66]. In addition, nanogrids are most likely to be the first commercial deployment of nanoelectronic circuits [67].

CMOL consists of a set of nanogrids fabricated on top of traditional CMOS, with the CMOS being used, among other things, for signal restoration and current drive, nanogrid addressing, and to communicate signals into and out of the nanogrids. The nanogrids themselves will generally have more specialized computation such as a memory, which augments the computation being performed by the CMOS.

A *nanogrid* consists of a set of parallel wires, with another set of parallel wires fabricated on top of and orthogonal to the first set. Likharev has shown that such grids need not be laid out in perfect dimensions or alignment, but can be reproduced by using nanoimprint templates. Sandwiched between the two grids is a planar material made from a specific molecular structure such that, where the horizontal and vertical wires cross, a molecular switch is created. Several mechanisms have been identified to effect the desirable electrical properties where two nanowires cross.

The most important property is of a binary "latching switch" with two metastable internal states [68]. This nanoscale device can be programmed to either an "on" or an "off" state by using two sets of voltages. The lower set is used to read out the device to determine its state, while the higher set is used to change the state of the device. The programming voltages are used to switch the device between high- and low-resistance states. The lower read-out voltages are used to determine the resistance or

5) Researchers are investigating a number of molecular technologies based on computational paradigms other than charge, such as spintronics, quantum cellular automata, and DNA computing. However, as neural circuits operate on the principle of charge accumulation, charge-based computation seems a better match, although further study of these other technologies is required.

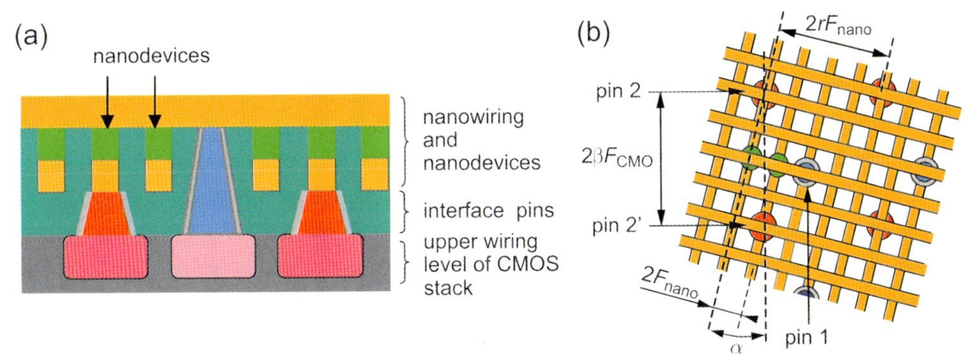

Figure 10.10 CMOL [1]. (a) A schematic side view. (b) Top view showing that any nanodevice may be addressed via the appropriate pin pair (e.g. pins 1 and 2 for the leftmost of the two shown devices, and pins 1 and 2′ for the rightmost device). Panel (b) shows only two devices; in reality, similar nanodevices are formed at all nanowire crosspoints. Also not seen on panel (b) are CMOS cells and wiring.

"state" of the molecule. Another required characteristic of these devices is rectification, where current flow is allowed only in one direction.

One of the most important characteristics of CMOL is the unique way in which the grids are laid at an angle with respect to the CMOS grid. Each nanowire is connected to the upper metal layers of the CMOS circuitry through a pin. In order for the CMOS to address each nanowire in the denser nanowire crossbar arrays, when it is fabricated, the nanowire crossbar is turned by an angle α, where α is the tangent of the ratio of the CMOS inter-pin distance to the nanogrid inter-pin distance. This technique allows the grid to be aligned arbitrarily with the CMOS and still have most nanowires addressable by some selection of CMOS cells. A nanowire is contacted by two CMOS cells, both of which are required to input a signal or read a signal. This basic connectivity structure is shown in Figure 10.10.

Although CMOL is not necessarily biologically inspired, it represents a promising technology for implementing such algorithms, as will be seen in the next section. CMOL uses charge-accumulation as its basic computational paradigm, which is also used by neural structures. Other nanoscale devices such as spin technologies do not implement a charge accumulation model, so such structures would have to emulate a charge-accumulation model, probably in digital.

10.3.3 An Example: CMOL Nano-Cortex

A high-level analysis has been performed of the implementation of a cortical column in CMOL. It is assumed that column operation is based on the Palm model discussed above. For this analysis, multiple column communication will be ignored and, for the sake of simplicity, a non-learning model will be assumed where the weights are computed off-line and downloaded into the nanogrid. Some typical values for the

Table 10.1 Typical values of parameters used for a cortical column analysis.

Parameter	Range	Typical Value I	Typical Value II
Hypercolumn node size	128 ∼ 128 K	1 K	16 K
Weight matrix size (single-weight-bit)	$2^{14} \sim 2^{34}$ bits	2^{20} bits	2^{28} bits
Weight matrix size (multi-weight-bit)	$2^{18} \sim 2^{51}$ bits	2^{30} bits	2^{42} bits
Multi-weight-bits	7 ∼ 17 bits	10 bits	14 bits
No. of active nodes in hypercolumn	7 ∼ 17	16	16
Inner-product result bits (single-weight-bit)	3 ∼ 5 bits	5 bits	5 bits
Inner-product result bits (multi-weight-bit)	11 ∼ 21 bits	14 bits	18 bits

parameters used in the cortical column architectural analysis are listed in Table 10.1. These values represent typical numbers used by several different simulation models, in particular, Lansner and his group at KTH [69]. Related investigations have been conducted at IBM [70], where a mouse cortex-sized model has been simulated on a 32 K processor IBM BlueGene/L.

Four basic designs have been analyzed, as shown in Figure 10.11:

- All-digital CMOS
- Mixed-signal CMOS
- All-digital hybrid CMOS/CMOL
- Mixed-signal hybrid CMOS/CMOL.

For the CMOS designs and the CMOS portion of CMOL, a 22-nm process was assumed as a "maximally" scaled CMOS. To approximate the features for this process, a simple linear scaling of a 250-nm process was made. The results of this

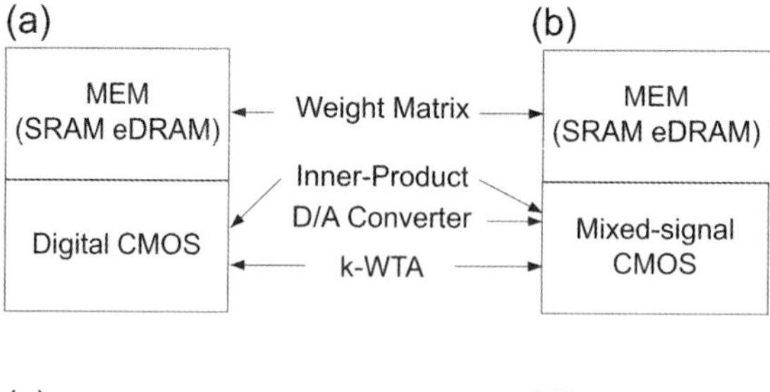

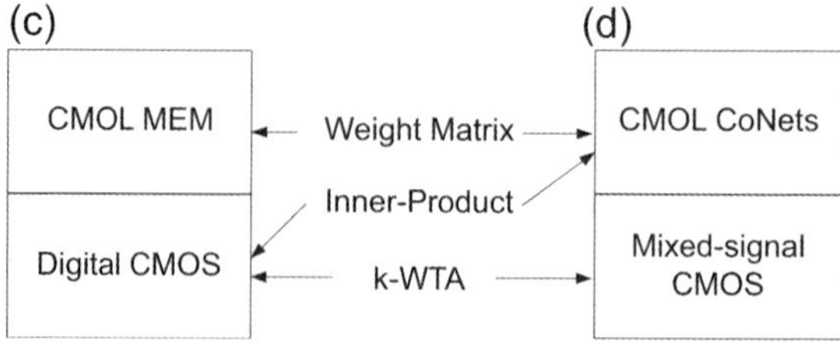

Figure 10.11 Architecture space - biologically inspired models. (a) All-digital CMOS; (b) mixed-signal CMOS; (c) all-digital hybrid CMOS/CMOL; (d) mixed-signal hybrid CMOS/CMOL.

Table 10.2 Analysis results.

Design		No. of column processors	Power (W)	Update rate (G nodes s^{-1})	Memory (%)
CMOS All Digital	1-bit eDRAM	6600	528	3072	2.9
CMOS Mixed-Signal	1-bit eDRAM	19 500	487	22 187	9.0
CMOL All Digital	1-bit CMOL Mm	4 042 752	317	4492	40
CMOL MS	1-bit CoNets	10 093 568	165	11 216	100

analysis, where the cost-performance for the four systems with the assumption of an 858 mm^2 die size (the maximum lithographic field size expected for a 22-nm process), are presented in Table 10.2.

With regards to Table 10.2, with the mixed signal CMOL it was possible to implement approximately 10 M columns, each having 1 K nodes, with 1 K connections each, for a total of 10 Tera-connections. In addition, this entire network can be updated once every millisecond – which is approaching biological densities and speeds, although of course with less functionality. Such technology could be built into a portable platform, with the biggest constraint being the high power requirements. Current studies include investigations into spike-based models [71] that should allow a significant lowering of the duty cycle and the power consumed.

Although real-time learning/adaptation was not included in the circuits analyzed here, deployed systems will need to be capable of real-time adaptation. It is expected that additional learning circuitry will reduce density by about two- to threefold. Neither has the issue of fault tolerance been addressed, although the Palm model has been found to tolerate errors, single 1 bits set to 0, in the weight matrix of up to 10%. For this reason, and given the excellent results of Likharev and Strukov [66] on the fault tolerance of CMOL arrays used as memory, it is expected that some additional hardware will be required to complement algorithmic fault tolerance, although this should not reduce the density in any significant way.

10.4 Summary and Conclusions

In this chapter, a brief and somewhat superficial survey has been provided of the specialized hardware developed over the past 20 years to support neurobiological models of computation. A brief examination was made of the current efforts and speculation made on how such hardware, especially when implemented in nanoscale electronics, could offer unprecedented compute density, possibly leading to new capabilities in computational intelligence. Biologically inspired models seem to be a better match to nanoscale circuits.

The mix of continued Moore's law scaling, models from computational neuroscience and molecular-scale technology portends a potential paradigm shift in how computing is carried out. Among other points, the future of computing is most likely not about discrete logic but rather about encoding, learning, and performing inference over stochastic variables. There may be a wide range of applications for such devices in robotics, in the reduction and compression of widely distributed sensor data, and power management.

One of the leading lights of the first computer revolution saw this clearly. At the IEEE Centenary in 1984 (*The Next 100 Years*; IEEE Technical Convocation), Dr. Robert Noyce, the co-founder of Intel and co-inventor of the Integrated Circuit, noted that:

> *Until now we have been going the other way; that is, in order to understand the brain we have used the computer as a model for it. Perhaps it is time to reverse this reasoning: to understand where we should go with the computer, we should look to the brain for some clues.*

References

1 K. Likharev, CMOL: Freeing advanced lithography from the alignment accuracy burden, in: The International Conference on Electron, Ion, and Photon Beam Technology and Nanofabrication '07, Denver, 2007.

2 W. S. McCulloch, W. H. Pitts, A logical calculus of the ideas immanent in nervous activity, *Bull. Mathematical Biophys.* 1943, **5**, 115–133.

3 K. Steinbuch, Die Lernmatrix, *Kybernetik* 1961, 1.

4 F. Rosenblat, *Principles of Neurodynamics: Perceptrons and the Theory of Brain Mechanisms*, Spartan, New York, 1962.

5 L. Minsky, M. A. S. Papert, *Perceptrons: An introduction to computational geometry*, MIT Press, Cambridge MA, 1988.

6 SOAR.Web Page. http://sitemaker.umich.edu/soar.

7 J. Hopfield, Neural networks and physical systems with emergent collective computational abilities. *Proc. Natl. Acad. Sci. USA* 1982, **79**, 2554–2558.

8 D. Rumelhart, G. Hinton, R. Williams, Learning internal representations by error propagation, *Nature* 1986, **323**, 533–536.

9 P. J. Werbos, *The Roots of Backpropagation: From Ordered Derivatives to Neural Networks and Political Forecasting*, Wiley-Interscience, 1994.

10 G. Palm, F. Schwenker, F. T. Sommer, A. Strey, Neural associative memories, in: A. Krikelis, C. C. Weems (Eds.), *Associative Processing and Processors*, IEEE Computer Society, Los Alamitos, CA, 1997, pp. 284–306.

11 A. Sandberg, A. Lansner, K.-M. Petersson, Ö. Ekeberg, Bayesian attractor networks with incremental learning, *Network: Computation in Neural Systems* 2002, **13** (2), 179–194.

12 S. Zhu, Associative memory as a primary component in cognition, PhD Dissertation (in preparation) CSEE Department, School of Science and Engineering, Oregon Health & Science University, Portland, OR.

13 D. Hammerstrom, Neural networks at work, *IEEE Spectrum* 1993, 26–32.

14 D. Hammerstrom, Working with neural networks, *IEEE Spectrum* 1993, 46–53.

15 J. Lazzaro, J. Wawrzynek, Speech recognition experiments with silicon auditory models, *Analog Integrated Circ. Signal Proc.* 1997, **13**, 37–51.

16 S. Haykin, B. Kosko (Eds.), Intelligent signal processing. *Proceedings of the IEEE*, Volume 86, IEEE Press 1989.

17 L. Rabiner, A tutorial on hidden Markov models and selected applications in speech recognition, *Proceedings IEEE* 1989, **77** (2), 257–286.

18 H. Bourlard, N. Morgan, *Connectionist Speech Recognition – A Hybrid Approach*, Kluwer Academic Publishers, Boston, MA, 1994.

19 J. L. Hennessy, D. A. Patterson, *Computer Architecture: A Quantitative Approach*, Morgan Kaufmann, Palo Alto, CA, 1991.

20 Intel. *IA-32 Intel Architecture Software Developer's Manual, Volume 1: Basic Architecture*. 2001 (cited 2001; Available from: http://developer.intel.com/design/pentium4/manuals/245470.htm)

21 C. Mead, *Analog VLSI and Neural Systems*, Addison-Wesley, Reading, Massachusetts, 1989.

22 M. A. Mahowald, *Computation and Neural Systems*, California Institute of Technology, 1992.

23 R. Douglas, Fifteen years of neuromorphic engineering: Progress, problems, and prospects, in: Proceedings, Brain Inspired Cognitive Systems – BICS2004, University of Stirling, Scotland, UK, 2006.

24 M. Holler, *et al.*, An electrically trainable artificial neural network (ETANN) with 10240 'floating gate' synapses, in: *Proceedings, International Joint Conference on Neural Networks*, IEEE, Washington DC, 1989.

25 I. Nestor, Ni1000 Recognition Accelerator – Data Sheet, 1996, 1–7. Available at: http://www.warthman.com/projects-intel-ni1000-TS.htm.

26 M. J. L. Orr, *Introduction to Radial Basis Function Networks*, Centre for Cognitive Science, University of Edinburgh, Edinburgh, 1996.

27 L. O. Chua, T. Roska, *Cellular Neural Networks and Visual Computing*, Cambridge University Press, 2002.

28 D. Balya, B. Roska, T. Roska, F. S. Werblin, A CNN framework for modeling parallel processing in a mammalian retina, *Int. J. Circuit Theory Applications* 2002, **30**, 363–393.

29 A. F. Murray, The future of analogue neural VLSI, in: *Proceedings Second International ICSC Symposium on Intelligent Systems for Industry*, The International ICSC Congress, June 26–29, Paisley, UK, 2001.

30 H. P. Graf, L. D. Jackel, W. E. Hubbard, VLSI implementation of a neural network model. *IEEE Computer* 1988, **21** (3), 41–49.

31 E. Culurciello, R. Etienne-Cummings, K. Boahen, An address event digital imager. *IEEE J. Solid-State Circuits* 2003, **38** (2), 505–508.

32 Y. N. Rao, D. Erdogmus, G. Y. Rao, J. C. Principe, Stochastic error whitening algorithm for linear filter estimation with noisy data, *Neural Networks Archive: Special issue: Advances in Neural Networks Research* 2003, **16** (5–6), 873–880.

33 T. Y. W. Choi, *et al.*, Neuromorphic implementation of orientation hypercolumns. *IEEE Trans. Circuits Systems II: Analog Digital Signal Proc.* 2005, **52** (6), 1049–1060.

34 D. Hammerstrom, A VLSI architecture for high-performance, low-cost, on-chip learning, in: *International Joint Conference on Neural Networks*, IEEE Press, San Diego, 1990.

35 D. Hammerstrom, A digital VLSI architecture for real-world applications, in: S. F. Zornetzer,*et al.* (Eds.), *An Introduction to Neural and Electronic Networks*, Academic Press, San Diego, CA, 1995, pp. 335–358.

36 J. L. Hennessy, D. A. Patterson, *Computer Architecture: A Quantitative Approach*, 3rd edn, Morgan Kaufmann, Palo Alto, CA, 2002.

37 U. Ramacher, W. Raab, J. Anlauf, U. Hachmann, J. Beichter, N. Brüls, M. Weißling, E. Schneider, R. Männer, J. Gläß, Multiprocessor and memory architecture of the neurocomputer SYNAPSE-1, in: *Proceedings, World Congress on Neural Networks*, INNS Press, Portland, Oregon, Volume 4, pp. 775–778, 1993.

38 R. Means, L. Lisenbee, Extensible linear floating point SIMD neurocomputer array processor, in: *Proceedings, International*

Joint Conference on Neural Networks, IEEE Press, Seattle, Washington, 1991.

39 J. Wawrzynek, K. Asanovic, B. Kingsbury, J. Beck, D. Johnson, N. Morgan, Spert-II: A vector microprocessor system. *IEEE Computer* 1996, 79–86.

40 L. Bengtsson, *et al.* The REMAP Reconfigurable Architecture: a Retrospective, in: A. R. Omondi, J. C. Rajapakse (Eds.), *FPGA Implementations of Neural Networks*, Springer-Verlag, 2006.

41 V. Mountcastle, *Perceptual Neuroscience – The Cerebral Cortex*, Harvard University Press, Cambridge, MA, 1998.

42 V. B. Mountcastle, An organizing principle for cerebral function: the unit model and the distributed system, in: G. M. Edelman, V. B. Mountcastle (Eds.), *The Mindful Brain*, MIT Press, Cambridge, MA, 1978.

43 V. Braitenberg, A. Schüz, *Cortex: Statistics and Geometry of Neuronal Connectivity*, Springer-Verlag, Berlin, 1998.

44 C. Johansson, M. Rehn, A. Lansner, Attractor neural networks with patchy connectivity, *Neurocomputing* 2006, **69**, 627–633.

45 R. Hecht-Nielsen, A theory of thalamocortex, in: R. Hecht-Nielsen, T. McKenna (Eds.), *Computational Models for Neuroscience – Human Cortical Information Processing*, Springer, London, 2003.

46 R. R. Granger, Engines of the brain: The computational instruction set of human cognition. *AI Magazine* 2006, **27** (2), 15–32.

47 R. Granger, *et al.*, Non-Hebbian properties of LTP enable high-capacity encoding of temporal sequences. *Proc. Natl. Acad. Sci. USA* 1994, **91**, 10104–10108.

48 D. George, J. Hawkins, Invariant pattern recognition using Bayesian inference on hierarchical sequences, in: *Proceedings, 2005 IEEE International Joint Conference on Neural Networks*, Volume 3, pp. 1812–1817, 2005.

49 S. Grossberg, Adaptive resonance theory, in: *The Encyclopedia of Cognitive Science*, Macmillan Reference Ltd, London, 2003.

50 A. Lansner, A. Holst, A higher order Bayesian neural network with spiking units. *Int. J. Neural Systems* 1996, **7** (2), 115–128.

51 C. Johansson, A. Lansner, Towards cortex-sized artificial nervous systems, in: *Knowledge-Based Intelligent Information and Engineering Systems KES'04*, WelTec-Springer, Wellington, New Zealand, 2004.

52 M. Arbib, Towards a neurally-inspired computer architecture. *Natural Computing* 2003, **2** (1), 1–46.

53 Y. Roudi, A. Treves, An associative network with spatially organized connectivity, *J. Stat. Mech.: Theor. Exp.* 2004, **2004**, P07010.

54 A. Renart, N. Parga, E. T. Rolls, Associative memory properties of multiple cortical modules, *Network: Comput. Neural Syst.* 1999, **10**, 237–255.

55 N. Levy, D. Horn, E. Ruppin, Associative memory in a multimodular network, *Neural Computation* 1999, **11**, 1717–1737.

56 J. A. Anderson, P. Allopenna, G. S. Guralnik, D. Scheinberg, J. A. Santini, S. Dimitriadis, B. B. Machta, B. T. Merritt, Programming a parallel computer: The Ersatz Brain Project, in: W. Duch, J. Mandziuk, J. M. Zurada (Eds.), *Challenges to Computational Intelligence*, Springer, Berlin, 2006 (in press).

57 J. Bailey, A VLSI interconnect strategy for biologically inspired artificial neural networks, PhD Thesis, Computer Science/ Engineering Department, Oregon Graduate Institute, Beaverton, OR, 1993.

58 J. Bailey, D. Hammerstrom, Why VLSI implementations of associative VLCNs require connection multiplexing, *Proceedings, International Conference on Neural Network*, 1988, pp. 173–180.

59 D. Hammerstrom, The connectivity requirements of simple association, or How many connections do you need? in: *IEEE Conference on Neural Network Information Processing*, IEEE Press, 1987.

60 E. Means, D. Hammerstrom, Piriform model execution on a neurocomputer, in:

International Joint Conference on Neural Networks, Seattle, WA, 1991.

61 K. A. Boahen, Point-to-point connectivity between neuromorphic chips using address events. *IEEE Trans. Circuits Systems II - Analog Digital Signal Proc.* 2000, **47** (5), 416–434.

62 J. P. Lazzaro, J. Wawrzynek, A multi-sender asynchronous extension to the address-event protocol, in: *Proceedings, 16th Conference on Advanced Research in VLSI*. IEEE Computer Society, Washington, DC, USA, p. 158, 1995.

63 D. Hammerstrom, J. Bailey, Neural-model, computational architecture employing broadcast hierarchy and hypergrid, point-to-point communication. US Patent No. 4,983,962, issued January 8, 1991.

64 J. H. Lee, K. K. Likharev, *CMOL CrossNets as Pattern Classifiers*, Stony Brook University, Stony Brook, 2005.

65 K. K. Likharev, D. V. Strukov, CMOL: Devices, circuits, and architectures, in: G. Cuniberti, *et al.* (Eds.), *Introducing Molecular Electronics*, Springer, Berlin, 2004.

66 D. B. Strukov, K. K. Likharev, Defect-tolerant architectures for nanoelectronic crossbar memories. *J. Nanosci. Nanotechnol.* 2007, **7**, 151–167.

67 G. S. Snider, R. S. Williams, Nano/CMOS architectures using a field-programmable nanowire interconnect, *Nanotechnology* 2007, **18**, 035204.

68 K. K. Likharev, D. B. Strukov, CMOL: Devices, Circuits, and Architectures, in: G. Cuniberti, *et al.* (Eds.), *Introduction to Molecular Electronics*, Springer, Berlin, pp. 447–477, 2004.

69 A. Lansner, *et al.* Detailed simulation of large scale neural networks, in: J. M. Bower (Ed.), *Computational Neuroscience: Trends in Research 1997*, Plenum Press, Boston, MA, 1997, pp. 931–935.

70 R. Ananthanarayanan, D. S. Modha, *Anatomy of a Cortical Simulator*, in: *Super Computing 2007 (SC07)*, EEE Press, Reno, Nevada, 2007.

71 W. Maass, Computing with spiking neurons, in: W. Maass, C. M. Bishop (Eds.), *Pulsed Neural Networks*, MIT Press, A Bradford Book, Cambridge, MA, 1999.

72 M. Hänggi, Available from: http://www.ce.unipr.it/pardis/CNN/cnn.html.

11
Nanowire-Based Programmable Architectures

André DeHon

11.1
Introduction

Today, chemists are demonstrating bottom-up synthesis techniques which can construct atomic-scale switches, field-effect devices, and wires (see Section 11.2). While these are key components of a computing system, it must also be understood if these can be assembled and organized into useful computing devices. That is, can arbitrary logic be built from nanowire building blocks and atomic-scale switches?

- Do we have an adequate set of capabilities to build logic?
- How do we cope with the regularity demanded by bottom-up assembly?
- How do we accommodate the high defect rates and statistical assembly which accompany bottom-up assembly techniques?
- How do we organize and interconnect these atomic-scale building blocks?
- How do we address nanowires from the lithographic scale for testing, configuration, and IO?
- How do we get logic restoration and inversion?
- What net benefit do these building blocks offer us?

The regular synthesis techniques can be used to assemble tight-pitch, parallel nanowires; this immediately suggests that programmable crossbar arrays (Section 11.4.1) are built as the key building blocks in these architectures. These crossbar arrays can be used as memory cores (Section 11.5), wired-OR logic arrays (Section 11.6.1), and programmable interconnect (Section 11.6.3) – memory, logic, and interconnect – all of which are the key components needed for computation.

The length of the nanowires must be limited for yield, performance, and logical efficiency. Consequently, the nanowires are organized into a collection of

Nanotechnology. Volume 4: Information Technology II. Edited by Rainer Waser
Copyright © 2008 WILEY-VCH Verlag GmbH & Co. KGaA, Weinheim
ISBN: 978-3-527-31737-0

modest-sized, interconnected crossbar arrays (Section 11.6.3). A reliable, lithographic-scale support structure provides power, clocking, control, and bootstrap testing for the nanowire crossbar arrays. Each nanowire is coded so that it can be uniquely addressed from the lithographic support wires (Section 11.4.2). With the ability to address individual nanowires, individual crosspoints can be programmed (Section 11.8) to personalize the logic function and routing of each array and to avoid defective nanowires and switches (Section 11.7).

As specific nanowires cannot, deterministically, be placed in precise locations using these bottom-up techniques, stochastic assembly is exploited to achieve unique addressability (Section 11.4.2). Stochastic assembly is further exploited to provide signal restoration and inversion at the nanoscale (Section 11.4.3). Remarkably, starting from regular arrays of programmable diode switches and stochastic assembly of non-programmable field-effect controlled nanowires, it is possible to build fully programmable architectures with all logic and restoration occurring at the nanoscale.

The resulting architectures (Section 11.6) provide a high-level view similar to island-style field-programmable gate arrays (FPGAs), and conventional logic mapping tools can be adapted to compile logic to these arrays. Owing to the high defect rates likely to be associated with *any* atomic-scale manufacturing technology, all viable architectures at this scale are likely to be post-fabrication configurable (Section 11.7). That is, while nanowire architectures can be customized for various application domains by tuning their gross architecture (e.g. ratio of logic and memory), there will be no separate notion of custom atomic-scale logic.

Even after accounting for the required, regular structure, high defect rates, stochastic assembly, and the lithographic support structure, a net benefit is seen from being able to build with nanowires which are just a few atoms in diameter and programmable crosspoints that fit in the space of a nanowire junction. Mapping conventional FPGA benchmarks from the Toronto20 benchmark set [1], the designs presented here should achieve one to two orders of magnitude greater density than FPGAs in 22 nm CMOS lithography, even if the 22 nm lithography delivers defect-free components (Section 11.10).

The design approach taken here represents a significant shift in design styles compared to conventional lithographic fabrication. In the past, reliance has been placed on virtually perfect and deterministic construction and complete control of features down to a minimum technology feature size. Here, it is possible to exploit very small feature sizes, although there is no complete control of device location in all dimensions. Instead, it is necessary to rely on the statistical properties of large ensembles of wires and devices to achieve the desired, aggregate component features. Further, post-fabrication configuration becomes essential to device yield and personalization.

This chapter describes a complete assembly of a set of complementary technologies and architectural building blocks. The particular ensemble presented is one of several architectural proposals which have a similar flavor (Section 11.11) based on these types of technologies and building blocks.

11.2
Technology

11.2.1
Nanowires

Atomic-scale nanowires (NWs) can be engineered to have a variety of conduction properties, from insulating to semiconducting to metallic. The composition of a NW can be varied along its axis and along its radius, offering powerful heterostructures to provide both controllable devices and interconnect integrated into a single structure.

Seed catalysts are used to control the diameter of a NW during the composition process and constrain the growth to a small region [2, 3].

In addition, semiconducting can be doped during the growth process by controlling the mix of elements in the ambient environment [4]. This can produce conducting NWs with heavy doping and field effect controllable NWs with a suitably light doping [5].

The doping profile or material composition can change along the length of a NW by controlling the ambient process environment over the time [6–8]. This leads to properties such as gateable and not-gateable regions within a single NW (Figure 11.1). After the axial growth, the NW's surface can be used as a substrate for atomic layer growth to produce a radial material composition (Figure 11.2), for example with SiO_2 as an insulator and spacer [9–11].

In order to increase the conductivity of NWs beyond heavily doped semiconductors, nickel silicide (NiSi) can be generated by coating selected regions with nickel and subsequently annealing the area [12].

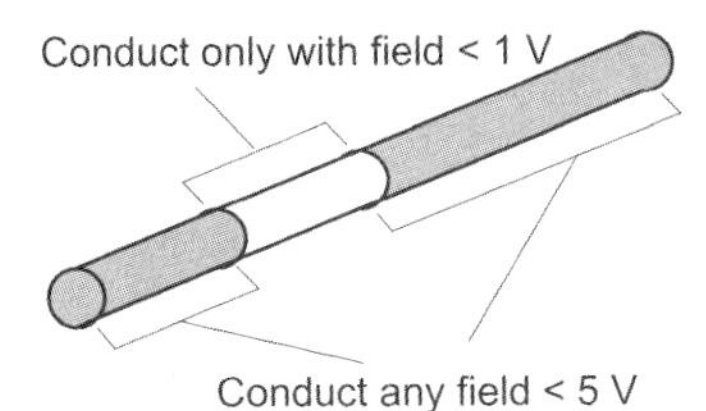

Figure 11.1 Axial doping profile places selective gateable regions in a nanowire.

Figure 11.2 Radial doping profile.

11.2.2 Assembly

Langmuir–Blodgett (LB) flow techniques can be used to align a set of NWs into a single orientation, close pack them, and transfer them onto a surface [11, 13]. The resulting wires are all parallel with nematic alignment. By using wires with an oxide sheath around the conducting core, the wires can be packed tightly. The oxide sheath defines the spacing between conductors and can, optionally, be etched away after assembly. The LB step can be rotated and repeated so that multiple layers of NWs are obtained [11, 13] such as crossed NWs for building a crossbar array or memory core (see Section 11.4.1).

11.2.3 Crosspoints

Many technologies have been demonstrated for non-volatile, switched crosspoints. Common features include:

- resistance which changes significantly between on and off states;
- the ability to be made rectifying;
- the ability to turn the device on or off by applying a large voltage differential across the junction;
- the ability to operate at a low voltage differential without switching the device state; and
- the ability to be placed within the area of a crossed NW junction.

Chen *et al.* [14] demonstrated a nanoscale Ti/Pt-[2]rotaxane-Ti/Pt sandwich which exhibits hysteresis and non-volatile state storage showing an order of magnitude resistance difference between on and off states for several write cycles. With 1600 nm^2 junctions, the on resistance ($R_{ondiode}$) was roughly 500 $K\Omega$, and the off resistance ($R_{offdiode}$) 9 $M\Omega$. After an initial burn-in step, the state of these devices can be switched at ± 2 V and read at ± 0.2 V. The basic hysteretic molecular memory effect is not unique to the [2]rotaxane, and the junction resistance is continuously tunable [15]. The exact nature of the physical phenomena involved is the subject of active investigation.

In conventional very large-scale integration (VLSI), the area of an SRAM-based programmable crosspoint switch is much larger than the area of a wire crossing. A typical, CMOS switch might be $2500\lambda^2$ [16], compared to a $5\lambda \times 5\lambda$ bottom-level metal wire crossing, making the crosspoint 100-times the area of the wire crossing. Consequently, the nanoscale crosspoints offer an additional device size reduction beyond that implied by the smaller NW feature sizes. This particular device size benefit reduces the overhead for configurability associated with programmable architectures [e.g. FPGAs, programmable logic arrays (PLAs)] in this technology compared to conventional CMOS.

11.2.4 Technology Roundup

It is possible to create wires which are nanometers in diameter and which can be arranged into crossbar arrays with nanometer pitch. Crosspoints which both switch conduction between the crossed wires and store their own state can be placed at every wire crossing without increasing the pitch of the crossbar array. NWs can be controlled in FET-like manner, and can be designed with selectively gateable regions. This can all be done without relying on ultrafine lithography to create the nanoscale feature sizes. Consequently, these techniques promise smaller feature sizes and an alternate – perhaps more economical – path to atomic-scale computing structures than top-down lithography. Each of the capabilities previously described has been demonstrated in a laboratory setting as detailed in the reports cited. It is assumed that, in future, it will be possible to combine these capabilities and to scale them into a repeatable manufacturing process.

11.3 Challenges

In the top-down lithographic model, a minimum, lithographically imageable feature size is defined, and devices are built that are multiples of this imageable feature size (e.g. half-pitch). Within the limits of this feature size, the size of features and their relative location to each other in three dimensions – both in the two-dimensional (2-D) plane of each lithographic layer and with adequate registration between layers – could be perfectly specified. This provided complete flexibility in the design of circuit structures as long as the minimum imageable and repeatable feature size rules were adhered to.

When approaching the atomic-scale, it becomes increasingly difficult to maintain this model. The precise location of atoms becomes relevant, and the discreteness of the underlying atoms begins to show up as a significant fraction of feature size. Variations occur due to statistical doping and dopant placement and interferometric mask patterning. Perfect repeatability may be extremely difficult or infeasible for these feature sizes.

These bottom-up approaches, in contrast, promise finer feature sizes that are controlled by physical phenomena but do not promise perfect, deterministic alignment in three dimensions. It may be possible to achieve good repeatability of certain types of small feature sizes (e.g. NW diameters) and correlation of tiny features within a single NW using axial and radial composition, but there may be little correlation from NW to NW in the plane or between NW planes. This may prompt the question of whether it would be reasonable to forego the perfect correlation and complete design freedom in three dimensions in order to exploit smaller feature sizes. The techniques summarized here suggest that this is a viable alternative.

11.3.1
Regular Assembly

The assembly techniques described above (see Sections 11.2.2 and 11.2.3) suggest that regular arrays can be built at tight pitch with both NW trace width and trace spacing using controlled NW diameters. While this provides nanometer pitches and crosspoints that are tens of nanometers in area, it is impossible to differentiate deterministically between features at this scale; that is, one particular crosspoint cannot be made different in some way from the other crosspoints in the array.

11.3.2
Nanowire Lengths

Nanowire lengths can be grown to hundreds of microns [17] or perhaps millimeters [18] in length. However, at this high length to diameter ratio, they become highly susceptible to bending and ultimately breaking. Assembly puts stresses along the NW axis which can break excessively long NWs. Consequently, a modest limit must be placed on the NW lengths (tens of microns) in order to yield a large fraction of the NWs in a given array. Gudiksen *et al.* [19] reported the reliable growth of Si NWs which are over 9 μm long, while Whang et al. [11, 20] demonstrated collections of arrays of NWs of size 10 μm × 10 μm. Even if it was possible physically to build longer NWs, the high resistivity of small-diameter NWs would force the lengths to be kept down to the tens of microns range.

11.3.3
Defective Wires and Crosspoints

At this scale, wires and crosspoints are expected to be defective in the 1 to 10% range:

- NWs may break along their axis during assembly as suggested earlier, and the integrity of each NW depends on the ~100 atoms in each radial cross-section.
- NW to microwire junctions depend on a small number of atomic scale bounds which are statistical in nature and subject to variation in NW properties.
- Junctions between crossed NWs will be composed of only tens of atoms or molecules, and individual bond formation is statistical in nature.
- Statistical doping of NWs may lead to high variation among NWs.

For example, Huang *et al.* [13] reports that 95% of the wires measured had good contacts, while Chen *et al.* [21] reported that 85% of crosspoint junctions measured were usable. Both of these were early experiments, however, and the yield rates would be expected to improve. Nonetheless, based on the physical phenomena involved it is anticipated that the defect rates will be closer to the few percent range than the minuscule rates frequently seen with conventional, lithographic processing. Consequently, two main defect types may be considered:

- *Wire defects:* a wire is either functional or defective. A functional wire has good contacts on both ends, conducts current with a resistance within a designated range, and is not shorted to any other NWs. Broken wires will not conduct current. Poor contacts will increase the resistance of the wire, leaving it outside of the designated resistance range. Excessive variation in NW doping from the engineered target can also leave the wire out of the specified resistance range. It can be determined if a wire is in the appropriate resistance range during testing (see Section 11.8.1) and arranged not to use those which are defective (see Section 11.7.1).
- *Non-programmable crosspoint defects:* a crosspoint is programmable, non-programmable, or shorted into the on state. A programmable junction can be switched between the resistance range associated with the on-state and the resistance range associates with the off-state. A non-programmable junction can be turned off, but cannot be programmed into the on-state; a non-programmable junction could result from the statistical assembly of too few molecules in the junction, or from poor contacts between some of the molecules in the junction and either of the attached conductors. A shorted junction cannot be programmed into the off-state. Based on the physical phenomena involved, non-programmable junctions are considered to be much more common than shorted junctions. Further, it is expected that fabrication can be tuned to guarantee that this is the case. Consequently, shorted junctions will be treated like a pair of defective wires, and both wires associated with the short will be avoided.

Currently, the bridging of adjacent NWs is considered NOT to be a major defect source. Radial shells around the (semi) conducting NW cores prevent the shorting of adjacent NWs. At present, there is insufficient experience to determine if variations in core shell thickness, imperfect planar NW alignment, or other effects may, nonetheless, lead to bridging defects between adjacent NWs. If such bridging were to occur, it could make a pair of NWs indistinguishable, perhaps effectively giving two addresses to the NW pair. These bridged NW pairs could be detected and avoided but their occurrence would necessitate slightly more complicated testing and verification algorithms than those detailed in Section 11.8.

After describing the building blocks and architecture, the way in which the two main defect types within the architecture are accommodated is described in Section 11.7.

11.4 Building Blocks

By working from the technological capabilities and within the regular assembly requirements, it is possible to construct a few building blocks which enable the creation of a wide range of interesting programmable architectures.

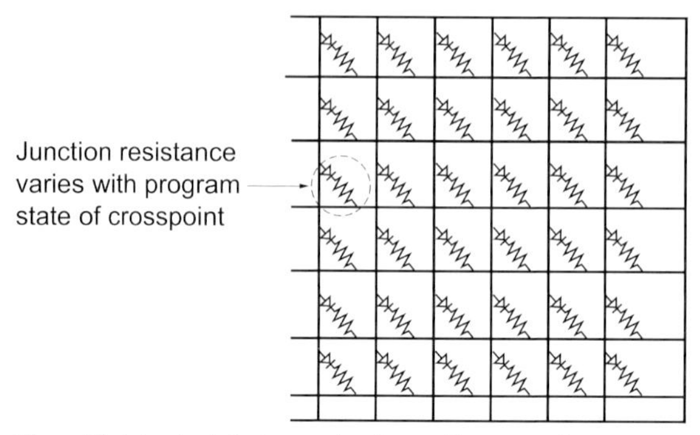

Figure 11.3 Logical diode crossbar formed by crossed nanowires.

11.4.1 Crosspoint Arrays

As suggested in Section 11.2.2 and demonstrated by Chen *et al.* [21] and Wu *et al.* [22], assembly processes allow the creation of tight-pitch arrays of crossed NWs with switchable diodes at the crosspoints (see Figure 11.3). Assuming for the moment that contact can be made to individual NWs in these tight-pitch arrays (see Section 11.4.2), these arrays can serve as:

- memory cores
- programmable, wired-OR planes
- programmable crossbar interconnect arrays.

11.4.1.1 Memory Core

As noted in Section 11.2.3, by applying a large voltage across a crosspoint junction, the crosspoint can be switched into a high or low resistance state. Consequently, if the voltage on a single row and a single column line can be set to desired voltages, each of the crosspoints can be set to a particular conduction state. It is further noted that the system can operate at a lower voltage without resetting the crosspoint. Consequently, a crosspoint's state can be read back by applying a small, test voltage to a column input and observing the current flow, or rate of charging, of a row line to tell if the crosspoint has been set into a high or low resistance state.

11.4.1.2 Programmable, Wired-OR Plane

When a method of programming the crosspoints into high or low resistance states has been effected, the OR logic can be programmed into a crosspoint array. Each row output NW serves as a wired-OR for all of the inputs programmed into the low resistance state. Consider a single row NW, and assume for the moment that the means is available to pull a non-driven NW down to ground. Now, if any of the column NWs which cross this row NW are connected with low resistance crosspoint junctions and are driven to a high voltage level, the current into the column NW will be able to flow into the row NW and charge the row NW up to a higher voltage value (see O1, O3, O4, and O5 in Figure 11.4). However, if none of the connected column NWs is high,

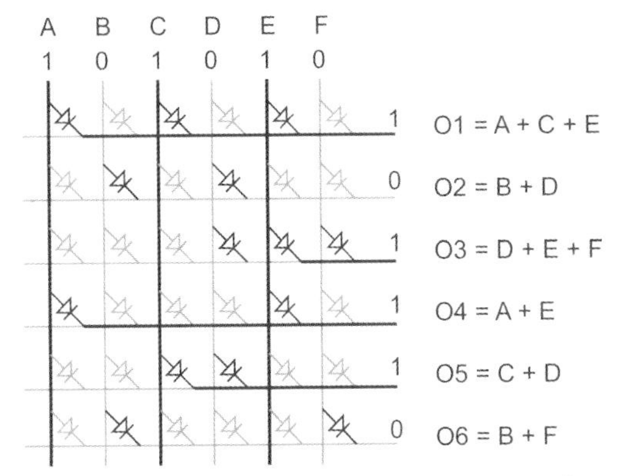

Figure 11.4 Wired-OR plane operation. Programmed on crosspoints are shown in black; off crosspoints are shown in gray. Dark lines represent a nanowire (NW) pulled high, while light lines remain low. Output NWs are marked dark, starting at the diode that pulls them high, in order to illustrate current flow; the entire output NW would be pulled high in actual operation.

the row NW will remain low (see O2 and O6 in Figure 11.4). Consequently, the row NW effectively computes the OR of its programmed inputs.

The output NWs do pull their current directly off the inputs and may not be driven as high as the input voltage. Hence, these outputs will need restoration (see Section 11.4.3).

11.4.1.3 Programmable Crossbar Interconnect Arrays

A special use of the Wired-OR programmable array is for interconnect. That is, if a restriction is introduced to connecting a *single* row wire to each column wire, the crosspoint array can serve as a crossbar switch. This allows any input (column) to be routed to any output (row) (e.g. see Figure 11.5). This structure is useful for

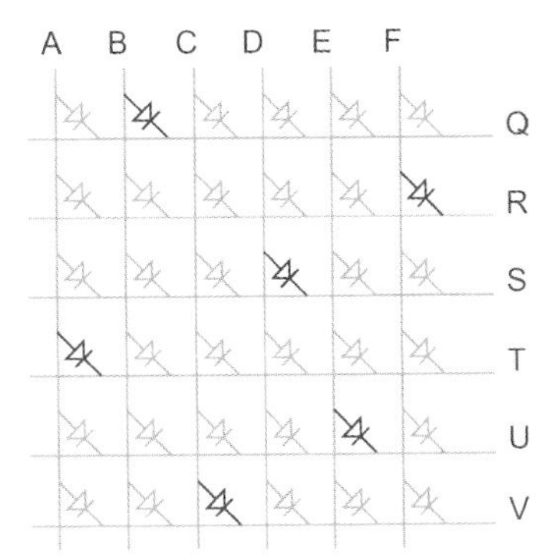

Figure 11.5 Example of crossbar routing configuration. Programmed on crosspoints are shown in black; off crosspoints are shown in gray. Here, the crossbar is shown programmed to connect A → T, B → Q, C → V, D → S, E → U, and F → R.

post-fabrication programmable routing to define a logic function and to avoid defective resources (see Section 11.3.3).

11.4.2 Decoders

A key challenge is bridging the length scale between the lithographic-scale wires that can be created using conventional top-down lithography and the small-diameter NWs that can be grown and assembled into tight-pitch arrays. As noted above, it must be possible to establish a voltage differential across a single row and column NW to write a bit in the tight-pitch NW array. It must also be possible to drive and sense individual NWs to read back the memory bit. By building a decoder between the coarse-pitch lithographic wires and the tight-pitch NWs, it is possible to bridge this length scale and to address a single NW at this tight pitch [23–26].

11.4.2.1 NW Coding

One way to build such a decoder is to place an address on each NW using the axial doping or material composition profile described previously. In order to interface with lithographic-scale wires, address bit regions are marked off at the lithographic pitch. Each such region is then either doped heavily so that it is oblivious to the field applied by a crossed lithographic-scale wire, or is doped lightly so that it can be controlled by a crossed lithographic scale wire. In this way, the NW will only conduct if all of the lithographic-scale wires crossing its lightly doped, controllable regions have a suitable voltage to allow conduction. If any of the lithographic-scale wires crossing controllable regions provide a suitable voltage to turn off conduction, then the NW will not be able to conduct.

It should be noted that each bit position can only be made controllable or non-controllable with respect to the lithographic-scale control wire; different bit positions cannot be made sensitive to different polarities of the input. Consequently, the addresses must be encoded differently from the dense, binary address normally used for memories. One simple way to generate addresses is to use a dual-rail binary code. That is, for each logical input bit, the value and its complement are provided. This results in two bit positions on the NW for each logical input address bit – one for the true sense and one for the false sense. To code a NW with an address, either bit position is simply coded to be sensitive to exactly one sense of each of the bit positions (see Figure 11.6). This results in a decoder which requires $2\log_2(N)$ address bits to address N NWs.

Denser addressing can be achieved by using $N_a/2$-hot codes; that is, rather than forcing one bit of each pair of address lines to be off and one to be on, it is simply required that half of the address bits, N_a, be set to a voltage which allows conduction, and half to be set to a voltage that prevents conduction. This scheme requires only $1.1\log_2(N) + 3$ address bits [24].

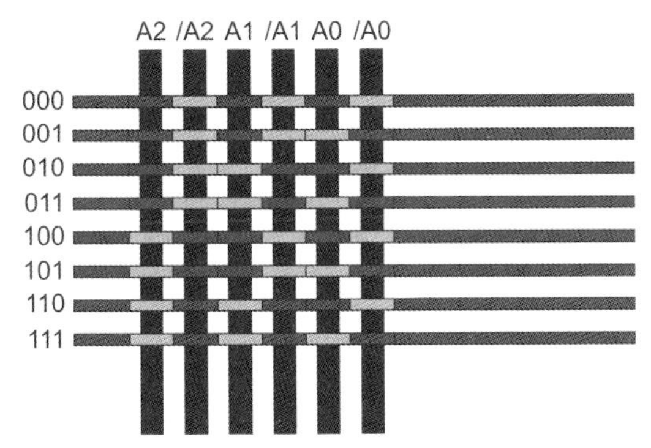

Figure 11.6 Dual-rail address coding.

11.4.2.2 Decoder Assembly

If each NW in the array has a unique address in the selected coding scheme, then each individual NW in the array can be uniquely addressed. However, the NW assembly techniques do not allow particular NWs to be placed in particular locations – it can only be arranged to create a tight-pitch parallel ensemble of a collection of NWs.

Instead, it appears that if the code space for the NWs is made large compared to the size of the array to be addressed, it can be guaranteed statistically (with arbitrarily high probability) that every NW in an array has a unique address. Starting with a very large number of NW codes, the NWs can be mixed up before assembly so that a random selection occurs of which NW codes go into each of the crosspoint arrays being assembled. As long as the array formed is sufficiently small compared to the code space for the NWs, strong guarantees can be provide that each array contains NWs with unique codes [24]. It transpires that there is no need for a large number of address bits in order to guarantee this uniqueness. For example, the $N_a/2$-hot codes need a total of only $\lceil 2.2\log_2(N)\rceil + 11$ bits to achieve over a 99% probability that all NWs in an array will have unique addresses. If a few duplicates are tolerable, the codes can be much tighter [25, 26].

Wires can be coded to tolerate misalignment during assembly. Hybrid addressing schemes which segment the collection of NWs in an array into lithographic-scale contact groups can be used to reduce the size of the NW codespace needed (for further details see Ref. [24]).

11.4.2.3 Decoder and Multiplexer Operation

Now it is known that uniquely coded NWs can be assembled into an array, it can be seen how the decoder operates. First, assume that all the NWs are either precharged or weakly pulled to a nominal voltage. The desired NW address is then applied to the lithographic-scale address lines. The desired drive voltage is also applied to a common line attached to all the NWs. If the selected address is present in the array, it will allow conduction from the common line into the array charging up the selected NW. All other NWs will differ in at least one bit position, and will thus be disabled by the address lines. Consequently, only the selected NW is charged strongly to the

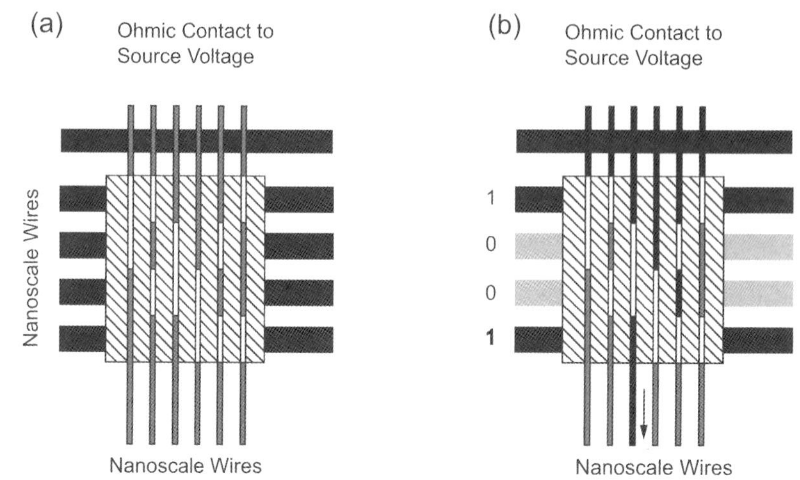

Figure 11.7 Coded NW decoder. (a) Decoder configuration: white NW regions are coded and controllable, while gray regions are not controllable and acts as wires. (b) Dark represents lines driven high; light gray shows lines low or undriven. Only the coding on the third line matches the applied address (1 0 0 1) and allows conduction. All other cases have a high address voltage crossing a lightly doped region, which prevents conduction.

voltage driven on the common line, and all other NWs are held at the nominal voltage (see Figure 11.7).

It should be noted that there is no directionality to the decoder, and consequently this same unit can serve equally well as a multiplexer. That is, when an address is applied to the lithographic-scale wires it allows conduction through the addressing region for only one of the NWs. Consequently, the voltage on the common line can be sensed rather than driven. Now, the one line which is allowed to conduct through the array can potentially pull the common line high or low. All other lines have a high resistance path across the lithographic-scale address wires and will not be able to strongly effect the common line. This allows a single NW to be sensed at a time (see Figure 11.8) as there is a need to read out the crosspoint state, as described in Section 11.4.1.1.

11.4.3
Restoration and Inversion

As noted in Section 11.4.1.2, the programmable, wired-OR logic is passive and non-restoring, drawing current from the input. Further, OR logic is not universal, and to build a good composable logic family an ability will be required to isolate inputs from output loads, restore signal strength and current drive, and invert signals.

Fortunately, NWs can be field-effect controlled, and this provides the potential to build FET-like gates for restoration. However, in order to realize these ways must be found to create the appropriate gate topology within the regular assembly constraints (see Section 11.3.1).

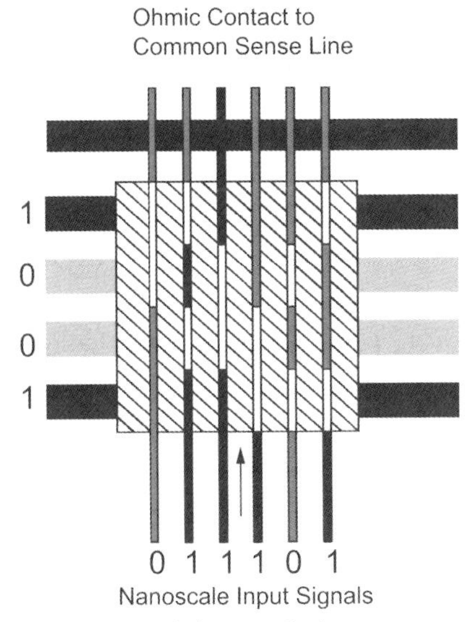

Figure 11.8 Coded NW multiplexer operation.

11.4.3.1 NW Inverter and Buffer

If two NWs are separated by an insulator, perhaps using an oxide core shell (see Section 11.2.1), then the field from one NW can potentially be used to control the other NW. Figure 11.9 shows an inverter which has been built using this basic idea. The horizontal NW serves as the input and the vertical NW as the output. This

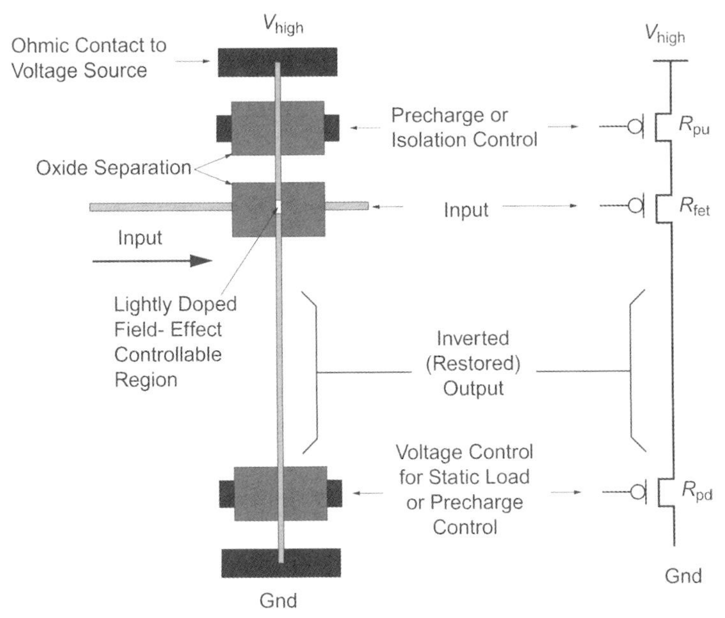

Figure 11.9 NW inverter.

gives a voltage transfer equation:

$$V_{\text{out}} = V_{\text{high}}\left(\frac{R_{\text{pd}}}{R_{\text{pd}} + R_{\text{fet}}(\text{input}) + R_{\text{pu}}}\right) \tag{11.1}$$

For the sake of illustration, the vertical NW has a lightly doped p-type depletion mode region at the input crossing forming a FET controlled by the input voltage (R_{fet}(Input)). Consequently, a low voltage on the input NW will allow conduction through the vertical NW ($R_{\text{fet}} = R_{\text{onfet}}$ is small), and a high input will deplete the carriers from the vertical NW and prevent conduction ($R_{\text{fet}} = R_{\text{offfet}}$ is large). As a result, a low input allows the NW to conduct and pull the output region of the vertical NW up to a high voltage. A high input prevents conduction and the output region remains low. A second crossed region on the NW is used for the pull down (R_{pd}); this region can be used as a gate for predischarge, so the inverter is pulled low before the input is applied, then left high to disconnect the pulldown voltage during evaluation. Alternately, it can be used as a static load for PMOS-like ratioed logic. By swapping the location of the high- and low-power supplies, this same arrangement can be used to buffer rather than invert the input.

Note that the gate only loads the input capacitively, and consequently current isolation is achieved at this inverter or buffer. Further, NW field-effect gating has sufficient non-linearity so that this gate provides gain to restore logic signal levels [27].

11.4.3.2 Ideal Restoration Array

In many scenarios, there is a need to restore a set of tight-pitch NWs such as the outputs of a programmable, wired-OR array. To do this, the approach would be to build a restoration array as shown in Figure 11.10a. This array is a set of crossed NWs

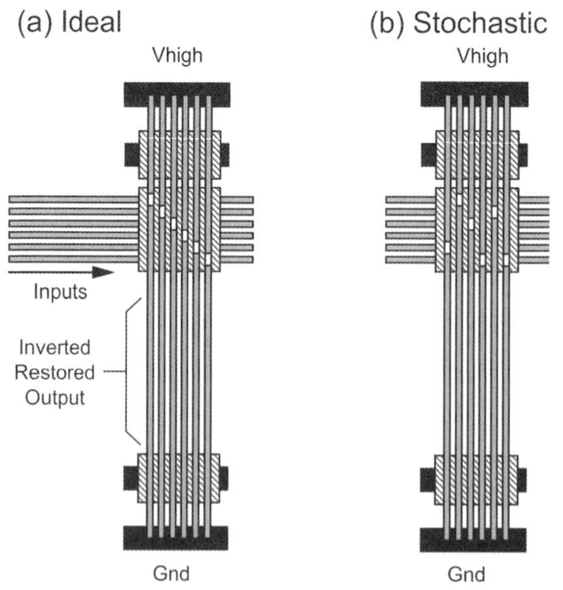

Figure 11.10 Restoration array.

which can be assembled using NW assembly techniques. If each of the NWs was sensitive to all of the crossed inputs, the result would be that all of the outputs would actually compute the NOR of the same set of inputs. To avoid computing a redundant set of NORs and instead simply to invert each of the inputs independently, these NWs are coded using an axial doping or material composition profile. In this way, each NW is field-effect sensitive to only a single NW, and hence provides the NW inversion described for a single one of the crossed NWs and is oblivious to the inputs of the other NWs.

The only problem here is that there is no way to align and place axially doped NWs so that they provide exactly this pattern, as the assembly treats all NWs as identical.

11.4.3.3 **Restoration Array Construction**

Although the region for active FETs is a nanoscale feature, it does not require small pitch or tight alignment. As such, there may be ways to mask and provide material differentiation along a diagonal as required to build this decoder.

Nonetheless, it is also possible to stochastically construct this restoration array in a manner similar to the construction of the address decoder. That is, an assembly is provided with a set of NWs with their restoration regions in various locations. The restoration array will be built by randomly selecting a set of restoration NWs for each array (see Figure 11.10b).

Two points differ compared to the address decoder case.

- The code space will be the same size as the desired restoration population.
- Duplication is allowed.

The question then is how large a fraction of the inputs will be successfully restored for a given number of randomly selected restoration NWs? This is an instance of the Coupon Collector Problem [28]. If the restoration array is populated with the same number of NWs as inputs, the array will typically contain restoration wires for 50–60% of the NW inputs. One way to consider this is that the array must be populated with 1.7- to 2-fold as many wires as would be hoped to yield due to these

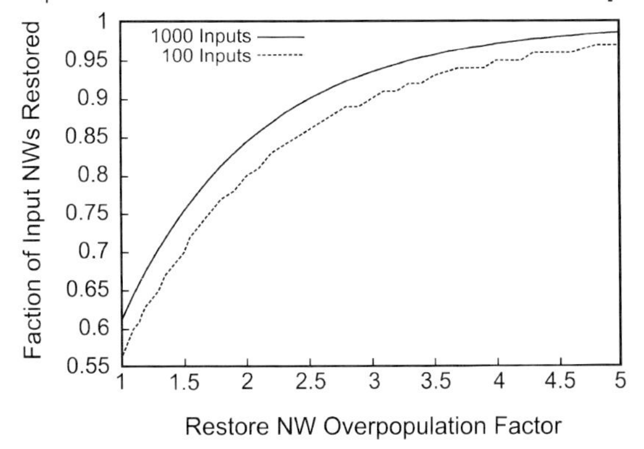

Figure 11.11 Fraction of input NWs restored as a function of restoration overpopulation.

stochastic assembly effects. If the number of restoration wires is increased relative to the number of input NWs, then a higher fraction of the inputs can be restored (as shown in Figure 11.11). For further details on these yield calculations, see Refs. [26, 29].

11.5
Memory Array

By combining the crosspoint memory cores with a pair of decoders, it is possible to build a tight-pitch, NW-based memory array [30]. Figure 11.12 shows how these elements come together in a small memory array, which is formed using crossed, tight-pitch NWs. Programmable diode crosspoints are assembled in the NW–NW crossings, while lithographic-scale address wires form row and column addresses. Write operations into the memory array can be performed by driving the appropriate write voltages onto a single row and column line. Read operations occur by driving a reference voltage onto the common column line, setting the row and column addresses, and sensing the voltage on the common row read line.

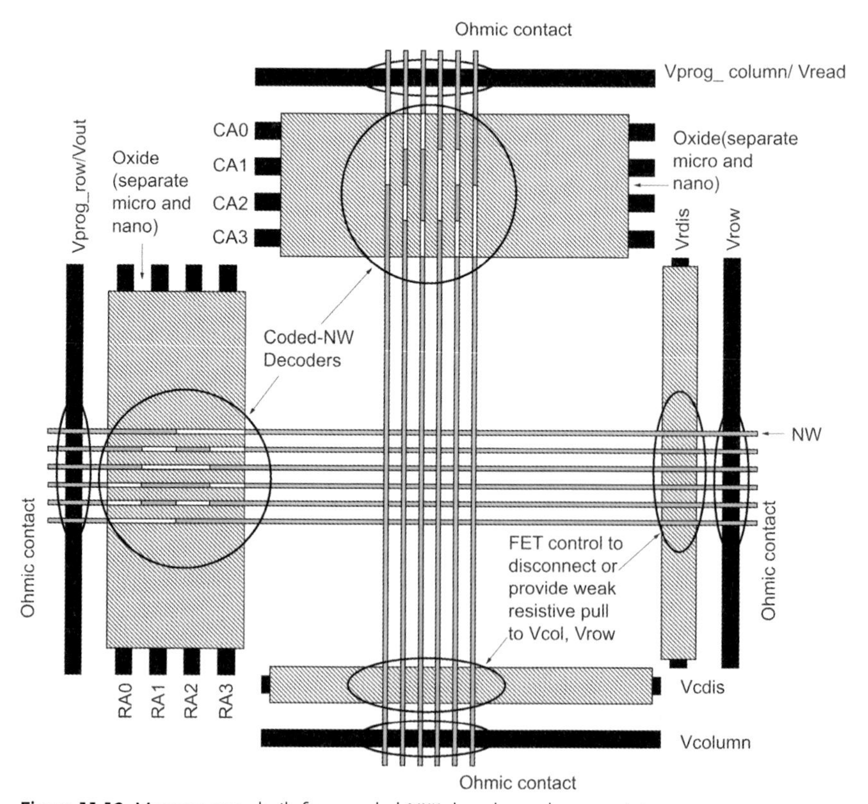

Figure 11.12 Memory array built from coded NW decoder and crosspoint memory core.

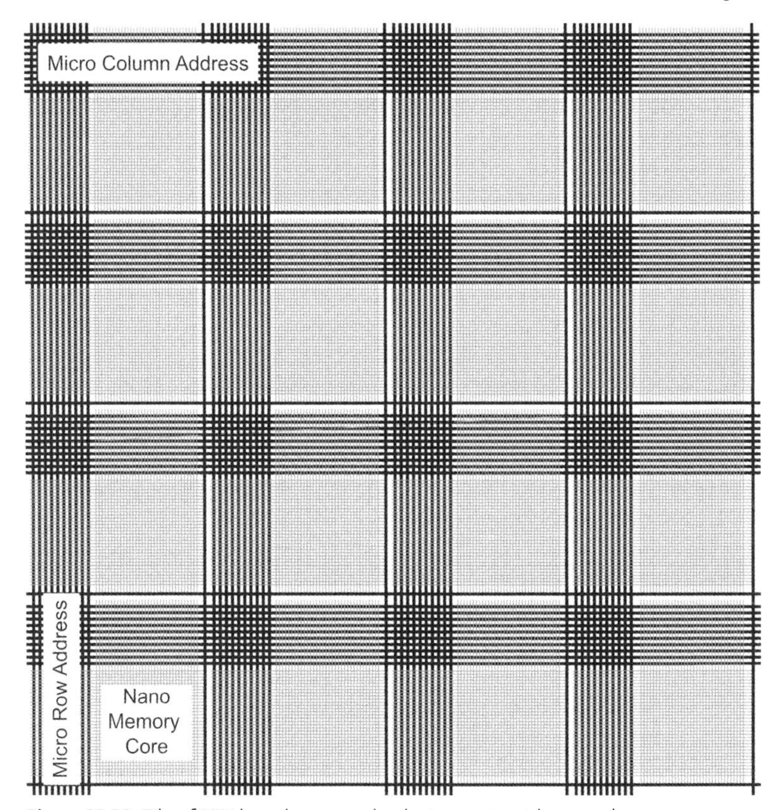

Figure 11.13 Tile of NW-based memory banks to construct large-scale memory.

Limitations on reliable NW length and the capacitance and resistance of long NWs prevent the building of arbitrarily large memory arrays. Instead, the large NW memories are broken up into banks similar to the banking used in conventional DRAMs (see Figure 11.13). Reliable, lithographic-scale wires provide address and control inputs and data inputs and outputs to each of the NW-based memory banks. The expected yield would be only a fraction of the NWs in the array due to wire defects. Error-correcting codes (ECC) can be used to tolerate non-programmable crosspoint defects. After accounting for defects, ECC overhead, and lithographic control overhead, net densities on the order of 10^{11} bits cm^{-2} appear achievable, using NW pitches of about 10 nm [29].

11.6 Logic Architecture

By combining the building blocks introduced in Section 11.4 it is possible to construct complete, programmable logic architectures with all logic, interconnect, and restoration occurring in the atomic-scale NWs. Diode crosspoints organized into Wired-OR logic arrays provide programmable logic, field-effect restoration arrays provide gain and signal inversion, and the NWs themselves provide interconnect among arrays. Lithographic scale wires provide a reliable support infrastructure which allows device testing and programming (see Section 11.8), addressing individual NWs using the decoders introduced in Section 11.4.2. Lithographic-scale wires also provide power and control logic evaluation.

11.6.1 Logic

Figure 11.14 shows a simple PLA created using the building blocks from Section 11.4 and first introduced by DeHon and Wilson [31]. The design includes two interconnected logic planes, each of which is composed of a programmable Wired-OR array, followed by a restoration array. It should be noted here that two restoration arrays are actually used – one providing the inverted sense of the OR-term logic and one providing the non-inverted buffered sense. This arrangement is similar to conventional PLAs where the true and complement sense of each input is provided in each PLA plane. Since Wired-OR logic NWs can be inverted in this nanoPLA, each plane effectively serves as a programmable NOR plane. The combination of the two coupled NOR–NOR planes can be viewed as an AND–OR PLA with suitable application of DeMorgan's laws and signal complementation.

11.6.1.1 **Construction**

The entire construction is simply a set of crossed NWs as allowed by the regular assembly constraints (see Section 11.3.1). Lithographic-scale etches are used to differentiate regions (e.g. programmable-diode regions for the Wired-OR). The topology allows the same NWs that perform logic or restoration to carry their outputs across as inputs to the array that follows it.

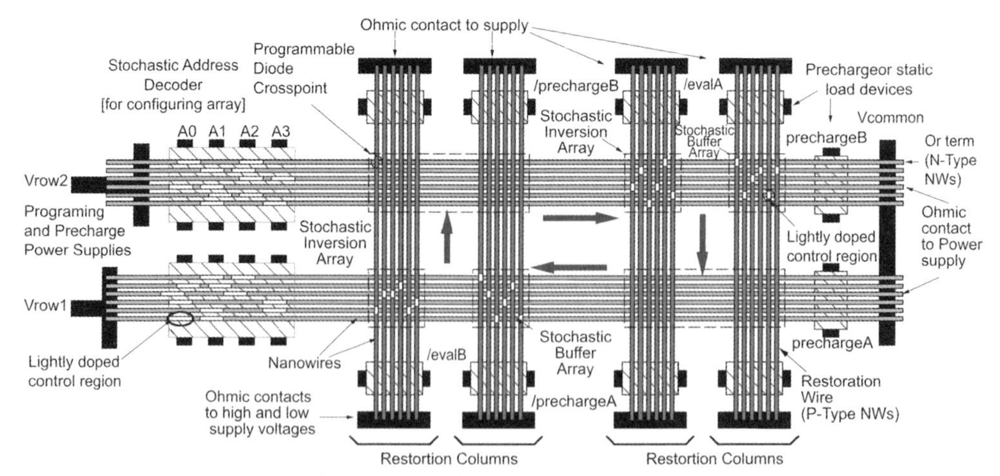

Figure 11.14 Simple nanoPLA block.

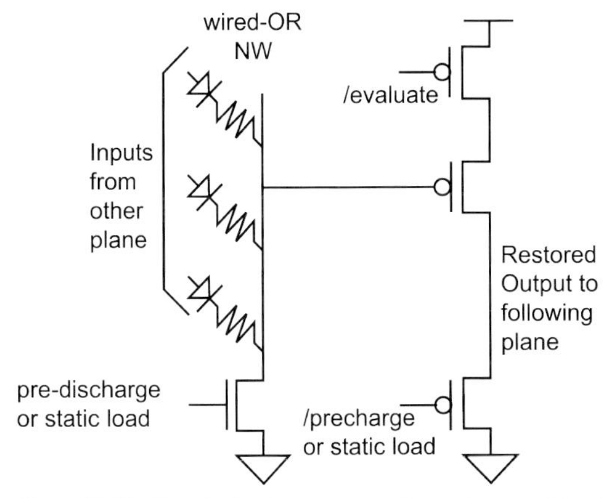

Figure 11.15 Rough circuit equivalent for each nanoPLA plane.

11.6.1.2 **Logic Circuit**

The logic gates in each PLA plane are composed of a diode-programmable Wired-OR NW, followed by a field-effect buffer or inverter NW (see Figure 11.15). The field-effect stage provides isolation as there is no current flow between the diode stage and the field-effect stage output. That is, the entire OR stage is capacitively loaded rather than resistively loaded. The OR stage simply needs to charge up its output which provides the field for the field-effect-based restoration stage. When the field is high enough (low enough for P-type NWs) to enable conduction in the field-effect stage, the NW will allow the source voltage to drive its output.

11.6.1.3 **Programming**

At the left-hand side of Figure 11.14 a decoder is formed (as introduced in Section 11.4.2) using the vertical microscale wires A0 to A3. These lithographic-scale wires allow the selection of individual NWs for programming. Each usable vertical restoration NW is driven by a horizontal NW. Consequently, decoders are only needed to address the horizontal NWs (see Section 11.8).

11.6.2
Registers and Sequential Logic

With slight modification as to how the control signals on the identified logic stages are driven, this can be turned into a clocked logic scheme. An immediate benefit is the ability to create a finite-state machine out of a single pair of PLA planes. A second benefit is the ability to use precharge logic evaluation for inverting restoration stages.

11.6.2.1 **Basic Clocking**

The basic nanoPLA cycle shown in Figure 11.14 is simply two restoring logic stages back-to-back (see Figure 11.16). For the present clocking scheme, the two stages are evaluated at altering times.

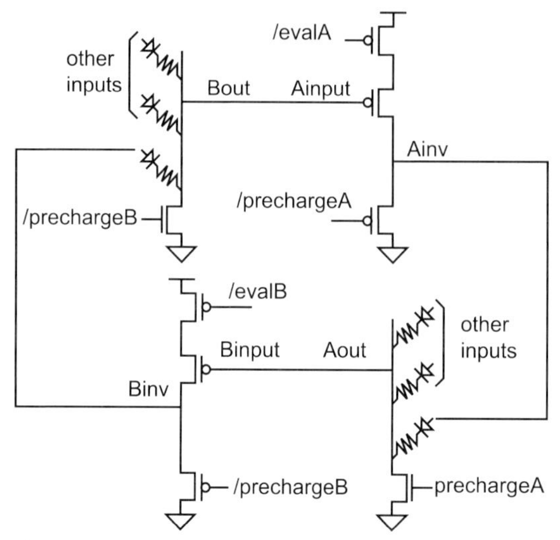

Figure 11.16 Precharge clocked INV-OR-INV-OR (NOR-NOR, AND-OR) cycle.

First, it should be noted that if all three of the control transistors in the restoring stages (restoring precharge and evaluate and diode precharge; e.g. evalA and prechargeA in Figure 11.16) are turned off, there is no current path from the input to the diode output stage. Hence, the input is effectively isolated from the output. As the output stage is capacitively loaded, the output will hold its value. As with any dynamic scheme, eventually leakage on the output will be an issue which will set a lower bound on the clock frequency.

With a stage isolated and holding its output, the following stage can be evaluated. It computes its value from its input, the output of the previous stage, and produces its result by suitably charging its output line. When this is done, this stage can be isolated and the succeeding stage (which in this simple case is also its predecessor) can be evaluated. This is the same strategy as two-phase clocking in conventional VLSI (e.g. Refs. [32, 33]).

In this manner, there is never an open current path all the way around the PLA (see Figures 11.16 and 11.17). In the two phases of operation, there is effectively a single register on any PLA outputs which feed back to PLA inputs.

11.6.2.2 **Precharge Evaluation**

For the inverting stage, the pulldown gate is driven hard during precharge and turned off during evaluation. In this manner, the line (A_{inv}) is precharged low and pulled it up only if the input (A_{input}) is low. This works conveniently in this case because the output will also be precharged low. If the input is high, then there is no need to pullup the output and it is simply left low. If the input is low, the current path is allowed to pullup the output. The net benefit is that inverter pulldown and pullup are both controlled by strongly driven gates and can be fast, whereas in a static logic scheme, the pulldown transistor must be weak, making pulldown slow compared to pullup. Typically, the weak pulldown transistor would be set to have an order of magnitude

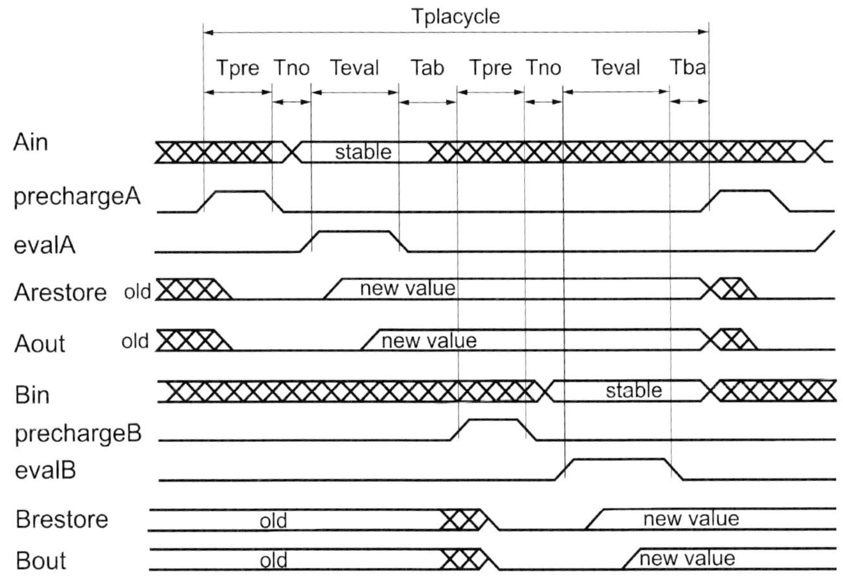

Figure 11.17 Clocking/precharge timing diagram.

higher resistance than the pullup transistor so this can be a significant reduction in worst-case gate evaluation latency.

Unfortunately, in the buffer case the weak pullup resistor can neither be precharged to high nor turned off, and so there are no comparable benefits there. It is possible that new devices or circuit organizations will eventually allow precharge buffer stages to be built.

11.6.3 Interconnect

It is known from VLSI that large PLAs do not always allow the structure which exists in logic to be exploited. For example, an n-input XOR requires an exponential number of product terms to construct in the two-level logic of a single PLA. Further, the limitation on NW length (see Section 11.3.2) bounds the size of the PLAs that can reasonably be built. Consequently, in order to scale up to large-capacity logic devices, modest size nanoPLA blocks must be interconnected; these nanoPLA blocks are extended to include input and output to other nanoPLA blocks and then assembled into a large array (see Figure 11.18), as first introduced by DeHon [34].

11.6.3.1 Basic Idea

The key idea for interconnecting nanoPLA blocks is to overlap the restored output NWs from each such block with the wired-OR input region of adjacent nanoPLA blocks (see Figure 11.18). In turn, this means that each nanoPLA block receives inputs from a number of different nanoPLA blocks. With multiple input sources and outputs routed in multiple directions, this allows the nanoPLA block also to serve as a

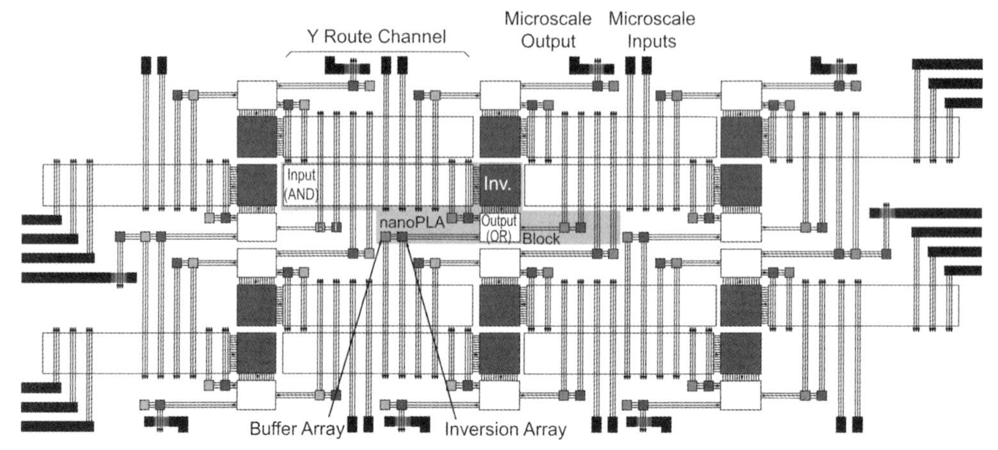

Figure 11.18 nanoPLA block tiling with edge IO to lithographic scale.

switching block. By arranging the overlap appropriately, Manhattan routing can be supported, thereby allowing the array of nanoPLA blocks to be configured to route signals between any of the blocks in the array.

11.6.3.2 NanoPLA Block

- *Input wired-OR region.* One or more regions of programmable crosspoints serves as the input to the nanoPLA block. Figures 11.18 and 11.19 show a nanoPLA block design with a single such input region. The inputs to this region are restored output NWs from a number of different nanoPLA blocks. The programmable crosspoints

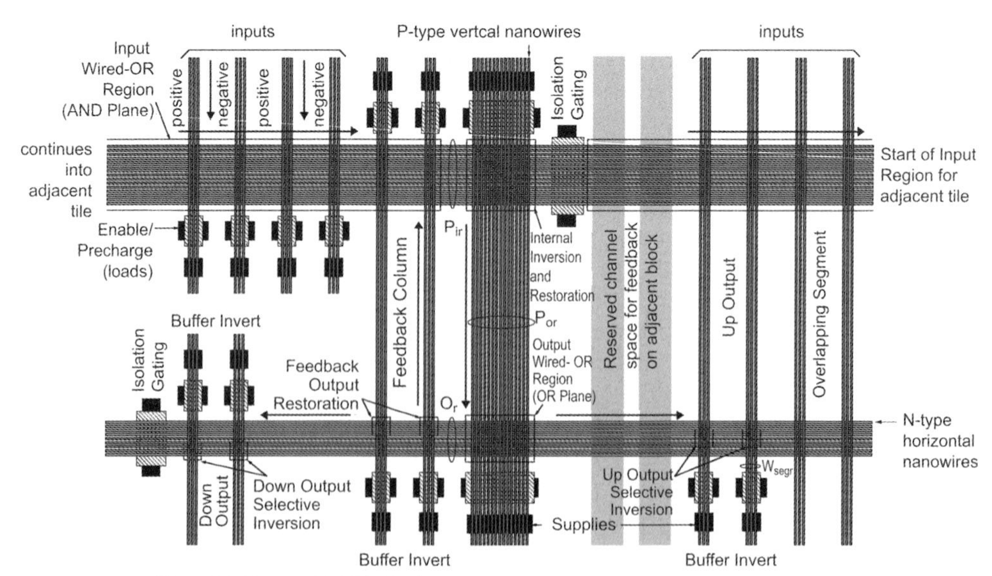

Figure 11.19 nanoPLA block tile.

allow those inputs to be selected which participate in each logical product term (PTERM) building a wired-OR array, as in the base nanoPLA (see Section 11.6.1).

- *Internal inversion and restoration array.* The NW outputs from the input block are restored by a restoration array. The restoration logic is arranged at this stage to be inverting, thus providing the logical NOR of the selected input signals into the second plane of the nanoPLA.
- *Output OR plane.* The restored outputs from the internal inversion plane become inputs to a second programmable crosspoint region. Physically, this region is the same as the input plane. Each NW in this plane computes the wired-OR of one or more of the restored PTERMs computed by the input plane.
- *Selective output inversion.* The outputs of the output OR plane are then restored in the same way as the internal restoration plane. On this output, however, the selective inversion scheme introduced in Section 11.6.1 is used. This provides both polarities of each output, and these can then be provided to the succeeding input planes. This selective inversion plays the same role as a local inverter on the inputs of conventional, VLSI PLA; here it is placed with the output to avoid introducing an additional logic plane into the design. As with the nanoPLA block, these two planes provide NOR–NOR logic. With suitable application of DeMorgan's laws, these can be viewed as a conventional AND–OR PLA.
- *Feedback.* As shown in Figures 11.18 and 11.19, one set of outputs from each nanoPLA block feeds back to its own input region. This completes a PLA cycle similar to the nanoPLA design (see Section 11.6.1). These feedback paths serve the role of intracluster routing similar to internal feedback in conventional Island-style [35] FPGAs. The nanoPLA block implements registers by routing signals around the feedback path (Section 11.6.2.1). The signals can be routed around this feedback path multiple times to form long register delay chains for data retiming.

11.6.3.3 **Interconnect**

- *Block outputs.* In addition to self feedback, output groups are placed on either side of the nanoPLA block and can be arranged so they cross input blocks of nanoPLA blocks above or below the source nanoPLA block (see Figure 11.18). Like segmented FPGAs [36, 37], output groups can run across multiple nanoPLA block inputs (i.e. Connection Boxes) in a given direction. The nanoPLA block shown in Figure 11.19 has a single output group on each side, one routing up and the other routing down. It will be seen that the design shown is sufficient to construct a minimally complete topology.
- Since the output NWs are directly the outputs of gated fields: (i) an output wire can be driven from only one source; and (ii) it can only drive in one direction. Consequently, unlike segmented FPGA wire runs, directional wires must be present that are dedicated to a single producer. If multiple control regions were coded into the NW runs, conduction would be the AND of the producers crossing the coded regions. Single direction drive arises from the fact that one side of the

gate must be the source logic signal being gated so the logical output is only available on the opposite side of the controllable region. Interestingly, the results of recent studies have suggested that conventional, VLSI-based FPGA designs also benefit from directional wires [38].

- *Y route channels.* With each nanoPLA block producing output groups which run one or more nanoPLA block heights above or below the array, the result is vertical (Y) routing channels between the logic cores of the nanoPLA blocks (see Figure 11.18). The segmented, NW output groups allow a signal to pass a number of nanoPLA blocks. For longer routes, the signal may be switched and rebuffered through a nanoPLA block (see Figure 11.20). Because of the output directionality, the result is separate sets of wires for routing up and routing down in each channel.
- *X routing.* While Y route channels are immediately obvious in Figure 11.18, the X route channels are less apparent. All X routing occurs through the nanoPLA block. As shown in Figure 11.19, one output group is placed on the opposite side of the nanoPLA block from the input. In this way, it is possible to route in the X direction by going through a logic block and configuring the signal to drive a NW in the

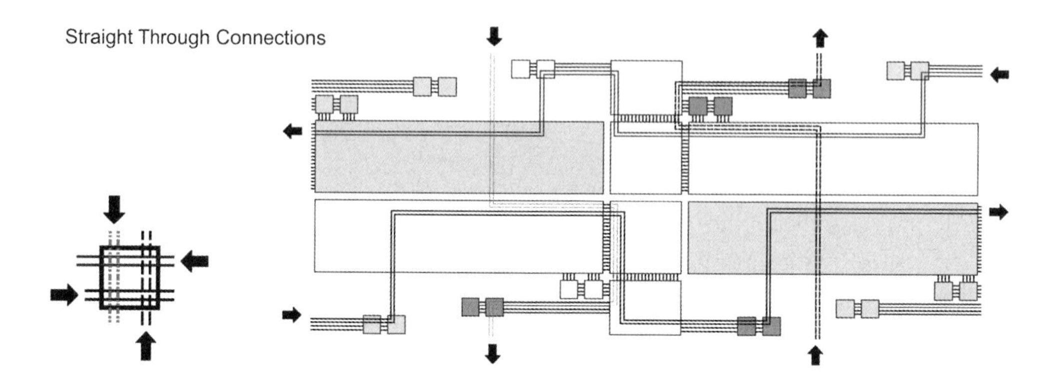

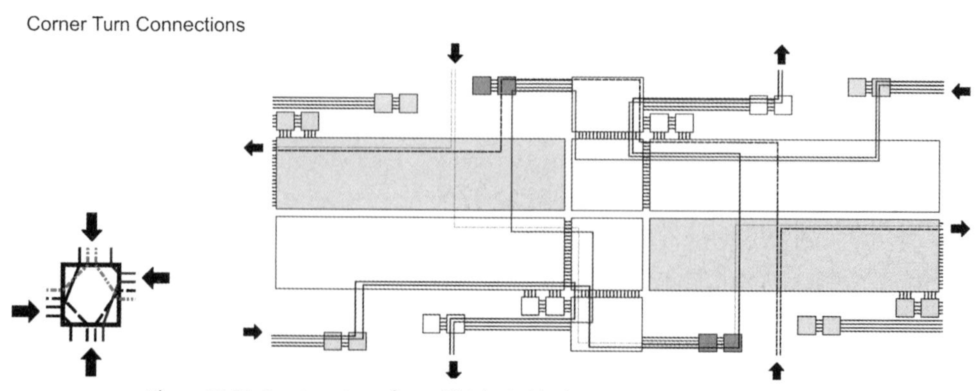

Figure 11.20 Routing view of nanoPLA logic block.

output group on the opposite side of the input. If all X routing blocks had their inputs on the left, then it would be possible only to route from left to right. To allow both left-to-right and right-to-left routing, the orientation of the inputs is alternated in alternate rows of the nanoPLA array (see Figures 11.18 and 11.20). In this manner, even rows provide left-to-right routing, while odd rows allow right-to-left routing.

- *Relation to Island-style Manhattan design.* Logically viewed, this interconnected nanoPLA block is very similar to conventional, Island-style FPGA designs, especially when the Island-style designs use directional routing [38]. As shown in Figure 11.20, there are X and Y routing channels, with switching to provide X–X, Y–Y, and X–Y routing.

11.6.4 CMOS IO

These nanoPLAs will be built on top of a lithographic substrate. The lithographic circuitry and wiring provides a reliable structure from which to probe the NWs to map their defects and to configure the logic (see Section 11.8).

For input and output to the lithographic scale during operation, IO blocks can be provided to connect the nanoscale logic to lithographic-scale wires, in much the same way that lithographic-scale wires are connected to bond pads on FPGAs. The simplest arrangement resembles the traditional, edge IO form of a symmetric FPGA with inputs and outputs attached to NWs at the edges of the routing channels (see Figure 11.18).

NW inputs can easily be driven directly by lithographic-scale wires. As the lithographic-scale wires are wider pitch, a single lithographic wire will connect to a number of NWs. With the lithographic wire connected to the NWs, the NW crosspoints in the nanoPLA block inputs can be programmed in the same way they are for NW inputs.

It is possible to connect outputs in a similar manner. Such a direct arrangement could be particularly slow, as the small NWs must drive the capacitance of a large, lithographic-scale wire. Alternately, the NWs can be used as gates on a lithographic-scale field-effect transistor (FET) (see Figure 11.21). In this manner, the NWs are only

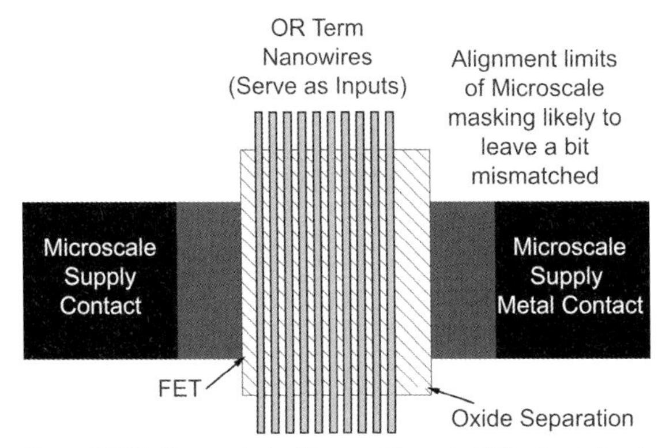

Figure 11.21 Nanoscale to lithographic-scale FET output structure.

loaded capacitively by the lithographic-scale output, and only for a short distance. The NW thresholds and lithographic FET thresholds can be tuned into comparable voltage regions so that the NWs can drive the lithographic FET at adequate voltages for switching. As shown, multiple NWs will cross the lithographic-scale gate. The OR-terms driving these outputs are all programmed identically, allowing the multiple-gate configuration to provide strong switching for the lithographic-scale FET.

11.6.5 Parameters

The key parameters in the design of the nanoPLA block are shown in Figure 11.22, where:

- W_{seg} is the number of NWs in each output group.
- L_{seg} is the number of nanoPLA block heights up or down which each output crosses; equivalently, the number of parallel wire groups across each Y route channel in each direction. In Figure 11.1 $L_{seg} = 2$, and this is maintained throughout the chapter.
- F is the number of NWs in the feedback group; for simplicity, $F = W_{seg}$ is maintained throughout the chapter.
- P is the number of logical PTERMs in the input (AND) plane of the nanoPLA logic block.
- O_p is the number of totals outputs in the OR plane. As each output is driven by a separate wired-OR NW, $O_p = 2 \times W_{seg} + F$ for the nanoPLA block focused on in this chapter, with two routing output groups and a feedback output group.
- P_p is the number of total PTERMs in the input (AND) plane. As these are also used for route-through connections, this is larger than the number of logical PTERMs in each logic block.

$$P_p \leq P + 2 \times W_{seg} + F \tag{11.2}$$

That is, in addition to the P logical PTERMs, one physical wire may be needed for each signal that routes through the array for buffering; there will be at most O_p of these.

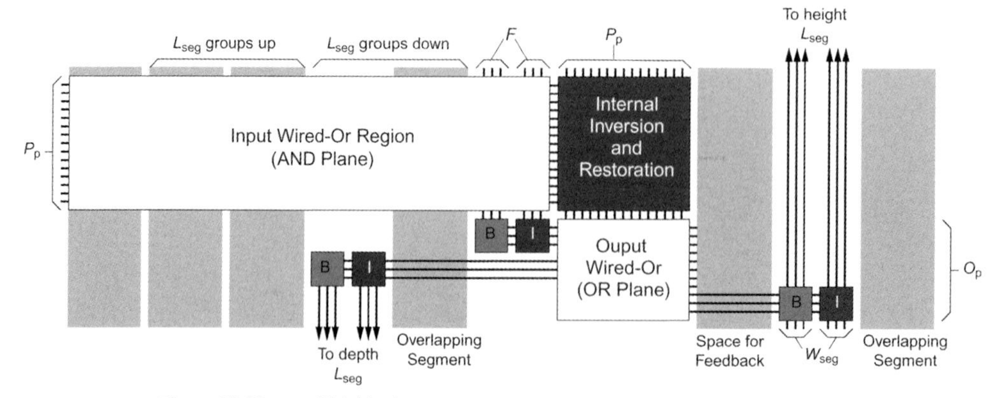

Figure 11.22 nanoPLA block parameters.

Additionally, the number and distribution of inputs [e.g. one side (as shown in Figure 11.22), from both sides, subsets of PTERMs from each side], the output topology (e.g. route both up and down on each side of the array), and segment length distributions could be parameterized. However, in this chapter attention is focused on this simple topology, with $L_{seg} = 2$. Consequently, the main physical parameters determining nanoPLA array size are W_{seg} and P_p.

11.7 Defect Tolerance

As noted in Section 11.3.3, it is likely that a small percentage of wires are defective and crosspoints are non-programmable. Furthermore, stochastic assembly (see Sections 11.4.2.2 and 11.4.3.3) and misalignment will also result in a percentage of NWs which are unusable. Fortunately, NWs are interchangeable and the crosspoints are small. Consequently, spare NWs can be provisioned into an array (e.g. overpopulate compared to the desired P_p and W_{seg}), NWs can be tested for usability (see Section 11.8.1), and the array configured using only the non-defective NWs. Further, a NW need not have a perfect set of junctions to be usable (see Section 11.7.4).

11.7.1 NW Sparing

Tolerating wire defects is a simple matter of provisioning adequate spares, separating the good wires from the bad, and configuring the nanoPLA blocks accordingly. For a given PLA design, each block should have a minimum number of usable wires (P_p and W_{seg}). As there will then be wire losses, the physical array is designed to include a larger number of physical wires to ensure that the yield of usable wires is sufficient to meet the logical requirements.

Using the restoration scheme described in Section 11.4.3, wires work in pairs. A horizontal OR-term wire provides the programmable computation or programmable interconnect, and a vertical restoration wire provides signal restoration and perhaps inversion. A defect in either wire will result in an unusable pair. Consequently, each logical OR-term or output will yield only when both wires yield. Let P_{wire} be the probability that a wire is not defective; then, the probability of yielding each OR-term is:

$$P_{OR} = (P_{input\text{-}wire} \times P_{restore\text{-}wire}). \tag{11.3}$$

An M-choose-N calculation can then be performed to determine the number of wires that must physically populate (N) to achieve a given number of functional wires (M) in the array. The probability of yielding exactly i restored OR-terms is:

$$P_{yield}(N, i) = \left(\binom{N}{i} (P_{OR})^i (1 - P_{OR})^{N-i} \right). \tag{11.4}$$

That is, there are $\binom{N}{i}$ ways to select i functional OR-terms from N total wires, and the yield probability of each case is: $(P_{OR})^i(1 - P_{OR})^{N-i}$. An ensemble is yielded with M items whenever M or more items yield, so the system yield is actually the cumulative distribution function:

$$P_{\text{M of N}} = \sum_{M \leq i \leq N} \left(\binom{N}{i} (P_{OR})^i (1 - P_{OR})^{N-i} \right) \qquad (11.5)$$

Given the desired probability for yielding at least M functional OR-terms, $P_{\text{M of N}}$, Eq. (11.5) provides a way of finding the number of physical wires, N, that must be populated to achieve this. For the interconnected nanoPLA blocks, the product terms (P_p) and interconnect wires (W_{seg}) will be the Ms in Eq. (11.5), and a corresponding pair of raw numbers N will be calculated to determine the number of physical wires that must be placed in the fabricated nanoPLA block. Here, P_r will be used to refer to the number of raw product term NWs needed to assemble, and W_{segr} to the number of raw interconnect NWs. Figure 11.23 illustrates how much larger N needs to be than $M = 100$ for various defect rates and yield targets.

11.7.2
NW Defect Modeling

A NW could fail to be usable for several reasons:

- The NW may make poor electrical contact to microwires on either end (let P_c be the probability the NW makes a good connection on one end).
- The NW may be broken along its length (let P_j be the probability that there is no break in a NW in a segment of length L_{unit}).
- The NW may be poorly aligned with address region (wired-OR NWs) or restoration region (restoration NW) (let P_{ctrl} be the probability that a the NW is aligned adequately for use).

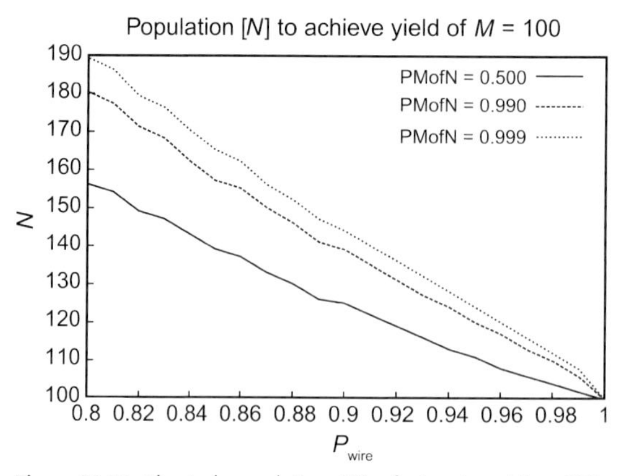

Figure 11.23 Physical population (N) of wires to achieve 100 restored OR-terms (M).

Consequently, the base NW yield looks like:

$$P_{\text{wire}} = (P_{\text{C}})^2 \times (P_{\text{j}})^{L_{\text{wire}}/L_{\text{unit}}} \times P_{\text{ctrl}} \qquad (11.6)$$

Typically, $P_c = 0.95$ (after Ref. [5] and $P_j = 0.9999$ with $L_{unit} = 10$ nm (after Ref. [19]; see also Refs. [27, 34]). P_{ctrl} can be calculated from the geometry of the doped regions [24]. P_{wire} is typically about 0.8.

11.7.3 Net NW Yield Calculation

A detailed calculation for NW population includes both wire defect effects and stochastic population effects. Starting with a raw population number for the NWs in each piece of the array, it is possible to:

- calculate the number of non-defective wired-OR wires within the confidence bound [Eqs. (11.6) and (11.5)];
- calculate the number of those which can be uniquely addressed using the following recurrence:

$$\begin{aligned} P_{\text{different}}(T, N, u) = & \left(\frac{T-(u-1)}{T}\right) \times P_{\text{different}}(T, N-1, u-1) \\ & + \left(\frac{u}{T}\right) \times P_{\text{different}}(T, N-1, u) \end{aligned} \qquad (11.7)$$

where T is the number of different wire types (i.e. the size of the address space), N is the raw number of nanowires populated in the array, and u is the number of unique NWs in the array.

- calculate the number of net non-defective restored wire pairs within the confidence bound [Eqs. (11.3), (11.6), and (11.5)];
- calculate the number of uniquely restored OR terms using Eq. (11.7); in this case, T is the number of possible restoration wires rather than the number of different NW addresses.

These calculations indicate how to obtain P_r and W_{segr} to achieve a target P_p and W_{seg}.

11.7.4 Tolerating Non-Programmable Crosspoints

As will be seen in Table 11.1, PLA crosspoint arrays are typically built with approximately 100 net junctions. If were demanded that all 100 crosspoint junctions on a NW were programmable in order for the NW to yield, then an unreasonably high yield rate per crosspoint would be required. That is, assuming a crosspoint is programmable with probability P_{pgm} and a NW has N_{junc} input NWs – and hence crosspoint junctions

Table 11.1 Area minimizing nanoPLA design points (Ideal Restoration, with $W_{litho} = 105$ nm, $W_{fnano} = W_{dnano} = 10$ nm); area ratios estimate how much larger 22 nm lithographic FPGAs would be compared to the mapped nanoPLA designs.

Design	P_p	W_{seg}	Area ratio
alu4	60	8	340
apex2	54	15	39
apex4	62	7	210
bigkey	44	13	69
clma	104	28	30
des	78	25	26
diffeq	86	21	32
deip	58	18	59
elliptic	78	27	27
ex1010	66	9	290
ex5p	67	18	390
frisc	92	34	17
misex3	64	8	150
pdc	74	13	360
s298	79	15	110
s38417	76	22	32
seq	72	18	69
spla	68	12	630
tseng	78	25	20

– then the probability that all junctions on a NW are programmable is

$$P_{\text{pgmwire}} = (P_{\text{pgm}})^{N_{\text{junc}}} \tag{11.8}$$

To have $P_{\text{pgmwire}} \geq 0.5$, P_{pgm} would need to be >0.993. However, as was noted in Section 11.3.3, the non-programmable crosspoint defect rates would be expected to be in the range of 1 to 10% ($0.9 \leq P_{\text{pgm}} \leq 0.99$).

As introduced by Naeimi and DeHon [39], it is apparent that a NW with non-programmable crosspoints can still be used to implement a particular OR-term, as long as it has programmable crosspoints where the OR-term needs on-programmed junctions. Furthermore, as the array has a large number of otherwise interchangeable NWs (e.g. 100), it is possible to search through the array for NWs that can implement each particular OR-term.

For example, if a logic array (AND or OR plane) of a nanoPLA has defective junctions (as marked in Figure 11.24), the OR-term $f = A + B + C + E$ can be assigned to NW $W3$, despite the fact that it has a defective (non-programmable) junction at ($W3$, D); that is, the OR-term f is compatible with the defect pattern of NW $W3$.

As the number of programmed junctions needed for a given OR-term is usually small (e.g. 8–20) compared to the number of inputs in an array (e.g. 100), the probability that a NW can support a given OR-term is much larger than the probability that it has no junction defects. Assuming that C is the fan-in to the OR-term, and assuming random

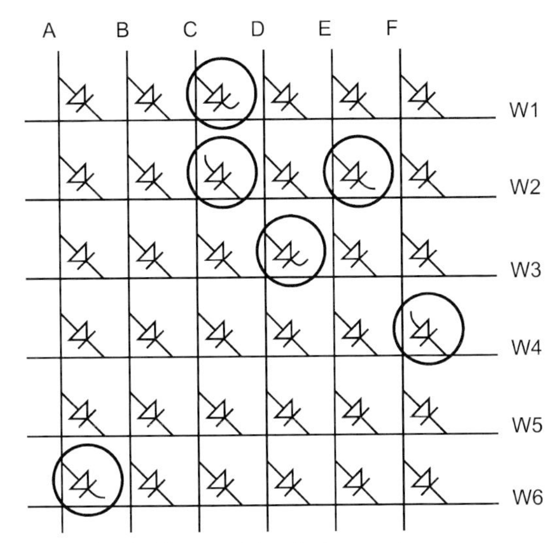

Figure 11.24 OR array with defective junctions.

junction defects, the probability that the NW can support the OR-term is

$$P_{\text{support}}(C) = (P_{\text{pgm}})^{C}. \tag{11.9}$$

For example, in a 100 NW array, if $P_{\text{pgm}} = 0.95$, $P_{\text{support}}(13) \approx 0.51$, and $P_{\text{pgmwire}} \approx 0.006$. Furthermore, as multiple NWs can be used in an array to find a compatible match, failure to map a NW will only occur if there are no compatible NWs in the array.

$$P_{\text{match}} = (C,\ N_{\text{wire}}) = (1 - (1 - P_{\text{support}}(C)^{N_{\text{wire}}}). \tag{11.10}$$

Hence, the probability of failing to find a match for the $C = 13$ OR-term in a 100 NW array is $[1 - P_{\text{match}}(13, 100) \leq 10^{-31}]$. Alternately, this means we have a 99% chance of finding a match after checking only 8 NWs ($P_{\text{match}}(13, 8) > 0.99$).

Naeimi and DeHon [39] developed the analysis and mapping strategy in greater detail for tolerating non-programmable crosspoints. DeHon and Naeimi [30] further expanded the mapping strategy to the interconnected nanoPLAs described in Section 11.6.3, and showed that non-programmable defect rates of up to 5% could be accommodated, with no additional overhead.

11.8 Bootstrap Testing

11.8.1 Discovery

Since addressing and restoration is stochastic, there is a need to discover the live addresses and their restoration polarity. Further, as the NWs will be defective it is vital to identify those NWs which are usable and those which are not. Here, the restoration columns (see Figures 11.14 and 11.19) are used to help identify useful addresses. The gate side supply (e.g. the top set of lithographic wire contacts in Figure 11.10) can be driven to a high value, after which a voltage is sought on the opposite supply line (e.g. the bottom set of lithographic wire contacts in Figure 11.10; these contacts are marked V_{high} and Gnd in Figure 11.10, but will be controlled independently as described here during discovery). There will be current flow into the bottom supply only if the control associated with the p-type restoration wire can be driven to a sufficiently low voltage. The process is started by driving all the row lines high, using the row precharge path. A test address is then applied and the supply (V_{row} in Figure 11.14) is driven low. If a NW with the test address is present, only that line will now be strongly pulled low. If the associated row line can control one or more wires in the restoration plane, the selected wires will now see a low voltage on their field-effect control regions and enable conduction from the top supply to the bottom supply. By sensing the voltage change on the bottom supply, the presence of a restored address can be deduced. Broken NWs will not be able to effect the bottom supply. NWs with excessively high resistance due to doping variations or poor contacts will not be able to pull the bottom supply contact up quickly enough. As the buffering and inverting column supplies are sensed separately it will be known whether the line is buffering, inverting, or binate.

No more than $O((P_p)^2)$ unique addresses are needed to achieve virtually unique row addressing [24], so the search will require at most $O((P_p)^2)$ such probes. A typical address width for the nanoPLA blocks is $N_a = 14$, which provides 3432 distinct 7-hot codes, and a typical number of OR-terms might be 90 (see Table 11.1). Hence, 3432 addresses may need to be probed to find 90 live row wires.

When all the present addresses in an array and the restoration status associated with each address are known, logic can be assigned to logical addresses within each plane, based on the required restoration for the output. With logic assigned to live addresses in each row, the address of the producing and consuming row wires can now be used to select and program a single junction in a diode-programmable OR plane.

11.8.2 Programming

In order to program any diode crosspoint in the OR planes (e.g. Figure 11.14), one address is driven into the top address decoder, and the second address into the bottom. The stochastic restoration performs the corner turn, so that the desired programming voltage differential is effectively placed across a single crosspoint. The voltages and control gating on the restoration columns are then set to define which programmable diode array is actually programmed during a programming operation. For example, in Figure 11.14 the ohmic supply contacts at the top and bottom are the control voltages; the signals used for control gating are labeled with precharge and eval signal names. To illustrate the discovery and programming process, DeHon [29] presents the steps involved in discovering and programming an exemplary PLA.

11.8.3
Scaling

It should be noted that each nanoPLA array is addressed separately from its set of microscale wires ($A0$, $A1$, ... and V_{row}, V_{bot}, and V_{top}; see Figure 11.14). Consequently, the programming task is localized to each nanoPLA plane, and the work required to program a collection of planes (e.g. Figure 11.18) only scales linearly with the number of planes.

11.9
Area, Delay, and Energy

11.9.1
Area

From Figures 11.19 and 11.22 the basic area composition of each tile can be seen. For this, the following feature size parameters are used:

- W_{litho} is the lithographic interconnect pitch; for example, for the 45-nm node, $W_{litho} = 105$ nm [40].
- W_{dnano} is the NW pitch for NWs which are inputs to diodes (i.e. Y route channel segments and restored PTERM outputs).
- W_{fnano} is the NW pitch for NWs which are inputs to field-effect gated NWs; this may be larger than W_{dnano} in order to prevent inputs from activating adjacent gates and to avoid short-channel FET limitations.

The tile area is computed by first determining the tile width, TW, and tile height, TH:

$$TW = (3 + 4(L_{seg} + 1)) \times W_{litho} + (P_{or} + 4(L_{seg} + 1)W_{segr}) \times W_{dnano} \quad (11.11)$$

$$TH = 12 \times W_{litho} + (O_r + P_{ir}) \times W_{fnano} \quad (11.12)$$

$$AW = (N_a + 2) \times W_{litho} \quad (11.13)$$

$$Area = (AW + TW) \times TH \quad (11.14)$$

where P_{or}, P_{ir}, O_r, and W_{segr} (shown in Figure 11.19) are the raw number of wires needed to populate in the array in order to yield P_p restored inputs, O_p restored outputs, and W_{seg} routing channels (see Section 11.7.3). The two 4s in TW arise from the fact that there are $L_{seg} + 1$ wire groups on each side of the array ($2\times$), and each of those is composed of a buffer/inverter selective inversion pair ($2\times$). A lithographic spacing is charged for each of these groups as they must be etched for isolation and controlled independently by lithographic scale wires. The 12 lithographic pitches in TH account for the three lithographic pitches needed on each side of a group of wires for the restoration supply and enable gating. As segmented wire runs end and begin between the input and output horizontal wire runs, these three lithographic pitches are paid for four-fold in the height of a single nanoPLA block: once at the bottom top of the block (see Figure 11.19).

N_a is the number of microscale address wires needed to address individual, horizontal nanoscale wires [24]; for the nanoPLA blocks in these studies, N_a is typically 14 to 20. Two extra wire pitches in the AddressWidth (AW) are the two power supply contacts at either end of an address run.

11.9.2
Delay

Figures 11.16 and 11.17 show the basic nanoPLA clock cycle, $T_{placycle}$. The component delays shown in Figure 11.17 (e.g. nanowire precharge and evaluation times) are calculated based on the NW resistances and capacitances, the crosspoint resistances, and the nanowire FET resistances [29]. NW resistance and capacitance can be calculated based on geometry and material properties using the NW lengths, which are roughly multiples of the tile width, TW, and tile height, TH, identified in the previous section. If simply heavily doped silicon nanowires are used, the NW resistances can be close to 10 MΩ, and this results in nanoPLA clock cycle times in the tens of nanoseconds. However, if the regions of the NW which do not need to be semiconducting are converted selectively – that is, everything except the diode crosspoint region and the field-effect restoration region – into a nickel silicide (NiSi) [12], the NW resistances can be reduced to the 1 MΩ range. As a result, the nanoPLA clock cycle is brought down to the nanosecond region. This selective conversion can be performed as a lithographic-scale masking step and, with careful engineering, subnanosecond nanoPLA cycle times may be possible. As long as the NW resistance is in the 1 MΩ range, it will dominate the on-resistance of both the field-effect gating in the restoration NW (R_{onfet}) and diode on-resistances ($R_{ondiode}$) in the 100 KΩ range.

11.9.3
Energy and Power

The nanoPLAs will dissipate active energy, charging and discharging the functional and configured NWs.

$$E_{NW} = \frac{1}{2} C_{Wire} V^2. \tag{11.15}$$

As noted in the previous section, C_{wire} can be computed from the material properties and geometry. To tolerate variations in NW doping, it is likely the operating voltage will need to be 0.5 to 1 V.

The raw E_{NW} can be discounted by the fraction of NWs typically used in a routing channel or a logic array, F. This tends to be 70–80% with the current tools and designs. When using the selective inversion scheme, both polarities of most signals will typically be driven to guarantee a close to 50% activity factor, A.

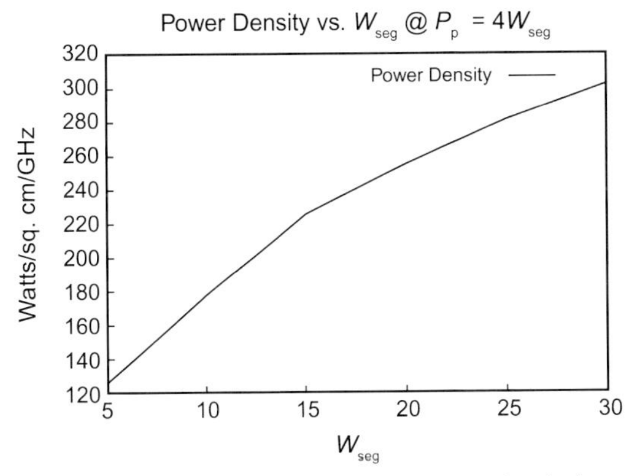

Figure 11.25 Power density as a function of W_{seg} for ideal restore, stochastic address case with $W_{litho} = 105$ nm, $W_{fnano} = 10$ nm, $W_{dnano} = 10$ nm.

Assuming an operating frequency of f, the power for a nanoPLA tile is

$$P_{array} = \sum_{\text{all NWs}} (A \times F \times E_{NW} \times f). \tag{11.16}$$

The power density is then

$$P_{density} = \frac{P_{array}}{Area}. \tag{11.17}$$

Here, *Area is* the area for the tile as calculated in Eq. (11.14).

Figure 11.25 shows the power density associated with interconnected nanoPLAs, and suggests that the designs may dissipate a few hundred Watts per cm^2. In typical designs, compute arrays would be interleaved with memory banks (see Section 11.5), which have much lower power densities. Nonetheless, this suggests that power management is as much an issue in these designs as it is in traditional, lithographic, designs.

11.10 Net Area Density

Recent developments in technology suggest that it is possible to build and assemble 10 nm-pitch NWs with crosspoints at every NW–NW crossing. To use these, it is necessary to pay for lithographic addressing overhead, to use regular architectures, and tolerate defects. In order to understand the net benefits, the characteristics of composite designs are analyzed. As an example, conventional FPGA benchmarks are mapped from the Toronto 20 benchmark suite [1] to NW logic with $W_{litho} = 105$ nm (45 nm roadmap node) and $W_{fnano} = W_{dnano} = 10$ nm. This provides a count of nanoPLA blocks and the logical P_p and W_{seg} parameters identified in Section 11.6.5, and these calculations can then be used for yield and statistical assembly (see Section 11.7) to compute physical nanowire population, and the area equations in Section 11.9.1 to compute composite area. Subsequently, the resultant minimum area obtainable is compared with the nanoPLA designs to lithographic 4-LUT FPGAs at the 22 nm node [40]. As shown in Table 11.1, and further detailed in Ref. [29], the routed nanoPLA designs are one to two orders of magnitude smaller than 22 nm lithographic FPGAs, even after accounting for lithographic addressing overhead, defects, and statistical addressing.

The datapoints in Table 11.1 are based on a number of assumptions about lithographic and nanowire pitches and statistical assembly. DeHon [29] also examined the sensitivity to these various parameters, and showed that the statistical restoration assembly costs a factor of three in density for large arrays, while the cost of statistical addressing is negligible. If the diode pitch (W_{dnano}) could be reduced to 5 nm, another factor of almost two in area could be saved. Moreover, if the lithographic support were also reduced to the 22 nm node ($W_{litho} = 45$ nm), a further three-fold factor in density advantage would be gained compared to the data in Table 11.1.

11.11 Alternate Approaches

During recent years, several groups have been studying variants of these nanowire-based architectures (see Table 11.2). Heath *et al.* [41] articulated the first vision for constructing defect-tolerant architectures based on molecular switching and bottom-up construction. Luo *et al.* [42] elaborated the molecular details and diode-logic structure, while Williams and Kuekes [23] introduced a random particle decoder scheme for addressing individual NWs from lithographic-scale wires. These early designs assumed that diode logic was restored and inverted using lithographic scale CMOS buffers and inverters.

Goldstein and Budiu [43] described an interconnected set of these chemically-assembled diode-based devices, while Goldstein and Rosewater [44] used only two-terminal non-restoring devices in the array, but added latches based on resonant-tunneling diodes (RTDs) for clocking and restoration. Snider *et al.* [45] suggested nanoFET-based logic and also tolerated non-programmable crosspoint defects by matching logic to the programmability of the device.

Strukov and Likharev [46] also explored crosspoint-programmable nanowire-based programmable logic and used lithographic-scale buffers with an angled topology and nanovias so that each long NW could be directly attached to a CMOS-scale buffer. Later, Snider and Williams [47] built on the Strukov and Likharev interfacing concept and introduced a more modest design which used NWs and molecular-scale switches only for interconnect, performing all logic in CMOS.

These designs all share many high-level goals and strategies, as have been described in this chapter. They suggest a variety of solutions to the individual technical components including the crosspoint technologies, NW formation, lithographic-scale interfacing, and restoration (see Table 11.2). The wealth of technologies

Table 11.2 Comparison of NW-based logic designs.

Design source	Component					
	Crosspoint technology	NW	Logic	Litho ↔ NW	Restoration	Reference(s)
HP/UCLA	Molecular switch diode	Imprint lithography	Nanoscale wired-OR	Random particles	CMOS	22, 41
CMU nanoFabric	Molecular switch diode	NanoPore templates	Nanoscale wired-OR	–	RTD latch	43, 44
SUNY CMOL	Single-electron transistor	Interferometric lithography	Nanoscale wired-OR	Offset angles	CMOS	46
HP FPNI	Molecular switch diode	Imprint lithography	CMOS (N)AND	Offset angles	CMOS	47
This chapter	Switchable diode	Catalyst NWs	Nanoscale wired-OR	Coded NWs	NW FET	–

and construction alternatives identified by these and other research groups has increased the general confidence that there are options to bypass any of the challenges that might arise when realizing any single technique or feature in these designs.

11.12 Research Issues

While the key building blocks have been demonstrated as previously cited, considerable research and development remains in device synthesis, assembly, integration, and process development. At present, no complete fundamental understanding of the device physics at these scales is available, and a detailed and broader characterization of the devices, junctions, interconnects, and assemblies is necessary to refine the models, to better predict the system properties, and to drive architectural designs and optimization.

The mapping results outlined in Section 11.10 were both area- and defect-tolerance driven. For high-performance designs, additional techniques, design transformations, and optimizations will be needed, including interconnect pipelining (e.g. Ref. [48]) and fan-out management (e.g. Ref. [49]).

In Section 11.7 it was noted that high defect rates could be tolerated when the defects occurred before operation. However, new defects are likely to arise during operation, and additional techniques and mechanisms will be necessary to detect their occurrence, to guard the computation against corruption when they do occur, and rapidly to reconfigure around the new defects.

Further, it is expected that these small feature devices will encounter transient faults during operation. Although the exact fault rates are at present unknown, they are certainly expected to exceed those rates traditionally seen in lithographic silicon. This suggests the need for new lightweight techniques and architectures for fault identification and correction.

11.13 Conclusions

Bottom-up synthesis techniques can be used to produce single nanometer-scale feature sizes. By using decorated NWs – for example, by varying composition at the nanometer scale, both axially and radially – the key nanoscale features may be built into the NWs. Moreover, the NWs can be assembled at tight, nanoscale pitch into dense arrays, contacted to a reliable, lithographic-scale infrastructure, and individually addressed from the lithographic scale. The aggregate set of synthesis and assembly techniques appears adequate for the building of arbitrary logic at the nanoscale, even if the only programmable elements are non-restoring diodes.

Bottom-up self-assembly demands that highly regular structures are built that can be differentiated stochastically for addressing and restoration. NW field-effect gating provides signal restoration and inversion while keeping signals at the dense, nanoscale pitch. Post-fabrication configuration allows the definition of deterministic computation on top of the regular array, despite random differentiation and high rates of randomly placed defects. When these NWs are assembled into modest-sized interconnected PLA arrays, it is estimated that the net density would be one-to-two orders of magnitude higher than for defect-free lithographic-scale FPGAs built in 22 nm CMOS. This should provide a pathway by which to exploit nanometer-pitch devices, interconnect, and systems without pushing lithography into providing these smallest feature sizes.

Acknowledgments

The architectural studies into devices and construction techniques which emerge from scientific research do so only after close and meaningful with the physical scientists. These studies have been enabled by collaboration with Charles M. Lieber and his students. The suite of solutions summarized here includes joint investigations with Helia Naeimi, Michael Wilson, John E. Savage, and Patrick Lincoln.

These research investigations were funded in part by National Science Foundation Grant CCF-0403674 and the Defense Advanced Research Projects Agency under ONR contracts N00014-01-0651 and N00014-04-1-0591.

Any opinions, findings, and conclusions or recommendations expressed in this material are those of the authors, and do not necessarily reflect the views of the National Science Foundation or the Office of Naval Research.

Christian Nauenheim and Rainer Waser helped to produce this brief chapter as a digested version of Ref. [29].

References

1 V. Betz, J. Rose, FPGA place-and-route challenge, 1999. Available at http://www.eecg.toronto.edu/~vaughn/challenge/challenge.html.

2 A. M. Morales, C. M. Lieber, A laser ablation method for synthesis of crystalline semiconductor nanowires. *Science* 1998, **279**, 208–211.

3 (a) Y. Cui, L. J. Lauhon, M. S. Gudiksen, J. Wang, C. M. Lieber, Diameter-controlled synthesis of single crystal silicon nanowires. *Appl. Phys. Lett.* 2001, **78** (15), 2214–2216. (b) Y. Cui, Z. Zhong, D. Wang, W. U. Wang, C. M. Lieber, High performance silicon nanowire field effect transistors. *Nano Lett.* 2003, **3** (2), 149–152.

4 Y. Cui, X. Duan, J. Hu, C. M. Lieber, Doping and electrical transport in silicon nanowires. *J. Phys. Chem. B* 2000, **104** (22), 5213–5216.

5 Y. Huang, X. Duan, Y. Cui, L. Lauhon, K. Kim, C. M. Lieber, Logic gates and computation from assembled nanowire building blocks. *Science* 2001, **294**, 1313–1317.

6 M. S. Gudiksen, L. J. Lauhon, J. Wang, D. C. Smith, C. M. Lieber, Growth of nanowire superlattice structures for nanoscale photonics and electronics. *Nature* 2002, **415**, 617–620.

7 Y. Wu, R. Fan, P. Yang, Block-by-block growth of single-crystalline Si/SiGe

superlattice nanowires. *Nano Lett.* 2002, **2** (2), 83–86.

8 M. T. Björk, B. J. Ohlsson, T. Sass, A. I. Persson, C. Thelander, M. H. Magnusson, K. Depper, L. R. Wallenberg, L. Samuelson, One-dimensional steeplechase for electrons realized, *Nano Lett.* 2002, **2** (2), 87–89.

9 L. J. Lauhon, M. S. Gudiksen, D. Wang, C. M. Lieber, Epitaxial core-shell and core-multi-shell nanowire heterostructures. *Nature* 2002, **420**, 57–61.

10 M. Law, J. Goldberger, P. Yang, Semiconductor nanowires and nanotubes. *Annu. Rev. Mater. Sci.* 2004, **34**, 83–122.

11 D. Whang, S. Jin, C. M. Lieber, Nanolithography using hierarchically assembled nanowire masks. *Nano Lett.* 2003, **3** (7), 951–954.

12 Y. Wu, J. Xiang, C. Yang, W. Lu, C. M. Lieber, Single-crystal metallic nanowires and metal/semiconductor nanowire heterostructures. *Nature* 2004, **430**, 61–64.

13 Y. Huang, X. Duan, Q. Wei, C. M. Lieber, Directed assembly of one-dimensional nanostructures into functional networks. *Science* 2001, **291**, 630–633.

14 (a) D. Chen, J. Cong, M. Ercegovac, Z. Huang, Performance-driven mapping for cpld architectures. *IEEE Trans. Comput.-Aided Des. Integr. Circuits Syst.* 2003, **22** (10), 1424–1431. (b) Y. Chen, G.-Y. Jung, D. A. A Ohlberg, X. Li, D. R. Stewart, J. O. Jeppesen, K. A. Nielsen, J. F. Stoddart, R. S. Williams, Nanoscale molecular-switch crossbar circuits. *Nanotechnology* 2003, **14**, 462–468.

15 D. R. Stewart, D. A. A. Ohlberg, P. A. Beck, Y. Chen, R. S. Williams, J. O. Jeppesen, K. A. Nielsen, J. F. Stoddart, Molecule-independent electrical switching in pt/organic monolayer/ti devices. *Nano Lett.* 2004, **4** (1), 133–136.

16 A. DeHon, Reconfigurable architectures for general-purpose computing. AI Technical report 1586 (oct.), MIT Artificial Intelligence Laboratory, Cambridge, MA, 1996.

17 Y. Wu, P. Yang, Germanium nanowire growth via simple vapor transport. *Chem. Mater.* 2000, **12**, 605–607.

18 B. Zheng, Y. Wu, P. Yang, J. Liu, Synthesis of ultra-long and highly-oriented silicon oxide nanowires from alloy liquid. *Adv. Mater.* 2002, **14**, 122.

19 M. S. Gudiksen, J. Wang, C. M. Lieber, Synthetic control of the diameter and length of semiconductor nanowires. *J. Phys. Chem. B* 2001, **105**, 4062–4064.

20 D. Whang, S. Jin, Y. Wu, C. M. Lieber, Large-scale hierarchical organization of nanowire arrays for integrated nanosystems. *Nano Lett.* 2003, **3** (9), 1255–1259.

21 Y. Chen, D. A. A Ohlberg, X. Li, D. R. Stewart, R. S. Williams, J. O. Jeppesen, K. A. Nielsen, J. F. Stoddart, D. L. Olynick, E. Anderson, Nanoscale molecular-switch devices fabricated by imprint lithography. *Appl. Phys. Lett.* 2003, **82**, 10, 1610–1612.

22 W. Wu, G.-Y. Jung, D. Olynick, J. Straznicky, Z. Li, X. Li, D. Ohlberg, Y. Chen, S.-Y. Wang, J. Liddle, W. Tong, R. S. Williams, One-kilobit cross-bar molecular memory circuits at 30-nm half-pitch fabricated by nanoimprint lithography. *Appl. Physics A* 2005, **80**, 1173–1178.

23 S. Williams, P. Kuekes,Demultiplexer for a molecular wire crossbar network. United States Patent Number 6,256,767, 2001.

24 A. DeHon, P. Lincoln, J. Savage, Stochastic assembly of sublithographic nanoscale interfaces. *IEEE Trans. Nanotech.* 2003, **2** (3), 165–174.

25 B. Gojman, E. Rachlin, J. E. Savage, Decoding of stochastically assembled nanoarrays, in: *Proceedings of the International Symposium on VLSI*, Lafayette, USA, IEEE Computer Society, 2004.

26 A. DeHon, Law of large numbers system design, in: S. K. Shukla, R. I. Bahar (Eds.), *Nano, Quantum and Molecular Computing: Implications to High Level Design and Validation*, Kluwer Academic Publishers, Boston, MA, Chapter 7, pp. 213–241, 2004.

27 A. DeHon, Array-based architecture for FET-based, *nanoscale electronics. IEEE Trans. Nanotech.* 2003, **2** (1), 23–32.

28 F. G. Maunsell, A problem in cartophily. *The Math. Gazette* 1937, **22**, 328–331.

29 A. DeHon, Nanowire-based programmable architecture. *ACM J. Emerging Technol. Comput. Systems* 2005, **1** (2), 109–162

30 A. DeHon, H. Naeimi, Seven strategies for tolerating highly defective fabrication. *IEEE Design Test Comput.* 2005, **22** (4), 306–315.

31 A. DeHon, M. J. Wilson, Nanowire-based sublithographic programmable logic arrays, in: *Proceedings International Symposium on Field-Programmable Gate Arrays*, Napa Valley, CA, IEEE Publishers, pp. 123–132, 2004.

32 C. Mead, L. Conway, *Introduction to VLSI Systems*, Addison-Wesley, 1980.

33 N. H. E Weste, D. Harris, *CMOS VLSI Design: A Circuits and Systems Perspective*, 3rd edn., Addison-Wesley, 2005.

34 A. DeHon, Design of programmable interconnect for sublithographic programmable logic arrays, in: *Proceedings International Symposium on Field-Programmable Gate Arrays*, Monterey, CA, ACM Publishers, pp. 127–137, 2005.

35 V. Betz, J. Rose, A. Marquardt, *Architecture and CAD for Deep-Submicron FPGAs*, Kluwer Academic Publishers, Norwell, MA, 1999.

36 S. Brown, M. Khellah, Z. Vranesic, Minimizing FPGA interconnect delays. *IEEE Des. Test Comput.* 1996, **13** (4), 16–23.

37 V. Betz, J. Rose, FPGA routing architecture: Segmentation and buffering to optimize speed and density, in: *Proceedings International Symposium on Field-Programmable Gate Arrays*, Monterey, CA, ACM Publishers, pp. 59–68, 1999.

38 G. Lemieux, E. Lee, M. Tom, A. Yu, Directional and single-driver wires in FPGA interconnect, in: *Proceedings International Conference on Field-Programmable Technology*, Brisbane, Australia, IEEE Publishers, pp. 41–48, 2004.

39 H. Naeimi, A. DeHon, A greedy algorithm for tolerating defective crosspoints in NanoPLA design, in: *Proceedings IEEE International Conference on Field-Programmable Technology*, Brisbane, Australia, IEEE Publishers, pp. 49–56, 2004.

40 ITRS, International technology roadmap for semiconductors. http://public.itrs.net/Files/2001ITRS/, 2001.

41 J. R. Heath, P. J. Kuekes, G. S. Snider, R. S. Williams, A defect-tolerant computer architecture: Opportunities for nanotechnology. *Science* 1998, **280** (5370), 1716–1721.

42 Y. Luo, P. Collier, J. O. Jeppesen, K. A. Nielsen, E. Delonno, G. Ho, J. Perkins, H.-R. Tseng, T. Yamamoto, J. F. Stoddart, J. R. Heath, Two-dimensional molecular electronics circuits. *ChemPhysChem* 2002, **3** (6), 519–525.

43 S. C. Goldstein, M. Budiu, NanoFabrics: Spatial computing using molecular electronics, in: *Proceedings International Symposium on Computer Architecture*, Gothenburg, Sweden, ACM Publishers, pp. 178–189, 2001.

44 S. C. Goldstein, D. Rosewater, Digital logic using molecular electronics. *IEEE ISSCC Digest Tech. Papers* 2002, 204–205

45 G. Snider, P. Kuekes, R. S. Williams, CMOS-like logic in defective, nanoscale crossbars. *Nanotechnology* 2004, **15**, 881–891.

46 D. B. Strukov, K. K. Likharev, CMOL FPGA: a reconfigurable architecture for hybrid digital circuits with two-terminal nanodevices. *Nanotechnology* 2005, **16** (6), 888–900.

47 G. Snider, R. S. Williams, Nano/CMOS architectures using a field-programmable nanowire interconnect. *Nanotechnology* 2007, **18** (3).

48 W. Tsu, K. Macy, A. Joshi, R. Huang, N. Walker, T. Tung, O. Rowhani, V. George, J. Wawrzynek, A. DeHon, HSRA: High-speed, hierarchical synchronous reconfigurable array, in: *Proceedings International Symposium on Field-Programmable Gate Arrays*, Monterey, CA, ACM Publishers, pp. 125–134, 1999.

49 H. J. Hoover, M. M. Klawe, N. J. Pippenger, Bounding fan-out in logical networks. *J. Assoc. Comput. Machinery* 1984, **31** (1), 13–18.

12
Quantum Cellular Automata

Massimo Macucci

12.1
Introduction

The concept of quantum cellular automata (QCA) was first proposed by Craig Lent and coworkers [1] at the University of Notre Dame in 1993, as an alternative based on bistable electrostatically coupled cells to traditional architectures for computation. Overall, the QCA architecture probably represents the proposal for an alternative computing paradigm that has been developed furthest, up to the experimental proof of principle [2]. As will be discussed in the following sections, its strengths are represented by the reduced complexity (in particular for the implementation based on ground-state relaxation), extremely low power consumption, and potential for ultimate miniaturization; its drawbacks are the extreme sensitivity to fabrication tolerances and stray charges, the difficulty in achieving operating temperatures reasonably close to room temperature, the undesired interaction among electrodes operating on different cells, and the very challenging control of dot occupancy.

The initial formulation of the QCA architecture relied on the relaxation to the ground state of an array of cells: computation was performed enforcing the polarization state of a number of "input" cells and then reading the state of a number of "output" cells, once the array had relaxed down to the ground state consistent with the input data. Such an approach is characterized (as will be discussed in the following) by a simple – at least in principle – layout, but suffers from the presence of many states very close in energy to the actual ground state. This leads to an extremely slow and stochastic relaxation, which may lead to unacceptable computational times.

The slow and unreliable evolution of the ground-state relaxation approach was addressed with the introduction of a modified QCA architecture based on clocked cells [3], which can exist in three different conditions, depending on the value of a clock signal:

- The "locked" condition corresponds to having tunneling between dots inhibited, and therefore the cell can be used to "drive" nearby cells.

Nanotechnology. Volume 4: Information Technology II. Edited by Rainer Waser
Copyright © 2008 WILEY-VCH Verlag GmbH & Co. KGaA, Weinheim
ISBN: 978-3-527-31737-0

- The "null" condition corresponds to having no electrons in the cell and therefore no polarization.
- The "active" condition is the one in which the cell adiabatically reaches the polarization condition resulting from that of the nearby cells.

The clocked QCA architecture solves the problem of unreliable evolution and allows data pipelining, but introduces a remarkable complication: the clock signal must be distributed, with proper phases, to all the cells in the array. Unless a "wireless" technique for clock distribution could be devised (some proposals have been made in this direction, but a definite solution is yet to be found), one of the most attractive features of QCA circuits – the lack of interconnections – would be lost.

Current research is focusing on the possibility of implementing QCA cells with molecules [4] or with nanomagnets [5], in order to explore the opportunities for further miniaturization (molecular cells) and for overcoming the limitations imposed by Coulomb coupling (nanomagnetic cells). However, these technologies do not seem to be suitable for fast operation: highly parallel approaches could make up for the reduced speed, but this would further complicate system design and the definition of the algorithms.

Although the basic principle of operation is sound, the above-mentioned technological difficulties and the reliability problems make practical application of QCA technology unlikely, at least in the near future. Nevertheless, the QCA concept remains of interest and the subject of lively research, because of its innovation potential and because it opens up a perspective beyond the so far unchallenged three-terminal device paradigm for computation.

In this chapter, an overview of the QCA architecture will be provided, with a discussion of its two main formulations: the ground-state relaxation approach and the clocked QCA approach. In Section 12.2 the issue of operation with cells with more than two electrons will also be addressed, as well as the details of intercell interaction. Section 12.3 will focus on the various techniques that have been developed to model QCA devices and circuits, while Section 12.4 will be devoted to the challenges facing the implementation of QCA technology. Actual physical implementations of QCA arrays will be addressed in Section 12.5, and an overlook for the future will be presented in Section 12.6.

12.2
The Quantum Cellular Automaton Concept

12.2.1
A New Architectural Paradigm for Computation

An early proposal for an architecture based on interacting quantum dots was formulated by Bakshi *et al.* in 1991 [6]: these authors considered parallel elongated quantum dots, defined "quantum dashes" (see Figure 12.1), each of which should have an occupancy of one electron. Their basic argument was that, once the electron

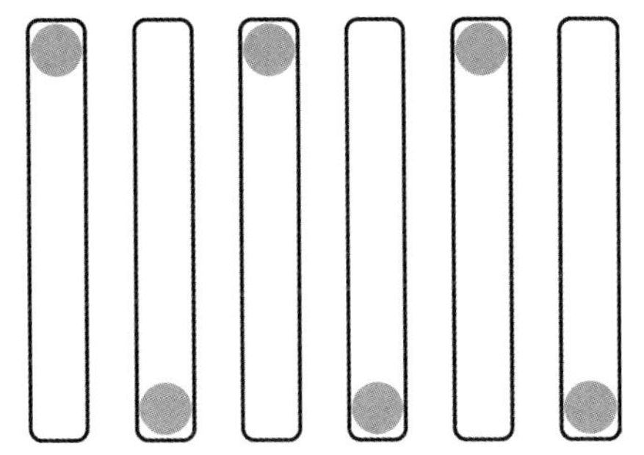

Figure 12.1 Series of elongated quantum dots (quantum dashes) with the hypothesized anti-ferroelectric ordering.

in the first dash was confined into one end of the dash, the electron in the next dash would be confined into the opposite end, as a result of electrostatic repulsion. This configuration would propagate along the line of dashes, leading to a sort of anti-ferroelectric ordering that could then be exploited for the implementation of more complex functions. This initial concept, however, had a serious problem, consisting in the fact that the localization of electrons along the chain of dashes would soon decay (See Figure 12.2), because the electrostatic repulsion due to an electron localized at the end of a dash is not sufficient to significantly localize the electron in the nearby dash, the probability density of which would just be slightly displaced. The Notre Dame group realized that this problem could be solved with the insertion of a barrier in the middle of the dash: in this way, the electron wave function must be localized on either side of the barrier and the electrostatic interaction from the electron in the nearby dash is sufficient to push the electron into the opposite half of the dash (Figure 12.3). This concept can be easily understood considering a two-level system subject to an external perturbing potential V [7]. The Hamiltonian of such a system in second quantization reads:

$$\hat{H}=\sum_{i=1,2} n_i E_i + t(b_1^\dagger b_2 + b_2^\dagger b_1) + \sum_{i=1,2} n_i q V_i, \tag{12.1}$$

where n_1 and n_2 are the occupation numbers of levels 1 and 2, respectively, $b_i^\dagger$ and b_j are the creation and annihilation operators for levels i and j, t is the coupling between

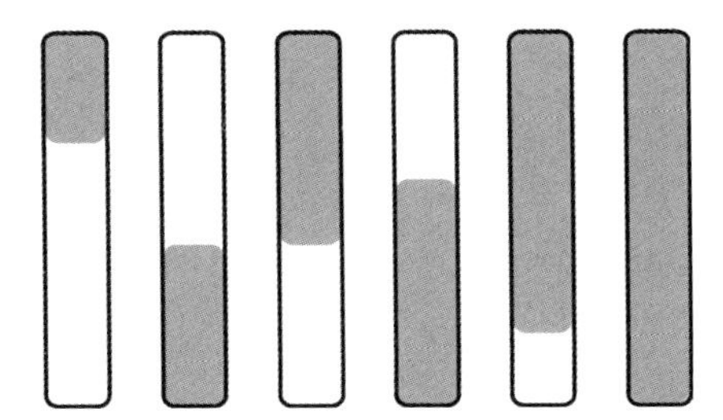

Figure 12.2 Sketch of the actual electron density within a chain of dashes.

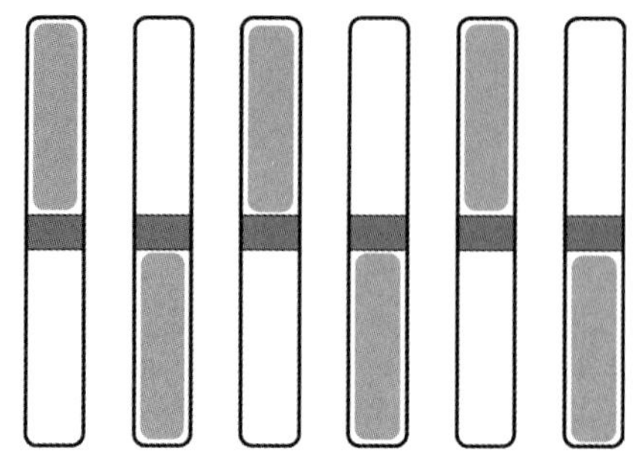

Figure 12.3 Chain of dashes with the inclusion of potential barriers: electrons are now localized on either side of the barriers and an anti-ferroelectric ordering is achieved.

the two levels and V_i is the external perturbing potential at the location of the i-th level. The creation operator $b_i^\dagger$ applied to a state with $(n-1)$ electrons yields a state with n electrons, thereby "creating" an electron in state i, while the annihilation operator b_j applied to a state with n electrons yields a state with $(n-1)$ electrons, thereby "destroying" an electron from state j. For example, the application of $b_i^\dagger b_2$ transfers an electron from level 2 to level 1. Each level can be associated with one of the sides into which the dash is divided by a potential barrier: if the barrier is exactly in the middle of the dash, $E_1 = E_2$ and the value of t depends on the height and thickness of the barrier; the higher and the thicker the barrier, the smaller t will be. A sketch of the potential profile is provided in Figure 12.4, where the dots represent the locations of the two levels 1 and 2. If $E_1 + V_1$ is chosen as the energy reference and ε is defined as $E_2 + V_2 - E_1 - V_1$, the Hamiltonian can be represented in the basis ($|0\rangle$ $(n_1=1,\ n_2=0)$, $|1\rangle$ $(n_1=0,\ n_2=1)$) simply as (computing the matrix elements between the basis states)

$$\mathbf{H} = \begin{pmatrix} 0 & t \\ t & \varepsilon \end{pmatrix}. \tag{12.2}$$

The state $|0\rangle$ corresponds to having an electron in level 1 and no electron in level 2, while $|1\rangle$ corresponds to the situation with level 1 empty and an electron in level 2. The eigenvalues of this representation can be easily computed:

$$e_1 = \frac{1}{2}\left(\varepsilon - \sqrt{\varepsilon^2 + 4t^2}\right) \quad e_2 = \frac{1}{2}\left(\varepsilon + \sqrt{\varepsilon^2 + 4t^2}\right). \tag{12.3}$$

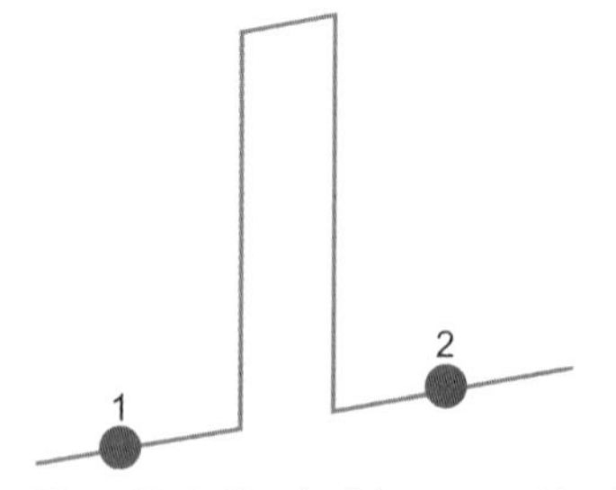

Figure 12.4 Sketch of the potential landscape defining the two levels, 1 and 2, separated by a barrier.

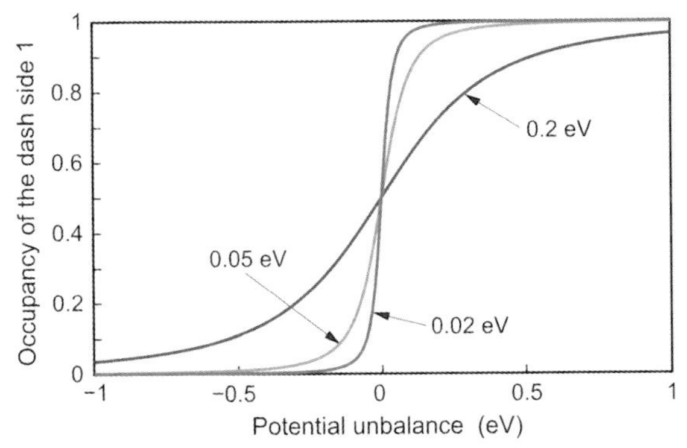

Figure 12.5 Occupancy of one of the levels of a two-level system, as a function of the potential unbalance resulting from an external applied electric field, for different values of the coupling parameter t.

The unbalance term ε in this case depends only on the external potential produced by the electron in the nearby dash: it will vary between a negative and a positive value, depending on the position of the electron.

The occupancy of the first level – that is, of the dash side labeled with 1 – will be given by the square modulus of the corresponding coefficient of the ground-state eigenvector, which can be computed along with the eigenvalues. Such a quantity is plotted in Figure 12.5 as a function of the unbalance ε for different values of the coupling energy t. It is apparent that, for low values of t (and therefore for an opaque barrier), the electron moves abruptly from one level to the other, as the external field is varied. Therefore, it is strongly localized even for very small values of such a field, while for high values of t (and therefore for a transparent or inexistent barrier) a very smooth transfer of the probability density from one level to the other are needed and large values of the perturbing field are required to achieve some degree of localization. Thus, the introduction of a barrier in the middle of the dash creates a sort of bistability – that is, a strongly non-linear response to external perturbations, which is at the heart of QCA operation and allows the regeneration of logic values. From the wire of dashes with barriers of Figure 12.3, the next step is represented by joining two adjacent dashes to form a square cell, which is the basis of the Notre Dame QCA architecture. The square cell allows the creation of two-dimensional (2-D) arrays, as shown in Figure 12.6, which can implement any combinatorial logic function. In an isolated cell the two configurations or polarization states, with the electrons along one or the other diagonal, are equally likely. However, if an external perturbation is applied, or in the presence of a nearby cell with a fixed polarization, one of them will be energetically favored. The two logic states, 1 and 0, can be associated with the two possible polarization configurations, as shown in Figure 12.7, where the solid dots represent the electrons. If a linear array of dots is created, enforcing the polarization corresponding to a given logic state on the first cell will lead to the propagation of the same polarization state along the whole chain, in a domino fashion. Such a linear array is usually defined a "binary wire", and can be used to propagate a logic variable

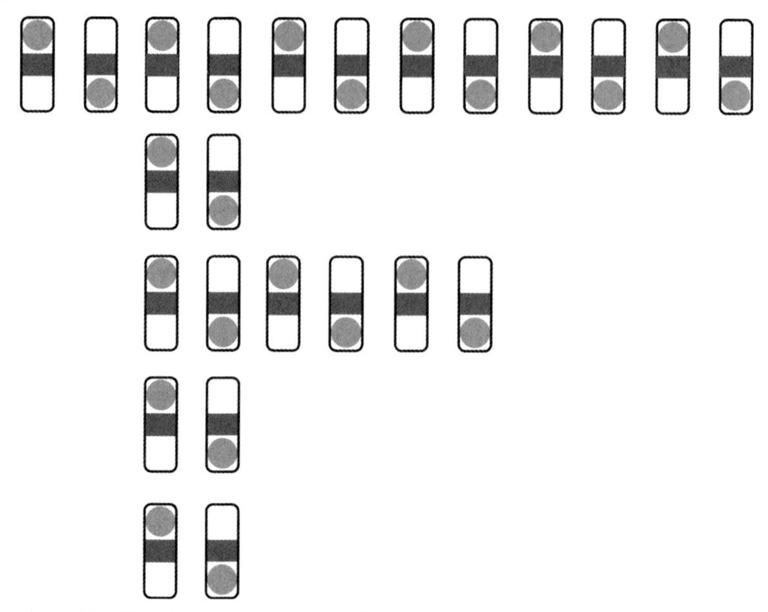

Figure 12.6 Two-dimensional array of cells made up of two adjacent dashes: this is the basis of the QCA architecture.

across a circuit: here, the strength and, at the same time, the weakness of the QCA architecture is noticed. Indeed, signal regeneration occurs during propagation along the chain, as a result of the non-linear response of the cells, but the transfer of a logic variable from one location in the circuit to a different location may require a relatively large number of cells. In other words, there are no interconnects, but the number of elementary devices needed to implement a given logic function may become much larger than in a traditional architecture.

The basic gate in QCA logic is represented by the majority voting gate, which is shown in Figure 12.8. Cells A, B and C are input cells, whose polarization state is enforced from the outside (here and in the following such "driver" cells are represented with a double boundary), while cell Y is the output cell, the polarization of which represents the result of the calculation. On the basis of a full quantum calculation or of simple considerations based on a classical electrostatic model, it is possible to show that the logic value at the output will correspond to the majority of the input values. Thus, for example, if A = 1, B = 0, C = 0, then Y = 0, or, if A = 1, B = 0, C = 1, the output will be Y = 1. From the majority voting gate it is

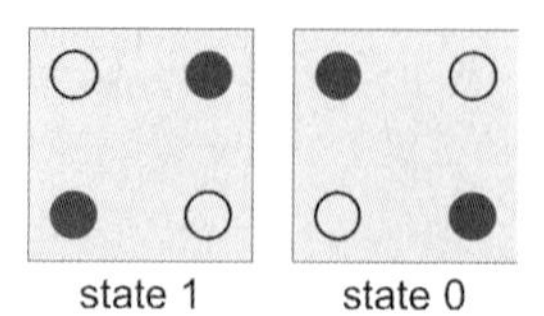

Figure 12.7 Basic configurations of a QCA cell with two electrons.

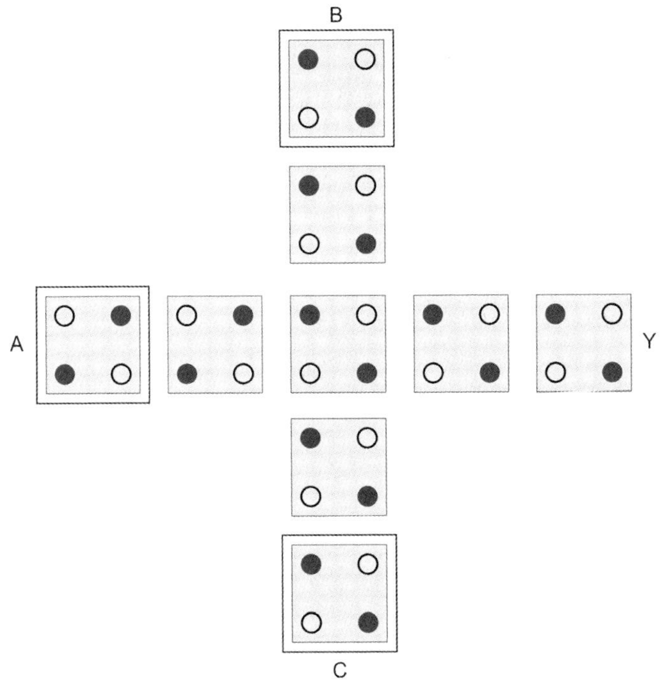

Figure 12.8 Layout of a majority voting gate.

straightforward to derive a two-input AND and OR gate: if A = 1, B and C will be the inputs of an OR gate, while if A = 0, B and C will represent the inputs of an AND gate. In order to be able to create an arbitrary combinatorial network, there is also a need for the NOT gate: this is just slightly more complex, and can be implemented with the layout shown in Figure 12.9 [15].

A generic logic function can thus be obtained with a 2-D array of cells: a number of cells at the perimeter of the array will be used as input cells, by enforcing their polarization condition with properly biased gates, and another group of perimetral

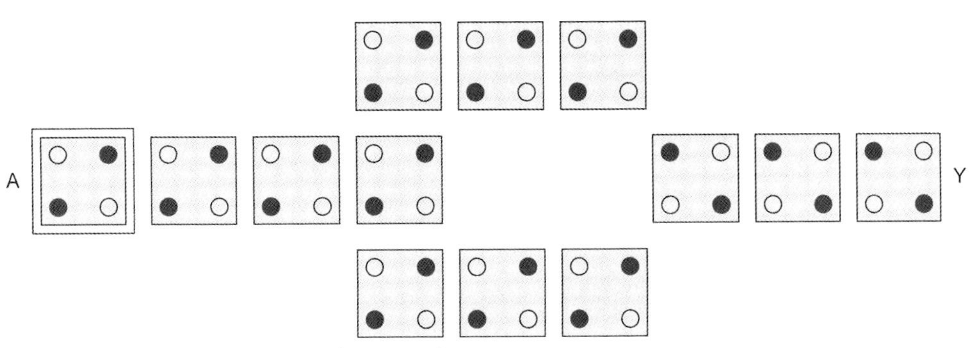

Figure 12.9 Layout for a NOT gate in the QCA architecture.

cells will act as outputs. Once the input values have been enforced, the array is allowed to evolve into the ground state and, when this has been reached, the state of the output cells corresponds to the result of the computation.

As the number of cells in the array increases, its energy spectrum – that is, the set of energies corresponding to all the possible configurations – becomes more complex, with a large number of configurations that have energies very close to the actual ground state. As a result, the evolution of the array may become stuck in one of these configurations for a long time, thus leading to a very slow computation. Furthermore, due to the appearance of states that are very close in energy to the ground state, the maximum operating temperature decreases as the number of cells is increased. In particular, it has been shown with entropy considerations [8] or by means of an analytical model [9] that, for the specific case of a binary wire, the maximum operating temperature falls logarithmically with the number of cells.

12.2.2
From the Ground-State Approach to the Clocked QCA Architecture

The above-mentioned problems severely limit the applicability of ground-state computation to real-life situations, and make the evolution of a large QCA system unreliable. To solve such problems, a modified architecture was proposed by Lent and coworkers [3], inspired, in its implementation with metal tunnel junctions, to the parametron concept introduced by Korotkov [10]. The so-called "clocked QCA architecture" derives from the concept of adiabatic switching [3]: based on the adiabatic theorem [11], it is possible to show that if the Hamiltonian of a system is made to evolve slowly from one initial form H_i to a final form H_f, a particle starting in the n-th eigenstate of H_i will be carried over into the n-th eigenstate of H_f. Thus, starting with particles in the ground state of the system, they will never leave the ground state during the evolution of the Hamiltonian, thereby preventing the previously mentioned problems of trapping into metastable states.

To implement this concept of adiabatic switching, the confinement potential defining the cell must be variable in time. In practice, the barriers separating the dots are modulated by an external potential, representing the clock signal. When the barriers are low, the cell is in the "null" state and has no polarization. In contrast, when the barriers reach their maximum height the cell is in the "locked" state and its polarization cannot be modified as a result of the action of the nearby cells (the electrons are prevented from tunneling between dots). It is only during the "active" phase, when the barriers have an intermediate height, that the polarization state can change according to the that of the nearby cells.

Attention should now be focused on the particular implementation of a clocked cell that has been experimentally realized [12]. This consists of a six-dot cell as proposed by Tóth and Lent [13], developed from the metal-island cells used in the first experimental demonstration of QCA action [2]. The barriers are represented by tunnel junctions obtained by including a very thin dielectric (usually created by oxidation) between two metallic electrodes (usually aluminum), and the quantum dots are replaced with metal islands. The six-island clocked cell is represented in

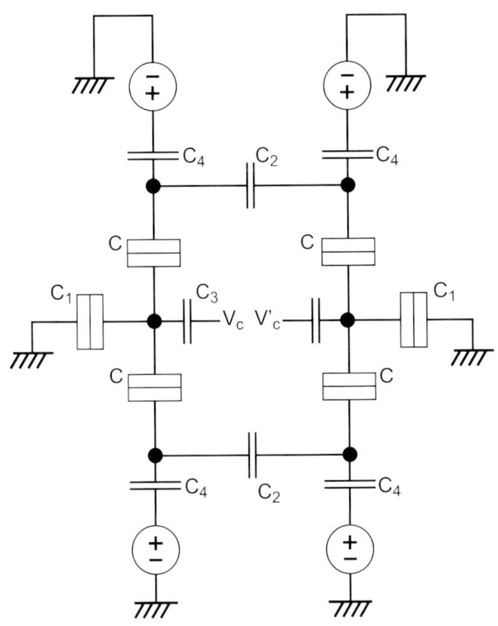

Figure 12.10 Clocked cell for implementation with tunnel junctions: tunneling between the dots of each half cell is controlled by the voltages applied to the middle dots through the C_3 capacitors.

Figure 12.10. Tunneling is possible only between the upper and the lower dot of each half of the cell (it can be shown that this does not limit in any way the logic operation of the cell) and the barrier height between the active islands is controlled by means of the potential applied to the central island. If the potential on the central island is low, then the electron will be trapped there (null state). As the potential on the central island is raised, a condition will be reached in which the electron can tunnel into one of the active dots, the one that is favored at the time by the potential created by the nearby cells (active state). Finally, as the potential on the central dot is further raised, the electron will be trapped in the dot into which it has previously tunneled, even if the polarization of the other cells is reversed (locked state).

Ideally, the computation should evolve with a cell in the locked state driving the next cell that moves from the null state to the locked state, going through the active state. When the state of a cell must be the result of that of more than one neighboring cell (as in the case of the central cell of a majority voting gate), all the cells acting on it should be at the same time in the locked state. The sequence of clock phases would allow the information to travel along the circuit in a controlled way, thus achieving a deterministic evolution and eliminating the uncertainty about the time when the calculation is actually completed that plagues the ground-state relaxation scheme. Furthermore, since the flux of data is steered by the clock, it would also possible to have data pipelining: new input data could be fed into the inputs of the array as soon as the previous data have left the first level of logic gates and moved to the second. Ideally, within this scheme each cell should be fed a different clock phase with respect to its nearest neighbors, which would imply an extremely complex wiring structure. Such a solution has been adopted in the experiments performed so far to demonstrate the principle of operation of clocked QCA logic [12]. However, in large-scale circuits it would forfeit one of the main advantages of the QCA architecture – that is, the simplicity deriving from the lack of interconnections. In order to address this problem, it has been proposed to divide the overall QCA array into "clocking zones": such regions consist of a number of QCA cells and would be subject to the same clock phase and evolve together while in the active state (similarly to a small array operating according to the ground-state relaxation principle). They would then be locked all at the same time, in order to drive the following clocking zone. This would reduce the complexity of the required wiring, and has been proposed in particular for the implementation of QCA circuits at the molecular scale, where it is impossible to provide different clock phases to each molecule, as the wires needed for clocking would be much larger than the molecules themselves! There are many difficulties involved, however, because each clocking zone is affected by the problems typical of ground-state relaxation (although on a smaller scale), and the clock distribution is still extremely challenging. For example, conducting nanowires have been suggested as a possible solution to bring the clock signal to regions of a molecular QCA circuit, but achieving uniformity in the clocking action of a nanowire on many molecular cells is certainly a very challenging task.

Notwithstanding all of these difficulties, the clocked scheme appears to be the only one capable of yielding a reasonably reliable QCA operation in realistic circuits, as will be discussed in the following sections.

12.2.3
Cell Polarization

At this point it is necessary to provide a rigorous definition of cell polarization, in order to be able to describe quantitatively the interaction between neighboring cells and to determine whether cells with an occupancy of more than two electrons could possibly be used. Indeed, according to the initial definition of cell polarization given by the Notre Dame group, the operation of cells with more than two electrons would not be possible. Their original definition of cell polarization was

$$P = \frac{\rho_1 + \rho_3 - \rho_2 - \rho_4}{\rho_1 + \rho_2 + \rho_3 + \rho_4}, \qquad (12.4)$$

where ρ_i is the charge in the i-th dot (dots are numbered counterclockwise starting from the one in the upper right quadrant). With such an expression, as soon as the number of electrons increases above two, full polarization can no longer be achieved, as the maximum possible value for the numerator is 2. There can be at most a difference of two electrons between the occupancy of one diagonal and that of the other, since configurations with a larger difference would require an extremely large external electric field (to overcome the electrostatic repulsion between electrons).

Starting from the observation that a QCA cell is overall electrically neutral, because of the presence of ionized donors, of the positive charge induced on the metal electrodes, and of the screening from surface charges, Girlanda *et al.* [14] proposed a different expression for the polarization of a cell, which is more representative of its action upon the neighboring cells. Indeed, neutralization occurs over an extended region of space; thus, although the global monopole component of the field due to a cell is zero, there can be some effect on the neighboring cells associated with the total number of electrons. However, in practical cases this turns out to be negligible compared to the dipole component associated with the charge unbalance between the two diagonals. The alternative expression for cell polarization introduced in Ref. [14] reads

$$P = \frac{\rho_1 + \rho_3 - \rho_2 - \rho_4}{2q}, \tag{12.5}$$

where q is the electron charge. Use of this expression is supported by semiclassical electrostatic considerations and by detailed quantum simulations [14], and leads to the conclusion that QCA action can be observed whenever the cell occupancy is of $4N + 2$ electrons, where N is the integer. This means that control of the occupancy of the dots is less stringent than previously expected, but is still quite difficult.

12.3 Approaches to QCA Modeling

12.3.1 Hubbard-Like Hamiltonian

The first approach to QCA simulation was developed by the Notre Dame group [15], based on an occupation number, Hubbard-like formalism. Within such an approach the details of the electronic structure of each quantum dot are neglected, and a few parameters are used to provide a description of the dots and their interaction. Although based on a few phenomenological parameters, this technique has been successful in providing a good basic understanding of the operation of QCA cells.

The occupation number Hamiltonian for a single, isolated cell reads

$$\begin{aligned} H_0 &= \sum_{i,\sigma} E_{0,i}, n_{i,\sigma} + \sum_{i>j,\sigma} t(b^{\dagger}_{i,\sigma} b_{i,\sigma} + b^{\dagger}_{j,\sigma} b_{i,\sigma}) \\ &+ \sum_{i} E_{Qi} n_{i,\uparrow} n_{i,\downarrow} + \sum_{i>j,\sigma,\sigma'} V_Q \frac{n_{i,\sigma} n_{j,\sigma'}}{|\vec{R}_i - \vec{R}_j|}, \end{aligned} \tag{12.6}$$

where $E_{0,i}$ is the ground-state energy of the i-th dot (assumed to be isolated), $b^{\dagger}_{j,\sigma}$ and $b_{j,\sigma}$ are the creation and annihilation operators, respectively, for an electron in the j-th dot with spin σ, $n_{j,\sigma}$ is the number operator for electrons in the i-th dot with spin σ, t is the tunneling energy between neighboring dots, V_Q is equal to $e^2/(4\pi\varepsilon)$ (e is the electron charge and ε is the dielectric permittivity), E_{Qi} is the on-site charging energy for the i-th dot [16], and $\vec{R}_i$ is the position of the i-th dot center. The tunneling energy t cannot be computed directly, and must be evaluated with some approximation. A commonly used approximation consists in assuming t to be equal to half of the level-splitting resulting because of the presence of a barrier of finite height between the dots. In the presence of a driver cell in a given polarization state, the above-written Hamiltonian must be augmented with a term that expresses the electrostatic contribution from such a cell:

$$H_{int} = \sum_{i \in \text{cell1}} \sum_{j \in \text{cell2}} V_Q \frac{\rho_{j,2} - \bar{\rho}}{|\vec{R}_{j,2} - \vec{R}_{i,1}|}, \tag{12.7}$$

where $\rho_{j,2}$ is the number of electrons in the j-th dot of the driver cell, ρ is the average number of positive neutralizing charges per dot, $\vec{R}_{j,2}$ and $\vec{R}_{i,1}$ are the positions of the j-th dot of the driver cell (cell 2) and of the i-th dot of the driven cell (cell 1), respectively.

Diagonalization of the total Hamiltonian can be performed easily using an occupation number representation: $|n_{1,\uparrow}, n_{1,\downarrow}; n_{2,\uparrow}, n_{2,\downarrow}; n_{3,\uparrow}, n_{3,\downarrow}; n_{4,\uparrow}, n_{4,\downarrow}\rangle$.

The dimension of the basis, considering only two-electron states, would be 256 but, on the basis of spin considerations [17], the number of basis vectors required for the determination of the ground state is just 16.

A representation of the complete Hamiltonian on such a basis consists in a sparse matrix with only four non-zero off-diagonal elements in each row. Eigenvalues and eigenvectors can be obtained numerically, and the ground state of the driven cell will be

$$|\psi_0\rangle = \sum_{i=1}^{16} \alpha_i |i\rangle, \tag{12.8}$$

where α_i is the i-th element of the eigenvector, corresponding to the lowest eigenvalue, and $|i\rangle$ is the i-th element of the basis used for the representation of the Hamiltonian. The average charge in each dot is given by

$$\rho_i = \langle \psi_0 | n_{i\uparrow} + n_{i\downarrow} | \psi_0 \rangle, \tag{12.9}$$

from which the cell polarization can then be computed. In Figure 12.11 the polarization of the driven cell computed as a function of the polarization of the driver cell is reported; that is, the cell-to-cell response function. Calculations have been performed for a cell with four dots located at the vertices of a square with a 24 nm side. The dots have a diameter of 16 nm, except for one, the diameter of which varies between 15.94 and 16.06 nm, and the separation between the centers of the driver and driven cells is 32 nm. Material parameters for GaAs have been considered, with an effective mass $m^* = 0.067\ m_0$ and a relative permittivity $\varepsilon_r = 12.9$; furthermore the tunneling energy t has been assumed to be 0.1×10^{-3} eV.

In the case of identical dots (all with the same 16-nm diameter), the response function is symmetric, while just an extremely small variation in the diameter of a dot leads to strong asymmetry and eventually to failure of operation, with the driven cell always stuck in the same state for any value of the polarization of the driver cell. It appears that such a sensitivity to geometric tolerances is a very serious practical

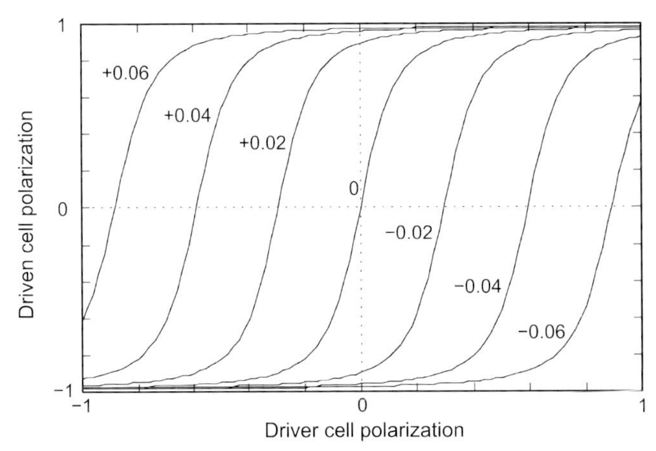

Figure 12.11 Cell-to-cell response function for cells with a separation of 32 nm, an interdot distance of 24 nm, and dots with a diameter of 16 nm. The different curves correspond to an error on the diameter of the lower left dot varying between −0.06 nm and 0.06 nm.

problem, but cannot be fully gauged with the occupation-number Hamiltonian approach, because it is not possible to directly relate the diameter of idealized quantum dots to actual geometric quantities, such as the size of the gates defining the quantum dots.

In order to be able to provide reliable estimates of the acceptable errors on actual geometric parameters, a more realistic model is needed which takes into account the detailed structure of the cell. To this purpose, the approach described in the following subsection has been developed.

12.3.2
Configuration–Interaction

More traditional, iterative self-consistent approaches, such as the Hartree technique or techniques based on the local density functional approximation (LDA), perform very poorly in the simulation of an active QCA cell, in particular in the region around zero polarization. The main problem with iterative self-consistent methods is that, in the application to QCA problem, they tend to become unstable, as a result of the strong degeneracies and of the marked bistability of the system. One effective technique to treat a realistic model for a QCA cell consists in configuration–interaction. This method is very well known in the field of molecular chemistry [18], and has found significant application also for treating semiconductor quantum dots [19, 20].

While, for example, the Hartree–Fock method consists in finding an optimized Slater determinant representing the single-determinant solution for the many-body Schrödinger equation (i.e. the Slater determinant that minimizes the ground-state energy), in the configuration–interaction picture the many-particle wave function is expressed as a linear combination of Slater determinants. In principle, if the basis of determinants were infinite, the solution would be exact; however, in practice a finite basis must be considered, which introduces some degree of approximation, depending on the number of elements and on how good their choice is in terms of the actual solution.

The application of configuration–interaction to the analysis of QCA cells is presented in Ref. [21]: the Hamiltonian for a cell is written as

$$\begin{aligned}\hat{H} = & -\frac{\hbar^2}{2m^*}\nabla_1^2 - \frac{\hbar^2}{2m^*}\nabla_2^2 + V_{\mathrm{con}}(\vec{r}_1) + V_{\mathrm{con}}(\vec{r}_2) + V_{\mathrm{driv}}(\vec{r}_1) + V_{\mathrm{driv}}(\vec{r}_2) \\ & + \frac{1}{4\pi\varepsilon}\frac{e^2}{|\vec{r}_1 - \vec{r}_2|} - \frac{1}{4\pi\varepsilon}\frac{e^2}{\sqrt{|\vec{r}_1 - \vec{r}_2|^2 + (2z)^2}} - \frac{1}{4\pi\varepsilon}\frac{e^2}{2z},\end{aligned} \tag{12.10}$$

where $\hbar$ is the reduced Planck constant, m^* is the electron effective mass, V_{con} is the bare confinement potential (due to the electrodes, the ionized donors, the charged impurities, and the bandgap discontinuities), V_{driv} is the Coulomb potential due to the charge distribution in the neighboring driver cell, e is the electron charge, the last two terms include the effects of the image charges (since, for simplicity, a Dirichlet boundary condition is assumed at the surface and at an infinitely far away conducting substrate), and z is the distance of the 2DEG from the surface of the heterostructure where the boundary condition is enforced.

A matrix representation of this Hamiltonian is derived by computing the matrix elements of Eq. (12.10) between the elements of the basis of Slater determinants, and is then diagonalized, obtaining the ground-state energy as the lowest eigenvalue and the ground-state wave function as a linear combination of the basis elements with coefficients corresponding to the elements of the associated eigenvector.

This technique does not have convergence problems, as it is intrinsically a one-shot method and allows the consideration of a realistic confinement potential, obtained from a detailed numerical calculation. However, if the intention was to introduce more realistic boundary conditions for the semiconductor surface, or in general to provide a more refined treatment of the electrostatic problem, going beyond the method of images, the problem would, computationally, be very intensive. This is because, in order to compute the matrix elements of the Hamiltonian, the complete Green's function of the Poisson equation between each pair of points in the domain would be needed.

If an occupancy of only two electrons per cell were to be considered, the actual two-electron wave function is very close to the Slater determinant constructed from the one-electron wave functions of the isolated dots. Therefore, the size of the basis needed to obtain a good configuration–interaction solution is small, of the order of 100 determinants. Instead, if there are more than four electrons per cell (and thus more than one electron per dot), the strong electron–electron interaction determines a significant deformation of the wave functions and therefore a large number of basis elements constructed from the single-electron orbitals is needed. For example, in the case of a six-electron cell, more than 1000 determinants are necessary. As the number

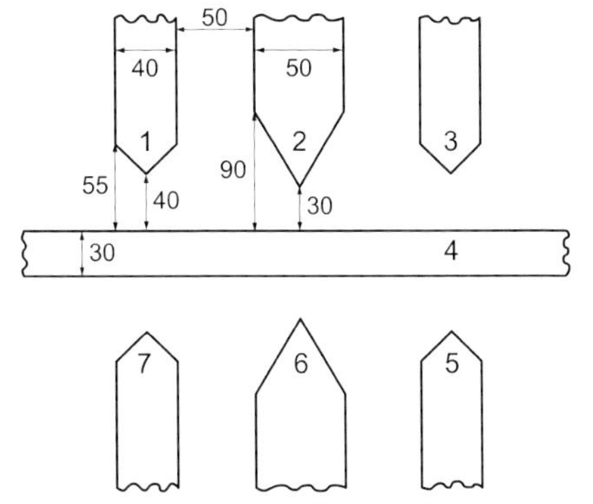

Figure 12.12 Gate layout for the definition of a working QCA cell (all measures are in nanometers).

of electrons is raised, there is a combinatorial increase in the size of the basis, and consequently the problem soon becomes intractable from a computational point of view.

Notwithstanding the above-mentioned limitations on the way that Coulomb that interaction can be included, and on the maximum number of electrons that can be considered, configuration–interaction has been very successfully applied to the simulation of QCA systems. In fact, it has allowed the demonstration that, for a semiconductor implementation, an array of holes (defining the quantum dots) in a depletion gate held at constant potential cannot possibly be fabricated with the required precision. Alternative gate arrangements, such as those shown in Figure 12.12, are possible [21], and have been used in the experimental demonstration of QCA action in GaAs/AlGaAs heterostructure-based devices [22]. However, they imply severe technological difficulties and the need for adjustment of individual gate voltages to correct for geometrical imperfections and for the unintentional asymmetries introduced by the presence of nearby cells [23].

12.3.3 Semi-Classical Models

Quantum models of QCA cells are needed to describe the bistable behavior of the single cell, and also to provide information on the technological requirements needed to obtain successful QCA operation. They are, however, too complex (from a computational point of view) to be applied to the analysis of complete QCA circuits consisting of a large number of cells. The time required to complete a simulation of a circuit made up of just a few tens of cells would be prohibitive. Therefore, a multiscale approach is needed, which is structured in a way similar to that of traditional microelectronics, where circuit portions of increasing complexity are treated with models based on progressively more simplified physical models.

It should be noted that the effect at the core of QCA action is purely classical – that is, it is the Coulomb interaction between electrons. As long as electrons are strongly localized, they behave substantially as classical particles, and a semi-classical model, based on the minimization of the electrostatic energy, can capture most of the behavior of a QCA circuit.

If the only point of interest is to determine the ground-state configuration of an array of QCA cells and in computing the energy separation ΔE between the first excited state and the ground state, then a simple electrostatic model can be used. The quantity ΔE is essential to determine the maximum operating temperature of the circuit: it must be at least a few tens of kT (where k is the Boltzmann constant and T is the absolute temperature); otherwise, the system will not remain stably in the ground state. The basic electrostatic model developed in Ref. [9] relies on a cell model in which the charge of the two electrons is neutralized either by positive charges of value $e/2$ located in each dot, or by image charges located at a distance from the dots and representing the effect of metal electrodes or of Fermi level pinning at the semiconductor surface. Although a cell can be in the two configurations with the electrons aligned along one of the diagonals, other configurations are also possible. However, in most cases they are not energetically favored. A more complete model must also introduce such configurations, corresponding to the two electrons occupying the dots along one of the four sides of the cell. While the configurations with the electrons on the diagonals are associated with the logical values 1 and 0, the other configurations do not correspond to any logical value and are thus indicated with X in Figure 12.13, where all possible configurations are represented.

The total electrostatic energy is given by [24]:

$$E = \sum_{i \neq j} \frac{q_i q_j}{4\pi\varepsilon_0 \varepsilon_r r_{ij}} \tag{12.11}$$

If, for the sake of simplicity, the neutralizing charge is considered to be located directly in each dot (in an amount $e/2$), the total charge in each dot can assume only two values: either $+e/2$ or $-e/2$, thereby leading to

$$q_i q_j = \frac{1}{4} e^2 \operatorname{sgn}(q_i q_j) \tag{12.12}$$

If the distance between the dots is expressed in terms of the ratio $R = d/a$ and of the electron positions, the following can be written:

$$E = \frac{e^2}{4a} \frac{1}{4\pi\varepsilon_0 \varepsilon_r} \sum_{i \neq j} \frac{s_{ij}}{\sqrt{(n_{ij} R + l_{ij})^2 + m_{ij}^2}}, \tag{12.13}$$

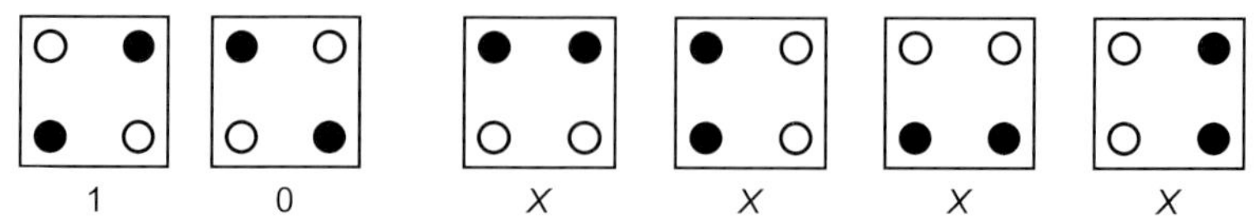

Figure 12.13 Possible configurations of a four-dot cell with two electrons. The configurations marked with X do not correspond to a well-defined logic state.

where $n_{ij} \in \{0, \ldots, N_{cell} - 1\}$ is the separation, in terms of number of cells, between the cell with dot i and the cell with dot j, $s_{ij} \in \{-1, 1\}$ is the sign of $q_i q_j$, $l_{ij} \in \{-1, 0, 1\}$ and $m_{ij} \in \{0, 1\}$, is the position of dots i and j within the relative cells. The quantity l_{ij} is equal to 0 if both the i and the j dots are on the left side or on the right side of the cell, to -1 if dot i is on the right side and dot j is on the left side, and to 1 if dot i is on the left side and dot j is on the right. Analogously, m_{ij} is equal to 0 if both dots i and j are on the top or on the bottom of a cell, to 1 if one dot is on the top and the other is on the bottom.

The most direct approach consists of computing the energy associated with each possible configuration by means of the direct evaluation of Eq. (12.13) and choosing the configuration that corresponds to the minimum energy. With this procedure the complete energy spectrum for the circuit is also obtained; that is, the energy values corresponding to all possible configurations. However, such a method soon becomes prohibitively expensive from a computational point of view, as the number of configurations to be considered is 6^N, where N is the number of active cells (i.e. of cells whose polarization is not enforced from the outside, as in the case of the driver cells). As the number of cells is increased, a simplified model can be used in which only the two basic states of a cell are considered, thus reducing the total number of configurations down to 2^N. This does not introduce significant errors, as long as the X states are unlikely (which is in general true), except for the case of intercell separation equal to or smaller than the cell size, when X states with both electrons vertically aligned may occur.

An example of application of the semi-classical simulation technique with six states per cell is shown in Figure 12.14, where the maximum operating temperature of a binary wire is reported as a function of the number of cells, for a 60%, 90% and 99% probability of obtaining the correct logical output, assuming an interdot distance of 40 nm, an intercell separation of 100 nm and GaAs material parameters. It should

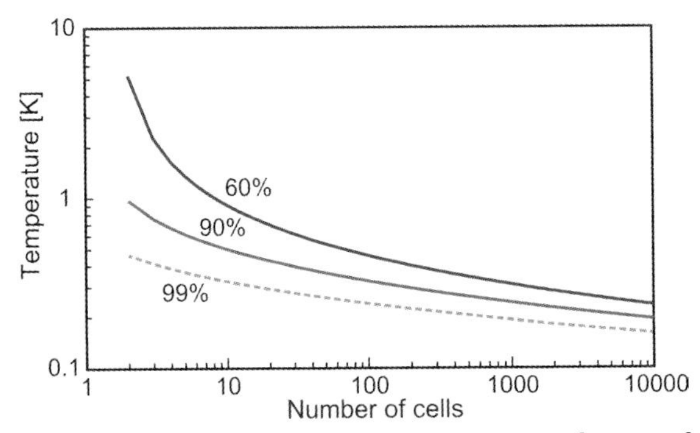

Figure 12.14 Maximum operating temperature, as a function of the number of cells, for a binary wire made up of GaAs cells with an interdot separation of 40 nm and an intercell separation of 100 nm. The maximum operating temperature has been computed for a 99%, 90% and 66% probability of obtaining the correct logical output.

be noted that the probability of obtaining the correct logical output is in general larger than the probability of being in the ground state, as there are also a number of excited states in which the polarization of the output cell has the correct value.

It is apparent that, even with the simplification down to just two states per cell, large circuits cannot be simulated with the semi-classical approach just described, because of the exponential increase in the time required to perform a complete exploration of the configuration space. This has led to the development of techniques based on an incomplete, targeted exploration of the configuration space, such as that described in the following subsection.

12.3.4 Simulated Annealing

The concept of simulated annealing derives from that of thermal annealing, whereby a material is brought into a more crystalline and regular phase by heating it and allowing it to cool slowly. Analogously, in simulated annealing the aim is to reach the ground state of the system, starting from a generic state at a relatively high temperature, and then to perform a Monte Carlo simulation in which at each step an elementary transition within a cell (chosen at random) is accepted with a probability P depending on the energy E_{old} of the system before the transition, and on the energy E_{new} after the transition:

$$P = \begin{cases} 1 & \text{if} \quad E_{new} \leq E_{old} \\ \exp[-(E_{new} - E_{old})/kT] & \text{if} \quad E_{new} > E_{old} \end{cases} \tag{12.14}$$

It is apparent that, in this way, the evolution of the system is steered along a path of decreasing energy, whilst at the same time trapping in a local minimum is prevented in most cases by the non-zero probability of climbing to a higher energy configuration. This procedure is iterated many times, gradually decreasing the temperature, until convergence to a stable configuration is achieved [17].

The application of simulated annealing to QCA circuits was originally proposed for their operation [25], and has since been applied to their modeling [26]. This has allowed the analysis of circuits with a number of cells of the order of 100 with limited computational resources and with just a few hours of CPU time. With large circuits, the simulated evolution of the circuit may occasionally become stuck in a local energy minimum, which would then be erroneously assumed as the ground state. The probability of this happening can be minimized by performing the equivalent of "thermal cycling". Once a stable state has been reached, the temperature is raised again, driving the circuit into an excited state, and a new annealing run is performed, reaching a new stable state. If the whole procedure is repeated several times, there is a better chance of reaching the ground state. It is possible to show that the probability P of the computational procedure stopping in the ground state is given by $P = 1 - (1 - P_0)^m$, where P_0 is the probability of reaching the ground state without cycling, and m is the number of cycles. With this technique it is possible to reliably simulate QCA circuits with a few hundreds of cells.

12.3.5
Existing Simulators

A number of simulators have been developed to study both the static and dynamic behaviors of QCA circuits. One of the first available was AQUINAS (A Quantum Interconnected Network Array Simulator, from the Notre Dame group), where cells are modeled within a Hartree–Fock approximation and the time-dependent Schrödinger equation is solved with the Crank–Nickolson algorithm. Relatively large systems can be handled, as a result of an approximation consisting in the representation of the state of a single cell with a simplified two-dimensional basis [27]. NASA researchers have added to AQUINAS capabilities for the statistical analysis of data, thus creating TOMAS (Topology Optimization Methodology using Applied Statistics) AQUINAS [28].

A static simulator for the determination of the ground state of a QCA circuit on the basis of a classical electrostatic model has been developed by the group in Pisa, and is currently available on the Phantoms Computational Hub (http://vonbiber.iet.unipi.it). The simulator has been named QCAsim, and operates according to the approach described in Section 12.3.3. In general, it can compute the ground-state configuration of a generic array of cells via a complete exploration of the configuration space, assuming for each cell six possible configurations for the two electrons. It is possible to specify whether neutralization charges should be included and in which positions (on the same plane as the electrons, on a different plane, as image charges, etc.).

The group in Pisa has also developed a dynamic simulator, MCDot (also available on the Phantoms HUB). This was conceived specifically for the QCA implementation based on metal tunnel junctions, and is therefore based on the Orthodox Theory of the Coulomb Blockade [29] with the addition of cotunneling effects treated to first order in perturbation theory [30]. The operation of such a code will be described in more detail in Section 12.4.3 while discussing limitations for the operating speed. Although the code was originally developed for circuits with metallic tunnel junctions, its range of applicability can be easily extended to different technologies, extracting appropriate circuit parameters and defining an equivalent circuit. For example, it has been successfully applied to the simulation of silicon-based QCA cells [17]. To this purpose, linearized circuit parameters can be determined from three-dimensional simulations around a bias point and then used in MCDot. The most challenging part of the parameter extraction procedure is represented by the capacitances and resistances of the tunneling junctions obtained by defining a lithographic constriction in silicon wires [31]: the detailed geometry and the actual distribution of dopants cannot be known exactly, and resort to experimental data is often necessary.

Recently, another simulator has been developed at the University of Calgary, QCADesigner (http://www.qcadesigner.ca). This uses a two-state model for the representation of each cell, derived from the theory developed by the Notre Dame group. QCADesigner is meant to be an actual CAD (computer-aided design) tool, applicable to the design of generic QCA circuits and with the capability for testing their operation with a targeted or exhaustive choice of input vectors.

12.4 Challenges and Characteristics of QCA Technology

12.4.1 Operating Temperature

As mentioned previously, the maximum achievable operating temperature is one of the main challenges in QCA technology. Indeed, the energy separation ΔE between the first excited state and the ground state of the system must be much larger than the thermal energy, if disruption of the operation as a result of thermal fluctuations is to be prevented. Unfortunately, the magnitude of the dipole interaction between cells is very small, of the order of millielectronvolts for cells with a size of a few tens of nanometers, and is further reduced by the screening action of nearby conducting electrodes and surfaces. This is why, with currently available technologies, the operation of a QCA circuit is not conceivable at temperatures beyond 10–20 K, and has so far been demonstrated only in the 100 mK range.

An increase in operating temperature requires an increase in the strength of the dipole interaction between cells, which can be achieved by reducing the size of the cell, by decreasing the dielectric permittivity of the material in which cells are embedded, or by resorting to a new type of interaction. As far as the dielectric permittivity is concerned, for semiconductor implementations silicon is more promising than gallium arsenide, because silicon dots can be defined by etching and be embedded in silicon oxide, which has a permittivity of 3.9 (much smaller than that of gallium arsenide, which is about 12). In Figure 12.15 the maximum operating temperature is plotted as a function of cell size for the silicon–silicon oxide and the gallium arsenide–aluminum gallium arsenide material systems. While for the GaAs/AlGaAs system the variation of the permittivity between the two materials is small, and can be neglected in approximate calculations, for silicon it is assumed that most of the electric field lines go through silicon-oxide (which encompasses the dots on all sides) and therefore its permittivity is used in the calculations. It is apparent that the

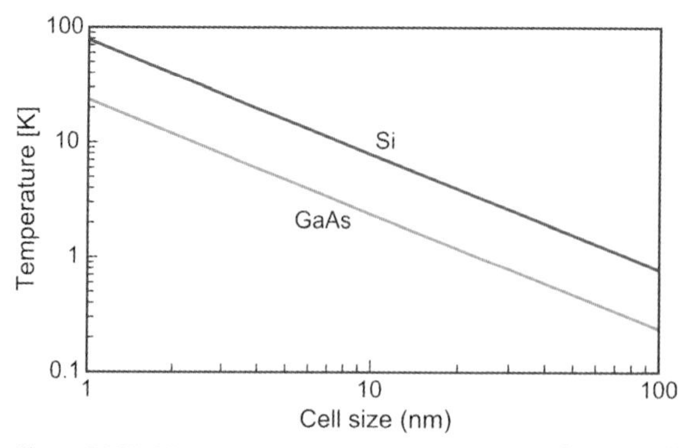

Figure 12.15 Maximum operating temperature as a function of the interdot separation within the cell, for gallium arsenide and silicon material systems.

stronger electrostatic interaction in silicon dots makes them suitable for relatively higher operating temperatures.

From Figure 12.15 it is however clear that an extremely small feature size would be needed to achieve operation at temperatures that are easily attainable.

An interaction that could allow QCA operation at room temperature is the one between nanomagnets characterized by a bistable behavior [17] (this will be discussed further in the next section). The magnetic interaction can be made strong enough to allow proper behavior of the circuit up to room temperature, but the achievable data processing speed is probably very low, of the order of a few kilohertz. On the other hand, a magnetic QCA circuit could exhibit an extremely reduced power dissipation, which could make it of interest for specific applications where speed is not a major issue, while keeping power consumption low is essential.

12.4.2 Fabrication Tolerances

The issue of fabrication tolerances has been introduced previously, and is probably the major limitation of the QCA concept, particularly for its implementation with semiconductor technologies. Detailed simulations [21] have shown that an approach based on the creation of an array of cells with dots defined by means of openings in a depletion gate cannot possibly lead to a working circuit. This is due to the fact that even extremely small errors in the geometry of such holes are sufficient to perturb the value of the confinement energy for the corresponding quantum dot to make it permanently empty or permanently occupied, no matter what the occupancy of the nearby dots is. Although shrinking the size of the cell the electrostatic interaction energy is increased, the above-mentioned problem becomes even more serious, due to the larger increase of the quantum confinement energy. An evaluation was made of the precision that could be achieved with state-of-the-art lithographic techniques and compared with the requirements for proper operation of a QCA circuit [32]. An array of square holes was obtained on a metal gate by means of electron beam lithography (Figure 12.16), after which the contour of the holes was extracted from a scanning

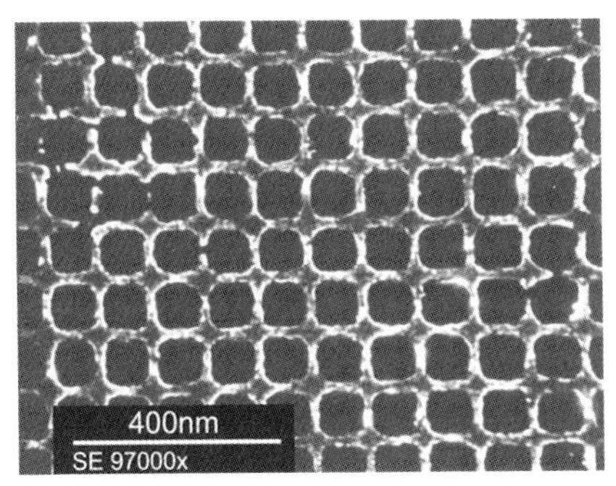

Figure 12.16 Scanning electron microscope image of the "hole array" that has been defined with state-of-the-art electron beam lithography for the purpose of evaluating the achievable precision.

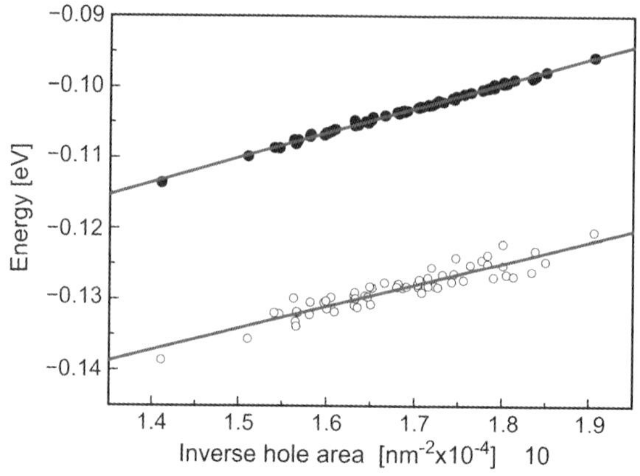

Figure 12.17 Scattering plot of the ground-state energy of single isolated quantum dots (closed circles) or of dots included in a cell (open circles) as a function of the inverse hole area.

electron microscope image and a solution of the Schrödinger equation was performed for the confinement potential obtained from each group of four holes (corresponding to a cell). The results showed a significant variance for the values of the confinement energy, as shown in Figure 12.17, where the ground-state energy for single dots and for four-dot cells is reported as a function of the inverse area of the holes defining them. From the almost linear dependence of the energy on the inverse area, it can be deduced that the local irregularities on the contour do not play an essential role, while the overall area is quite critical. It is clear that there is a dispersion of about 4 meV around the average value, while, from configuration–interaction calculations, it is shown that the allowed dispersion would be only 3 μeV, more than three orders of magnitude smaller.

Sensitivity to fabrication tolerances is ultimately the consequence of the same issue preventing operation at higher temperatures – that is, the smallness of the electrostatic interaction between cells. Imperfections are expected to play a role also with molecular-scale QCA circuits because, although molecules are in principle identical, once they are attached to a substrate any defects and stray charges from the substrate will disrupt the symmetry of the cells.

12.4.3
Limitations for the Operating Speed

The maximum operating speed of QCA circuits is ultimately limited by the dynamics of the evolution toward the ground state (if the ground-state relaxation paradigm is used), or by the tunneling rate between quantum dots. First, consider a non-clocked circuit, such as that represented for the case of a binary wire in Figure 12.18. The polarization state of the first cell is switched by inverting the bias voltages, and the cells of the wire will follow; however, according to a time evolution that may be rather

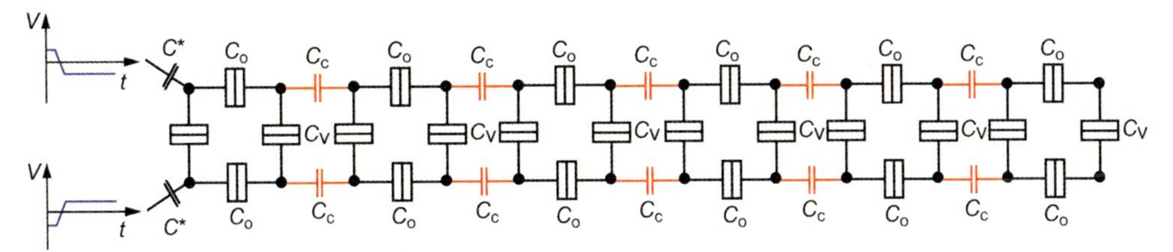

Figure 12.18 Equivalent circuit of a non-clocked binary wire; the voltages applied at the left end enforce the polarization state of the driver cell.

complex and involved. In particular, the presence of states that are very close in energy to the ground state, although corresponding to different configurations, will increase the time required for settling.

It is possible to obtain estimates of the time required for completion of the computation in a QCA array by performing simulations with a Monte Carlo approach. A Monte Carlo simulator specifically devised for QCA circuits was presented in Ref. [33]. This is based on the Orthodox Coulomb Blockade theory: the transition rate between two configurations differing by the position of one electron (which has tunneled through one of the junctions) and by a free energy variation ΔE can be expressed as

$$\Gamma = \frac{1}{e^2 R_T} \frac{\Delta E}{1 - \exp\frac{-\Delta E}{kT}}, \tag{12.15}$$

where R_T is the tunneling resistance of the junction.

Such a quantity is computed for all possible transitions, after which one of the transitions is chosen with a probability proportional to the corresponding rate. This procedure is repeated for each elementary time step into which the simulation period is divided, and the time-dependent currents in the branches of the circuit can be calculated from the contribution of the electron transitions.

This simulator can be used for the analysis of both clocked and unclocked circuits [17], as well as of a wide variety of single-electron circuits. By applying it to a six-cell binary wire with capacitances of the order of a few attofarads (values that are within the reach of lithographic technologies in the near future [33]), relaxation times of the order of 0.1 ms have been obtained; these are quite large, considering the extremely advanced technology needed for the fabrication of such devices. The reason for the slow operation is in the stochastic relaxation process, which brings the system to the ground state through a rather irregular path in the configuration space.

The Monte Carlo simulator MCdot can also be applied to the simulation of clocked QCA circuits. It has, for example, been used for the investigation of a clocked binary wire, as represented in Figure 12.19. The capacitance values are of the order of a few attofarads [33] and, at a temperature of 2.5 K, clock periods down to 0.1 μs can be achieved, as can be deduced from Figure 12.20, where the probability P_{co} of obtaining the correct logical state is plotted as a function of the clock interval for the second

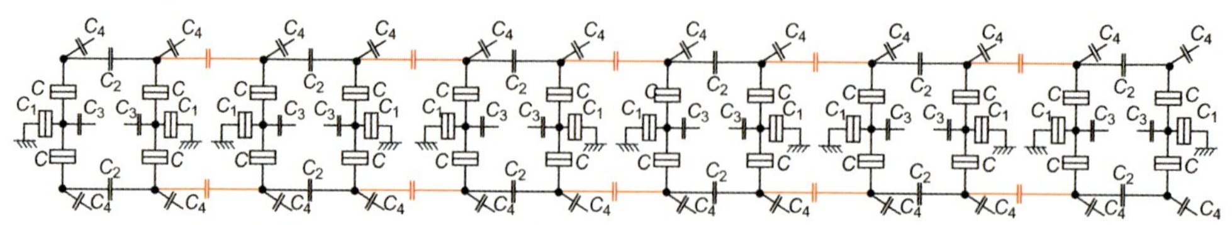

Figure 12.19 Equivalent circuit of a clocked binary wire; the C_4 capacitors are connected to voltage sources providing the bias required for proper operation, while the C_3 capacitors are connected to the clock signals.

(circles) and the last (squares) cells in the chain. As the tunneling resistances are assumed to be 200 kΩ, the resulting *RC* constant is of the order of 10^{-12} s. There is therefore a difference of about five orders of magnitude between the *RC* time constant of the circuit and the minimum clock period. This is due to a series of reasons [17]: the average time an electron takes to tunnel through a 200-kΩ junction with a 1.5-mV voltage is around 20 ps; furthermore the time during which the cell is active is only about one-tenth of the actual ramp duration; the active time, to be reasonably sure of regular operation, must be at least ten times the tunneling time; a clock period is made up of four ramps; and the intercell coupling is about five times smaller than the intracell coupling, which involves a further slow-down. Taken together, all of these effects lead to the above-mentioned reduction of the speed by five orders of magnitude with respect to the *RC* time constant, and make QCA technology not very suitable for high-speed applications.

On the other hand, the relatively slow operation of QCA circuits further limits their power dissipation, and in particular makes the power dissipated in capacitor charging–discharging negligible, as will be discussed in the next subsection.

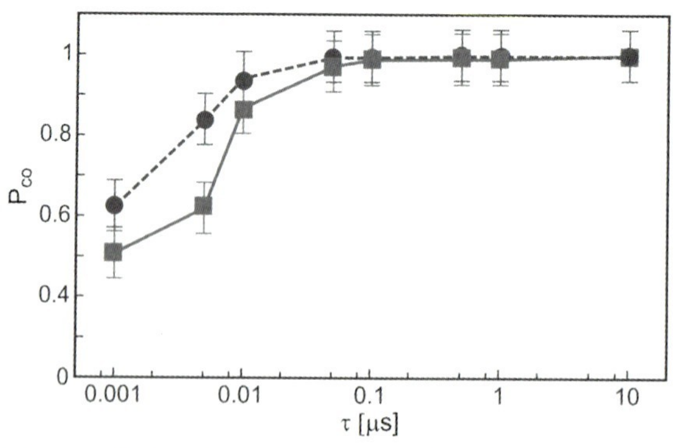

Figure 12.20 Probability (P_{co}) of correct logical output for the second (circles) and the third (squares) cell in a clocked binary wire as a function of the clock period. The *RC* constants of the circuit are of the order of 10 ps.

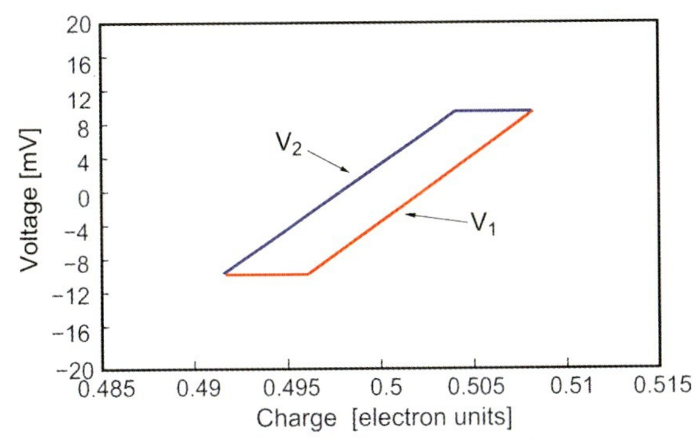

Figure 12.21 Representation of the voltages applied to a driver cell as a function of the charge flowing in the leads: the area inside the parallelogram corresponds to the work performed by the voltage sources on the QCA circuit.

12.4.4
Power Dissipation

One of the most attractive features of QCA circuits is represented by the limited power dissipation, which results mainly from the fact that there is no net transfer of charge across the whole circuit: electrons move only within the corresponding cell. The energy dissipated can be computed by integrating over the $V_i - Q$ plane [34] (Figure 12.21), where Q is the charge transferred from the source, for each external voltage source V_i and taking the algebraic sum of the results. The voltage V_2 is varied linearly until the unbalance is reversed: up to this point the charge variation corresponds to charging of the equivalent capacitance seen by V_2; then, some time after the new bias condition has been established, the electrons in the cell will tunnel, thus leading to a charge variation at constant voltage, which is represented by the horizontal segment. It is this tunneling event that makes the switch operation irreversible: without it, the area comprised between the two curves would be zero, as the voltage would simply be reversed across an equivalent capacitor, without changing its magnitude.

The energy dissipation depends on the voltage unbalance that is applied to the input cells, and that is reversed when the input data change: the larger the unbalance, the faster the switch, but the larger the dissipation, too. In the case of a single driven cell, for a typical unbalance of a few millivolts the power dissipation for a single switching event is about 10^{-22} J [35]. When considering a binary wire, the energy dissipated when the polarization of the driver cell is reversed, followed by all the other cells, and then increases very slowly as the number of cells is increased: the value for a six-cell wire is just 1% larger than that for a three-cell wire. This is due to the fact that the external voltage sources that provide energy are directly connected only to the driver cell, and the electrostatic action of the electrodes of the driver cell decays rapidly when moving along the chain. This leads to a very marginal contribution to the dissipated energy from the cells further away but, at the same time, is the fundamental reason for the above-mentioned slow and irregular switching of a long chain.

So far, the energy loss associated with the capacitor charging process has not been included. If the unbalance reversal were abrupt, such an energy loss would be (as well known) equal to the electrostatic energy stored in the capacitor, and therefore it would represent the main contribution to dissipation. However, due to the other limitations in switching speed, there is no reason to perform such a switching with very steep ramps. It transpires from calculations, applying the expressions typically used in adiabatic logic [36], that the energy loss in capacitor charging performed with a speed compatible with the response of the circuit is negligible with respect to that due to electron tunneling in all practical cases [35].

For the case of clocked circuits the simulation is more complex and must be performed over a complete clock cycle; the conclusion is however similar, as far as a single cell is concerned: about 6×10^{-22} J dissipated in a clock cycle for the above-mentioned typical circuit parameters. In the clocked case, however, the dissipated energy is supplied by the clock electrodes directly to each cell (or clocking zone, in the case of groups of cells sharing the same clock phase), and therefore there is a linear increase in the energy dissipation as the number of cells is increased, contrary to what happens with the unclocked circuits. Indeed, with the clocked architecture there is an improvement in terms of speed and regularity of operation, but the power consumption in increased. Also in this case, it is possible to show that the contribution from the energy loss associated with capacitor charging is negligible, because the clock ramps can be much slower than the relevant time constants in the circuit.

Overall, the dissipation for a switching event of a single QCA cell is four orders of magnitude smaller than that projected for an end-of-the-roadmap MOS transistor by the ITRS Roadmap. However, the transistor operates at 300 K, while the simulations have been performed, for the clocked case, at 2.5 K. Cooling down to such a temperature requires energy, which can be estimated on the basis of the efficiency of a Carnot cycle refrigerator [37]. Inclusion of the energy lost for cooling reduces the ratio of the energy dissipated in a transistor to that dissipated in a QCA cell by two orders of magnitude. The advantage of the QCA cell still remains two orders of magnitude, but a fair comparison would require a relatively large effort, as a larger number of QCA cells is in general needed than transistors in order to obtain the same logic function. Furthermore, the energy savings that can be obtained in CMOS adiabatic logic should also be taken into consideration.

12.5 Physical Implementations of the QCA Architecture

12.5.1 Implementation with Metallic Junctions

The implementation of QCA circuits with metal islands connected by tunnel junctions was introduced in the previous sections. Tunnel junctions with extremely small area (and therefore very small capacitance) can be fabricated between slightly overlapping electrodes, on top of a silicon oxide substrate, using the shadow mask evaporation technique. The QCA array in this case corresponds to a single-electron circuit, with tunnel junctions, capacitors, and voltage sources. From a technological point of view, a circuit with metallic junctions is relatively simple to implement, but has the major drawback, with currently available fabrication capabilities, of yielding capacitances no smaller than a few hundred attofarads [17], thus making operation possible only at temperatures of 100 mK or lower.

With such a technique, several QCA circuits have been demonstrated by the Notre Dame group: the basic driver cell-driven cell interaction [2], the operation of a clocked cell [12], a clocked QCA latch [38], and power gain in a QCA chain [34].

The problems connected with the undesired influence of each electrode on the proper balance of the cells via stray capacitances have been solved with a clever experimental scheme, based on an initial evaluation of the capacitance matrix among all electrodes. Once this is known, when the voltage of one electrode is varied, the voltages applied to all the other electrodes can be corrected in such a way as to compensate for the effects deriving from undesired capacitive couplings.

Although this technology has been very successful in the experimental demonstration of QCA operation, it appears very difficult to scale it down in order to increase the operating temperature so that it can be applied to large-scale circuits.

12.5.2
Semiconductor-Based Implementation

There are two main semiconductor implementations of QCA technology that have been attempted: one based on the Si/SiO_2 material system, and the other on the GaAs/AlGaAs material system. As previously stated, silicon has the advantage of the reduced permittivity of SiO_2, which allows the operating temperature to be raised but the fabrication of nanostructures in GaAs (defined by means of depletion gates) is more developed and tested.

For the approach based on GaAs/AsGaAs, a high-mobility, two-dimensional electron gas (2DEG) is formed at the heterointerface, and the quantum dots forming a cell are obtained by means of electrostatic depletion performed with properly shaped metal gates (Figure 12.22). In the experiments conducted to date, there is a hint of QCA effect, but it has not been possible to obtain full cell operation due to the too-small value of the capacitance coupling the upper with the lower dots across the barrier created by the horizontal electrode.

As the 2DEG is a few tens of nanometers below the surface, it is not possible to effectively define (at the 2DEG level) features that are significantly smaller; this also implies that dots cannot be made very close to each other, which represents a limitation on the maximum achievable interdot capacitance.

The advantage of GaAs technology is that tunnel barriers between dots can be finely tuned (contrary to what happens with the silicon–silicon oxide material system; see Section 12.5.3) by adjusting the bias voltage applied to the split gates defining them. In the cell represented in Figure 12.22, tunneling can occur between the top dots and

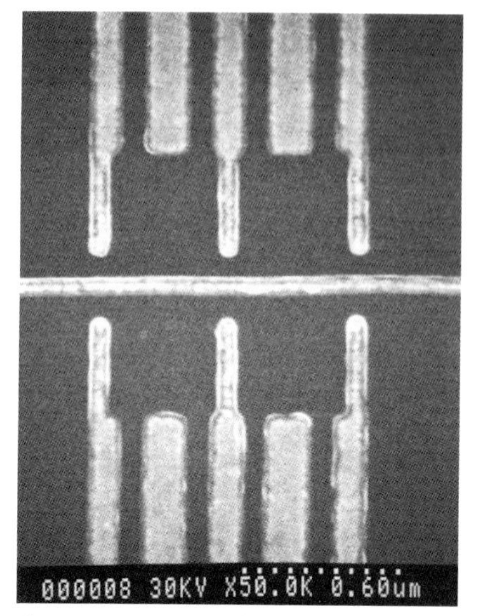

Figure 12.22 Scanning electron microscope image of the gate layout used for the investigation of the feasibility of a QCA cell in the GaAs/AlGaAs material system.

between the bottom dots, but not between one of the top dots and one of the bottom dots. This is not a problem, however, as the two configurations, with the electrons aligned along either diagonal, can still be achieved, and thus cell operation is unaltered.

A series of experiments has been performed on the prototype GaAs cell, operating, for example, the bottom part, while using one of the split gates in the top part as a non-invasive charge detector [39, 40], to monitor the motion of electrons between the two bottom dots. The outer quantum point contacts (QPC) in the bottom part are pinched off, to guarantee that the total number of electrons remains constant. Therefore, as a result of a variation of the voltage applied to the plunger gate of the dot at the bottom left (the shorter gate in the middle of the dot region), it is possible to observe motion of an electron from one dot to the other: as the plunger gate becomes more negative, an electron is moved from left to right. It has been observed [39] that the motion of an electron between the two bottom dots causes a shift of the Coulomb blockade peaks relative to one of the upper dots by about 20% of the separation between two consecutive peaks: this coupling is estimated to be sufficient to determine a reverse motion of an electron between the upper dots (i.e. the basic QCA effect).

The gate layout used for this experiment can also be applied to the implementation of a binary wire, but not for general logic circuits, because lateral branching is not possible due to the presence of the leads reaching each gate. It should also be pointed out that, even for the implementation of a simple binary wire, a careful balancing procedure would be needed, because even the finite length of the wire may be sufficient to create a fatal unbalance for all the cells, except for the one in the middle [23].

The other semiconductor implementation that has been attempted is based on silicon dots embedded in silicon oxide. As mentioned above, this material system has

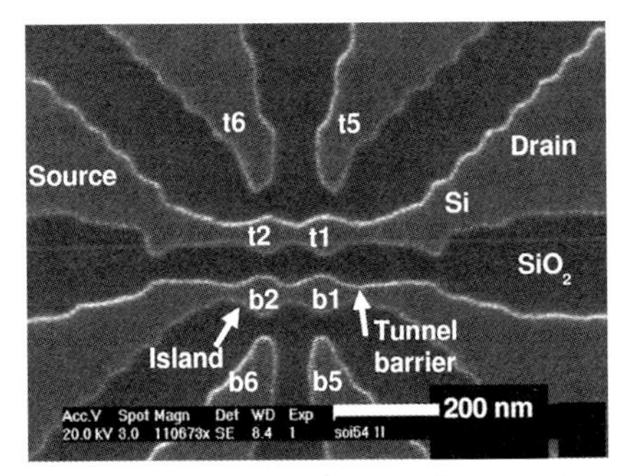

Figure 12.23 Scanning electron microscope image of a prototype QCA cell fabricated with the silicon/silicon oxide material system.

the advantage of the lower permittivity of silicon oxide with respect to gallium arsenide. However, although smaller feature sizes are achievable, control of the tunnel barriers is quite difficult as they are obtained by lithographically defining a narrower silicon region between two adjacent dots [31]. A prototype silicon QCA cell was fabricated at the University of Tübingen starting from a SOI (Silicon-On-Insulator) substrate, defining the structure by means of electron-beam lithography and reactive ion etching. The lithographically defined features are then further shrunk by means of successive oxidations [41]. It can be seen in Figure 12.23 that the tunneling junctions (between the two upper and the two lower dots) have been obtained by creating a narrower region, with a cross-section small enough that it does not allow propagation of electrons at the Fermi energy.

Such tunnel junctions are not easily controllable and, depending on the value of the Fermi energy and on the distribution of charged impurities, they may contain multiple barriers. However, it has been shown [17, 42] that, by properly tuning the back-gate voltage and the bias voltages applied to the gates, it is possible to achieve a condition in which both junctions contain a single barrier. Clear control of dot occupancy by means of the external gates has been demonstrated, by monitoring the conductance of the upper and lower double-dot systems. A clear demonstration of the QCA effect has not yet been possible due to the limited capacitive coupling between the upper and the lower double dots. However, simulations have shown [17] that a modified layout, with reduced spacing between the upper and the lower parts, should allow the observation of cell operation at a temperature of 0.3 K (probably up to 1 K), which is definitely higher than that required for metal dots and for GaAs. Unfortunately, also in this case, the basic layout used for the experiments should be significantly modified to make it suitable for complex circuits.

12.5.3 Molecular QCA

Another possible approach to the implementation of QCA circuits, as pioneered by Lent [4], is based on single molecules: this would satisfy the ultimate miniaturization requirement (a single molecule for a single computational function) and possibly reduce the precision constraints, exploiting the fact that molecules are identical by construction. Furthermore, approaches to fabrication based on self-assembly could be envisioned, which would significantly decrease fabrication costs.

The molecular QCA concept relies on molecules containing four (or possibly two) sites where excess electrons can be located and which are separated by potential barriers. It has been demonstrated that potential barriers do exist at the molecular level and that they do lead to bistability effects [7]: a simple example is represented by a methylene group ($-CH_2-$) placed between two phenyl rings.

For the implementation of a complete cell, several candidate molecular structures have been proposed, such as metallorganic mixed-valence compounds containing four ruthenium atoms that represent the four dots. However, investigations are continuing to determine whether sufficient coupling is achievable between cells, because the screening action of the electronic clouds of the ligand atoms may determine too large a suppression of the electrostatic interaction. Furthermore, the problem of attaching the molecules to a substrate, in order to create properly structured arrays, is still only partially solved. In particular, the presence of imperfections or unavoidable stray charges at the surface of the substrate may create asymmetries preventing correct QCA operation, notwithstanding the identity of all molecules.

A simple molecule that has been proposed by the Notre Dame group as a model system for half of a cell is the so-called Aviram molecule [43], in which the two dots are represented by allyl groups at the ends of an alkane chain. Quantum chemistry calculations have shown that some bistability effects can be obtained, as well as sufficient electrostatic interaction between neighboring molecules, although, due to the reactivity of the allyl groups and to the difficulty to attach this molecule to a substrate, it does not seem a likely candidate for experiments.

Whilst overall the molecular approach seems the most appropriate solution for the implementation of the QCA concept in the long term, many problems – some of which are fundamental in nature – remain unsolved, such as finding a reliable way to assemble molecular arrays, managing the effect of stray charges, and determining whether the interplay of molecular sizes and screening effects will allow reasonably high operating temperatures.

12.5.4
Nanomagnetic QCA

To date, implementations of the QCA concept have been considered that rely on an electrostatic interaction between dots within a cell and between neighboring cells. It is also possible, as mentioned in the introduction, to exploit other forms of interaction, which may be less susceptible to the effects of temperature and of imperfections. One such alternative solution is represented by nanomagnetic QCA circuits. The concept is rather simple: an array of elongated single-domain dots obtained from properly chosen magnetic material will relax into an antiferromagnetic ordering, and it will be possible to drive the evolution of the system with an external clock consisting in an oscillating magnetic field that also supplies the energy needed for power gain along the circuit. The first experimental investigation into the possibility of propagating the magnetic polarization along a chain of nanomagnets was performed by Cowburn and Welland [44], who managed to show the operation of a chain of magnetic nanodots.

The specific nanomagnetic approach to QCA circuits has been investigated mainly by Csaba and Porod, who have determined that an energy difference between the ground state and the first excited state of 150 kT at room temperature can be achieved with elongated nanomagnets that are manufacturable with existing technologies. However, there is also a relatively high barrier (100 kT) between the two states, and therefore at room temperature the system would be stable in both configurations. Thus, a pure ground-state relaxation scheme is not applicable, and resort must be made to the above-mentioned oscillating clock field. Such a field is used to turn the magnetic moments of all nanomagnets into a neutral state, from which they can relax into the ground state (as long as they remain in the instantaneous ground state).

A chain with an even number of cells (including the driver cell) will act as an inverter, as antiferromagnetic ordering is present. Thus, implementation of the NOT gate is straightforward; the majority voting gate can be implemented [17] in a way similar to that for electrostatically coupled QCA systems, with three binary wires converging on a single cell, from which another binary wire representing the output departs. Therefore, a generic combinatorial network can be realized in a way quite similar to what has been seen for other QCA technologies.

Simulators for nanomagnetic QCA circuits have been developed by Csaba and Porod [45], in which the complete micromagnetic equations are solved numerically. It has also been noticed that, for dot sizes below 100 nm, a significant simplification can be used – the single-domain approximation – in which the magnetization condition of the dots can be represented by means of single vectors instead of vector fields. Such an approximation is valid because magnetic dots below a size of 100 nm operate as single domains. The equations governing the evolution of single domains can be written as a system of ordinary differential equations and may then be recast into the form of a standard SPICE model. This allows efficient and easy simulation of relatively complex architectures of nanomagnetic QCAs, and has made it possible to show that logic signal restoration and power gain can be achieved, at the expense of the external oscillating magnetic field.

12.5.5 Split-Current QCA

An alternative approach to QCA implementation has been proposed by Walus *et al.* [46], who suggested a QCA cell in which tunneling of electrons between the dots does not take place; rather, the interaction is between tunneling currents that flow vertically through a double resonant tunneling diode structure. The cross-section of the cell proposed by Walus *et al.* is shown in Figure 12.24: the lower quantum well region of the double-resonant tunneling structure is partitioned into four dots in the horizontal plane.

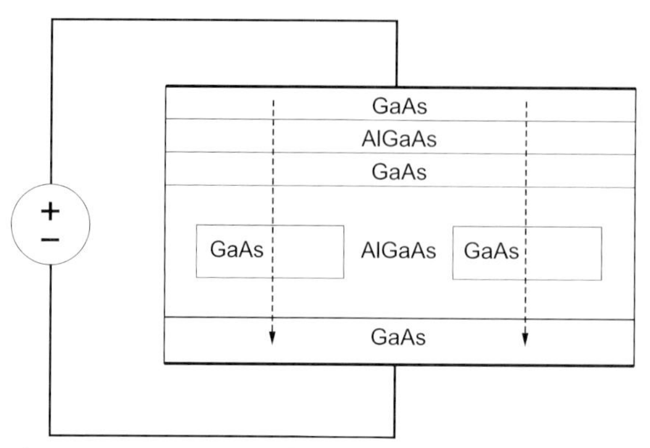

Figure 12.24 Cross-section of the split-current QCA cell, based on four parallel double-resonant tunneling structures.

The current flows mainly through these four dots, and the actual value of the current through each dot is strongly dependent on the position of the alignment of the energy levels in the upper and lower GaAs wells. Starting from a situation where the upper and lower levels are aligned, the flow of current will create a charge density that, in turn, will perturb the position of the resonant levels in the nearby dots: the larger the current, the greater the induced level shift. Therefore, for an isolated cell, the rest condition will be with current flowing mainly through either pair of antipodal dots (i.e. the dots that are furthest from each other), so that the resonant level shift is minimized. If another, driver, cell is placed next to it, with a well-defined polarization, the same polarization state will be obtained, as a result of Coulomb interaction between dots. Thus, operation similar to that of previously discussed electrostatically coupled cells will be achieved. The authors of Ref. [46] suggest that clocking is also possible, by controlling the voltage applied in the vertical direction across the resonant tunneling structures.

Although interesting in principle, this approach forfeits one of the main advantages of the QCA architecture, namely the potentially extremely low power consumption, since non-zero currents flowing in the vertical direction through the resonant tunneling diodes are always present, except for the "null" phase of the clock.

12.6 Outlook

The QCA concept has been the subject of significant research activity throughout the past decade, leading to results of general interest in the field of nanoelectronics. The practical implementation of QCA circuits is, however, still elusive, because of a few major problems connected with the weakness of the proposed cell-to-cell interaction mechanisms and with the extreme sensitivity to fabrication tolerances. Novel concepts are being explored, in particular aimed at the ultimate miniaturization, with cells consisting of single molecules, or aimed at an increase of inter-cell interaction, with cells made up of single-domain nanomagnets.

It is possible that these will lead to applications in niche markets in which extremely low power consumption is essential and high data processing speed is not a requirement. On the basis of the limited achievable switching speed and of the functional density not expected (if realistically evaluated) to be much higher than that achievable with CMOS technology, it is unlikely that QCA circuits will find application in large-scale integration. The QCA concept, however, can be at the basis of applications that go beyond its original purpose, for example in the field of metrology, where some of its weaknesses, such as the extreme sensitivity to external charges, may become important assets.

Overall, in the – so far unsuccessful – quest for a technology capable of succeeding CMOS, QCA have represented a very interesting diversion. Although such an approach may be too bold in relation to existing and near-future technological capabilities, it has contributed a wealth of novel understanding about the ultimate limitations of computation at the nanoscale.

References

1 C. S. Lent, P. D. Tougaw, W. Porod, *Appl. Phys. Lett.* 1993, **62**, 714.

2 I. Amlani, A. O. Orlov, G. L. Snider, C. S. Lent, G. H. Bernstein, *Appl. Phys. Lett.* 1998, **72**, 2179.

3 C. S. Lent, P. D. Tougaw, *Proc. IEEE* 1997, **85**, 541.

4 C. S. Lent, *Science* 2000, **288**, 1597.

5 G. Csaba, A. Imre, G. H. Bernstein, W. Porod, V. Metlushko, *IEEE Trans. Nanotechnol.* 2002, **1**, 209.

6 P. Bakshi, D. A. Broido, K. Kempa, *J. Appl. Phys.* 1991, **70**, 5150.

7 M. Girlanda, M. Macucci, *J. Phys. Chem. A* 2003, **107**, 706.

8 C. S. Lent, P. D. Tougaw, W. Porod, in: Proceedings of the Workshop on Physics and Computation - Physcomp, Dallas, Texas, November 17–20, IEEE Computer Press, pp. 1–13, 1994.

9 C. Ungarelli, S. Francaviglia, M. Macucci, G. Iannaccone, *J. Appl. Phys.* 2000, **87**, 7320.

10 A. N. Korotkov, *Appl. Phys. Lett.* 1995, **67**, 2412.

11 D. J. Griffths, *Introduction to Quantum Mechanics*, Prentice-Hall, Englewood Cliffs, NJ, 1994.

12 A. O. Orlov, I. Amlani, R. K. Kummamuru, R. Ramasubramaniam, G. Toth, C. S. Lent, G. H. Bernstein, G. L. Snider, *Appl. Phys. Lett.* 2000, **77**, 295.

13 G. Toth, C. S. Lent, *J. Appl. Phys.* 1999, **85**, 2977.

14 M. Girlanda, M. Governale, M. Macucci, G. Iannaccone, *Appl. Phys. Lett.* 1999, **75**, 3198.

15 P. D. Tougaw, C. S. Lent, *J. Appl. Phys.* 1994, **75**, 1818.

16 C. A. Stafford, S. Das Sarma, *Phys. Rev. Lett.* 1994, **72**, 3590.

17 M. Macucci (Ed.), *Quantum Cellular Automata: Theory, Experimentation and Prospects*, Imperial College Press, London, 2006.

18 R. McWeeny, *Methods of Molecular Quantum Mechanics*, Academic Press, London, 1989.

19 G. W. Bryant, *Phys. Rev. Lett.* 1987, **59**, 1140.

20 M. Brasken, M. Lindberg, D. Sundholm, J. Olsen, *Phys. Rev. B* 2000, **61**, 7652.

21 M. Governale, M. Macucci, G. Iannaccone, C. Ungarelli, J. Martorell, *J. Appl. Phys.* 1999, **85**, 2962.

22 S. Gardelis, C. G. Smith, J. Cooper, D. A. Ritchie, E. H. Linfield, Y. Jin, *Phys. Rev. B* 2003, **67**, 033302.

23 M. Girlanda, M. Macucci, *J. Appl. Phys.* 2002, **92**, 536.

24 J. D. Jackson, *Classical Electrodynamics*, Wiley, New York, 1962.

25 M. Akazawa, Y. Amemiya, N. Shibata, *J. Appl. Phys.* 1997, **82**, 5176.

26 M. Macucci, G. Iannaccone, S. Francaviglia, B. Pellegrini, *Int. J. Circul. Theoret. Appl.* 2001, **29**, 37.

27 P. D. Tougaw, C. S. Lent, *J. Appl. Phys.* 1996, **80**, 4722.

28 C. D. Armstrong, W. M. Humphreys, A. Fijany, The design of fault-tolerant quantum dot cellular automata based logic, in: *Proceedings, 2nd International Workshop on Quantum Dots for Quantum Computing and Classical Size Effect Circuits*, University of Notre Dame, August 7–9, 2003. Also available at: http://www.cambr.uidaho.edu/symposiums/symp11.asp.

29 V. Averin, K. K. Likharev, in: B. L. Altshuler, P. A. Lee, R. A. Webb (Eds.), *Mesoscopic Phenomena in Solids*, Elsevier, Amsterdam, 1991.

30 L. R. C. Fonseca, A. N. Korotov, K. K. Likharev, A. A. Odinstov, *J. Appl. Phys.* 1995, **78**, 3238.

31 R. Augke, W. Eberhardt, C. Single, F. E. Prins, D. A. Wharam, D. P. Kern, *Appl. Phys. Lett.* 2000, **76** 2065.

32 M. Macucci, G. Iannacone, C. Vieu, H. Launois, Y. Jin, *Superlatt. Microstruct.* 2000, **27**, 359.

33 L. Bonci, M. Gattobigio, G. Iannaccone, M. Macucci, *J. Appl. Phys.* 2002, **92**, 3169.

34 R. K. Kummamuru, J. Timler, G. Toth, C. S. Lent, R. Ramasubramaniam, A. O. Orlov, G. H. Bernstein, G. L. Snider, *Appl. Phys. Lett.* 2002, **81**, 1332.

35 L. Bonci, M. Macucci, in: Proceedings of the European Conference on Circuit Theory and Design, Cork, Ireland, vol. II, p. 239. Also available at: http://ieeexplore.ieee.org/xpl/RecentCon.jsp?punumber=10211.

36 R. C. Merkle, *Nanotechnology* 1993, **4**, 21.

37 International Technology Roadmap for Semiconductors (ITRS), 2005 Edition Semiconductor Industry Association.

38 A. O. Orlov, R. K. Kummamuru, R. Ramasubramaniam, G. Toth, C. S. Lent, G. H. Bernstein, G. L. Snider, *Appl. Phys. Lett.* 2001, **78**, 1625.

39 M. Field, C. G. Smith, M. Pepper, D. A. Ritchie, J. E. F. Frost, G. A. C. Jones, D. G. Hasko, *Phys. Rev. Lett.* 1993, **70**, 1311.

40 G. Iannaccone, C. Ungarelli, M. Macucci, E. Amirante, M. Governale, *Thin Solid Films* 1998, **336**, 145.

41 C. Single, R. Augke, F. E. Prins, D. P. Kern, *Semicond. Sci. Technol.* 1999, **14**, 1165.

42 C. Single, F. E. Prins, D. P. Kern, *Appl. Phys. Lett.* 2001, **78**, 1421.

43 A. Aviram, *J. Am. Chem. Soc.* 1988, **110**, 5687.

44 R. P. Cowburn, M. E. Welland, *Science* 2000, **287**, 1466.

45 G. Csaba, W. Porod, *J. Comput. Electronics* 2002, **1**, 87.

46 K. Walus, A. Budiman, G. A. Jullien, *IEEE Trans. Nanotech.* 2004, **3**, 249.

13
Quantum Computation: Principles and Solid-State Concepts

Martin Weides and Edward Goldobin

... how can we simulate the quantum mechanics? ... Can you do it with a new kind of computer - a quantum computer? It is not a Turing machine, but a machine of a different kind. R. P. Feynman, 1982 [1]

13.1
Introduction to Quantum Computing

For half a century the conventional electronic information processing based on Boolean logic and using a von Neumann-type architecture has been very successful in solving many numerical problems, and its computational power is still increasing. The term *Neumann-type architecture* describes a device, which implements a so-called *Turing machine* by a specifying sequential architectures of information processing. In short, a *Turing machine* contains a program (software), a finite state control, a tape (memory) and a read-write tape-head [2]. It can be proven that conventional computers are equivalent to a Turing machine.

However, there are – and will continue to be – some restrictions for conventional computation, as will be seen below. Most of today's electronics is based on devices with digital logics, apart from the very specialized analog computers, which can solve for example differential equations up to a certain size. Both the feature size and energy consumption per logic step of conventional computers have been much reduced in the recent decades, and will continue to do so for some more years [3]. The *total* energy dissipation per unit area is still increasing due to increasing packaging density. A power density of $\sim$100 W cm^{-2} can be regarded as a reasonable limit, given by the thermal conductivity of materials and the geometry for setting up temperature gradients. Note that a common kitchen heating plate at full power has as a tenth of this power density. This limit will be reached in a few decades.

If one day the energy dissipation per a single logic step is equal to $k_BT\ln 2$, where $k_B = 1.38 \times 10^{-23}$ J K^{-1} is the Boltzmann constant and T the theoretical limiting

Nanotechnology. Volume 4: Information Technology II. Edited by Rainer Waser
Copyright © 2008 WILEY-VCH Verlag GmbH & Co. KGaA, Weinheim
ISBN: 978-3-527-31737-0

value of temperature, the device has to use a so-called reversible logic (see Section 13.2). The implementation of reversible computing demands a precise control of the physical dynamics of the computation machine to prevent a partial dissipation of the input information (i.e. energy) into the form of heat. One type of reversible computer is the *quantum computer* which, by definition, relies on the time-reversible quantum mechanics.

A quantum computer is a device for information processing that is based on distinctively quantum mechanical phenomena, such as quantization of physical quantities, superposition and entanglement of states, to perform operations on data. The amount of data in a quantum computer is measured in quantum bits or *qubits*, whereas a conventional digital computer uses binary digits, in short: *bits*. The quantum computation relies on the quantum information carriers, which are single quantities, whereas the conventional computer uses a huge number of information carriers. In addition, the quantum devices may be much more powerful than the conventional devices, as a different class of (quantum) algorithm exploiting quantum parallelism can be implemented. Some specific problems cannot be solved efficiently on classical computers because the computation time would be astronomically large. However, they could be solved in reasonable time by quantum computers. This emerging technology attracts much attention and effort, both from the scientific community and industry, although the possibility of building such a device with a large number ($\gg$10) of qubits is still not answered.

For a more detailed introduction to classical and quantum computation, the interested reader is referred to a great selection of excellent textbooks such as, for example, Feynman Lectures on Computation [2] and others [4, 5].

13.1.1 The Power of Quantum Computers

In recent years there has been a growing interest in quantum computation, as some problems, which are practically intractable with classical algorithms based on digital logic, can be solved much faster by massive parallelism provided by the superposition principle of quantum mechanics.

In theoretical computer science, all problems can be divided into several classes of complexity, which represents the number of steps of the most efficient algorithm needed to solve the problem. The class **P** consists of all problems that can be solved on a Turing machine in a time which is a *polynomial* function of the input size. The class **BPP** (Bound-error, Probabilistic, Polynomial time) is solvable by a *probabilistic* Turing machine in polynomial time, with a given small (but non-zero) error probability for all instances. It contains all problems that could be solved by a conventional computer within a certain probability. It comprises all problems of **P**. The class **NP** consists of all problems whose solutions can be *verified* in polynomial time, or equivalently, whose solution can be *found* in polynomial time on a non-deterministic machine. Interestingly, no proof for $\mathbf{P} \neq \mathbf{NP}$ – that is, whether **NP** is the same set as **P** – has yet been found. Thus, the possibility that $\mathbf{P} = \mathbf{NP}$ remains, although this is not believed by many computer scientists.

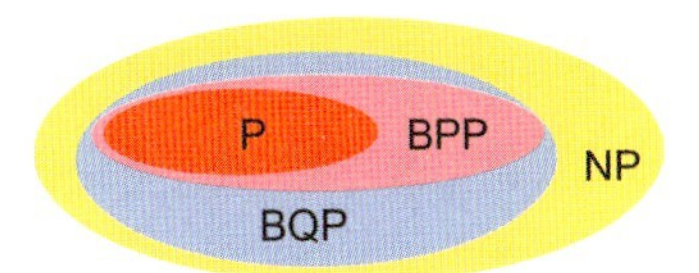

Figure 13.1 Diagram of complexity classes. A quantum computer can solve **BQP** (Bounded error, Quantum, Polynomial time) problems, whereas digital logic can just solve problems from the **P** (polynomial) class efficiently.

The class of problems that can be efficiently solved by *quantum computers*, called **BQP** (bounded error, quantum, polynomial time), is a strict superset of **BPP** (see Figure 13.1). For details on complexity classes, see Nielsen and Chuang [4].

The most important problems from the **BQP** class that cannot be solved efficiently by conventional computation (i.e. they do not belong to **P** or **BPP** class) but by quantum computation, are summarized below.

13.1.1.1 Sorting and Searching of Databases (Grover's Algorithm)

Quantum computers should be able to search unsorted databases of N elements in $\sim\sqrt{N}$ queries, as shown by Grover [6], rather than the linear classical search algorithm which takes $\sim N$ steps of a conventional machine (see Figure 13.2). This speedup is considerable when N is getting large.

13.1.1.2 Factorizing of Large Numbers (Shor's Algorithm)

A quantum algorithm for the factorization of large numbers was proposed by Shor [7], who showed that quantum computers could factor large numbers into their prime factors in a polynomial number of steps, compared to an exponential

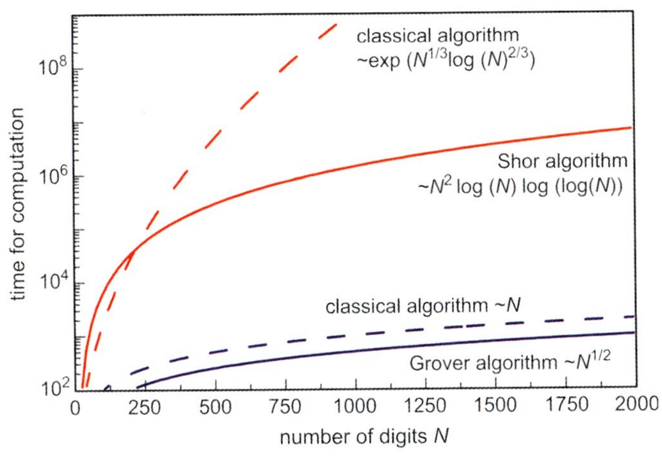

Figure 13.2 (In red): Factorization of a N digit number. (In blue): searching an unsorted database of N elements. Time requested for classical (dashed line) versus quantum (continuous line) algorithm. Note the logarithmic scale of the ordinate.

number of steps on classical computers (see Figure 13.2). The difficulty of prime factorization is a cornerstone of modern public key cryptographic protocols. The successful implementation of Shor's algorithm may lead to a revolution in cryptography.

13.1.1.3 Cryptography and Quantum Communication

Contrary to the classical bit, an arbitrary quantum state cannot be copied (no-cloning theorem; see Section 13.3.4) and may be used for secure communication by means of quantum key distribution or quantum teleportation. By sharing an entangled pair of qubits (so-called EPS-pair after a famous paper from Einstein, Podolski and Rosen [8]), signals can be transmitted that cannot be eavesdropped upon without leaving a trace; that is, performing a measurement and thereby destroying the entangled state.

All of these three important challenges for quantum computations have been implemented on NMR qubits or photons; that is, their feasibility has been proven using a small set of qubits.

13.2 Types of Computation

Here, information theory and logic will be briefly reviewed.

13.2.1 Mathematical Definition of Information

From statistical thermodynamics it is known that the entropy S of a system is defined as

$$S = k_B \ln\Omega$$

where Ω is the number of possible configurations or microscopic states. If an ideal gas is considered in a given volume and compressed under *isothermal* (T=const.) conditions, the mean translational kinetic energy of the gas is not changed. However, its entropy is reduced, because fewer possible positions are available for the gas atoms in the new volume. Conservation of the total energy leads to a dissipation of thermal energy

$$\Delta W = T\Delta S$$

to the outside thermal reservoir.

In information theory the situation is similar, the aim being to measure information [9]. The number Ω of possible configurations of a binary code with m elements is

$$\Omega = 2^m.$$

A stored bit can be in one of two states. If these states have the same probability (i.e. $\Omega = 2$), the minimum entropy associated with this bit is

$$S = k_B \ln \Omega = k_B \ln 2.$$

In case that the number of bits m is reduced during the computation process, for example by logic gates with less output than input bits, the energy dissipation per lost bit is given by

$$\Delta W = T \Delta S = k_B T \ln 2.$$

This is the so-called *Landauer principle* [10], which states that erasure is not thermodynamically reversible. Any loss of information inherently leads to a (minimum) energy dissipation of this amount per reduced bit in the case of an isothermal operation, as the entropy of the systems changes. In contrast, to prevent the thermal energy destroying the stored information, the minimum power for storage of information is $k_B T \ln 2$. The *energy consumption* in computation is closely linked to the *reversibility* of the computation.

13.2.2 Irreversible Computation

In all switching elements with more input than output bits, a loss of information occurs upon the information processing. For example, the Boolean function AND is defined as

$$(0,0) \to 0; \quad (0,1) \to 0; \quad (1,0) \to 0; \quad (1,1) \to 1\,.$$

The 0 output of this conventional two-valued gate can be caused by three possible input signal configurations. This type of computing is called *irreversible*.

13.2.3 Reversible Computation

Reversible computation needs to be based on switching elements which do not lose information. An example is the NOT gate, which is a single bit in- and output

$$(0) \to (1); \quad (1) \to (0)$$

or the controlled NOT (CNOT) gate, which makes the XOR operation reversible $(x, y) \to (x, x\ XOR\ y)$:

$$(0,0) \to (0,0); \quad (0,1) \to (0,1); \quad (1,0) \to (1,1); \quad (1,1) \to (1,0)\,.$$

The input can always be deduced from the output.

Now, this should be considered from the physical point of view. The classical two-state system is prepared and stored in the two stable states 0 or 1, respectively. It can be characterized by a particle placed in a double-well potential (see Figure 13.3a). To start the controlled reversible switching of a bit, some definite energy must be fed into the system to overcome the energy barrier separating the two minima of the double well potential. The energy is then available to perform a switching in an adjacent minimum of the potential, and remains available for subsequent switching processes

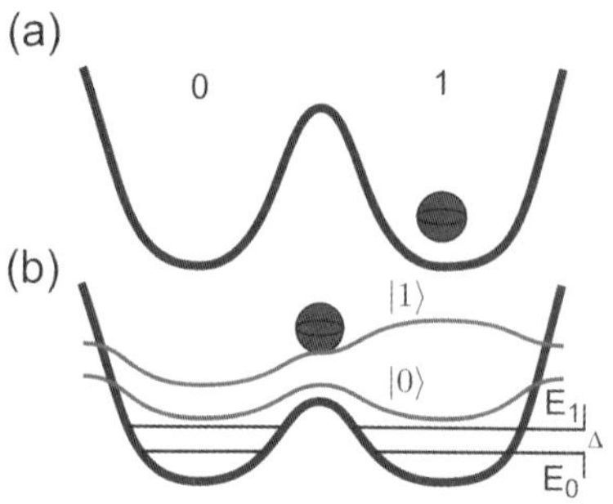

Figure 13.3 Representation of information in (a) a classical computer by a bit with two states (0, 1) and (b) a quantum computer by a quantum-mechanical two-level systems. $|0\rangle$, $|1\rangle$ denote the two quantum states.

or to switch the system back into its original state. Thus, both forward- and backward-switching processes have the same probabilities and the net speed of the computational process is zero. The flow of information must be determined by the gates and an adequate feedback prevention must be built into the logic gates.

13.2.4
Information Carriers

In classical information-processing systems the information transfer is based on the flow of particles, that is, electronic charges. A huge number of information carriers is involved for fault-tolerant operations. The information processing itself can be done by either reversible or irreversible logic gates.

A pure reversible information transfer could be implemented by means of discrete stationary states of a microscopic system which interacts by fields to move the information and to set the logic states. For example, the Quantum Cellular Automata (QCA) (see Chapter 12) are based on elementary cells with two stable states (0 or 1), which can be toggled by fields emerging from neighboring cells. There is no flow of charges or particles, as single atomistic entities (electrons, electron spin, small molecules) solely change their position in a potential well. In principle, an ideal QCA circuit would operate in the thermodynamic limit of information processing and at the same time is still calculating with conventional Boolean logic.

13.3
Quantum Mechanics and Qubits

In quantum computation the information is represented by the quantum properties of particles, so-called *qubits*, and its operations are devised and built by quantum mechanisms. The information content of a single qubit is obtained by a measurement process. An ideal quantum mechanical measurement of a single qubit can only measure one degree of freedom, and returns either 0 or 1. The information encoded in a quantum mechanical system is described by a vector in the Hilbert space. The components of the Hilbert space vector denote probability amplitudes related to the outcome of certain measurements. Physical pure states are unit-norm vectors in the Hilbert space. Any observable of a system – that is, any quantity which can be measured in a physical experiment – should be associated with a self-adjoint linear operator. The measurable values are the eigenvalues of this operator and their probability is related to the projection of the physical state on the respective subspace. To keep the norm of the physical state fixed the operator should be unitary; that is, all eigenvalues of the operator matrix are complex numbers having absolute value 1.

Shortly, the aim of this subchapter is to summarize the important features of quantum mechanics. A good introduction can be found in various textbooks, for example in Ref. [11].

- *Quantization of states:* The observable quantities do not vary continuously but come in discrete *quanta*. In real systems they can be represented by electric charge, spin, magnetic flux, phase of electromagnetic wave, chemical structure, and so on.
- *Superposition of states:* The linear combination of two or more eigenstates results in a quantum superposition. If the quantity is measured, the state will randomly collapse onto one of the values in the superposition with a probability proportional to the square of the amplitude of that eigenstate in the linear combination. This superposition of states makes quantum computation qualitatively more powerful than classical one, because if we have $\alpha|0\rangle + \beta|1\rangle$ we simultaneously make computations with $|0\rangle$ and $|1\rangle$, roughly speaking.
- *Entanglement of states:* Two spatially separated and non-interacting quantum systems (qubits) that have been interacting may have some locally inaccessible information in common. An entangled state cannot be written as a direct product of two states from two subsystem Hilbert spaces. The entanglement of states makes quantum cryptography possible.

13.3.1 Bit versus Qubit

In a classical computer the information is encoded in a sequence of bits, having two distinguishable and stable states, for example, as a particle placed in a double-well potential, which are conventionally read as either a 0 or a 1 (Figure 13.3a). Apart from some similarities with a classical bit the qubit is overall very different. As in case of the bit, two distinguishable states – that is, different eigenstates of an operator – are needed. For example, a spin $\frac{1}{2}$ particle has two possible states ($\uparrow$ or $\downarrow$), or a photon can be polarized either vertically or horizontally. The state of a qubit may be expressed by basis states (or vectors) of the Hilbert space. The Dirac, or the so-called bra-ket, notation is used to represent them. This means that the two logical basis states, so-called eigenstates, are conventionally written as $|0\rangle$ and $|1\rangle$ (pronounced: "ket 0" and "ket 1").

Unlike the bit (being either 0 or 1) the qubit is not necessarily in the $|0\rangle$ or $|1\rangle$ state, but it can be rather in a superposition of both states. A particle representing a qubit is described by a quantum-mechanical wave function and can tunnel under the barrier which separates two wells (i.e. the two states). As a consequence, a quantum system in a double-well potential has the two lowest energy states $|0\rangle = \frac{1}{\sqrt{2}}(|\uparrow\rangle + |\downarrow\rangle)$ and $|1\rangle = \frac{1}{\sqrt{2}}(|\uparrow\rangle - |\downarrow\rangle)$. Here $\uparrow$ and $\downarrow$ are two classical states, like the states 0 and 1 in the double-well potential (Figure 13.3a). In the ground state $|0\rangle$ the wave function is symmetric, whereas for the first excited state $|1\rangle$ it is antisymmetric, Figure 13.3b). A system prepared in a superposition state exhibits coherent oscillations between the two wells, and the measuring probability for finding the particle in each well oscillates with frequency $\omega = \Delta\,\hbar^{-1}$, where Δ is the splitting of the lowest energy level (see Figure 13.3b).

13.3.2
Qubit States

A pure qubit state is a linear superposition of both eigenstates. This means that the qubit can be represented as a linear combination of $|0\rangle$ and $|1\rangle$:

$$|\psi\rangle = \alpha|0\rangle + \beta|1\rangle, \tag{13.1}$$

where α and β are probability amplitudes and can, in general, be complex.

When measuring this qubit in the standard eigen-basis, the probability of outcome $|0\rangle$ is $|\alpha|^2$ and of $|1\rangle$ is $|\beta|^2$. Because the absolute squares of the amplitudes are equal to probabilities, α and β must be constrained by the equation $|\alpha|^2 + |\beta|^2 = 1$. The Eq. (13.1) could be rewritten as

$$|\psi\rangle = |\psi(\varphi,\theta)\rangle = e^{-i\varphi/2}\cos\frac{\theta}{2}|0\rangle + e^{i\varphi/2}\sin\frac{\theta}{2}|1\rangle.$$

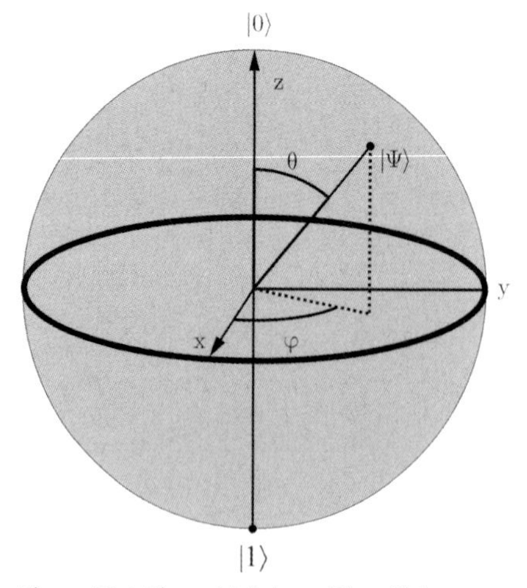

Figure 13.4 The qubit $|\psi\rangle = \alpha|0\rangle + \beta|1\rangle$ is represented as point (θ,φ) on a unit sphere, the so-called Bloch sphere. θ and φ are defined by $\alpha = e^{-i\varphi/2}\cos(\theta/2)$ and $\beta = e^{i\varphi}\sin(\theta/2)$. $|\psi\rangle$ is represented by the unit vector $[\cos(\varphi)\sin(\theta), \sin(\varphi)\sin(\theta), \cos(\theta)]$, called the Bloch vector.

The common phase factor resulting from the complex nature of α and β was neglected. The state of a single qubit can be represented geometrically by a point on the surface of the Bloch sphere (see Figure 13.4). A single qubit has two degrees of freedom φ, θ. A classical bit can only be represented by two discrete values 0 or 1. Note that the two complex numbers α, β in Eq. 13.1 in fact correspond to four numbers: $\mathrm{Re}(\alpha)$, $\mathrm{Im}(\alpha)$, $\mathrm{Re}(\beta)$, $\mathrm{Im}(\beta)$. However, these numbers are not independent; they are linked by the unity norm and the physical irrelevant common phase factor can be neglected. Any two-level quantum physical system can be used as a qubit; for example, the electron charge, polarization of photons, the spin of electrons or atoms and the charge, flux or phase of Josephson junctions could be used for implementation of qubits.

13.3.3
Entanglement

The entanglement of qubits is a subtle non-local correlation that has no classical analog. It allows a set of qubits to express a higher correlation than is possible in classical systems. As an example, consider the two entangled qubits A and B

$$|\Phi\rangle = |\phi\rangle_A |\psi\rangle_B = |\phi_A \psi_B\rangle = |\phi\psi\rangle = \frac{1}{\sqrt{2}}(|00\rangle + |11\rangle).$$

The first system is in state $|\phi\rangle_A$ and the second in state $|\psi\rangle_B$. Both systems have the two basis vectors $|0\rangle$ and $|1\rangle$. When measuring state $|\Phi\rangle$ the outcomes $|00\rangle$ and $|11\rangle$ have equal probabilities. It is impossible to attribute to either system A or system B a definite pure state as their states are superposed with one another. It is seen that if a 0 state is measured in A, then there will be an obligatory 0 state when measuring B. So A and B are not independent, they are *entangled*.

Entanglement allows multiple states to be acted on simultaneously, unlike classical bits that can only have one value at a time.

A number of entangled qubits taken together is a *qubit register*, with basis states of the form $|x_1 x_2 ... x_n\rangle$. An n-qubit register has a 2^n dimensional space, being much larger than a classical n-bit register.

Entanglement of states in quantum computing has been referred to as *quantum parallelism*, as the state can be in a quantum superposition of many different classical computational paths which can all proceed concurrently.

13.3.4
Physical State

The quantum Hamiltonian operator $\hat{H}$ generates the time evolution of quantum states and applied to the state vector yields the observable corresponding to the total energy of the system. The eigenvectors of $\hat{H}$, denoted by $|x\rangle$, provide an orthonormal basis for the Hilbert space of the system. The equation

$$\hat{H}|x\rangle = E_x|x\rangle.$$

yields the spectrum of allowed energy levels of the system, given by the set of eigenvalues E_x. Since $\hat{H}$ is a Hermitian operator, the energy is always a real number.

The time evolution $|\psi(t)\rangle$ of quantum states is given by the time-dependent Schrödinger equation:

$$\hat{H}\,|\psi(t)\rangle = i\hbar\frac{\partial}{\partial t}|\psi(t)\rangle.$$

If $\hat{H}$ is independent of time this equation can be integrated to obtain the state at any time:

$$|\psi(t)\rangle = \exp\left(-\frac{i\hat{H}t}{\hbar}\right)|\psi(0)\rangle,$$

where $|\psi(0)\rangle$ is the state at some initial time ($t = 0$).

13.3.4.1 Measurement

A measurement of a quantum state inevitably alters the system, as it projects the state onto the basis states of the measuring operator. Only if the state is already the eigenstate of the measuring operator then the state does not change. Thus, the superposition of states collapse into one or the other eigenstate of the operator, defined by the probabilities amplitudes. The precise amplitudes (α and β of a single qubit) can be found by multiple recreation of the superposition and subsequent measurements.

13.3.4.2 No-Cloning Theorem

Unlike the classical bit, of which multiple copies can be made, it is not possible to make a copy (clone) of an unknown quantum state [12, 13]. This so-called *no-cloning theorem* has profound implications for error correction of quantum computing as well as quantum cryptography. For example, such as it prevents the use of classical error correction techniques on quantum states, no backup copy of a state to correct subsequent errors could be made. However, error correction in quantum computation is still possible (see Section 13.5 for details). The no-cloning theorem protects the uncertainty principle in quantum mechanics, as the availability of several copies of an unknown system, on which each dynamical variable could be measured separately with arbitrary precision would bypass the Heisenberg uncertainty principle $\Delta x \Delta p$, $\Delta E \Delta t \geq \hbar/2$.

13.4 Operation Scheme

Quantum bits must be coupled and controlled by gates in order to process the information. At the same time, they must be completely decoupled from external influences such as thermal noise. It is only during well-defined periods that the control, write and readout operations take place to prevent an untimely readout.

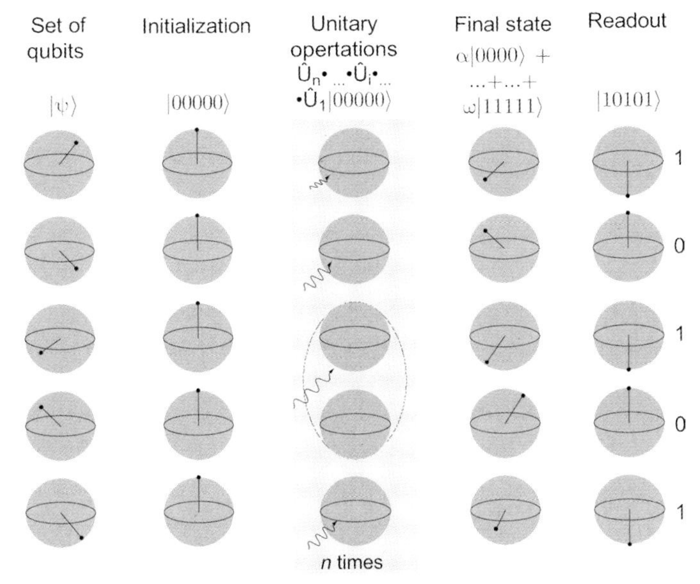

Figure 13.5 Operation scheme of quantum computation. After the system is prepared as a quantum register, controlled unitary operations $\hat{U}$ on single or entangled qubits are performed by gate operations in a controlled manner $\hat{U} = \exp(-i\hat{H}t/\hbar)$ ($\hat{H}$: Hamiltonian, t: time). The readout is done by projection onto basis states, yielding probability distributed Boolean values.

13.4.1 Quantum Algorithms: Initialization, Execution and Termination

The operation scheme of quantum computation is depicted in Figure 13.5. To start an quantum algorithm the qubit register must be initialized in some specified well-defined state, for example by a dissipative process to the ground state $|00...0\rangle$.

Then, the computation is done by an appropriate sequence of applied unitary operations $\hat{U} = \exp(-i\hat{H}t/\hbar)$ with Hamiltonian $\hat{H}$ to the qubits. The actual mechanism – that is, electromagnetic waves, voltages or magnetic fields of well-defined energy/amplitude and time t – depends strongly on the type of qubit. In each step of the algorithm, the qubit vector is modified by multiplying it by a unitary matrix. The components of the matrix are determined by the physics of the device. The unitary character of the matrix ensures that the matrix is invertible, making the computation reversible.

For a given algorithm the operations will always be done in exactly the same order. Since there is no way to read the qubit state before the final destructive readout measurement, there are no conditional statements such as 'IF... THEN...'. However, there are conditional gate operations such as the controlled NOT (CNOT) gate.

After termination of the algorithm the qubit vector must be read out by measurement. Quantum mechanics ensure that the measurement will destroy the qubit vector by projection onto the eigenstate of the corresponding observable, and only a probability distributed n-bit vector is obtained.

Even when neglecting the decoherence sources during the unitary transformations, the experimental readout schemes can never be perfectly efficient. Thus, it should be possible to repeat the measurement to enhance the probability of the obtained results by a majority polled output.

13.4.2 Quantum Gates

Once a quantum register is initialized the qubits must be manipulated in order to process the information by quantum gates, just as in case of the classical logic gates for conventional digital information processing. Quantum logic gates are represented by unitary matrices, as they are reversible, unlike many classical logic gates. However, some universal classical logic gates, such as the *Toffoli* gate, also provide reversibility, and can be directly mapped onto quantum logic gates. Mostly quantum gates operate on spaces of one or two qubits, thus written as matrices the quantum gates can be described by 2×2 or 4×4 matrices with orthonormal rows.

Examples of a single qubit operation are the *Hadamard* gate, which puts the initialized qubit in superposition state, represented by the Hadamard matrix

$$\hat{H}_{\text{Hadamard}} = \frac{1}{\sqrt{2}} \begin{bmatrix} 1 & 1 \\ 1 & -1 \end{bmatrix},$$

and the CNOT gate for two qubit operations, defined as

$$\hat{H}_{\text{CNOT}} = \begin{bmatrix} 1 & 0 & 0 & 0 \\ 0 & 1 & 0 & 0 \\ 0 & 0 & 0 & 1 \\ 0 & 0 & 1 & 0 \end{bmatrix} = \begin{bmatrix} \mathbf{1} & \mathbf{0} \\ \mathbf{0} & \mathbf{X} \end{bmatrix},$$

with $\mathbf{1}$ the identity matrix and $\mathbf{X}$ the first Pauli matrix.

Both, in classical and quantum computation, all possible operations can be reduced to a finite set of *universal* gates, which can be used to construct the specific algorithm of the information processing. To achieve universality for classical reversible gates, three-bit operations are needed, whereas in the quantum regime only one- and two-qubit gate operations are sufficient. This underlines the versatile character of quantum logic.

13.5 Quantum Decoherence and Error Correction

Decoherence is the mechanism by which the information encoded in the superposed and entangled qubits register degrades over time. For example, dephasing caused by fluctuations of the energy level of two quantum mechanical states gives an additional phase proportional to the energy change of the superposition states. With increasing number of qubits the computation power, as well as the probability for decoherence, will increase, and the need for decoherence control becomes predominant. This could be done by the implementing of quantum error correcting gates. In general, the sources of error can be: (i) non-ideal gate operations; (ii) interaction with environment causing relaxation or decay of phase coherence; and (iii) deviations of the quantum system from an idealized model system.

In classical computers every bit of information can be re-adjusted after every logical step by using non-linear devices to re-set the information bit to the 0 or 1 state. Contrary in a quantum system, no copy can be made of a qubit state without projecting it onto one of its eigenstates and thus destroying the superposition state.

Quantum information processing attracted much attention after Shor's surprising discovery that quantum information stored in a superposition state can be protected against decoherence. The single qubit is encoded in multiple qubits, followed by a measurement yielding the type of error, if any, which happened on the quantum state. With this information the original state is recovered by applying a proper unitary transformation to the system. This stimulated much research on quantum error correction, and led to the demonstration that arbitrarily complicated computations could be performed, provided that the error per operation is reduced below a certain error threshold.

By repeated runs of the quantum algorithm and measurement of the output, the correct value can be determined to a high probability. In brief, quantum computations are probabilistic.

13.6 Qubit Requirements

Di Vincenzo [14] listed criteria that any implementation for quantum computers should fulfill to be considered as *useful*:

- A scalable physical system with well-characterized qubits. A quantum computer consisting of a few qubits is not sufficient for useful computation.
- The ability to initialize the state of all qubits to a simple basis state.
- Relative long decoherence times τ_{dec}, much longer than the gate-operating time. To observe quantum-coherent oscillations the requirement $\tau_{\text{dec}}\Delta \gg \hbar$ must be fulfilled, and the fidelity loss per single quantum gate operation should be below some criteria.
- A universal set of quantum gates is needed to control the quantum system.
- A qubit-specific measurement capability to perform a readout and to transfer the information to conventional computers.

13.7 Candidates for Qubits

Various quantum mechanical two-level systems have been examined as potential candidates for qubits. In this overview, the aim is to concentrate on the first system where the feasibility of quantum computation was demonstrated, and highlight some potential *solid-state* systems. The solid-state technology is easily scalable and can be integrated into conventional electric circuits. At the time of writing [15] there is no notable clear favorite for quantum computing, because on one hand the known systems must be improved, while on the other hand new promising candidates for quantum computing hardware are continuously appearing.

13.7.1 Nuclear Magnetic Resonance (NMR)-Based Qubits

Atomic nuclei are relatively isolated from the environment and thus well protected against decoherence. Their spins can be manipulated by properly chosen radio frequency irradiation. Elementary quantum logic operations have been carried out on the spin $\frac{1}{2}$ nuclei as qubits by using nuclear magnetic resonance (NMR) spectroscopy [16]. This system is restricted by the low polarization and the limited numbers of resolvable qubits. The nuclei (see Figure 13.6) were in molecules in a solution, and a magnetic field defined the two energy-separated states of the nuclei. These macroscopic samples with many nuclear qubits provide massive redundancy, as each molecule serves as a single quantum computer. The qubits interact through the electronic wave function of the molecule. Quantum calculations with up to seven qubits have been demonstrated. For example, the Shor algorithm was implemented to factorize the number 15 [17].

13.7.2 Advantages of Solid-State-Based Qubits

Large number of qubits could be assembled using existing solid-state technology. The main drawback is that the accuracy of the devices is not as high as established NMR-based qubits described above. Solid-state quantum computers encounter a new set of problems, as the spatial inhomogeneity during processing causes differences between nominally identical devices. Digital logic tolerates imperfection by restoring a signal to its intended value, whereas quantum computing forsakes this

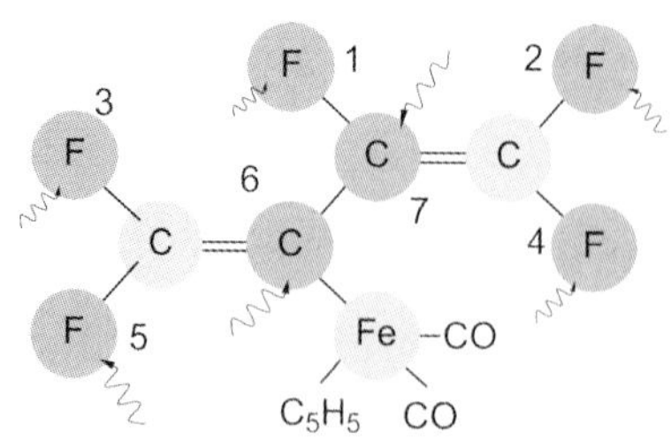

Figure 13.6 One molecule acts as seven-qubit system. The nuclei of five fluorine and two carbon atoms in the molecule form seven nuclear spin qubits (light gray), which are programmed by radiofrequency pulses and can be detected by nuclear magnetic resonance (NMR) spectroscopy [16].

methodology. Until now the imprecision of conventional solid-state devices prevents the storage of more than a few bits in a single device. A solid-state quantum computing device should be cooled down to a few mK to prevent thermal noise, caused for example, by phonons and destroying the coherent superposition of states.

The computations are performed by microwave pulses, that decrease the barrier separating the two states and brings the system into quantum realm. Subsequent pulses manipulate the Hamiltonian, *viz.* the probability weighting of the state vector to perform useful operations.

The spread between qubit parameters could be compensated by individually chosen biases, managed by a conventional computer. However, leaving the reduced coherence time due to additional noise aside, this would complicate the layout of a quantum computer, as at least one additional wire per qubit is needed.

Fortunately, it transpires that quantum error correction can even take care of error caused by defective implementation of a quantum gate, in case that the error is not larger than 10^{-4} per operation [18].

13.7.3 Kane Quantum Computer

In 1998, Kane [19] suggested imbedding a large number of nuclear spin qubits, made up by isotopically pure ^{31}P (nuclear spin 1/2) donor atoms, at a moderate depth in a ^{28}Si (nuclear spin 0) crystal to build a quantum computer (see Figure 13.7). When placed in a magnetic field, two distinguishable states for the spin 1/2 nuclei of the P atoms appear. By a controllable overlap of the wave functions of electrons bound to this atom in the deliberately doped semiconductor, some interaction may take place between adjacent nuclear qubits. Voltages applied to electrodes (gate A) placed above the phosphorus atoms can modify the strength of the electron–nuclear coupling. Individual qubits are addressed by the A-gates to shift their energies in and out of resonance with an external radiofrequency. The strength of the qubit–qubit coupling by overlapping wavefunctions is controlled by electrodes placed midway between the P atoms (gate B). The operation principle is similar to the NMR-based qubits in a macroscopic molecule solution, except that the nuclei are addressed by potentials rather than by the radiofrequency pulses.

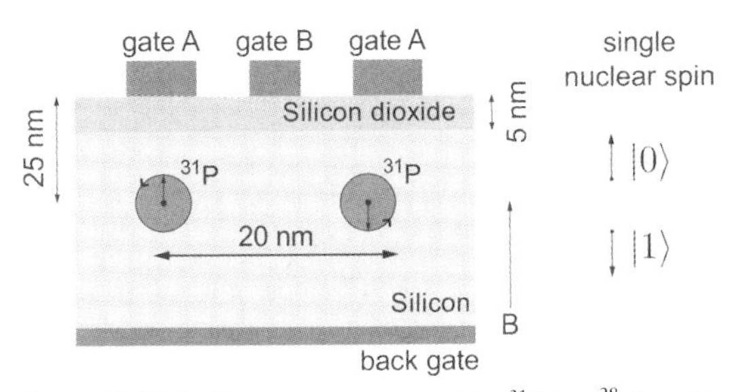

Figure 13.7 The Kane quantum computer: ^{31}P in a ^{28}Si matrix are controlled by gate electrodes (A, B) placed above them [19].

13.7.4 Quantum Dot

Beside the nucleus, the spin and charge of electrons may also be used to construct a double degenerate system. The electron spin-based qubits have the advantage that the Hilbert space consists of only two spin states, which strongly suppresses the leakage of quantum information into other states. In addition, the spin is less coupled to the environment than the charge, which results in longer decoherence times.

Quantum dots of 5 to 50 nm are thin, semiconducting multilayers on the substrate surface, where confined electrons with discrete energy spectra appear. By using Group III–V compound semiconductor materials such as GaAs and InAs of different lattice sizes, two-dimensional electron gas systems with high electron mobility can be constructed. To use these as qubits, their quantized electron number or spin is utilized [20]. The switching of the quantum state can be achieved either by optical or electrical means. In case of the spin-based quantum dot the electrons are localized in electrostatically defined dots (see Figure 13.8). The coupling between the electron spins is made via the exchange interaction, and the electron tunneling between the dots is controlled by a gate-voltage (gate B in Figure 13.8).

13.7.5 Superconducting Qubits

Superconductivity is a macroscopic quantum mechanical state, in which electrons with opposite spin and momenta form the so-called Cooper pairs and an energy gap in the quasi-particle spectrum appears. The interaction between the electrons is mediated by phonons. The superconducting qubits are based on *Josephson junctions* – that is, two superconductors separated by a tunnel barrier. The supercurrent that

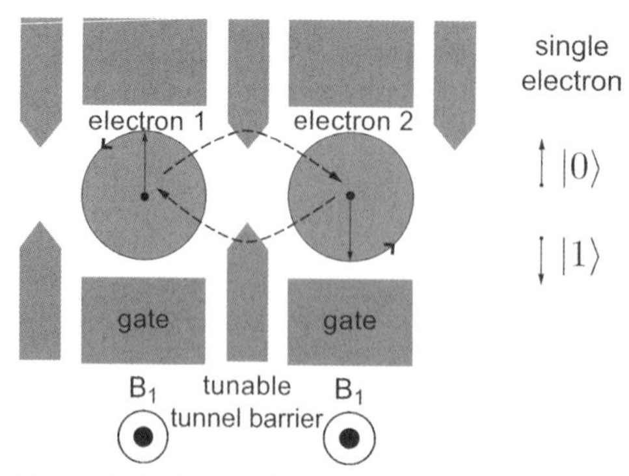

Figure 13.8 Scheme of spin-based quantum dots as a qubit system. The coupling of spins of localized electrons is formed by exchange interaction due to gate-voltage-controlled tunneling. Two electrons are localized in the regions defined by the gates. Single spin operations are performed by local magnetic fields [20].

crosses the weak link is the Josephson supercurrent. For details on superconductivity and the Josephson effect, see Refs. [21, 22].

Until now, several possible systems differing by 2e have been described for constructing a superconducting qubit. In the *charge* qubit a coherent state with a well-defined charge of individual Cooper pairs is used, while the *flux* qubit employs two degenerate magnetic flux states and the *phase* qubit is based on the phase difference of superconducting wavefunctions in two electrodes for quantum computation [23, 24].

13.7.5.1 Charge Qubits

The basis states of a charge qubit are two charge states. The qubit is formed by a tiny ~100-nm superconducting island – a Cooper pair box. Thus the charging electrostatic energy of a Cooper pair dominates in comparison with all other energies. An external gate voltage controls the charge, and the operation can be performed by controlling the applied gate voltages V_g and magnetic fields. A superconducting reservoir, coupled by a Josephson junction to the Cooper pair box, supplies the neutral reference charge level (see Figure 13.9). The state of the qubit is determined by the number of Cooper pairs which have tunneled across the junction. The readout is performed by a single-electron transistor attached to the island (not shown). By applying a voltage to this transistor, the dissipative current through the probe destroys the coherence of charge states and its charge can be measured [25].

13.7.5.2 Flux Qubits

A flux qubit is formed by a superconducting loop (1 μm size) interrupted by several Josephson junctions with well-chosen parameters (c.f. Figure 13.10) [26]. To obtain a double-well potential, as in Figure 13.10, either an external flux $\Phi_0/2$ or a π junction is needed. By including a π junction [27–29] in the loop the persistent current,

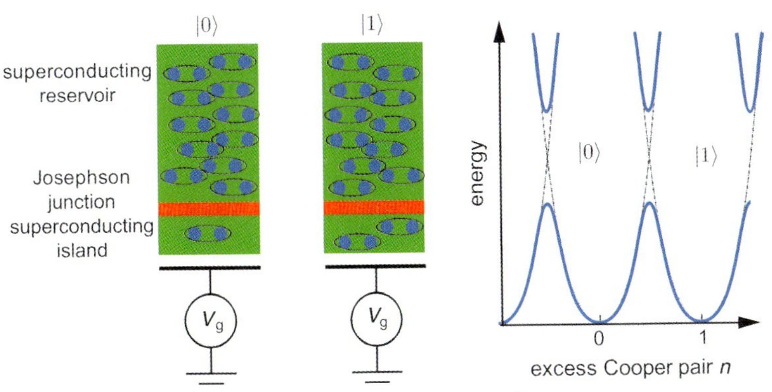

Figure 13.9 The charge qubit is formed by a Cooper pair box (CPB), separated from the superconducting reservoir (top) by a Josephson tunnel junction [25]. The basis states |0⟩ and |1⟩ differ by the number of excess Cooper pairs *n* on the CPB.

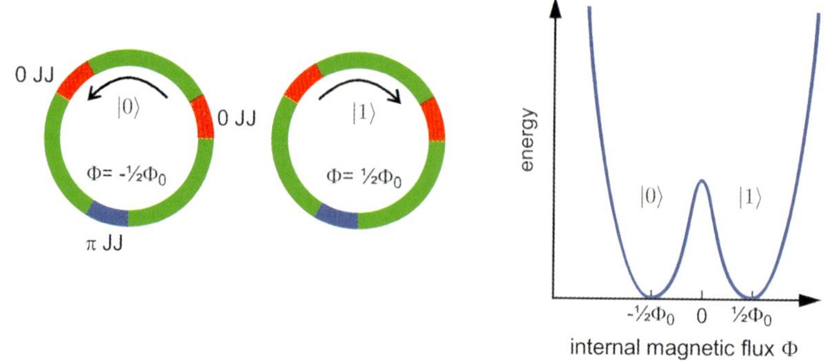

Figure 13.10 The basis states |0⟩ and |1⟩ of the flux qubit are determined by the direction of a persistent current circulating in three-junctions qubit [26]. The basis states |0⟩ and |1⟩ differ by the direction of the current in the superconducting loop. The π-Josephson junction (JJ) self-biases the qubit to the working point and, thus, substitutes an external magnetic flux.

generating the magnetic flux, may spontaneously appear and flow continuously, even in absence of an applied magnetic field [30]. The basis states of the qubit are defined by the direction of the circulating current (clockwise or counter-clockwise). The currents screen the intrinsic phase shift of π of the loop, such that the total flux through the loop is equal to $\pm\Phi_0/2$, i.e. half a magnetic flux quanta. The two energy levels corresponding to the two directions of circulating supercurrent are degenerate. If the system is in the quantum mechanical regime (low temperature to suppress thermal contributions) and the coupling between the two states (clockwise/counter-clockwise current flow) is strong enough (*viz.* the barriers are low), the system can be in the superposition of clockwise and counter clockwise states. This *quiet* qubit [31], is expected to be robust to the decoherence by the environment because it is self-biased by a π Josephson junction. Note that this flux-qubit device with a π junction is an optimization of the earlier scheme, where the phase shift of π was generated by an individual outer magnetic field of $\pm\Phi_0/2$ for each qubit [26].

The readout could by made by an additional superconducting loop with one or two Josephson junctions (i.e. SQUID loop) that is inductively coupled to the qubit.

To process the input and output of flux qubits an interface hardware based on the rapid single flux quantum (RSFQ) circuits could be used. These well-developed superconducting digital logics work by manipulating and transmitting a single flux quanta. In fact, this logic overcomes many problems of conventional CMOS-logics as it has a very low power consumption, an operating frequency of several hundred GHz, and is compatible with the flux qubit technology [32].

13.7.5.3 **Fractional Flux Qubits**

One variation of a flux π qubit is a fractional flux qubit. At the boundary between a 0 and a π coupled Josephson junction (i.e. a 0–π JJ) a *spontaneous* vortex of supercurrent

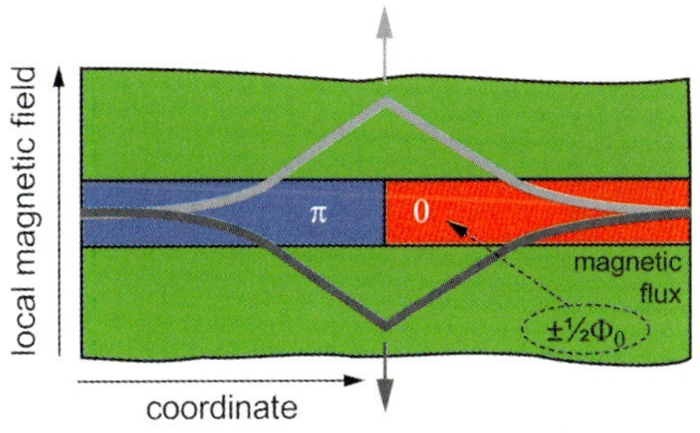

Figure 13.11 Sketch of a 0–π JJ with circulating supercurrent around 0–π phase boundary. The magnetic flux is equal to half of a flux quantum Φ_0 (semifluxon).

may appear under certain circumstances. Depending on the length L of the junction, the supercurrent carries a half-integer flux quantum $\Phi_0/2$ (called *semifluxon*) or fractions of it. Figure 13.11 depicts the cross-section of a symmetric 0–π *long* JJ. Classically, the semifluxon has a degenerate ground state of either positive or negative polarity, that corresponds to clockwise and counter-clockwise circulation of supercurrent around the 0–π boundary. The magnetic flux localized at the 0–π boundary is $\Phi_0/2$ and represents two degenerate classical states [33].

0–π Josephson junctions with a spontaneous flux in the ground state are realized with various technologies. The presence of spontaneous flux has been demonstrated experimentally in *d*-wave superconductor-based ramp zigzag junctions [34], in long Josephson 0–π junctions fabricated using the conventional Nb/Al-Al_2O_3/Nb technology with a pair of current injectors [35], in the so-called tricrystal grain-boundary long junctions [36–38] or in SIFS Josephson junctions [39] with a *step-like* ferromagnetic barrier. In the latter systems the Josephson phase is set to 0 or π by choosing a proper F-layer thicknesses d_F. The advantages of this system are that it can be prepared in a multilayer geometry (allowing topological flexibility) and it can be easily combined with the well-developed Nb/Al-Al_2O_3/Nb technology.

A *single* semifluxon ground state is double degenerate with basis flux states $|\uparrow\rangle, |\downarrow\rangle$. It transpires that the energy barrier scales proportionally to the junction length L, and the probability of tunneling between $|\uparrow\rangle$ and $|\downarrow\rangle$ decreases exponentially for increasing L [40]. Hence, a single semifluxon will always be in the classical regime with thermal activated tunneling for long junctions. As a modification, a junction of finite, rather small length L may be considered. In this case, the barrier height is finite and approaches zero when the junction length $L \to 0$. At this limit, the situation is not really a semifluxon, as the flux Φ present in the junction is much smaller than $\Phi_0/2$.

A 0–π–0 Josephson junction (see Figure 13.12) has *two* antiparallel coupled semifluxons for a distance a larger than the critical distance a_c [40]. The ground state of this system is either $|\uparrow\downarrow\rangle$ or $|\downarrow\uparrow\rangle$. For symmetry reasons both states are degenerate. The tunnel barrier can be made rather small, which results in a rather strong coupling with appreciable energy level splitting due to the wave functions overlap. Estimations show that this system can be mapped to a particle in a double-well potential, and thus can be used as a qubit like other Josephson junctions-based

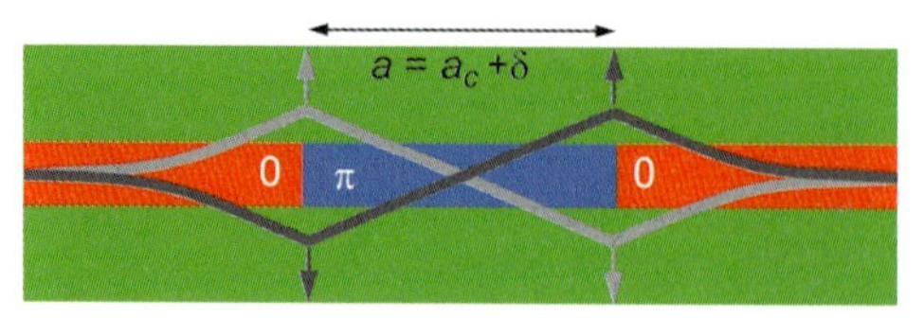

Figure 13.12 The two basis states are $|\uparrow\downarrow\rangle$ and $|\downarrow\uparrow\rangle$ of two coupled fractional vortices in a long linear 0–π–0 Josephson junction.

qubits. Thus, the 0–π–0 junctions are supposed to show the motion of a point-like particle in a double-well potential, and may be used as the basis cell of a fractional flux qubit.

13.8 Perspectives

Today, quantum computer algorithms allow to solve some specific **NP** problems, including factoring, sorting, calculating discrete logarithms, and simulating quantum physics. However, conventional computers will still be needed for addressing the quantum computer, for databases and for applications where either high computation power is not needed or sufficient coherence could not be provided, as in mobile devices. Quantum computers could be used as coprocessors for specific tasks such as en/decryption. The use of quantum computers may have several consequence, as today's data security algorithms could stay secure only if the keylength exceeds the storage capacity of quantum computers. The simulations of quantum mechanics might contribute to a variety of scientific and practical applications based on physics, chemistry, biology, medicine and related fields.

References

1 R. P. Feynman, *Int. J. Theoret. Physics* 1982, **21**, 467.

2 R. P. Feynman, *Feynman Lectures on Computation*, Addison-Wesley, Reading, MA, 1996.

3 International Technology Roadmap for Semiconductors, http://www.itrs.net/.

4 M. A. Nielsen, I. L. Chuang, *Quantum Computation and Quantum Information*, University Press, Cambridge, 2000.

5 J. Stolze, D. Suter, *Quantum Computation*, Wiley-VCH, 2004.

6 L. K. Grover, Proceedings 28th Annual ACM Symposium on the Theory of Computing, Philadelphia, Pennsylvania, ACM Publishers, New York, 1996.

7 P. W. Shor, *Proceedings, 35th Annual Symposium on the Foundations of Computer Science*, IEEE Computer Society Press, Los Alamitos, CA, 1994.

8 A. Einstein, B. Podolsky, N. Rosen, *Phys. Rev.* 1935, **47**, 777.

9 R. W. Hamming, *Coding and Information Theory*, Prentice-Hall, 1986.

10 R. Landauer, *IBM J. Res. Dev.* 1961, **5**, 183.

11 C. Cohen-Tannoudji, B. Diu, F. Laloë, *Quantum Mechanics*, Wiley-VCH, 1977.

12 W. K. Wootters, W. H. Zurek, *Nature* 1982, **299**, 802.
13 D. Dieks, *Phys. Lett. A* 1982, **92**, 271.
14 D. P. DiVincenzo, *Fortschr. Phys. Prog. Physics* 2000, **48**, 771.
15 GDEST , *EU/US Workshop on Quantum Information and Coherence*, December 8–9, Munich, Germany, 2005.
16 L. M. K. Vandersypen, I. L. Chuang, *Rev. Mod. Phys.* 2004, **76**, 1037.
17 L. M. K. Vandersypen, M. Steffen, G. Breyta, C. S. Yannoni, M. H. Sherwood, I. L. Chuang, *Nature* 2001, **414**, 883.
18 R. W. Keyes, *Appl. Phys. A* 2003, **76**, 737.
19 B. E. Kane, *Nature* 1998, **393**, 133.
20 D. Loss, D. P. DiVincenzo, *Phys. Rev. A* 1998, **57**, 120.
21 W. Buckel, R. Kleiner, *Superconductivity. Fundamentals and Applications*, Wiley-VCH, 2004.
22 A. Barone, G. Paterno, *Physics and Applications of the Josephson Effect*, John Wiley & Sons, 1982.
23 A. Ustinov, in: R. Waser (Ed.), *Nanoelectronics and Information Technology - Advanced Electronic Materials and Novel Devices*, Wiley-VCH, 2005.
24 Y. Makhlin, G. Schön, A. Shnirman, *Rev. Mod. Phys.* 2001, **73**, 357.
25 Y. Nakamura, Y. A. Pashkin, J. S. Tsai, *Nature* 1999, **398**, 786.
26 J. E. Mooji, T. P. Orlando, L. Levitov, L. Tian, C. H. van der Wal, S. Lloyd, *Science* 1999, **285**, 1036.
27 L. Bulaevskii, V. Kuzii, A. Sobyanin, *J. Exp. Theoret. Physics Lett.* 1977, **25**, 290.
28 V. V. Ryazanov, V. A. Oboznov, A. Y. Rusanov, A. V. Veretennikov, A. A. Golubov, J. Aarts, *Phys. Rev. Lett.* 2001, **86**, 2427.
29 M. Weides, M. Kemmler, E. Goldobin, D. Koelle, R. Kleiner, H. Kohlstedt, A. Buzdin, *Appl. Phys. Lett.* 2006, **89**, 122511.
30 T. Yamashita, S. Takahashi, S. Maekawa, *Appl. Phys. Lett.* 2006, **88**, 132501.
31 L. B. Ioffe, V. B. Geshkenbein, M. V. Feigel'man, A. L. Fauchère, G. Blatter, *Nature* 1999, **398**, 679.
32 M. Siegel, in: R. Waser (Ed.), *Nanoelectronics and Information Technology - Advanced Electronic Materials and Novel Devices*, Wiley-VCH, 2005.
33 E. Goldobin, D. Koelle, R. Kleiner, *Phys. Rev. B* 2002, **66**, 100508.
34 H. Hilgenkamp, A. Ariando, H. J. H. Smilde, D. H. A. Blank, G. Rijnders, H. Rogalla, J. R. Kirtley, C. C. Tsuei, *Nature* 2003, **422**, 50.
35 E. Goldobin, A. Sterck, T. Gaber, D. Koelle, R. Kleiner, *Phys. Rev. Lett.* 2004, **92**, 57005.
36 J. R. Kirtley, C. C. Tsuei, K. A. Moler, *Science* 1999, **285**, 1373.
37 J. R. Kirtley, C. C. Tsuei, M. Rupp, J. Z. Sun, L. S. Yu-Jahnes, A. Gupta, M. B. Ketchen, K. A. Moler, M. Bhushan, *Phys. Rev. Lett.* 1996, **76**, 1336.
38 A. Sugimoto, T. Yamaguchi, I. Iguchi, *Physica C* 2002, **367**, 28.
39 M. Weides, M. Kemmler, H. Kohlstedt, R. Waser, D. Koelle, R. Kleiner, E. Goldobin, *Phys. Rev. Lett.* 2006, **97**, 247001.
40 E. Goldobin, K. Vogel, O. Crasser, R. Walser, W. P. Schleich, D. Koelle, R. Kleiner, *Phys. Rev. B* 2005, **72**, 054527.

Index

Nanotechnology. Volume 4: Information Technology II. Edited by Rainer Waser
Copyright © 2008 WILEY-VCH Verlag GmbH & Co. KGaA, Weinheim
ISBN: 978-3-527-31737-0

p

q